Geographic Information 1992/3

Geographic Information 1992/3

The Yearbook of the Association for Geographic Information

Edited by
James Cadoux-Hudson
Geonex (UK) Ltd

and

Ian Heywood
University of Salford

Taylor & Francis
London • Washington, DC

UK	Taylor & Francis Ltd., 4 John Street, London WC1N 2ET
USA	Taylor & Francis Inc., 1900 Frost Road, Suite 101, Bristol, PA 19007

A catalogue record for this book is available from the British Library
ISBN 0-7480-0046-X

Cover design by Amanda Barragry based on illustration by Mike Beswick

Typeset by The Right Type Limited, Twickenham, Middlesex

Printed in Great Britain by Burgess Science Press, Rankine Road, Basingstoke, Hampshire

Contents

Preface

The 1992 edition of the AGI Yearbook represents a major change in direction from previous volumes. To assist us with the production of this year's volume we established an editorial panel consisting of nine section editors each with a responsibility for reviewing and commenting on GIS activities in their specialist field. These section editors, drawn from the commercial and academic GIS community, have been appointed (should they wish to continue!) for a period of three years. The rationale behind this strategy has been to (1) improve the coverage of the yearbook and (2) to try and maintain consistency in style and coverage between years.

Despite changes in the organizational structure of the yearbook our basic aim still remains the same, to provide a snap shot of what has happened how, why and where in the UK GIS community over the last year. Therefore, each part contains a detailed review paper written by the section editor covering the major issues applicable to their area of interest which have emerged over the last year. This is followed by a collection of papers which take a detailed look at either issues, events or applications which reflect the spirit of GIS activities in 1992.

Once again we are indebted to all those who have given up their time to make contributions in the form of articles, reports and reviews. Behind the scenes a special thanks must go to Beverly Heyworth of Salford University Business Services Ltd, who for the second year running, co-ordinated the compilation of the digital manuscript. In addition a special thanks must go to Steve Tomlinson who dealt with several digital manuscripts in particularly interesting formats.

On the production side, as editors, we owe a debt of gratitude to Richard Steele and Wendy Mould of Taylor & Francis who have trusted our decision to change the format of the AGI Yearbook and helped us co-ordinate the project.

Ian Heywood
James Cadoux-Hudson
May 1992

Notes on Section Editors

James Cadoux-Hudson

James Cadoux-Hudson is a chartered land surveyor. He gained a civil engineering degree at Nottingham University followed with 10 years' service in The Royal Engineers, much of which was in Military Survey and included many overseas projects in Europe, the Middle East, Africa and USA. James left Military Survey to run the survey department in Qatar where he introduced a GIS. Returning to UK he joined AT&T Istel for a time and then became a GIS consultant. Subsequent work has included the *AGI Yearbook*, work with DWH Associates and *Mapping Awareness*.

Heather Campbell

Heather Campbell is a Lecturer in the Department of Town and Regional Planning at the University of Sheffield. Her research interests include the organizational and institutional issues associated with the implementation of geographic information systems and the effective management of information in organizations.

Graham Clarke

Graham Clarke is a Lecturer in the Department of Geography at Leeds University. He is currently a Senior Consultant with Genemap, a consultancy company that is owned wholely by the University. Research interests include GIS, spatial modelling, and urban and social geography.

Chris E.H. Corbin

Chris Corbin has experience gained over the past 32 years of working with information technology working for computer manufacturers as well as organizations exploiting information technology within the UK nuclear and water industries. Since 1975 he has worked for Southern Water plc. Over the past 10 years this experience has included the successful implementation of a geographic information system within a corporate management information system for Southern Water.

Sarah Cornelius

Sarah Cornelius is a Lecturer in the Department of Environmental and Geographical Studies at Manchester Polytechnic. With a background in earth sciences and environmental technology she now teaches GIS and information technology to undergraduates in Geography and Environmental Management and on the Postgraduate Diploma/MSc in GIS run jointly with Huddersfield Polytechnic and Salford University. Current research interests include the application of GIS to environmental management and human and organizational impacts of GIS.

Anna Cross

Anna is a geography graduate who stayed on at Newcastle, in 1987, to specialize in geographical information systems (GIS). Her work involved the valuation of GIS as a useful epidemiological tool. In 1990, she was appointed as Research Associate, and more recently as co-ordinator, to the Centre for Research into Crime, Policing and the Community. This work involves the development of crime pattern analysis tools, GIS and currently the evaluation of urban crime projects being undertaken by Northumbria Police.

Dr Sara Finch

Sara Finch is responsible for developing business in geographic information systems (GIS) for Logica. This technology facilitates the handling of any spatial information and, through the use of common geographic references, allows disparate datasets to be combined and analysed. Sara gained her initial experience of GIS while employed by London University to carry out research and lecturing duties. During this time she published a number of papers and completed her PhD thesis on issues related to the capture and handling of digital map data and the use of GIS technology. She worked on GIS research projects for a variety of clients including Ordnance Survey, Natural Environment Research Council and the Greater London Council (GLC). She then joined Siemens to work as a consultant supporting the sales and marketing of the SICAD GIS product during its launch to the UK market. In 1988 Sara joined Logica, initially working as account manager for a number of clients in the central government sector. She moved to her current role in 1990.

David R. Green

David R. Green is currently Lecturer in Remote Sensing and Geographic Information Systems at the Centre for Remote Sensing and Mapping Science at Aberdeen University. Educated at the Universities of Edinburgh, Pennsylvania, and Toronto, he currently teaches in the MSc Environmental Remote Sensing Applications, Marine Resource Management, and Environmental Issues. Besides a strong commitment to GIS in secondary and higher education, his other research interests lie with the environmental applications of remote sensing and GIS, plant canopy reflectance modelling, the use of colour for cartographic output, and journalistic cartography. He is currently editor of the British Cartographic Society GIS Special Interest Group Newsletter, UK Associate Editor of GIS Europe, and a member of the AGI Organizing Committee and the Geographical Association (GA) ITWG on GIS in Secondary Education.

Dr Ian Heywood

Dr Ian Heywood is a lecturer in Geography at the University of Salford, England, specializing in Geographical Information Systems and Environmental Management. At present he is working in close collaboration with Tydac Technologies and in 1991/92 he has been on secondment to them as Director with responsibilities for education and training in the UK and Europe. His main research and development projects include mountain area GIS and training.

David Parker

David Parker heads the mapping information science research and teaching team in the Surveying Department at the University of Newcastle. Fundamental research concentrates on the capture, modelling and structuring of large-scale map and related data his application research applies this information for local authorities, the utilities, mineral exploration and the property profession.

Neil Stuart

Neil Stuart is a lecturer on Geographical Information Systems at the University of Edinburgh, Department of Geography. He gained a PhD from the University of Leeds for a thesis concerning the use of quadtree-based GIS for problems of land and water resource management. Neil lectures in the Undergraduate and Masters programmes in GIS at Edinburgh as well as running workshops on GIS principles and practices at sites worldwide. He is an editorial board member of the *International Journal of GIS* and book reviewer for several GIS journals. In addition to his educational responsibilities Neil conducts research in the use of GIS technology for a variety of land and water resource applications, with emphasis recently in developing countries.

Matthew Stuttard

Matthew Stuttard was a Lecturer on post-graduate courses in applied remote sensing at Silsoe College and also technical manager of the image processing and GIS facilities. He moved to Tydac Technologies where he was responsible for SPANS training and support in Europe. Matthew is now a project manager and consultant at Earth Observation Sciences, specializing in GIS applications using remotely sensed information.

Part I

The AGI in 1991/2

All enquiries regarding the AGI; matters of membership, publications, conferences, etc., should be directed to:

The Secretary,
The Association for Geographic Information,
12 Great George Street,
LONDON SW1P 3AD
Tel: 071 222 7000 Ext.226
Fax: 071 222 9430
Tlx: 915443 RICS G

1

AGI—Here now, where next

Peter Woodsford
AGI Chairman, 1991

Introduction

The third annual conference of the Association for Geographic Information, held at Birmingham in November 1991, was the culmination of a year of steady progress for the AGI. The event, with the theme "GIS—Here Now, Where Next?" attracted a combined delegate/visitor count of 2 300. Its success, at the premier national conference venue—the new International Conference Centre which will now be its permanent home—was particularly pleasing in the circumstances of economic recession, against which even the GIS industry has no immunity. The sessions of the conference focusing on the AGI and its review of the Chorley Report, together with the surrounding debates, produced several useful and important initiatives, including a constitutional review and a series of Ordnance Survey Roundtable Meetings. These are both discussed more fully below. At the Annual General Meeting, I was able to report a generally healthy financial state to members, along with a small increase in membership and a considerable increase in activities.

Activities

The work of the AGI is carried out mainly through its committees, which during 1991 numbered seven and involved approximately 175 participants. These activities are described elsewhere in this volume by Andrew Larner, the AGI Secretary. I will restrict myself to a few reflections on these activities and their growth and evolution.

Significant progress was made during 1991 in creating and sustaining a worthwhile and effective role for the AGI Council, in representing the interests of the geographic information community in both a consultative role and a pro-active lobbying role. However it comes as no surprise that five of the six constitutional issues to be reviewed relate to the Council; its structure, size, representation, voting procedure and the tenure of its officers. It is rightly perceived that improvements in these areas will contribute to the authority of the AGI and hence its ability to influence matters for the good. The action on the 1992 Chairman in ensuring that the constitutional review takes place speedily is commendable—such changes are important but are no panacea. They will not provide a substitute for a track record of effective action and involvement.

Standards activities are covered in Part 10 of this volume. They continue to be a major responsibility of the AGI, and the publication of NTF as a British Standard, under the title "BS 7567 Electronic transfer of Geographic Information (NTF) 1992" must rank as one of the Association's most important achievements to date. The need for standards activities is much broader than data transfer formats alone, and equally important is the role of AGI in promoting and co-ordinating other standardization efforts of fundamental practical importance such as the National Street Gazetteer and the Land and Property Gazetteer, both of which originate from Local Government Management Board working parties. Inputs from AGI will help ensure that these emerging standards are applicable as widely as possible, alongside lobbying to ensure they are given full place in legislation and steering through to British Standard status via the IST/36 Technical Committee. The start already made in the UK by AGI towards broadening the scope of standards activities means that we are well placed to make a major contribution to the recently established CEN Technical Committee 287, which will over the next two years define mandatory European standards for geographic information. All of these are important and legitimate activities for the AGI to pursue—the issue is how to accomplish them within available resources. In many of these areas the role of AGI is to evangelize and to demonstrate the advantages to be gained by standardization, to define the scope of the work required together with appropriate levels of prioritization, and to act as a catalyst for potential funding.

In other activities too, the key is to obtain maximum leverage from the limited resources AGI can deploy—both financial and the energy and enthusiasm of its participating members. Good recent examples include the Annual Awards Event organized by the Publicity Committee and the commissioning of review papers for publication by the AGI by supplying modest financing to young research workers.

The Association currently has an administrative Secretariat of two full-time and one part-time members. The flow of information both in and out of the Secretariat has increased enormously, yet there is still scope to increase its role as an information hub. The constitutional review is revisiting the issue of the appointment of a Director for the AGI. Such an appointment would provide continuity at a senior level and is the logical next step in the development of the Association.

The AGI has to balance its role as an umbrella organization with activities focused on special interest groups, and has deliberately set out to conduct these, where appropriate, in co-operation with other peer organizations. AGI has also set out to promote regional activities, to combat the inexorable pressure of London-centricity, and in particular to support the growth of the very active GIS communities in Scotland and Northern Ireland. As we move towards wider pan-European activities, the need for balancing pro-active regionally based activities will increase rather than diminish and this, I believe, is one of the more significant future challenges AGI has to meet in validating its role as an effective national umbrella organization. AGI will need to persist in promoting regional activities—it has much both to contribute and to gain from the smaller communities where there is great enthusiasm and some of the organizational problems are less severe.

The UK geographical information framework

This is not the place to record the full results of the Review by the AGI during 1991 of the 1987 Chorley Report. Suffice it to say that the exercise demonstrated the substantial measure of progress achieved against the Chorley recommendations, generated significant public debate and interest and focused attention on key areas where improvements to the health of the UK geographic information industry are still very neccessary. By all reasonable standards the Chorley Report must be judged a success in terms of its beneficial effects, but it will surprise no one that much remains to be done, particularly in the areas highlighted by Walter Smith in his Keynote Address at the 1991 AGI Conference (Smith, 1991). Nor should it surprise anyone that these areas are the most intractable ones, to do with the technically difficult subject of spatial referencing and with the commercially or politically sensitive areas of pricing, payback periods and access to information. Some new issues have come to prominence, including in particular the whole issue of Update, which I am sure will be of increasing importance as the whole subject matures, and others have evolved, so that demonstrator projects (never to my mind effectively realized in UK) have become cost/benefit authentications and there is a growing realization of the wider scope for the benefits of standardization exercises, beyond simply the transfer of data. Much of the efforts of the AGI in the next year or so will be directed to addressing the outstanding issues identified in its 1991 Chorley Review.

As a practical contribution to improving public access to geographic information, AGI has, since November 1991, been making freely available, under contract from the Department of the Environment, a database of spatially related datasets held by central government. The lessons to be learned from this service will need to be assessed in due course, and the service expanded to cover a wider range of datasets, either within the AGI or commercially.

The key role of the national mapping agencies needs no emphasis among any geographic information community. AGI has been pleased to take up the suggestion made by the minister responsible for the Ordnance Survey that it convene a series of round table meetings between Ordnance Survey and its users, to provide opportunities for Ordnance Survey to clarify and refine its policies and practices in relation to the availability and pricing of digital data in the light of the concerns and interests expressed by its user community. To date, these meetings have resulted in clarifications with respect to the operation of copyright, and alternative charging regimes have been explored, including more wholesale approaches. Users have highlighted the importance of predictability of charges—uncertainty as to the level of charges for data is a major obstacle. The talks have also given the user community a clear understanding of the parameters within which OS operates. The exercise has now to be brought to an effective and beneficial conclusion by:

- co-ordinating the user community position insofar as this is possible, presenting it to Ordnance Survey, and facilitating negotiations to enable Ordnance Survey to more effectively meet the requirements of the user community;

- if appropriate, using the AGI to advocate and press within government, the case for changes to any of the parameters within which the Ordnance Survey currently operates, where the effect of such parameters can clearly be shown to be to the detriment of the geographic information user community and the public good.

While the AGI does not seek such a role permanently, such a mediating function is of such value to all concerned that if it is successful, it may well merit repetition in the future.

The information industry is subject to an increasing level of complex legislation, at national and international levels, which is difficult to interpret when applied to electronic systems. Such legislation addresses topics including intellectual ownership, data protection, liability and freedom of information. Behind these issues, but closely related, are such considerations as public/private section interaction, formal quality assurance standards and codes of practice. It is vital that organizations in the geographic information industry, selling, providing or using information should understand how they may be affected by these issues; and are in a position to contribute to the ongoing debate. In an important new initiative prompted by the Chairman for 1992, Gurmukh Singh, AGI is establishing a limited life working party to examine these issues, to review existing and proposed legislation, taking into account the existing infrastructure of the information industry and the views of its organs, to report to the membership and recommend further action as appropriate.

International comparisons

A danger of a national association is that it becomes too inward-looking and parochial. It is therefore valuable to survey, albeit briefly, some of the major patterns discernable in other countries. In many respects the issues being faced are common to those we face in UK, with access to data, availability of data and pricing of data being key issues. It is pleasing to note how well the availability of large scale digital data from our Ordnance Surveys compares with other countries—and puzzling to reflect on the lack of 1:50 000 data cover compared with many other countries. It is a situation that must surely soon be rectified, one feels.

There is of course a major exception to the pattern of data accessibility and cost—in the United States data originating from federal government programmes is covered by Freedom of Information principles and treated in effect as a common good. The boost this gives to GIS technology originating from the United States, and to related business activity adding value to data is very substantial. This coverage will be felt around the world during 1992, as the marketing gets underway of the Digital Chart of the World, offering data coverage of the whole world at 1:1 000 000 scale, together with supporting PC software, at the very low price of US$200. By contrast in many European countries there are growing requirements on national mapping agencies to achieve an increasing level of cost recovery. The level of cost recovery already achieved by Ordnance Survey Great Britain is the highest anywhere in the world. The effects of demands to increase fees prematurely are potentially damaging and need to be resisted. Internationally there are many unresolved issues over the ownership of data, the terms under which it is traded and the liabilities which attach to the provision of data. There is a very cogent and informative international survey of these issues in a paper by Professor David Rhind (written before his appointment as the Director-General of the Ordnance Survey) and presented at the 1991 EGIS conference (Rhind, 1991). In it he concludes that we are approaching the time when

> all that will determine regional differences in up-take and efficient use of GIS are the conditions under which data are available, and the skills of the staff; the latter are determined indirectly in practice by the former.

There is a great deal at stake. Many observers have concluded, in the words of a

Dataquest report (Hale and Gartzen, 1991), that

> Underlying all comparisons, there is a competitive advantage in GIS industry development to be gained by countries that make their data freely available at low cost and encourage development of standards for data formats.

From an international perspective, therefore, there is every justification for a national association with the mission of the AGI. There is however considerable diversity in the nature of associations for the promotion of geographic information. In the United States, professional bodies and special interest groups such as URISA, ACSM and AM/FM are extremely well established and powerful, although it could be argued that none of them have the promotion of geographic information as their central purpose. They are complemented by a well-resourced and dirigiste Federal Geographic Data Committee which co-ordinates the provision of spatial data through a system of lead agencies responsible for different data categories. This committee also issues a biennial summary of GIS Activities in the Federal Government, with sufficient detail of content for clear trends to emerge. Interestingly the FGDC evolved from an earlier co-ordination committee which came into being as a result of an Office of Management and Budget investigation which highlighted a substantial amount of duplication in digitizing spatial information.

In Canada the Inter-Agency Committee on Geomatics (IAGC) co-ordinates the development in GIS technology and Geomatics (a useful umbrella word!) activities. It covers many of the responsibilities carried out in the UK by the AGI, including standards work, research and education, user needs and applications and the cataloguing of government databases. A service equivalent to that launched by AGI in 1991 on behalf of the Department of the Environment (the Tradeable Information Initiative) is well established in Canada. IAGC activities are funded by contributions from twelve government departments. In parallel with this government activity is a business association of firms operating in Canada—the Geomatics Industry Association of Canada. Collaboration between government and industry is very close—a visible sign being publications such as an overview of the industry, published by External Affairs and International Trade Canada and clearly aimed at increasing the nation's market share (Canada exports about C$100 million worth of Geomatics products and services per annum). The 1992 national GIS conference in Canada included in its opening ceremony the signing of a memorandum of understanding on collaboration in Geomatics between the Governments of Canada and Qatar, another indication of the explicit role of government.

Closer to home (and away from such Utopian fantasies!) the situation in Sweden is quite close to that in UK, with a national interest body, ULI, formed in 1986, and having many similarities to AGI. ULI is funded by corporate membership subscriptions, there being a greater element of government and user membership participation than we in AGI have so far achieved. The level of resourcing achieved for standards activities in Sweden is particularly impressive. In France there are two parallel organizations, with essentially common membership and management. CNIG is the official government body and AF3IG is its unofficial, non-government counterpart. With Gallic pragmatism, the membership pursues issues through the appropriate channel. There may be a lesson in that for the concerns raised as to whether AGI should have a more "official" makeup.

There is increasing demand for pan-European co-ordination for GIS. As a result European Commission Directorate XIII have charged a team of four people, including

Mike Brand, the first Chairman of AGI, to facilitate the formation of a Europe-wide umbrella organization for GIS. AGI will be active to ensure a full UK contribution to the resulting activity.

So what conclusions can be drawn from this brief international review? There is clearly considerable variety, reflecting political, economic and historical circumstances. The UK situation is characterized by the minimalist role of government, and the 1992 election result means that this is likely to continue. While this gives us many opportunities, the lack of the share of the resourcing that comes from governments in most other countries gives us difficulty in resourcing all the things that need to be done in such areas as standards and demonstrator systems—or perhaps it just makes us more efficient?! More subtly, the minimalist role of government makes it more difficult to get to grips with co-ordination issues, a situation likely to be exacerbated by trends towards privatization and agency status. Just where in the system does pressure have to be exerted and cases argued. Most fundamentally of all, how do we ensure that a broader, long-term view is taken on issues of data availability and cost within the context of a vision of what can be achieved nationally by the effective use of geographic information and the associated technologies?

A wider role for the AGI?

The Association is becoming more effective in a consultative role, representing the interests of the geographic information community in the consultative phase of new legislation and subsequent regulations. Examples include the Streetworks Bill and aspects of the Environmental Protection Act. It has been quick to point out to government the potential of geographic information and the danger of ignoring that potential. It has begun to exploit lobbying channels such as PITCOM, the parliamentary IT committee, and is currently pursuing vigorously the benefits to be gained from appropriate use of the National Street Gazetteer and the Land and Property Gazetteer in current and impending legislation. We are only beginning to find how to pursue such matters effectively, and much more activity will be required in the future. Persistent public pressure and lobbying is needed to develop better trading frameworks for data, to encourage and endorse sharing, on a joint venture, project-related basis, of costs and rewards between the public and private sectors and to promote models for spreading the costs of data to users. AGI also needs to generate pressure on data suppliers (or more pertinently their political paymasters) not to stifle the market by seeking too high a return too early. This can most helpfully be done by helping to articulate and enlarge the market demand.

Persistence and clarity are need in equal measure in the pursuit of the issues of common spatial referencing and to create conditions in which the benefits of GIS as an integrating technology can be realized. Somehow more cohesion and synergy has to be created between the activities of the major players—Ordnance Survey, Land Registry, Census and Post Office to name but a few. The thinking that is already evident over the future evolution of the Census represents an important opportunity, and the enthusiasm and momentum already generated by the Domesday 2000 project will need full support and participation from the AGI and all the other players to ensure the success the importance of the concept demands.

We still face a challenging task adjusting to the ever-increasing pace of technological advance and the AGI, the academic community and the industry have common cause in focusing attention and increasing awareness of this. How many of us have really come to terms with the effects of the seemingly ever increasing power of workstations or of a universally available, miniaturized GPS capability affordable to all? Or of powerful notebook computers? And Multimedia? I sometimes think we do ourselves a disservice by adopting the stance that the technical issues are all solved, really, and that the key issues are managerial and organizational, just because they are so intractable. Technology still has many potent solutions and stimulants to bring to the Geographic Infomation scene, and they will include surprises and unforseen developments as well as logical unfoldings of current techniques. AGI has an ongoing role, alongside vendors, consultants and the academic community, in expounding and promoting technological developments through its conferences, newsletters, events and publications.

Nor must we allow ourselves to become engrossed in government-related issues to the exclusion of all else. If the right conditions can be achieved and demonstrated, growth in the private sector, in commercial applications, will play an increasingly important part. The programme of lunchtime presentations to senior executives instituted by the 1992 Chairman, Gurmukh Singh, and the planned media coverage for GIS instigated by the AGI in the financial press and on television will increase awareness and growth in the commercial sector. Our industry needs more champions at senior levels within the potential user community. The 1992 emphasis on commercial applications will bring a welcome balance to the 1991 focus, arising from the review of the Chorley Report, on government-related activities.

AGI and its members are fortunate to be part of an activity and industry that is vigorous and growing—that still has its future before it. There is plenty for us all to do in improving that future and in realizing the full potential of geographic information and its associated technologies.

References

Hale, K. and Gartzen, P., 1991, "Geographic Information Systems: The European Terrain", *Geodetical Info Magazine* **5**, (9), September, pp. 24–7.

Rhind, D., 1991, "Data access, charging, and copyright and their implications for GIS", in *EGIS 91, Second European Conference on Geographical Information Systems*, Brussels, April 1991, Vol 2, Utrecht, Netherlands: EGIS Foundation, pp. 929–45.

Smith, W., 1991, "Keynote Address", AGI 91, Third National Conference, Birmingham, November 1991, Association for Geographic Information, also in *Mapping Awareness and GIS in Europe,* **6**, (1), Jan/Feb 1992, pp. 9–13.

2

AGI Secretariat and Committees

Council and management committee

The AGI met five times in 1991 and is due to meet a further five times before the 1992 Yearbook goes to print. The council structure is designed to represent each sector of the Geographic Information Community in the UK. Each major sector has three representatives, in addition, there is a "catch-all" not otherwise represented category. The council structure and the overall committee structure is under review in 1992. The management committee is comprised of the officers of the Council and two other council members. The management committee ensures the smooth running of the Association from week to week. The Council has also agreed to investigate the possibility of engaging a Director. Apart from reviewing the progress made by individual committees and monitoring the finances of the AGI, the Council was involved in the following items in 1991/92: The Computerized Street Register; Local Authority Potentially Contaminated Land Registers; Future of the Regional Research Laboratories; The Parliamentary Information Technology Committee (PITCOM); A review of the "Chorley Report".

The Council has commented on a number of pieces of legislation in 91/92 helping Civil Servants to frame regulations which will make the most effective use of geographic information. The most notable examples of legislation affected Registers of Potentially Contaminated Land. The Council used these two legislative provisions to point out the necessity of the standards spatial referencing in a presentation to PITCOM. MPs were informed of the developing Street and Land and Property Gazetteers and the importance of their enforced use by legislation. The Council also commented on the Regional Research Laboratory Initiative.

The key part of the Council's year was a review of the Chorley Report. The review attempted to create an update on the progress in meeting the recommendations of the committee. The findings of the review formed a part of the 91 Conference.

A major issue raised by the AGI membership in 1991 was Ordnance Survey Copyright and Pricing Policy. As a result of comments to the AGI and the Ordnance Survey of Great Britain the Minister responsible for the Ordnance Survey asked the AGI to facilitate a discussion between the users and OS. These discussions chaired by the AGI Council's 1991 Chairman are still ongoing. As a result of the concern over Copyright, the AGI 1992 Chairman, Gurmukh Singh, has secured the support of a legal firm specializing in Copyright. The legal advisors will be members of the AGI's committee reviewing legal restrictions for the geographic information market. The results of this investigation will be presented at the 1992 conference.

The 1991 Conference Committee, under the Chairmanship of Nick Pearce, organized the Third Annual Conference and Exhibition. The event saw over 100 per cent growth in attendance at the Exhibition and 11 per cent growth in the Conference, an overall attendance of around 3 000 against a backdrop of recession and questionmarks over the viability of two separate national events.

A new venture for the 1991 conference was the involvement of the many different organizations from different sections of the GIS community. This widespread involvement underlined the AGI's role as the umbrella organization for GIS in the UK. The other benefit of the community shaping its own conference was the variety of new faces among the speakers and the new experiences that were related.

Education, Training and Research Committee

The 1991 Education Training and Research Committee (ET&R), chaired by Mike Blakemore, has focused its activities on producing short review publications of relevance to different parts of the GIS industry. There has been a long lead time in commissioning authors and producing publications. The processes of editing and quality controlling the publications is new to the AGI and has relied upon the work of its part time committee members in the Publications Committee and Council. Only four publications have been produced to date, however, a large number will be delivered during 1992.

The 1992 ET&R Committee, chaired by David Unwin, is now starting to turn its attention to its involvement with the wider education and research community. Competitions for schools in attaining the targets set for GIS in the national curriculum is just one of the ideas that are being investigated.

Membership Committee

As well as gaining new members for the Association, the Membership Committee is responsible for a range of activities for the membership. In 1991 the Committee, chaired by Phil Jeanes, intentionally focused upon building up activities for members. To achieve this the committee introduced the concept of Special Interest Groups (SIGs). To date the Committee has overseen the creation of the Environment and the Emergency Services SIGs.

Another area of interest to the AGI membership has been Property and Asset Management. While a SIG has not yet been launched for this area, four regional seminars were held to bring the latest information to the membership. The seminars included a briefing on the "registers of potentially contaminated land" required by the Environmental Protection Act 1990 and the CIPFA recommendations on Asset Registers.

Most recently the Committee has been involved in the creation of a Scottish "Regional Interest Group" for the Grampian area.

Promotions and Publicity Committee

Promotions is a small but very active Committee under the Chairmanship of Michael Nicholson. The committee seeks to raise the profile of the Association of Geographic Information. The Committee is responsible for a constant stream of press releases and articles going out from the AGI. As a result of the AGI's press release service the press now seeks out the AGI's opinion. Information distributed by the committee has been taken up by, among others, the *Times* and the *Financial Times*.

The Promotions Committee has also been responsible for the creation of the AGI Awards scheme and presentation dinner. The awards, organized by Alastair Macdonald, have been in four categories: The Student of the Year; The Journalist of the Year; The Technology Award, and; The Industry Award. The presentation of the awards has become a key item on the Associations Calendar. The presentation dinner has developed a reputation of being a meeting of the geographic information "family" and is always a sell-out. The 1991 AGI awards dinner was held on HMS Belfast, this year the venue moves to Lords Cricket Ground.

Publications Committee

The Publications Committee, chaired by Nigel Sheath, is directly responsible for the Association's Yearbook and its Newsletter.

The organization of the Yearbook has been undertaken by the joint editors James Cadoux-Hudson and Ian Heywood. James and Ian have created a hierarchy of editors and authors who track subject areas. The objective of this reorganization of the yearbook is to ensure the quality of the articles and comprehensiveness of the Yearbook as a record of a 12 month period.

As with the Yearbook, the Newsletter has been changed. The emphasis is now placed upon short news items giving the latest developments in the geographic information community in the UK and overseas.

Standards Committee

Since the production of the last yearbook there have been many changes for the Standards Committee. The Standards Committee structure has changed, its work has increased and there has been an increased recognition that the AGI is the home of national standards for the geographic information community.

The AGI Standards Steering Committee and BSI/IST/36 is chaired by John Rowley. 1991 started for this Committee with the creation of an agreement with the British Standards Institution. This agreement created a BSI Technical Committee on Geographic Information (BSI/IST/36). The AGI Standards Committee's first job in its BSI role was to turn NTF, the UK *de facto* standard for transferring geographic information into a British Standard. This transformation has been achieved; NTF has now developed into BS7567.

The work of the Standards Steering Committee has been to create a long-term strategy for a basket of standards. The objective of this strategy is to create a truly open environment for geographic information which will ensure the widespread and effective use of GIS. The creation of this standards strategy has allowed the AGI to play an active part in the creation of a European-wide agenda for geographic information standards. The European Standards body CEN created a technical committee which mirrors the work BSI/IST/36 on a European scale. The foundations of the work at CEN were created from the contributions of the AGI's strategy document and a similar document prepared by the Nordic group.

The Standards Steering Committee is set to continue is work at BSI and CEN level. The next specifications to move to British Standard Status will be "The National

Street Gazetteer'' and ''The National Land and Property Gazetteer''. These standards are set to underpin the legislation for the Computerized Street Works Register and the Local Authority Registers of Potentially Contaminated Land.

Under the Steering Committee are five working committees. Group A, Implementation Standards, chaired by Chris Gower, deals with the development of mainstream IT standards to ensure that they will cope with geographic information. Group A's most notable work has been carried out by Steve Dowers of Edinburgh University, who is putting the geographic input into the development of the SQL Database standard at international level.

Group B, Definitions and Concepts, chaired by Andrew Larner, has been actively involved in a number of projects. The Group is responsible for the ''GIS Dictionary'' which proved invaluable at the first CEN standards meeting where the UK were the only country to have defined the terms they were using. Group B has also been involved as the AGI/BSI link to both the Gazetteer projects run under the auspices of the Local Authorities Geographic Information Advisory Group.

Group C, GIS Data, chaired by Rob Walker, has been actively looking at a variety of standards ranging from ''quality'' through to spatial data types in a joint working group with Group B. The spatial data types group (SDTG) is likely to prove one of the AGI's most important in the coming years. If successful, the SDTG will lead to the definition of a common set of data types which Group A can put forward into the SQL standard, so ensuring compatible and portable data as well as software. The Group also aims to ensure the development of standard data models which will aid the coherent development of GIS in all industry sectors. Under Group C is the Surveying Sub-Group which was created to monitor the work at CEN level on a European reference system.

Group D, in the past, dealt with development of NTF. NTF is the UK's transfer format, and it has involved input from over 150 people of a variety of backgrounds. The variety of input has ensured that the format will transfer the widest possible variety of geographic data, unlike formats which have been developed in one industrial sector. Part of the development of NTF has been to ensure that it takes into consideration these sector-spawned formats. NTF now allows the transfer layer dedicated to the DIGEST Topological Structure which is part of the Military Sectors transfer format.

The fifth group was set out to deal specifically with the problem of a referencing framework for the whole of Europe. This project is being considered for standardisation by CEN, the European standards body. The project in the UK is being managed for the AGI by the Royal Institution of Chartered Surveyors.

Secretariat

The AGI now employs two full-time and one part-time members of staff. Andrew Larner, AGI Secretary, is now supported by his personal assistant, Maxine Allison and Brian Shephard who looks after the bookkeeping.

The AGI Secretary has also been working on the key projects identified by the Council, from developing the links with PITCOM and BSI to bidding for the role of disseminating the data sets of tradable information held by local government. The Secretariat was successful in this bid and now runs a telephone enquiry service for anyone interested in the spatial information held by central government. The next stage of this project will be to expand it to other sectors which hold key spatial data sets.

Part II

GIS and the World

3

The world of GIS

Introduction

It does not require much in the way of imagination to realize that the application of GIS technology is now a global phenomenon. One only needs look at the sales literature of any one of a number of the top five GIS vendors to find the standard "where we have sold systems" type map. For most of these vendors there are only a few patches on the globe which remain blank. As a little aside, there would probably be an interesting research project associated with the diffusion of GIS products and the changing nature of political boundaries. For example, the recent changes in the former Soviet Union and the subsequent realignment of internal political boundaries means that a vendor, having sold or donated a single system into Moscow, can no longer legitimately shade in the whole of what was the USSR. However, it works both ways; those vendors who had made sales in the old West Germany now have every right to shade into what was East Germany as well!

In the same way as many of the GIS vendors see the GIS market place as global we as editors feel that, though the focus of the AGI Yearbook should be directed primarily at what is happening in the UK, reviewing initiatives, experiences and research around the world will help provide a different perspective on events in the UK. We originally started out with the idea of providing a short snapshot of what was happening where on a global scale country by country. However, it soon became apparent that this was an impossible task. Who should we get to write for each country? Would they provide a biased perspective? What should we include? What should we leave out? In the end we have decided to base the section around a set of papers which describe the activities taking place in a selected set of countries. Each year we will introduce new countries or new authors to provide a different geographical and personal perspective. We recognize that at any one time it will only provide a glimpse of what is happening where, but as the papers in this part show, such a candid view can be extremely enlightening. Before taking a brief look at the countries portrayed this year, a plea for next year. If you are reading this section sipping wine in Spain, drinking coffee in Colombia or eating passion-fruits in Hawaii and feel that you have something to say, please contact us and we would be more than happy to consider your views for next edition.

This year we have papers from the following six countries; Austria, Australia, Bahrain, the Commonwealth of Independent States, France and Canada. In each case we have left it up to the author to identify the key issues rather than write to a template provided by us. We hope you will agree that this has captured the spirit of the moment rather than simply provided a catalogue of facts. Several of the authors have tried to draw parallels with what they see happening in the UK and some of them should give us food for thought. For example Derrick Steel, when describing activities in Australia, states that, "Obviously, where the charges for single data sets are pitched far too high there can be little opportunity for developing value adding processes." This statement born out of the all too real experience of the pricing of data by public service agencies in Australia should give those of you currently trying to put a market value on data owned by your organization something to think about.

In Austria, Josef Strobl stresses that it is still the issue of where to find spatial data that is one of the biggest drawbacks to the uptake of the technology. From an education and training point of view he indicates that the new challenge is not for education about GIS but for more general spatial concepts such as map projections and pattern recognition. Basic geography for the executive looks like it will soon be on the cards. In Bahrain (as the opening quotation to the chapter by Bell and AbdulGhani from the *Gulf Daily News* illustrates) it is the costly stories of near disaster which have initiated a comprehensive Land Inventory Project. The project is designed to establish a country-wide database which will identify and catalogue all property parcels with their appropriate owners and attributes.

This chapter is followed by an enlightening piece by Alexander Koshkarev on GIS in the newly united Commonwealth of Independent States. Koshkarev points out that two major factors limit the uptake and development of GIS technology. First, the lack of modern hardware such as high-resolution monitors and workstations (which are still on the restricted COCOM list) and second the unreliability of ex-Soviet maps which contain deliberate distortions and blank spots. Against this somewhat pessimistic backdrop Koshkarev makes it clear that the challenges facing those involved with the development of GIS in the CIS are on a scale several orders of magnitude different to those in the UK.

In reviewing recent developments of GIS in France, Jean Denégre points out that over the last two decades the main thrust of GIS development has be related to Urban Planning and the development of Urban Data Banks. These early systems, like those in the UK, were mainly devoted to computer-assisted cartography. In the 1980s French decentralization laws helped to change the direction of GIS development into agricultural and rural planning. An interesting concept introduced by Dengre is the GIS Observatory which was established at the national level to act as a GIS watch-dog to provide both data and information on GIS and remote sensing. At the institutional level general co-ordination is assumed by the Conseil National de l'Information Geographique. This national body appears to be similar to the AGI in some respects; for example, its representatives are made up from the same wide range of interested parties which represent the GIS community. On the other hand, the council of CNIG is directly responsible to the government, from whom it also receives financial support for specific projects.

In reviewing the state of GIS in Canada, Rogerson focusses attention on the educational sector. At the outset we are reminded that CGIS, the Canadian GIS, developed in the early 1960s was arguably the world's first GIS. Since these early days Rogerson points out that the Canadian GIS industry, often described as "Geomatics", has not only kept pace with the rest of the world but in many cases has been responsible for a large proportion of the GIS development activity. For example, at least five of the 'household" GIS product brands are Canadian in origin. However, on the educational front, he remains a little less positive. Rogerson points out that while undergraduate, diploma and research programmes are thriving there are few, if any, comprehensive Graduate programmes on offer. In fact, he states "Many Canadian Students wishing to complete doctoral work in GIS go to universities in the United States where, perhaps ironically, they may well be supervised by a Canadian''.

However, despite this missing element, the picture Rogerson paints of Canada's GIS educational activities is both healthy and innovative.

4

GIS in Austria

Josef Strobl
Department of Geography, University of Salzburg, Austria

Giving a concise and thorough report on the state of Geographic Information Processing in Austria has already become a not-so-trivial task. Despite the country's small area and population, GIS development activity at research institutions like Joanneum Research, Forschungszentrum Seibersdorf and some university departments got GIS off to an early start many years ago and provided widespread awareness of GIS's capabilities. Today most popular commercial products are well-established in the market and some of them can show an impressive list of successful installations. The detailed picture over various sectors of business and levels of government shows quite marked differences in terms of GIS penetration and success.

The single most important factor in driving the GIS marketplace probably has been the introduction of regional GIS's into the governments of Austria's nine provinces. These installations are today at quite different stages of implementation and have been initiated from diverse departmental needs, giving these systems a different focus in their first stages. Nonetheless, future homogenization will be greatly supported by their common choice of ESRI's ARC/INFO as primary GIS software. These provincial activities in turn caused a notable effect in the private sector. Companies offering services ranging from primary data acquisition to regional planning and specialized modeling computations increasingly turn to GIS as a key tool.

GIS today is an indispensable instrument in the planning for tomorrow's high speed rail network, the siting of waste deposits, fine-tuning permits for extractive industries and basically all other major "spatial" projects, most of them requiring mandatory environmental impact studies. One of the major roles GIS was to fill in several such planning and siting studies was not only to provide spatial analysis and cartographic documents but, in an ongoing dialogue with citizen groups, the different options and the rationale behind recommendations were interactively examined at a GIS workstation.

Managing the already scarce spots of unspoiled environment as national parks or under other schemes of protection is another field of already heavily entrenched GIS application. The national parks "Hohe Tauern" and "Kalkalpen" already operate GIS installations at their respective administrative and/or research facilities. Several other projected national park areas in various parts of the country, ranging from the lowlands bordering on the CSFR up into high mountain areas, are under study by GIS-centered projects aiming at producing basic information for proposing borders around future parks and giving first clues to the future management of these sensitive spots.

GIS introduction in federal government and agencies shows a much different picture. While most resource- and environment-related departments have already embarked on big-ticket projects like managing Austria's extensive public forests, documenting bedrock geology and soils, some agencies are lagging far behind. This is in part due to split responsibilities between levels of government (e.g. in environment, planning, road network maintenance), and on the other hand stems from the fact that allocation of the required resources is hard to obtain in times of fierce inter- and intra-departmental budgetary competition.

Still, some huge federal data-centered projects like the conversion of Austria's property records and cadastre have made steady progress over the last years, already providing operational access to fully digital records in some parts of the country. The focus now gradually shifts to smaller scale topographical mapping, where strong demand for digital basemaps builds up from planners and a multitude of special applications like emergency services, vehicle routing etc. These latter application disciplines are still hampered in their progress towards GIS implementation by this lack of digitally available topographic base maps.

Infrastructure and utilities management has to be viewed on two levels. There is considerable sectoral division on the regional to national levels, with mostly publicly-owned companies providing communication, gas, electricity and some other services. On the community level services tend to be more integrated, with a strong push towards more organizational separation from government. Whereas all the bigger cities in the provincial capital league are already successfully operating facility management systems from various vendors, the battle for second-tier cities and then for the numerous smaller communities just heats up. Aside from sheer size, one difference is that most bigger cities went through the CAD-stage of automation, whereas most smaller local utilities have to start from scratch, that is paper records. In today's closer integration of relevant technologies, it is definitely a much easier start for them.

Since these GIS developments in communal facilities management are not only driven by technological advances but the pending general availability of the national property cadastre, issues like standards and co-operation are of course hot topics. Some initiatives (e.g. in the Tyrol) try to work out schemes of regional co-operation like "centers of competence", from where local communal needs shall be serviced. The most difficult indicator of ultimate success definitely will be the level of integration finally reached. Some projects undertaking an overall definition of communal GIS needs ranging from facilities management to property and population records, billing of services and construction regulation and permits have run into severe difficulties due to the sheer diversity of considerations. Still, local government is certainly the most important area of growth for the GIS industry in the immediate future.

The issue of standards on a national, inter-departmental level is probably even more difficult. A series of symposia at Vienna's technical university ('GeoLIS') have dedicated much effort towards raising awareness and promoting documentation and exchange of data sets between agencies. Unfortunately, there is still no single or central place to look for spatial information from various disciplines. The national institute of standards already regulates some basic forms of interchange, but efforts on a higher level still lack the basic research most countries are probably already eagerly awaiting.

GIS usage in market-related applications is still in its infancy. Only a few market research, retailing and distribution companies employ anything more sophisticated than simple mapping packages. This lagging may be explained by being a comparatively small, immobile and obviously clearly structured market. Tourism being one of Austria's economic mainstays, marketing of vacation spots is one of the most promising areas where geographic information processing is just now trying to link up to the "point of sale".

Projected growth figures for Austria's GIS market being in line with the general European estimates for the next few years obviously creates strong demand for qualified specialists. Neither public administration, private business nor the GIS industry can today fill all open positions for GIS specialists. Several universities are trying to gear courses of study towards GIS, most notably at the Technical Universities of Vienna and Graz, Vienna's Agricultural University and the Geography departments at Vienna, Salzburg and Klagenfurt. In addition, short courses and postgraduate courses are offered with primarily a technical focus. Still, with the spreading of GIS application into a host of non-specialist disciplines, much more emphasis on general "spatial information science" will be needed in many disciplines to educate tomorrow's end users.

Looking east, but also north and south beyond Austria's borders, the creation of Europe's "new democracies" has also brought along what most people in GIS industry view as mid to long term opportunities. Numerous projects are brought up, defined and supported out of Austria, thereby contributing to the strength and competence of the local GIS industry as such. Trying to anticipate Austria's position on the international GIS scene, this focus on the East will probably converge with a rapidly increasing involvement in EEC-initiated activities. These two recent developments should create strong transnational involvement in the furthering of GIS technology, methodology and application.

References

Hvllriegl, H.P., 1991, Regionale und Kommunale Informationssysteme. Stand der raumbezogenen Informationssysteme in den Bundesldndern und Gemeinden Osterreichs. Geo-Informations-Systeme 4/1991 20-26.

5

An Australian experience

Derrick Steel
Steel Information Management

Introduction

Australia has often been held aloft as a shining example of what can be achieved with GIS and LIS. This has to a large extent been attributed to its status as "an open information society". Australia, like the USA, is a federal system of government where individual states have a far-reaching autonomy, creating the equivalent of mini-countries within the overall umbrella of federal foreign policy, defence and revenue raising. Each state has developed with different historical constraints, different sets of priorities and different institutional structures, but basically each undertakes the same functions. Some of these institutional structures are conducive to the development of a statewide GIS/LIS, others are not.

The basis for Australian LIS

Let's take land administration as an example, as it is internationally recognized as the root function for the largest users of GIS/LIS. In Australia, each state has its own equivalent to HM Land Registry which registers land ownership and the sale and transfer of land. Many states, including Western Australia, South Australia and the Northern Territories have a complete register of all land ownership. The ownership of any parcel of land can be relatively easily traced through the land registry office, with no restrictions on access. Naturally, with such a complete record of land ownership it provides the ideal opportunity to raise taxes from landowners. This land tax, along with rates, is used as the basis for raising many of the states' revenue requirements.

What has this information got to do with GIS/LIS? For those involved in cost benefit analysis for GIS, often the most difficult part of a project is to convince the financial community of the many "intangible benefits". However, where those benefits include tracing landowners who collectively are avoiding paying millions of dollars in land tax you are on to a certain winner. This basis underpins most of the GIS/LIS development in Australia—revenue collection.

The payback multiples for LIS development for revenue collection in many of the states ran into double figures. In that success, however, lay a major problem. The holders of the purse-strings wanted similar payback multiples for any subsequent development of GIS/LIS.

LIS/GIS co-ordination committees

In an attempt to continue the ongoing development, most states established land information co-ordination committees. Their declared mission was to ensure that individual state government data sets were collected only once, up to a standard that satisfied all, and that each one was maintained by a recognized custodial or trustee agency.

All else being equal this strategy should work. However, information tends to be used as a most potent weapon, either against competing agencies or as a means of self-elevation. Over time these LIS committees developed to exclude federal government, local government and the commercial sector. State government departments in turn came to regard these sectors as their captive customers. Each "customer" was earmarked to provide substantial revenue in order to justify the states" optimistic cost benefit projections for value added information. What they failed to realize was that these other sectors were just as capable of competing against them in the acquisition and sale of information.

In general, the market studies undertaken were woefully inadequate, being based on the states" legislated monopoly in the provision of "building block" datasets, such as the National Grid system and land ownership polygons. Many failed to implement an effective marketing strategy, due to the many internal disagreements between state departments. It became easier to trade one off against another or simply go back to more cost-effective, traditional methods, of acquiring and supplying information.

Cost recovery

In many cases it was more cost effective to purchase the required hard copy documentation and digitize it into your own database. This avoided all the restrictions, costs, bureaucracy and uncertainty associated with the purchase of digital data from another agency.

Charges for such digital data varied from as little as 10 cents per ownership polygon in Western Australia to as much as A$2.50 per polygon in Victoria. This difference in price demonstrated the guesswork involved in the valuation of LIS data. None of the prices had any market justification, as there was no historical basis or in depth market analysis on which to base an equitable pricing structure.

As the major datasets were mainly held by the state government sector their valuation of data lacked any commercially based reasoning. In general, they worked on the assumption that if my dataset A is worth $A and your dataset B is worth $B. By combining the two with a GIS/LIS, as a value adding process, the combined datasets should be worth $2(A + B). In conclusion, we can both double our income by combining our data.

State information co-ordination committees were set up based on this version of the pyramid selling theory. The theory was that as all state government databases are progressively combined their value will be increased in accordance with the following progression:

$A; $B; .... $n
$2(A + B); .... $2 ((n-1) + n)

$3(A + B + C); .... $3((n-2) + (n-1) + n)

which leads to the equation:

$n(A + B .... n)
where n is the total number of state datasets.

However, this theory did not take into account the costs of new product development, marketing and distribution. Many of these costs are dependent on the size and profile of the target market, where marketing and distribution costs alone can be between two and five times the production costs.

A more realistic commercial assessment for the value adding of data can be applied to single datasets, that is:

$(A + B .... n) + v

where v, the added value, is positive where the data acquisition, administrative, marketing, distribution and maintenance costs are less than the income generated from sales.

Obviously, where the charges for the single datasets are pitched too high there can be little opportunity for developing value adding processes. This is the rock on which many of Australia's land information co-ordination committees are now foundering and to which many of the UK public sector agencies are also heading.

From my experience, working within a multidisciplinary land information consultancy in Australia, I believe that the public sector is best positioned to acquire, store and maintain data, but they are hopelessly structured to effectively market and distribute information. This function, I believe should be left to the commercial/professional sector who can assess the particular need and hence value of data, in terms of how it is to be used.

Public sector organizations on the whole tend to overvalue their information holdings, frightened of undercharging in case some other organization makes a profit at their (or should I say the public's) expenses. As a result, the sale and value adding of their information is often not feasible. This is especially so when there is inflexibility in the pricing policy.

The development of distributed databases

In Australia, the early development of large centralized databases was supported by the states in order to implement their combined market valuation theory for land information. However, large centralized databases are normally the product of a rigid and inefficient bureaucratic, autocratic or authoritarian society. In the case of Australia, Canada and Sweden, I suggest that they belong to the bureaucracy classification.

Where does that leave the Australian states today?

Many states have simply run out of funds or have failed to realize the projected benefits. They have often succeeded in simply converting a public bureaucracy into a computerized public bureaucracy. The LIS equivalent to the Big Bang has now been somewhat discredited.

However, perhaps their most significant development as an alternative to a centralized LIS is the "hub theory". This recognizes the growth of distributed databases across a wide range of public and private sector organizations.

Instead of maintaining a centralized database, the hub is a computerized referencing system that allows immediate access to other agencies databases and the authorized abstraction of the required information. This structure helps to promote the development of smaller, more flexible databases within a regional framework, easier to fund locally, and closer to the true custodians of the data. However, these local data providers have to be empowered with a standardized referencing system in order to become part of the larger framework of available information.

Who funds the establishment of the referencing system?

Many of these initiatives were unsuccessful due to the limited local budget. However, the main reason for lack of progress has been the historical rivalry between many local agencies who fiercely maintain their independence. Many have gone as far as marking out the equivalent of tribal territories within their information holdings, designed to prevent others from accessing their information other than by legislation.

The UK—An international view

Among the basic datasets for any LIS/GIS are:

- A standardized National Grid System,
- Land ownership polygons,
- Basic land use classification,
- Road and railway networks,
- River networks and water features.

In the UK, despite the wide acceptance of large scale OS topographical mapping, it has to a large extent been developed as a default option to the UK's poor public record of land registration maps. Although most organizations use the OS map as a background for their own information, in reality they are tying their information to the National Grid. In future, the agencies will have the opportunity to bypass OS topographical maps by using satellite positioning systems to give direct National Grid co-ordinates to their own features of interest. In many cases OS large scale topographic detail is superfluous to needs and in other instances important detail is not recorded.

As for UK land ownership polygons, the UK are near the bottom of the world's league table for maintaining registers of land. Just over half of landowners are recorded on the government register and less than a quarter of UK land by area has ownership registered. This unenviable position can be directly attributed to the archaic land laws drafted during the nineteenth century to protect landowner's privacy.

For this position to be dramatically reversed legislation is needed to introduce further compulsory registration. This legislation should target all land held by government; all land subject to inheritance; all land used to raise capital or mortgages and all land receiving some form of agricultural subsidy. It could be applied on a regional basis, with a view to completing the register within 20 years—Domesday 2010.

Summary

The cost of GIS/LIS hardware, software and training will continue to reduce with the advent of more user friendly PC/Workstation packages. Hence the future success of GIS/LIS on a national basis will largely depend on the availability and cost of data. Viewing the availability and cost of data in the UK is indeed depressing in comparison with not only Australia but with other developed countries in the world.

As such, the UK's future is dependent on the development of a strong political lobbying force able to influence parliamentary support for major changes in legislation. Perhaps the Lord Chancellor's office is currently in a mood to take on board such an initiative. In the meantime this delay will help ensure that the UK does not make the same mistakes as other countries. On the other hand is the UK in danger of missing the boat altogether?

6

Geographic Information Systems in Bahrain

J.F. Bell and Yousif AbdulGhani
Ministry of Housing, State of Bahrain

> International telecommunication services in areas of Bahrain were cut off for more than five hours after a contractor damaged one of the fibre optic cables. More than 4000 subscribers were affected. The contractor was trying to replace a palm tree using a mechanical digger . . . .
>
> *Gulf Daily News*, 8 December 1991

Origins

By its nature GIS is a rich tapestry woven from widely diverse threads. Each thread has its own origin and it is the coming together of this diversity which gives GIS its power and value. In Bahrain, the threads came from an Electricity Department visit to Singapore; the high cost and danger level generated by uncharted underground services; the realization of the potential value of having land ownership information which was co-ordinated both literally and metaphorically; and, as the prime mover, the surge in land values created by the coming of oil wealth which brought a change away from a rural life-style and the realization that land was a very limited resource. This chapter is concerned with the survey input to the weave in Bahrain.

In the late 1970s and early 1980s it became clear just how far the knowledge of land ownership had been allowed to slip away from the relatively high standard created in the mid-1920s when plane table teams from the survey of India conducted detailed boundary surveys of the non-urban areas. The root of title in Bahrain is an Amiri gift. Title Deeds, each signed with the personal authority of His Highness the Amir, had been drawn on the basis of graphic surveys of perimeter measurements using a cloth tape. Necklaces had grown up around the villages as owners received their gift lands of known dimension but poorly controlled shape and virtually unknown position. With a settled population, a largely agrarian economy and low land values the records and survey techniques in use sufficed for public needs. The emergence of an oil rich economy, with escalating land values reflecting limited supply, was accompanied by burgeoning development.

In 1978 a Survey Directorate was established in the Ministry of Housing, and in 1979 a revised Land Registration Law regularized what was in effect a Register of Deeds and introduced better documentation. A Ministerial Order the following year sought to restore the standard of survey. A series of ambitious master plans for the capital city of Manama and other major towns were, unfortunately, all too frequently incompatible. A highly commendable resolve to alleviate this situation rapidly resulted in an exponential growth in the workload of the newly formed Survey Directorate.

Problems began to emerge of duplicated ownerships which could not be easily resolved with the records available; of very expensive conflicts of interest as the various service authorities spent money based upon individual interpretations of the master plans; of land owners with expensive development loans they were unable to use because the exact site of their land parcel could not be readily indicated. Costly events were created by cases where successive graphic surveys extrapolated from an initial misaligned development obstructed essential roads or resulted in substantial building over expensive buried services. Accidents occurred, or were narrowly averted, as high pressure water mains or high tension cabling were encountered unexpectedly.

In April 1982, following upon an intensive review of the problems, a number of essential requirements were implemented by His Excellency the Minister of Housing.

- A Special Land Inventory Project (LIP) was created to identify all live deeds outside the town centres of Manama and Muharraq, and to place those deeds in their relative position in a series of 1:1 000 Cadastral Index Maps.

- A contract was placed for new 1:1 000, aerial mapping of the island. As a by-product of this mapping, the contractors supplied digital data tapes of the topographic information which were to become the foundation of the GIS systems.

- The island has been transformed onto the UTM projection and gone metric in the 1970s. It was decided to support the new mapping and field surveys by densifying and adjusting the existing traverse control network to the point that all future surveys could be expressed readily in UTM terms.

- A decision in principle had already been made to introduce computer power and this took definable shape in the course of the review. The original system consisted of:

  — One DEC VAX 11/730 with
  — Five MB RAM,
  — One 84 MB system disk,
  — Two 675 MB additional data disks,
  — One 1650 bpi magnetic tape drive,
  — One DEC LA-120 system console,
  — Two Intergraph dual screen workstations,
  — One Calcomp 1055 pen plotter with 0.1m accuracy and a paper size of 36.7″ x paper roll length four pens,
  — Versatic V80 screen plotter, plot size up to 11″ x 17″ 200 dots/inch,
  — Five alphanumeric video terminals,
  — Intergraph IGDS and DMRS nucleus software,
  — VAX-Fortran compiler.

- Other decisions dealt with a review and, where appropriate, revision of the relevant law; staff organization; job flow control; standardized practice and procedures and a recruitment and training plan aimed at a 10 year programme to transform a department without a single formally-qualified Bahrain surveyor into a competent organization made of at least 90 per cent local staff.

It is to the enormous credit of the vision of HE the Minister, to the unstinting and far-sighted support of senior management of the Ministry and to the small staff in post in 1982 that in due course all of the above came to pass. The present GIS system still has far to go but about 1986 the emphasis within the Survey Directorate changed from pressurized problem solving to positive impetus towards an ambitious but very worthwhile goal.

Strategy

The Ministry of Works, Power and Water (MWPW), with responsibility for the majority of the services authorities, had encountered similar problems and were readily persuaded of the potential benefits of a common system based upon the new mapping. The two ministries started down a familiar road of purchasing parallel computer systems with a view to future networking. The physical join still lies in the future but there is now a steady transfer of data by "wheelbarrow" between the systems.

The concept is the equally familiar layer cake. With a common base, the UTM grid, each contributor feeds his own positional and qualitative information into a separate layer. Only he or she may write to that layer and he or she is responsible for the accuracy and currency of the information it contains. Any other user may consult and draw down information from a slice through such layers as required. To date 14 layers have been identified and are at various stages of involvement, including:

- Survey Directorate contributes the base grid, topographic and cadastral boundary information;
- Planning Directorate, from the same ministry, is beginning to make its contribution of planning information of all kinds;
- MWPW contributes location information on electricity power cables, road reservation requirements, water supply and sewerage;
- Bahrain Telecommunications, with multicore and fibre-optic cables to protect, is one of the more advanced users;
- The Central Municipal Council is developing the facility to deal with local authority applications;
- The Land Registry (LRD), housed in the Ministry of Justice, is being actively wooed to bring in information on current owners and in due course take to a comprehensive picture of registered land;
- Finally, in a land where spelling of names has a certain heuristic quality, the Central Statistics Organization (CSO) offers a unique registration number for each resident which is of service to all.

The link to CSO will in due course give direct access to the wealth of data from the comprehensive National Census of 1991 which opens up new horizons of demographic mapping. Already there are major independent users who have seen the virtue of becoming computer compatible with the system. The major oil company (BAPCO), at least one major engineering consultancy firm and the local private sector survey firms are all at varying stages of joining the club.

The system offers problems which must be negotiated. Mutual currency of data, compatibility of level of accuracy, the provision of forward planning information which is not precisely located but is vital to ensure that land is reserved for future uses and, perhaps most difficult of all, the constraints of security are all matters which have been, or will need to be, hammered out. With mutual awareness of the potential value and a will to succeed none of these problems is thought to be insurmountable.

Practical considerations

A geographic database is conceptually scale free; in practice it is necessary to think though the sources and users to determine the structure which will most easily absorb and disgorge the information. In Bahrain the main competitors are a system based upon the sheet lines of the 1;1 000 standard map series, which favours the survey element, and a system based upon the registration districts which are used by MWPW, CSO and a widely accepted National Addressing Project. Cadastral users are ambivalent on this issue because the unique parcel numbering system which they use is almost a parallel of the registration districts. With the introduction of a relational database system the problems of non-regular boundaries which tend to haunt the more orthodox users are quickly disappearing. The transition from spaghetti to shape proceeds apace.

At the time of the introduction of the database, each parcel of land had four systems of reference. The LRD historically identify every transaction on the land by a case number made up of the year and the serial number of the entry in their Case Book Register. Recognizing the ephemeral nature of a transaction, they also gave each parcel as it was encountered a registration number under which all case numbers associated with that parcel could be collated. Originally this number also was year sequential but in 1979 changed to an ongoing serial irrespective of date of issue. Survey Directorate independently evolved a unique plot numbering system based upon registration districts and this was closely cross referenced to the relevant survey job numbers which were serial though each year. The first requirement was a mutually invert file to relate the elements of this witches' brew together with the owner's name into a system which would enable entry of any one to automatically identify the others.

Within the cadastral layer there are four components which must be accommodated.

- Surveys carried out by graphic means, which are essentially linked to topographic hard detail of any accuracy of about 0.3m, must be separated from instrumental surveys accurate to 0.05 m. The source of this problem is a slightly misguided but now irreversible legal requirement introduced in 1980 for survey dimensions to be expressed to a repeatable accuracy of 0.1m.

- Positional information must be supported by attribute information about the plot and its owner.

- To be of best value the database must retain archival information giving a full history of each individual parcel of land.

- Finally, as the management control system for jobs in work uses much of the information which will be required later, the job management systems is also integrated into the overall database.

The philosophy which is now accepted clearly developed over time. The initial windfall of computerized topographic data brought the planned introduction of a system forward by about five years. The result was a certain amount of activity which was implemented before it was thought through, but most of the product would later be utilized to considerable effect.

With the topographic information in place, attention turned to land ownership and boundary information.

Development

- The 1:1 000 Cadastral Index maps created by the LIP depicted all known Title Deeds outside the older part of the two main townships with 95+ per cent completeness and relational accuracy of 3-10 metres. These maps were digitized to become the "blue" layer in the database, a technique which stayed well within the purported accuracy of the source.

- New graphic surveys, keyed to hard topographic detail to an accuracy of about 0.3m were captured as a "green" layer used to supersede the blue wherever opportunity allowed.

- Finally, past and present instrumental surveys with corner points co-ordinated on the UTM grid to <0.05m were entered numerically as the "red" layer. Again red was used to supersede blue and green wherever opportunity allowed.

The steady growth of the red layer is a measure of the progress of the definitive cadastral layer in the database. The colours in use have altered but the concept of a progressive upgrading though three layers continues.

The process left a significant gap made up of those land parcels which were surveyed graphically before the introduction of the system and which cannot now be positioned to an accuracy better than the blue layer of the Index. The upgrading and inclusion of these parcels to an acceptable accuracy is one of the more difficult areas of current investigation.

In the meantime the LIP, which had completed the non-urban Cadastral Index Map series in 1985, moved on to a second phase. Older Title Deeds in the rural areas were normally defined by their dimensions along physical boundaries such as tracks, bands and irrigation channels. Rapid development frequently destroyed these landmarks and the geometry of the original surveys was quite inadequate to pin the relative index into absolute position on the grid. Using groups of deeds bounded by established hard boundaries and detailed research interpreting old air photography, obsolete mapping and every other available source, the pattern of deeds in each cell was resurrected and co-ordinated to form a key contribution to the red layer. The benefits to consistency of survey and to the easing of future surveys in these areas was enormous.

In parallel to the database, the department commissioned a customized job management system to control the flow of work though the sections and to provide source data for production and productivity reports, in 1984 for clients. The record was keyed by job number but it also contained the same unique plot number as the database as well as information on ownership at the time of survey, plot location, map sheet number, centroid, parcel area, survey date and the type of final documentation. This information was retained in the computer after the job was complete and archived. It provided the basis for an attribute for each new parcel. Attributes were completed retrospectively, first for past jobs and then for parcels which predated the Survey Directorate's job system.

As plot numbers are superseded by subsequent surveys the information is retained and becomes the archival section of the database which, by fore and back referencing, enables a full history of every defined piece of land. When the second generation of this control system was created in-house, it was not only considerably more sophisticated but it focused upon making the information which it was intended to retain readily available to the database.

Current status

The current hardware system at the Ministry of Housing comprises:

- One Intergraph 350 (DEC Microvax 3500) server with:
- 32 MB RAM,
- 1 × 280 MB hard disk,
- 1 × 559 MB hard disk,
- 2 x 675 MB hard disk,
- 1650/6250 bpi magnetic tape drive,
- 1 × DEC LA 120 console,
- 2 × Interact-200 dual screen graphic workstations,
- 1 × Interview-220 dual screen graphic workstation,
- 1 × Intepro-340 single screen graphic workstation,
- 1 × Interact 68K dual screen graphic workstation,
- 1 × Versatic V80 thermal monochrome screen dup device,
- 1 × OCE 3036 colour electrostatic plotter,
- 1 × HP Draftmaster-II pen plotter,
- 1 × Roland GRX-400 pen plotter,
- 1 × Interserve-200 plotting server,
- 16 × VT220 Alphanumeric terminals.

The cadastral database consists of the graphic and non-graphic files. At present the graphic files are Intergraphic IGDS design files based upon the topographic 1:1 000 scale mapping which can be viewed on the graphic workstations. The non-graphic files containing the metric information of the parcel (UTM co-ordinates, areas and accuracy of the beacons etc.) are indexed (ISAM) or direct access files. Currently the ministry is in the final stages of integrating the textual information with the graphical database in an integrated Land Record Management system.

Future plans

Space does not allow a description of the experiences of other users, or even a digression into the techniques, problems and developments involved in keeping the topographic layer up to a target currency of 15 months. There is scope for a number of papers on these issues alone. The way ahead, based upon recognition of what has been achieved so far, is indicated by a description of future planning.

The ministry is moving from a purely digital mapping system to an integrated Land Information System by marking the graphical, metric and location representation of the land with its textual attributes. Once this objective is achieved, LRD will join in to complement the updating of ownership information. In addition, work has already started to integrate the various computer systems within the Ministry of Housing into a network to enable wider access to the system and the freer, faster and more accurate movement of data across the network.

To cope with recent developments in the computer industry and to meet ongoing user demands for more computer power and resources, the ministry is considering migrating, and downsizing its topographical and cadastral digital databases to mix with UNIX and PC network platforms. At the same time, electronic links are planned between the various GIS partners in Bahrain, notably the Ministry of Housing, MWPW, CSO and the Central Municipal Council. These links will enable the various authorized users to share information, and communicate more efficiently. In addition, MWPW is planning a major upgrade and transition from VAX-VMS to UNIX, and installing a UNIX-based management system.

It cannot be overstressed that, as in similar circumstances in UK in the 1960s and 1970s, a key element to success has been the sympathetic, knowledgeable reaction of those in a position to provide the funding and the motivation. As a result Bahrain now has a working system rather than just a system which works. It was suggested by one who has been here and who played a significant role in the development, that this chapter might be sub-titled "out to chaos". In a country where one can have an idea today, develop it over the next few years or so, a more appropriate aphorism has to be the hackneyed "small is beautiful".

Acknowledgement

The authors are indebted to His Excellency Sheikh Khalid bin Abdulla Al Khalifa, the Minister of Housing, for his gracious permission to publish this account of the development towards a Geographical Information System in Bahrain.

7

Geographical Information Systems in the CIS: A critical view on the critical state

Alexander Koshkarev
Russian Academy of Sciences

Introduction

It is not an easy task to review the state-of-the-art in the field of Geographical Information Systems (GIS) and computer-assisted cartography (Automated Mapping Systems—AMS) in the former USSR and the states that have emerged upon its dispersal to combine into the Commonwealth of Independent States, or CIS. It is well-known that in the past the situation with modern information technologies in the USSR/CIS has been far from bright. However, in the first part of this chapter the author has attempted to list the fields that are the most promising for GIS progress. Whether these hopes are to come true is for the near future to decide. This future gives no promises of easy life; it will most probably be an intricate intertwining of positive shifts in CIS society, and of the problems and contradictions inherited from the dark past as well as developing anew in the course of the difficult transition from communist slavery to freedom. It is understood that the expectations of the author on the irreversible character of this process formulated in the second part of this chapter should be percieved just as critically as the author's perspective on the state-of-the-art of his subject.

This review may be of certain interest for a British reader who is probably not familiar with the Russian-language publications on the GIS/AMS, the more so that our few publications in English in the *International Journal on GIS* (Koshkarev *et al.*, 1989) and in *Mapping Sciences and Remote Sensing* (Koshkarev, 1990) have essentially become obsolete. However, these publications provide detailed regional inventories of the GIS in the USSR, including Moscow, Kazan, Georgia, Baltic Republics, Moldavia, Western Siberia and the Far East, that are not possible in this chapter because of space limitation.

Current state

One can name several institutions that are successfully introducing the GIS technology to many applications; there are Thematic Mapping Systems in the Moscow State University, Automated System for Forest Mapping (All-Union Lesproject Amalgamation—now the Research Center "Lesresurs"), State Centre "Priroda"

("Nature") of the Main Geodesy and Cartography Department (digital mapping and remotely sensed image data processing). GIS activity continues in the Institute of Geography of the Russian Academy of Sciences (IG RAS), including development of the Thematic Electronic Mapping System with advanced GIS-functionality, the initial phase of the GIS project for the glaciology application, and the GIS project of "An electronic atlas of Tadjikistan: Scenarios of environment, population and economy development for the twenty-first century" which has been planned under the auspices of MAB Unesco, the UN University, and enjoys Swiss and US assistance (Badenkov and Koshkariov, 1990). The latter will involve data capturing, analysis and simulation, using TerraSoft software tools on different spatial levels, from macro-territory to detailed scales. Spatial data handling and analysis have been used in particular for evaluating the environmental consequences of the 1988 Spitak earthquake in Armenia (Borunov et al., 1991). Development in the GIS field has become possible essentially due to better availability of relatively cheap IBM PC-compatible computers with advanced peripherals. Of other current GIS projects one should mention: the Electronic Atlas/Reference Information System of the Volga River Basin (multi-departmental programme) and the Geographic Retrieval Information System of the Russian Far East. The activities on the international arena include involvement of the USSR in the GRID GEMS UNEP programme and establishment of the Moscow regional centre in the GRID global network (on the initiative of the former Ministry of Environment of the USSR—Minpriroda SSSR).

The GIS and AMS problems were the subjects of discussion of several regional and all-Union workshops, conferences and seminars: Scientific Workshop on Problems of GIS Sciences (Tartu University, Estonia, 1983); the first training seminar on GIS, "The Creation and Operation of GIS" (Tartu, 1985); All-Union Conference on Cartography in the Epoch of the Scientific Technological Revolution: Theory, Methods, Practice with subsections: Cartography, GIS, Remote Sensing, Ways of Interaction, and Principles and Methods of Cartographic Modelling on the basis of Automated Complexes, Databases and Remotely-Sensed Information (Moscow, IG RAS, 1987); All-Union Conference on Eco-Informatics and Ecological Databases, ECO-INFO-87 (Moscow, Institute of Evolutional Morphology and Ecology of Animals); second and third Regional Workshops on GIS, Cartographic Modelling and Image Processing (Vladivostok, Far Eastern Department of the RAS, 1985, 1987); All-Union Interdepartmental Workshop on GIS Applications for Environmental Tasks (Moscow, Minpriroda SSSR, 1991). Discussion on the problems in the field of GIS/AMS is the main goal of the regular seminar on geographic information and automated cartography, established in 1989 by the Cartography Section of the Moscow Branch of the USSR (now Russian) Geographical Society. The seminar was publishing a periodical *GEO-INTERFACE* newsletter (editor: A. Koshkarev). Alas, this only periodical on the GIS/AMS had to be cancelled because of lack of financial support.

In the end of 1991 Minpriroda SSSR implemented an AMS and GIS software survey (PC-based only) under leadership of A. Koshkarev. The questionnaire distributed to several dozens of agencies in this country included the questions that were similar to those listed in the *GIS World* magazine for the annual *GIS Sourcebook* (1990) and the British Columbia Forest Resource Development Agreement (Ferguson, 1989). The first conclusions suggested by this inventory are as follows.

Of the foreign-made GIS software five systems are now most widespread in the CIS: pc ARC/INFO (ESRI, Inc.), TerraSoft (Digital Resource Systems Ltd.), MapInfo (MapInfo Corp.), EPPL7 (Minnesota Land Planning Agency) and IDRISI

(Clark University). All licenses for this software were basically obtained as the "humanitarian aid" to science. Except for the raster-based GIS EPPL7, that is relatively low-cost (but sufficiently powerful and is used in many research insitutions including the Soil Institute named after V.V. Dokuchaev, the Institute of Geography of the RAS, Research Center "Lesresurs", the State Research Centre "Priroda"), all other software means are in the phase of adaptation and experiment. Unfortunately, the situation is the same with the pc ARC/INFO Joint Venture of ESRI and IG RAS established with direct participation of J. Dangermond, President of ESRI, for distribution of this product in the CIS countries. (By today it has only been used by the Estonian geographers; Meiner et al., 1990.) In absence of the system of teaching and training courses and extreme shortage of experts familiar with this software, this venture has not yet produced the awaited results that could be expected from this powerful *GIS tool—de facto* GIS system standard.

Of the home-made AMS and GIS software products one should mention the Thematic Mapping System developed in IG RAS; the MAG package implemented by the Department of Cartography and GISs of the Moscow State University for digital elevation modelling and analogous to the latter, but more powerful DEM data handling system "Relief-Processor" of the Kharkov University; KAPRIZ—the computer-assisted mapmaking tool (èishinev), and its advanced version—"ECOKART" designed for automatic decision-making on the environmental management issues; the information programming complex "10 æhern" for support of the data bank for monitoring of ground water flow in the 45-km radius of the Chernobyl NPP (Soviet-German Joint Venture "Intercomputer", Kiev); the system for scenario simulation of the environmental processes "ATLANT"/"ARCHIMED" of the All-Union Institute for Systems Studies RAS, etc. Though they lack a complete set of functions for the GIS technology, these products often have several more advanced and efficient particular GIS operations (DEM processing, screen map designing, simulation) and in some cases are finished commercial software tools that start to spread in the domestic market of the GIS systems and to compete there with the foreign-made products that are not so readily available, are more expensive and relatively more complicated for a domestic user.

Development of in-house software and adaptation of foreign-made products is hampered by shortage of modern hardware (high-resolution monitors, mini-computer based workstations and so on), that are still in the COCOM list; such equipment obtained in violation of these restrictions via third countries is few, concentrated in "closed" institutions and has no significant influence on the general hardware level.

View on the future

The author believes he is obliged to conclude by describing the bright perspectives awaiting the GIS Science in new Russia, and the fruits of benefits and prosperity that it will bring to the society and the state in future. Today is in no sense an easy time. The collapse of the totalitarian regimes in the now independent states of the former USSR and of its Eastern-European satellites, the processes of democratic revival of their societies, removal from the political arena of the vicious organizations of communist orientation, more or less successful economic reforms establishing new relationships, demilitarization of the post-communist societies, end of the "empires of fear", freedom of information and the new information order, transition to civilized forms of relations with the outer world—all these fundamental and inspiring changes have interfered in a profound, and hopefully irreversible way into the patterns of modern life, science, production. They are to decide the routes of GIS progress in the current dynamic and unstable CIS environment.

The new economic situation includes economic reform, transition to market mechanisms and privatization of property. Five years ago, at the dawn of perestroika the author has supposed that

> . . . the promises on the full-scale deployment of GIS in many ways depend also on how fast and how completely a new economic environment will be created [and that] hopes for the supplying of geography and other agencies developing GIS technologies for widespread consumption with hardware and software seem quite illusory without the creation of the economic prerequisites, levers, and mechanisms ensuring the normal circulation of information flows, (Koshkarev, 1990, p. 195).

What was meant here, first and foremost, were the data and new technology markets.

The new legislation in the field of economy provides favourable opportunities for establishing firms and Limited companies with private and mixed capitals. The examples of such new ventures are: ''Nooinform'' (a branch of joint venture ''Face-to-Face''), small enterprise ''KIBERSO'', firm ''DeCart'' and a number of other smaller enterprises taking contracts for the GIS development, application and digital mapping. These companies are able to compete with the state monopolies such as the former USSR Geodesy and Cartography Department, that did practically nothing for the development of GIS/AMS technology in the CIS, or like its Russian successor—the Russian State Mapping Service ''Glavcartografia'', that claims (though with no sufficient grounds yet) to keep the monopolistic position in the GIS/AMS industry. One should mention that until now the CIS countiries had no system of digital mapping, and the GIS developers had to do the operation of map digitizing themselves, that is done everywhere in the world by the state surveys and topographic services, like the UK Ordnance Survey and USGS.

The economic reform that includes the large-scale privatization of land property may bring around a catastrophic increase of the number of land owners (by the beginning of 1992 in Russia there were already 14 million family farms; it is expected that by the end of this year about one third of 215 million hectares of lands of the collective and state farms will change to private hands). This would require establishment of land information systems for a real estate and ownership inventory, evaluation and management of the urban and agricultural applications. Another expected consequence of the reform is growth of unemployment; this suggests the need for a federal information service on employment opportunities. (Isn't it time to think of developing, in Moscow, a GIS similar to ''NOMIS'' developed and managed in the UK by Blakemore?)

Perestroika has revealed the extreme instability of the CIS society. Besides the ethnic conflicts, it has initialized a series of the world-level natural-technogenic catastrophes with horrifying environmental and ecological consequencies: the Chernobyl NPP catastrophe, the tragic Spitak earthquake in 1988 and several less impressive but numerous catastrophes, harmful exogenic processes, threat of extinction of the Aral Sea—the ''creeping'' super-catastrophe of the twentieth century, that fits perfectly into the *Guinness Book of World Records* in the chapter of ''achievements'' of the communist totalitarian regimes. To reduce the risk and damage of the catastrophes it is planned to establish a federal information system for management in extreme situations—the GIS "GEOCON" as part of the Russian State Programme "Safety" (evaluation of an area, prevention and mitigation of the consequences of large natural and technogenic catastrophes).

The new information order is the collapse of the empire of fear. One element of the general paranoia in the former USSR was the fear of truth long and vainly concealed from the surrounding world and the people of the country by the "iron curtain" and the regime of total secrecy. In January 1991, addressing the Workshop on GIS held by Minpriroda of the USSR (in the dramatic period when under cover of the crisis in the Persian Gulf, "Moscow" issued the order to attack the Lithuanian TV Centre with the special militia regiment, that resulted in loss of 14 lives), Yu. G. Puzachenko defined the Soviet Union as a "unique system that thrived on the basis of general concealment of information" and warned of the danger of continuing the formation of the domestic GIS as the "geo-misinformation systems". The gulf of secrecy that was evident until recently limiting access to the detailed (large-scale) maps of the USSR and remotely sensed images has gravely damaged the geo-sciences by blocking normal research work. Deliberate distortions of the state statistical records used for plotting large-scale maps on socio-economic themes has been equally harmful.

The problem of unreliability of the Soviet cartographic sources (deliberate distortions and "blankspots") has been described in the West by Blakemore (1990) and it is no secret that for a long time the instructions for the USSR cartographic service obliged it to use distorted base maps for small-scale thematic mapping. The consequences of this inaccuracy have not yet been fully appraised, though even the first attempts at digitizing the thematic maps have demonstrated the great difficulties that await the GIS developers in the future. M. Blakemore's catchwords "there are errors, damned errors, and maps" (1990, p. 108), acquire the literal and sinister meaning for cartography in the CIS.

References

Badenkov, Y. and Koshkariov, A.V., 1990, Development of new techniques for resource evaluation and development: An electronic atlas for Tajikistan—"Making it work", GIS'90 symposium procceedings, March 13-16, Vancouver, Canada, p. 67.

Blakemore, M., 1990, Cartography, *Progress in Human Geography,* **14**, 1, pp. 101–11.

Borunov, A.K, Koshkariov, A.V. and Kandelaki, V.V., 1991, Geoecological consequences of the 1988 Spitak earthquake (Armenia), *Mountain Research and Development,* **11**, 4, pp. 19–35.

Ferguson E.A., 1989, *A technical and operational comparison of geographic information systems as applied to the Canada forest industry,* Vancouver: FRDA.

Koshkarev, A.V., 1990, Cartography, Geographic Information and Their Interaction, *Mapping Sciences and Remote Sensing,* **27**, 3, pp. 185–98.

Koshkarev, A.V., Tikunov, V.S. and Trofimov, A.M., 1989, The current state and the main trends in the development of geographical information systems in the USSR, *International Journal on Geographical Information Systems,* **3**, 3, pp. 257–72.

Meiner, A., Saare, L., Roosaare, J. and Roose, A., 1990, Geographic information system for nature management in the industrial region of North-Estonia (USSR), *EGIS'90 Proceedings,* First European Conference on Geographical Information Systems, Amsterdam, Netherlands, April 10–13, Vol 2, pp. 756-61.

Mounsey, H. and Tomlinson, R., 1988, *Building Databases for Global Science,* London: Taylor & Francis.

The 1990 GIS Sourcebook, 1990, *Geographic Information System Technology in 1990,* 2nd ed, USA, Fort Collins: GIS World.

8

The development of geographical information and GIS in France

Jean Denégre
Conseil National de l'Information Geographique

Introduction

The current development of GIS in France is intensive, like in most developed countries, but it is probably being impeded by the development of geographic information itself in digital form. This situation leads to a need for an increasing co-operation between all partners, because the acquisition of the geographic information implies an enormous effort of design, production or conversion of existing data, as well as integrating sources, disseminating and up-dating.

The situation in France in 1991 can be characterized by the dynamics and the diversity of the territorial levels concerned (national, regional, urban etc.); the role of the French "Institut Géographique National" in producing reference data; the converging efforts in the sector of cadastral information between the Bureau du Cadastre, public utilities and local authorities; the specific contribution of satellite remote sensing for regional and national GIS; the increasing economic weight of the geographic sector in national activities; and the role of the "Conseil National de l'Information Géographique" (French National Council for Geographic Information) (CNIG) for co-ordination and standardization.

Territorial and technical aspects

It is interesting to note the increasing diversity of the territorial levels concerned by GIS. During the past two decades GISs in France were mostly developed at urban levels (and were often called "Urban Data Banks"). It should also be noted that the first systems like those in Lille or Marseilles were mainly devoted to computer-assisted cartography. Nowadays the objective is to build GIS devoted to multi-purpose management in association with all potential partners. Characteristic examples of such an evolution are given by the cities of Toulouse or Mulhouse, as well as the Urban Communities of Lyon or Strasbourg.

However, new territorial levels arose in conjunction with the French decentralization laws (1982–1983); departmental and regional authorities become involved with the GIS in the mid-1980s like the "Département de Vaucluse" or "Département de l'Hérault" in Southern France. Regional projects started a little later like in the Midi-Pyrénées or Nord-Pas de Calais (French regions group several "départements"; there are 22 regions in metropolitan France with 95 "départements"). Those GISs are characterized by rural and agricultural aspects, and land use monitoring plays a major role in them.

Of course, the national level is also concerned with the implementation of GIS projects in several administration services like the Ministries of Environment, Agriculture, Equipment or Interior. A common requirement with all those territorial levels is the availability of geographic reference data. The French "Institut Géographique National" started to build its Topographic Data Base (BD Topo, based on 1:25 000 photogrammetric surveys) and its Cartographic Data Base (BD Carto, based on the digitizing of existing 1:50 000 maps), as well as revising the specifications of the French geodetic network, in order to meet the requirements of local users.

Another source of reference data is provided by the Bureau du Cadastre (580 000 cadastral maps cover the entire national area, but only a few of them have been digitized). In 1991 an agreement was made between the Cadastre and major French public utility services (France Télécom, Electricité and Gaz de France, followed by Lyonnaise des Eaux and Générale des Eaux) in order to combine their efforts for digitizing cadastral maps. In parallel, the Bureau du Cadastre is now equipping itself with specialized hardware and software for managing digital maps.

A third aspect of reference data is provided by SPOT satellite imagery but this concerns large territories rather than small units. Therefore applications are developing more for regional levels, for instance in the sectors of agriculture or of environment. A programme for providing the entire national area with satellite-derived land cover data (in the framework of the European CORINE project) is now under development.

As a conclusion we can note that the development of GIS has become so complex that it appeared necessary to establish a "GIS Observatory" to record relevant events in France and elsewhere. That observatory is now being set up with both a data base and a regular bulletin publication focused on GIS and remote sensing.

Economic aspects

The increasing role of geographic information in the national economy is not really visible through the numbers describing budgets and staff devoted to its production or transformation. The enquiry made by the CNIG in 1988-1989 (concerning the 1987 numbers) gave the following general estimations:

- finances (all expenses included) : 5 700 MF (£570 M)
- personnel (all categories included) : 24 000 persons (1 987 numbers)

Within those estimations expenses correspond mainly to salaries, the part of investments being rather low (16 per cent of budgets as an average). Yet that part has recently been increased by the contribution of spatial activity (SPOT programme) where the French effort is rather original compared to other European countries.

It should be noted that the above estimations include all geographic information activities, i.e. conventional ones, and not only GIS developments. Anyway those numbers can be regarded as surprisingly low compared to those of the national economy: about 0.15 per cent of the Gross National Product (PNB) (3900 billion FF, i.e. £390 billion). According to the UN World Statistics (0.1 per cent) it seems that France is slightly higher than the world average. However, if we consider that situations are extremely contrasted between all nations, it means that France is probably not among the most advanced developed countries from this point of view. This could confirm the UNO conclusion (in 1974), placing France in the seventh position behind Germany, Switzerland, United Kingdom, Denmark, Luxembourg and Italy.

In contrast to the finance devoted to geographic information one should also estimate the benefits obtained by the national community. One of the main efforts made by the CNIG in France consists in developing theory and practice in that area. A first study, published in 1990 (*Utility and Value of Geographic Information*, Michel Didier, Economica Editions), has stated the basic principles of an economic analysis. A second study was undertaken in 1991 by the same author for compiling an "economic guide book" for GIS, based on the application of cost/benefit analysis for decision-makers.

Institutional aspects

In order to improve co-ordination, several organizations committees have recently been created in some areas: local authorities, utilities, state administrations (Ministries of Equipment, Agriculture, etc.) research laboratories (CASSINI), computer-service bureau (ATOLL). At the national level the task of general co-ordination is assumed by the "Conseil National de l'Information Géographique".

The mission of the CNIG, a consultative organization placed under government authority, was defined in 1985; it consists of contributing to the development of geographic information by taking into account the needs of public and private users. The CNIG is also responsible for examining and co-ordinating the production and dissemination programmes which are executed by the state or with its financial support.

Since it was set up in 1986, the CNIG is formed by representatives of:

- four local authorities,
- 13 concerned ministries,
- five geographic-information producers,
- two scientific experts.

In fact the representativity of CNIG is even larger since it includes manufacturers and private consultants within commissions and working groups. This allows the Council to be a place where co-operation can be discussed along three main axes:

- national level and local level,
- producer level and user level,
- public level and private level.

As a result of that co-operation, the CNIG has drawn-up specifications and/or standard projects, like:

1 Specifications for the large-scale topo-cadastral basic information (TCBI) which comprises a dense control-point network, digitized cadastral maps, and topographic information (derived from the BD Topo or from specific surveys);

2 Specifications for a new geodetic reference system;

3 Terminology and typology for image maps;

4 Standard project for an exchange format of digital geographic data (EDIGO).

That last standard was elaborated in 1990 after a study of existing standards, like NTF and DIGEST. A first draft was compiled in early 1991 and transferred to the French standards body (AFNOR). After a first public enquiry the draft version was revised and the first EDIGO version was published in spring 1992. In parallel it was proposed to the European authorities (Comité Européen de Normalisation—CEN) to establish a Technical Committee (TC) devoted to geographic information. This was accepted by other countries in October 1991. The first meeting of that TC 287 was held in Brussels in February 1992.

Other CNIG activities concern primarily geographic research (static and dynamic positioning techniques, development of using satellite images, instrumentation, etc.) as well as toponymy (harmonizing place-names on French maps) and design of a national digital reference survey: definition of land parcel limits with physical markers and the legal values to be attached to them.

A general information activity is assumed by the CNIG Newsletter, printed in 5 000 copies (some are sent to AGI for information). An additional extension of CNIG actions is provided by the non-official "Association Française pour l'Information et l'Instrumentation Géographiques" (Fi3G) which, in particular, organizes public conferences and seminars. Among them is the next International Forum for Geographic Instrumentation and Information (Fi3G), endorsed by AGI, AM/FM/GIS, CERCO, EARSeL, ICA, ISPRS, etc., which was held in Strasbourg (25-27 May 1992): it should be a *rendezvous* for many experts and European partners of geographic information.

Reference

Didier, M., 1990, *Utility and Value of Geographic Information*, Paris: Economica.

9

GIS and Canadian educators

Robert J. Rogerson
University of Lethbridge

Canada has long been known for its role in the early and continuing development of Geographical Information Systems (GIS). The case can be made that the Canadian government was the site of the first GIS, the Canadian GIS (CGIS), in the early 1960s, and certainly the first to be so called. Subsequently a vigorous high technology sector, with some focus on GIS, has developed right across Canada. GIS software developers and vendors are located in centres from New Brunswick to southwestern British Columbia, with perhaps the largest concentration in the Ottawa region. A substantial and detailed review of the GIS industry in Canada, which now tends to be described by the Canadian term ''Geomatics'', appears in Forrest 1991.

The involvement of the educational sector, in particular universities and colleges in Canada, is less well-known. As might be expected, however, in a country where GIS, or Geomatics, is an important commercial and governmental undertaking, the educational sector is also vigorously involved in GIS. There are four ways in which the educational sector supports GIS activities in Canada:

1. GIS operators and project support personnel commonly receive their initial skills and interest through an undergraduate education and/or technical training at the postgraduate or diploma level;

2. GIS developers and leaders are often those with graduate degrees and research experience in GIS in Canadian universities;

3. Research activities are conducted or supervised by faculty in support of core GIS developments or new GIS applications;

4. Conferences, Workshops and Training Courses involving GIS are often organized by, or utilize the facilities of a college or university.

No one of these sectors of involvement is more important than another, but the first activity tends to involve both universities and colleges, and more personnel in both teaching and as students. The second and third activities tend to be the exclusive domain of universities, and concern far smaller numbers. The fourth is probably the most visible in a public sense and often gives colleges and universities an opportunity to publicize their attainments in the other three activities.

GIS in Undergraduate and diploma programs

In most of Canada, excluding the Maritime Provinces, GIS tends to be best established in departments of Geography. Virtually all of the departments reporting in the annual Directory of the Canadian Association of Geographers, and those Canadian departments which also appear in the American Association of Geographers Guide to Departments of Geography, indicate some involvement in teaching GIS. Only five of 30 departments reporting in the latter publication in 1990–91 do not include GIS as a "Department Speciality", and four of those are known to the author to have subsequently made a commitment to GIS in the last year. GIS is increasingly a core course in undergraduate geography programs, rather like Quantitive Methods in the 1960s. Many geography faculty have commented that GIS is a very popular topic with students who identify it as "practical", and career directed, and the courses are virtually always full or even over-subscribed. Many faculty have become GIS converts and either teach core GIS courses or introduce GIS laboratories into existing geography courses. Advertisements for geography faculty positions reveal that GIS is a "demand" area.

A typical, well-developed series of undergraduate GIS courses appears in the Geography Program at the University of Western Ontario (UWO) (Walden and Pazner 1992). An introduction to the subfield appears at the first year in a course entitled "Computers and Geographic Space". This is followed by a substantial "Introduction to GIS" at the second year, and three further advanced undergraduate courses. Several other courses in the department, dealing with topics from Cartography and Remote Sensing to Location Theory also provide fundamental support to GIS.

Professional programs in Agriculture, Forestry, Wildlife Management, Landscape Architecture, Environmental Design and Survey Engineering also include courses in GIS, often tied into extensive laboratory activities and detailed student projects. Good examples of the latter are the Natural Resources Planning Studio and the Advanced Regional Planning Studio in the professional Landscape Architecture degree at the University of Montreal (UM). Dr Bernard Lafargue explains that his students become involved in computer-based spatial analysis and modelling using very large data bases composed of bio-physical data and cultural information.

Supporting undergraduate programs are an increasing number of GIS teaching laboratories with student computer workstations. These laboratories more often than not fall under the administration of the department or faculty involved in teaching GIS, rather than the general Computer Services units within universities. This may be because GIS tends to require proximal peripheral equipment, such as plotters, digitizers and scanners, large data bases (and hence full hard disks in a PC lab), and often GIS software requires computers which are rather specialized and more powerful than those used for spread sheet or word-processing activities which are typical of many undergraduate computer labs.

The Geography Department at the University of Victoria (UV) has three well-established GIS teaching and research labs with a spectacular array of high-quality equipment including a large number of IBM PS/2s and eleven RS/6000s. Another substantial IBM-based laboratory is in the Department of Geography at Queen's University (QU). A large laboratory in the Department of Applied Geography at Ryerson Polytechnic Institute (RPI) has over 20 PCs which run a commercial GIS software package, while the GIS laboratory at UWO has more than 20 Macintosh

IIsi colour workstations. Most GIS teaching laboratories tend to have ten or fewer personal computers which are usually accessed by students working in pairs. Some were established through a major investment by the university or supporting vendors, or both, but most have grown incrementally like that in Agriculture at MacDonald College of McGill University, or that in Geography and Water Resources at the University of Lethbridge (UL), both of which have half-a-dozen workstations running powerful commercial software.

There is no consensus on whether teaching labs should be PC-based, Macintosh-based or Unix workstation-based, nor indeed whether software should be Mac-based, DOS-based, OS/2-based or Unix-based. Similarly, home-grown software is used in GIS labs with some frequency, for instance at the University of Ottawa (UO) where David Douglas has developed a package caled XYNIMAP, and more typically educational software such as OSU Map-for-the-PC, IDRISI and Map II are commonly used. Commercial packages make an appearence even in introductory courses. Since 1990, Geography and other departments in forty universities and colleges in Canada acquired SPANS commercial GIS software, mostly through the Canadian Centre for GIS in Education, a not-for-profit corporation supported by INTERA-TYDAC Technologies, the developer of SPANS (Rogerson 1990). More than 50% of these use the SPANS software in undergraduate teaching. Arc/Info is also a common commercial software in widespread use in Canadian universities. In British Columbia PAMAP and Terrasoft are popular among educators as they are among local commercial and government users, while in New Brunswick and other Maritime provinces, CARIS is a popular local product. Most university departments tend to focus their energy on one commercial product, or two at the most, while in technical colleges, especialy those with specialized diploma programs in GIS, several commercial systems may be operational at any one time.

The forecasted shortage of trained personnel in GIS stimulated the College of Geographic Sciences in Lawrencetown to begin an intensive year-long postgraduate training program in GIS in 1985 (Dramowicz, Wightman & Crant 1992). Other well-equipped programs of this type also exist at Sir Sandford Fleming College (SSFC) in Lindsay, Ontario, and at the British Columbia Institute of Technology. Typically, these institutions expose students to cutting-edge technology in both hardware and software, often of the most advanced commercial type. Each of these institutions produces between 15 and 30 intensively-trained graduates each year. Smaller programs which focus more on training students straight from High School and which may provide a less intensive training experience, but to many more students, are present in all provinces: Cabot Institute in Newfoundland, Holland College in Prince Edward Island, Northern and Southern Alberta Institutes of Technology, and many others.

It is very likely that the post-graduate diploma area will show considerable growth in coming years as more students with a freshly-minted academic degree decide to gain practical skills in a postgraduate diploma course.

Such technical institutions are also a popular destination for a workforce returning to school to upgrade skills or for retraining. In this way SSFC in Lindsay, Ontario, has formed an education/training partnership with the Ontario Ministry of Natural Resources and in 1991-2 had ten professional and technical ministry employees enrolled in its eight-month "GIS Applications Specialist" program.

GIS Graduate Programs

Canada has a tradition of producing some of the leading GIS thinkers: one might expect then, that GIS graduate programs are large and well-developed in Canadian universities. This however is not so. Masters programs are widely-distributed throughout the universities although most have a relatively small enrolment of a handful of students. This is in keeping with the relatively small number of faculty in any one department who would claim to be GIS specialists. At the doctoral level the numbers are even less, and many well-established GIS units have none or only one doctoral student.

Those departments where there appears to be some growing strength in supervising doctoral students are not many. In the 1990-91 American Association of Geographers Guide, no PhDs completions which had a clear focus on core GIS or GIS applications are reported in Canada. Nevertheless, there are PhD students in a number of departments: Geography at UV, UBC, Calgary, UWO, Waterloo, Toronto (UT), UM and Sherbrooke; Forestry at UBC, Alberta, Laval University; Survey Engineering at UT, and University of New Brunswick(UNB); Resource Management Science at UBC; these are therefore likely to be the major Canadian source of PhD graduates in GIS-based programs for some time.

Many Canadian students wishing to complete doctoral work in GIS go to universities in the United States where, perhaps ironically, they may well be supervised by a Canadian. Most students who remain in Canada for doctoral work are absorbed by the high technology industry or by governments, such that several of the new openings for GIS academics in Canadian universities have been filled by recent immigrants or non-Canadians.

University Research in GIS

Vigorous research activity in GIS is pursued in Canadian universities, supported by government granting agencies (most notably Natural Science and Engineering Research Council—NSERC) or by contracts with GIS developers or GIS users.

Research is directed to core GIS concerns, such as software and algorithm development, and to GIS applications, particularly in environmental sciences. Of particular note in core GIS developments are the software packages developed for the Macintosh by Micha Pazner and his associates at UWO which range from Map II, to a flight simulator, a new-generation grid-based spatial processor and a Hypercard Stack development software for image map visualization and browsing. A major investment by the Forestry industry in Quebec and NSERC established a research chair and a Geomatics Research Centre at Laval University. Although the objective is to provide for long- term improvements in forestry GIS, much fundamental research is taking place, most significantly the development of a Voronoi diagram based dynamic interactive vector data structure.

Interest in core GIS development is also quite strong in western Canada, from Peter Keller's work on incorporating discrete and continuous space-time dynamics in GIS at UV, to Brian Klinkenburg's interest in fractals and dynamic data exchange in GIS at UBC, to Nigel Water's work on expert systems and computer mapping of multivariate data sets at UC.

Applied GIS research is directed to marine topics (UV, UNB and Memorial University of Newfoundland), sea ice (Waterloo, Memorial University), water resources and soil erosion and agriculture and global climatic change (UV, UBC, UL, Regina), land use change (Regina, UM, Brock University), integration with Remote Sensing (UC, Carleton University, Sherbrooke, Waterloo), geology and waste disposal (UV, UT and OU), urban and political decision making (OU) medicine (QU), market research (Ryerson), and forestry (UBC, Alberta, UNB).

Support for applied research activities comes from NSERC, and from contract work involving GIS developers, GIS consultants and industries which are users of GIS. This is a lucrative field for universities, but the intellectual level of involvement is sometimes rather slight, and most GIS faculty try to limit their consulting activities to a necessary minimum. Sometimes the objective of involvement is to earn funds to purchase more equipment for the teaching and research laboratories in the institution, or to support a financial need expressed by a graduate student.

Broad scale problems related to the adoption of GIS, the development of GIS curricula, and the organization of national GIS networks are much discussed by many of the educational institutions in Canada. UNB has provided substantial leadership from the survey engineering sector and has recently proposed a national spatial data infrastructure (McLaughlin and Nichols 1992) to enhance the accessibility, flow and use of georeferenced material in Canada.

Conferences, Workshops and Training Courses

GIS activities in university-based conferences are well- established, although the core GIS conferences in North America tend to be hosted by Hotel-based conference centres rather than universities, primarily because they are so large. Universities therefore tend to be the sites of more intimate gatherings, of small national groups such as Geographers or Cartographers, or regional meetings and workshops. These often provide a real service to the local GIS community. University of Regina, for instance, has played a major role in hosting Saskatchewan GIS meetings which are attended by government and private sector GIS workers, rather than by academics involved in GIS.

Universities and colleges have also become the site of commercial GIS training course. University of Lethbridge became an official SPANS training centre in 1991, and other SPANS training centres have developed or are being developed at Sir Sandford Fleming College, University of Montreal and Memorial University of Newfoundland. The experiment has been sufficiently successful that INTERA-TYDAC Technologies now runs all its Ottawa-based training courses through the Carleton University Geomatics Training Institute. Canadian university faculty have also been involved in offering training courses outside Canada supported by Canadian vendors.

Summary

Canadian educators have played a significant part in supporting the activities of software developers, consultants and governments in maintaining a significant position for Canada in GIS. They will continue to do so, indeed, we can expect that their involvement, particularly in supplying an educated and trained workforce, is likely to increase. GIS faculty in Canadian universities and colleges appear to be quite enterprising and unafraid of relatively close links with the private sector which spearheads GIS development in Canada.

Acknowledgements

The information for this paper was obtained by asking at least one person at each Canadian University and at several of the major colleges to report on GIS activities. Only 15 responses were received. Other materials were gleaned from Journals, Directories and Newsletters, and from the author's general knowledge of GIS activities in Canadian educational institutions. Thanks to all who contributed information, and apologies to those whose activities, for one reason or another, were not included.

References

Dramowicz, K., Wightman, J.F. and Crant, J.S., 1992, Addressing GIS Personnel Requirements: A Model for Education and Training. Computers, The Environment and Urban Systems, in press.

Forrest, D., 1991, IV. International GIS: Canada 1991–2 International GIS Sourcebook, GIS World Inc., Fort Collins, CO, USA, pp. 438–443

McLaughlin, J. and Nichols, S., 1992, Building the National Spatial Data Infrastructure. Computing Canada, January 6th 1992, p. 24

Rogerson, R.J., 1990, GIS in Education: a New Partnership, *GIS World,* **3**, 6, pp.60–62

Walden, S. and Pazner, M., 1992, Western Ontario's GIS Offerings Extensive, *GIS World, Vol 5, no. 2, March 1992, pp. 78–80.*

Part III

GIS and the Environment

10

GIS in environmental management

Sarah Cornelius
Manchester Polytechnic

Introduction

Attempting to manage our environment presents us with seemingly endless problems and conflicts. Not only do we need to understand the dynamics of our physical and living surroundings, how we interact with them and the problems we cause; but also the effects of the management policies we implement. The potential GIS applications in this area are huge—from monitoring bird populations; to modelling the impacts of new developments, or analysing the effects of changes in management practices. This potential has long been recognized and some of the well known pioneering GIS applications were in this area. Some more recent areas of application in the United Kingdom include environmental auditing and assessment, resource management and short term project applications. Some of these are discussed in other chapters (particularly that dealing with local authority use of GIS), so this chapter attempts to identify some of the main issues and problems being faced by those currently implementing GIS in environmental management. The papers which follow give more detail of some of the recent application areas, particularly those which incorporate environmental and human data and have had to consider some of the conflicts between humans and the environment. To begin with, let us review some of the areas of application which have been at the forefront of environmental GIS over the past year or so.

Areas of application of GIS in environmental management

Environmental auditing and environmental assessment

Environmental auditing is a relatively new field, particularly for many local authorities in the United Kingdom (in common with other areas of industry and commerce—for example my own institution has initiated its first environmental audit in the last twelve months). In the local authority sphere, however, it seems that many are still grappling with internal audits and state of the environment reports before attempting full audits of their regions. One of the most sophisticated environmental audits to have taken place, and one which has utilized GIS, was performed in 1990 by Lancashire County Council. More details of Lancashire County Council's green audit and other environmental areas of GIS application can be found in Derek Taylor's chapter elsewhere in this volume.

The costs of undertaking such an exercise are obviously great, and may deter some from following suit. However, as Tony Hams states in the Local Government Management Board's (LGMB) publication *GIS News* (autumn 1991), the benefits are "colossal". Such statements clearly need clarification and quantification!

Contaminated land registers

Publicly available registers of contaminated land have become a requirement of the 1991 Environmental Protection Act. The act calls for the collection and dissemination of information on contaminated land together with monitoring of existing sites. This is a new application area eminently suited to GIS and there has been discussion of contaminated land registers in such publications as *AGI News* and the Local Government Management Board's *GIS News*. However, as with many GIS application areas, little cost benefit information has been forthcoming. Interestingly, this particular area of GIS application has seen the development of one of the first thematic customized GIS products to help environmental managers: "Reclaim"—the register of contaminated land information manager—from AutoCad and GEO/SQL.

The AGI have aided the discussion on how to use GIS for the contaminated land register by providing a list of recommendations. In addition to suggesting that the information required should be held in digital form, and that use should be made of GIS, the AGI have suggested that the boundaries of each area of contaminated land should be held; that large scale map production should be possible; that standard reports and letters should be possible and public access should be allowed. On the more technical side the AGI recommend the use of "GIS standards" such as the postcode, national grid reference and National Transfer Format for transfer of information (see AGI newsletter, 1991). It is issues such as these which will be crucial to the success of a digital register and which offer an exciting chance to develop and achieve standards across local authorities on referencing, data exchange, quality, charges and standards, and on the communication of information to the public.

Resource management

Forestry has received much attention over the last year in the GIS arena. It is an area where GIS may be able to help with planning and design of forest stands and their management, and Davidson (1991) discusses some of the advantages of using GIS in forestry. These include the ability to generate landscape simulations to help with design, felling and restocking of forests. Aspinall (1991) also writes of the great potential of GIS for application to forestry, at different scales and for different tasks. One of the major drawbacks to this area of application in the United Kingdom is the conversion of forest stocking maps and other information to digital format. The Forestry Commission has recently initiated pilot GIS projects, encouraged by advances in scanning and computer aided digitizing techniques (see *Mapping Awareness,* **5** (10), 50). Again, more attention is being given to the costs and benefits of GIS before a wholesale commitment is made.

Tropical forests have also been subjected to the GIS treatment in recent years. For example the World Conservation Monitoring Centre are using satellite imagery and GIS to produce world coverage of tropical rainforests (Collins, 1991).

English Nature have been involved in GIS for some time, and, having implemented their system in 1989, could be regarded as one of the veterans of environmental management GIS in the UK. They have applied GIS to many applications (see, for example, Budd Chapter 11). Other British organizations with responsibility for environmental management are the National Parks. Their involvement with GIS is much more recent, with the Lake District National Park acquiring, in 1991, GIS hardware and software, to make use of the land cover database specially commissioned for the Parks (Bird, 1991) and to demonstrate GIS capabilities to other Parks before they take the plunge. Taking just these two examples, there are overlaps in data and application areas. English Nature are responsible for the management of National Nature Reserves, many of which occur within National Parks, and also for sites of special scientific interest (SSSIs). Much of the data the two are collecting must be in common. Others interested in our countryside (for example, the Macaulay Land Use Research Institute; see Miller *et al.,* Chapter 13) also use similar data. There is a clear need for some coordination—a central broker of environmental data, or at least a register of environmental data. Perhaps the list published in this volume will help.

The National Parks work frequently at the interface between humans and the environment. Much of the areas they cover are at risk from over-use and abuse. Many environmental management organizations, such as the National Rivers Authorities (NRA), are working at this sensitive interface. The NRAs, for instance, manage the conflicting interests of wildlife, consumers and recreators along waterways, estuaries and coasts. The paper later in this chapter, from Anglian Region NRA, focuses on one of these areas—coastal protection—and discusses the establishment of a strategic GIS helping protect the coastal population and the sensitive coastal environment.

Another issue to be faced by those with long term resource management GIS projects is that of time. The GIS market is stabilizing sufficiently to encourage those with responsibility for environmental management to take up the technology. However, there still needs to be long term standards for referencing of data, for exchange, and for update. Socio-economic data is rife with inconsistencies if compared over time. For example, administrative boundaries change, units for the collection of important longitudinal data sets such as the census of population have been altered over time, and seemingly stable statistics on subjects such as unemployment are rendered incomparable due to changes in the definition of "unemployment" over time. It can only be hoped that in the environmental field lessons can be learnt from these problems.

Project based environmental GIS

The above are all long term management GIS projects, but that is not the only type of GIS application area in environmental management. Smaller, shorter term projects are being set up by, for example, consultancies, research groups and campaigning bodies. These projects may cover issues such as environmental auditing, environmental impact assessment and pollution monitoring. During 1991–2 there has been a rapid growth in this area and an increase in consultancies in the United Kingdom, (for example Aspinalls, ERL Ltd and W.S. Atkins), and pressure groups (for example, Friends of the Earth) contributing to GIS conferences and advertising for GIS staff. There are plenty of examples of those who have been in the game a little longer, with much experience to offer. Posford Duvivier, who have contributed a later chapter to this part, are one such organization.

Short term projects have their own peculiar problems. The most obvious being the requirement for relatively swift acquisition and input of the project database. In GIS, where data input is seemingly never quick and easy, this presents a significant problem, particularly as data collected for a previous project will not necessarily be appropriate for the current one. However, the emergence over the last year of vendors of environmentally related data (e.g. Taywood Data Graphics providing SSSI boundaries) and improvements in the Ordnance Survey's digital database are improving the situation. Again, an environmental data directory or broker of digital environmental data would be an immense help to those needing to trace appropriate data fast. Many short term projects also make extensive use of data sources such as aerial photographs. The work on estuary management by Posford Duvivier (see Zisman later in this part) and that of the National Parks have relied on aerial photographs as an important data source.

Other requirements are for rapid responses, usually within a defined and limited time frame, the incorporation of qualitative information and the need for specialized models and techniques. A useful review of the issues faced when using GIS for project based work can be found in Vaughan 1991.

Issues

Data

Most environmental management applications require huge amounts of data from myriad sources. This has been illustrated by for example Hunting *et al.* (1991) who list 24 datasets which are available for the Cuilcagh Karst management project, Harrison *et al.* (1991) who list 21 different major data sets appropriate for a GIS demonstrator for Devon County Council and Archibold and Jeffries-Harris (1991) who discuss the "amazingly diverse" range of data that a GIS for the Scottish Marine Estate of the Crown Estate Office will have to consider. The problems in identifying potential data sets, then obtaining them, then integrating them are huge and could be assisted greatly by the existance of registers, brokers, and standards.

As already mentioned, environmental data is becoming available commercially at ever increasing rates, but is still rare and expensive. The boundaries of SSSIs in Great Britain can now be purchased, there are increasing amounts of satellite imagery available and more satellites due to come on line in the near future. Data on other natural resources (for instance ecological data forest extents, boundaries of important conservation sites etc.) are harder to come by. Much of these data are surely being produced by environmental GIS users, for example the land-use data for the National Parks (Bird, 1991). Let us hope that they will be made available to others.

Human-Environment Interactions

One important area of environmental GIS applications is that addressing human-environment interactions. To date, many GIS in the environmental field have been resource based (eg water or forestry) or region based (eg national parks). Less common have been applications addressing environmental problems and issues where humans are involved, for example those concerned with recreation and tourism, or management of areas where human impacts are substantial. All of the papers in this chapter attempt to address this imbalance by focusing on GIS projects where human-environment interactions have had to be considered. Much environmental data is manageable, the environment in general changes slowly and its behaviour may be predictable. Humans on the other hand are rarely predictable. If the weather is bad they may not visit a site, if they do will they change their mode of transport to that site? In an estuary environment, for example, human activities such as fishing and flying may be affecting the ecosystem. But how can this be interpreted quantitatively for incorporation in a GIS?

Cost/benefit analysis for environmental applications of GIS

In common with many other application areas, environmental management is lacking clear cost benefit justification. This is causing many new users to tread carefully, taking application of the technology pilot project by pilot project. However, the use of the technology in this field would seem to have massive benefits, allowing the integration of the many data sets which may be used for even the simplest management strategy. There is a need for publication, by organizations already using the technology, of the costs and benefits they have experienced. Small scale, thematic and large scale strategic applications all need to be evaluated, to reassure environmental management bodies (who are often underfunded, or even of charitable status) that the investment is worthwhile. Some standard methodologies are required, perhaps along the lines of those being developed for local authorities by Coopers and Lybrand (Buxton, 1991).

Future environmental applications of GIS

Environmental Interpretation and Education

Environmental interpretation is one area which has escaped GIS to a large extent to date, but the potential is huge. Just as the *BBC Domesday Disk* has been used in education with great success, more modern GIS based environmental systems have great potential. This is one area where multimedia and virtual-reality technologies should join with GIS to help promote understanding of our environment. As part of Manchester Polytechnic's Cuilcagh Mountain project a demonstration of an educational PC based system for use in the Marble Arch Caves visitor centre within the management area will be developed. There are a few other projects in this area but more are needed, and have enormous potential. Virtual-reality, multimedia and CDROM all offer exciting opportunities to extend GIS. At a recent Royal Geographical Society (RGS) "Planning a Small Expedition" seminar it was suggested that not so far into the future data resulting from expeditions to far away places might be collated at the RGS so that in addition to the present library containing books on all parts of the globe, there might be a "virtual-reality room" where one could go to experience these otherwise difficult to reach and imagine places. An exciting prospect!

GIS for fieldwork

If environmental GIS are to become truly responsive to changes in the environment there needs to be attention given to the use of GIS equipment in the field. Global Positioning System is already a reality, as are portable PCs capable of hosting the largest and most powerful PC GIS software. However, if you were to take such a PC into the mountains, I suspect it would be the first item to be dumped at the first vertical rock wall. Portable? You need to be pretty strong to port a portable 486 very far! But there are already smaller, lighter machines being produced; there are also portable roll-up digitizers, but these come without robust carrying cases and would not stand up to use in the field. More problematic is powering equipment in far away places. Batteries are probably heavier than the equipment itself, and there is a distinct lack of solar powered equipment to date.

Such issues may seem a luxury for those of us who want to use GIS in remote and exciting places on expeditions and for field work. However, the issues are real. There are parts of the developing world where portable equipment may be the only possible GIS solution, and again, for educational purposes in schools and colleges, portable equipment could help make geographical, environmental and related subjects come alive.

Completely integrated GIS

As we have seen from the applications discussed above, the data required for the solution of environmental problems are varied, and in many cases need to be collected quickly. Remotely-sensed and air photograph data are frequently the answer, but integration with other data, for instance boundaries, river networks and so on, is often crucial. Many of the difficulties faced by current environmental management GISs are being reduced as integrated vector and raster GISs become available. The third and temporal dimensions also need to be considered. A good environmental GIS package needs to have functions to cope with vectors, rasters, images (air photographs and remotely sensed), 3 dimensional data and time series data.

Global projects and campaigning with GIS

Global databases are being established for climate change research programmes, monitoring of tropical rainforests and the like. It is with such projects that perhaps some of the greatest challenges for GIS lie. Not only are the required datasets huge, but the functionality requirements are greater. For example, modelling of climate change, extrapolation forwards and backwards in time to see the effects of change and the evolution of our climate might be required. Environmental campaigners need similar tools, which might, for example, model the effects on pollution levels of an increase in car ownership for comparison with the effects of rigorous car curbing policies. The campaigners require more too, they need to be able to communicate their results effectively and persuasively.

Conclusions

The early 1990s have seen great developments in GIS and environmental management. New data, new GIS products and new application areas are being developed at an increasing rate. So what has changed since the publication of "Geographic Information 1991"? In "Managing our future environment: Some GIS issues" (Heywood, 1991) the issues discussed were largely related to data and the technical capabilities of GIS. Today these issues are still important but 1991–2 has seen the emergence of commercially available environmental information and discussions on the adoption of standards for environmental information. So, the data issues are being addressed. The user base is also expanding, shown for instance in the development of an Environment Special Interest Group at the AGI. However, it must be said that most environmental management GIS applications do not yet have very sophisticated technical requirements in terms of GIS functionality. Most users, and particularly the new users, are still at the inventory stage of the GIS lifecycle, as they were in 1991 (Heywood, 1991). There is much collection of data and conversion of data to digital form being undertaken and huge thematic and regional databases are under construction. The challenge for these large inventory databases will be to make the change to analysis and then on to management applications. For project based environmental GIS the inventory phase is less problematic since the amount of data required are smaller, and there are now good examples of analysis and management applications, some of which are illustrated in the following chapters. For thematic GIS applications, such as the contaminated land registers, special products have been developed to ease the burden on the user of identifying functions required, training skilled operators and analysts, and customizing GIS. An evaluation of the effectiveness and appropriateness of such packages is awaited with interest.

All environmental management GIS applications need to include data on humans. It is the humans who do the managing, the humans who may be doing damage to some of our natural resources, and the humans who must be managed to prevent further degradation. The five chapters which follow are examples of current GIS work where human impacts are being considered and GIS is being used in a management context rather than simply as an inventory tool. There is much to learn from these case studies; let us hope that their calls for better data, and better appreciation of the costs and benefits of GIS for such applications will be answered in response to their publication.

References

AGI, 1991, Newsletter, ISSN 0962 4163, Issue 11, October/November.

Archibald, I. and Jeffries-Harris, T., 1991, A vectorised spaghetti-hooped boundary 10,000km long. In *Proceedings of the AGI conference,* November 1991, Birmingham, 2.18.1–2.18.5.

Aspinall, R., 1991, GIS and Forestry, *Mapping Awareness* **5**(4) 12–14.

Bird, A.C., 1991, GIS based data on landcover change in the national parks of England and Wales. In *Proceedings of the AGI Conference, November,* 1991, Birmingham, 2.12.1–2.12.3.

Buxton, 1991, Geographic Information in Local Government—steering in the right direction, *Mapping Awareness,* **5**(2), 39–41.

Collins, M., 1991, Mapping the Tropical Forests, *Mapping Awareness,* **5**(10), 25–28.

Davidson, D.A., 1991, Forestry and GIS, *Mapping Awareness,* **5**(5), 43–45.

Haines-Young, R., and Ward, N., 1991, GIS in the development of tree and forest strategies. In Proceedings of Mapping Awareness '91. Conference held in London February 1991. pp. 165–175.

Harrison, A., Dunn, R. and Turton, P., 1991, Environmental GIS: Technology, Data and Policy. In *Proceedings of the AGI conference,* November 1991, Birmingham, 2.14.1–2.14.5.

Heywood, I., 1991, Managing our future environment: some GIS issues. In Cadoux-Hudson, J. and Heywood, I. (Eds), *Geographic Information 1991,* London: Taylor & Francis, pp. 75–80.

Hunting, C., Cornelius, S. and Gunn, J., 1991, The preparation of a GIS for the Cuilcagh Karst. In *Proceedings of Coras Eolas Tireolaiochta na hEireann, the first Irish Conference on Geographical Information Systems,* October 1991, Cork.

Vaughan, D., 1991, Project-based Environmental GIS. In *Proceedings of AGI Conference, November 1991,* Birmingham, 2.21.1–2.21.5.

11

GIS in a Multifunctional Organization

Paul Dowie
The National Rivers Authority, Anglian Region

Introduction

The National Rivers Authority (NRA) was established in September 1989, as a result of the 1989 Water Act, with the following mission:

> The National Rivers Authority will protect and improve the water environment. This will be achieved through effective management of water resources and by substantial reductions in pollution. The Authority aims to provide effective defence for people and property against flood from rivers and the sea. In discharging its duties it will operate openly and balance the interests of all who benefit from and use river, groundwaters, estuaries and coastal waters. The Authority will be business-like, efficient and caring towards its employees.

While specific aims are dealt with by the NRA's water resources, water quality, flood defence, conservation, fisheries, recreation and navigation functions, many require a multifunctional consideration and response.

Anglian is one of ten regions covering England and Wales. It is the largest geographically covering one fifth (27 000 km^2) of the area of England. Over one quarter of the Region is below high tide or river flood level and the Region is responsible for maintaining 7 000 km of flood defence works. Licensed discharges within the region number 26 887 and there are 9 700 water abstraction licences. Also in 1990/91 1 700 serious pollution incidents were dealt with. Anglian Region employs over 1 100 people located at Regional Headquarters in Peterborough, three Area and eleven District Offices.

Balancing interests

NRA Anglian Region acts at the interface between human activities and the natural environment. Much of its work seeks to manage public interests in relation to its main objectives as defined in its mission statement. Environmentally friendly solutions to the problems that are facing our multifunctional organization require a much greater understanding of the interactions between functions that are taking place. This requires an information based approach to be adopted.

The NRA is a map based organization. Over 90 per cent of the data used by the organization to carry out its business is spatially or geographically referenced. A recent internal survey of a cross-section of employees found that over 80 per cent said that maps were essential to their work with some engineers spending one to two man days per week using maps, and most considering current access to spatial information in need of improvement.

The multifunctional nature of the organization and the need for strategic as well as operational information is reflected in the character of spatial information requirements (see Table 1). Specific activities require individual functions to use a particular scale. For example, regional and national data is required to determine the effect of global warming and sea level rise on North Sea wind, wave and tide climates. These interact with local conditions to undermine east coast sea defences. This level requires large scale survey data.

Table 1: Scale of hard copy OS maps most frequently used (percentage of responses)

	Operations (Engineers)	Environment Conservation, District (Quality Officers, etc)	Total
1:1 250	2	5	7
1:2 500	19	7	26
1:10 000	16	10	26
1:25 000	0	5	5
1:50 000	12	24	36

Each of the functions have their specific aims, but to achieve these aims considerable consultation is required both between functions such as Engineering, Conservation and Fisheries as well as with external bodies and local interests. There is therefore a need to be able to generalize and transfer data effectively to enable data integration across functions.

This mix of requirements necessitates the provision of sophisticated spatial information processing which both structures the data and enables the sharing of information between functions. Geographical Information Systems, GIS, as a technology and tool aims to provide these facilities.

GIS at NRA Anglian Region

The wide scope for GIS, given the Region's spatial data-handling requirements, necessitated taking a long-term view. There was also a short term need to gain experience with GIS over a period of time. The recognition of the above led to the establishment of a five-year rolling capital programme for investment in GIS in 1989. GIS are seen as capital investments with a life cycle which require an overall framework but each application must be prioritized and economically justified in full, on their own merits. In this way experience is gained while system, data and staff resources can be effectively managed.

Application areas have been identified within a framework whose cornerstones are:

- the provision of central shared data resources;
- user ownership and quality control of data;
- customized user applications;
- development using Intergraph's Modular GIS Environment, MGE.

From the beginning the importance of data as a resource and user commitment and responsibility were recognized and are embodied within the above framework. While taking this strategic view of GIS, individual applications are developed according to an incremental philosophy:

Stage I Discussion with users as to what they think they want to use GIS for. This results in a user requirement document.
Stage II A demonstration version is developed in conjunction with the users.
Stage III This demonstration is discussed with the users to produce a refined user requirement statement.
Stage IV A full Project Appraisal with cost-benefit analysis is undertaken and presented to a multi-functional panel.
Stage V The full application is developed for a pilot area.
Stage VI The full application is developed for the whole region.

Currently GIS is being used for shoreline management, catchment planning, planning liaison, telemetry, abstraction licences and some specialized mapping.

Implementation

The following three case-studies represent the range of current experiences with GIS in Anglian Region.

GIS as a mapping tool

NRA Anglian Region requires basic information about its area to be able to undertake emergency planning and to provide maps for operational use by Area and District staff. The specialized Regional Maps project was initiated to meet these needs and utilizes digital cartography to provide maps at 1:50 000 scale of over 40 themes or data sets for the whole region. Information layers include, for example, flood areas, aquifer outcrops, Sites of Special Scientific Interest, rain gauge and sluice gate locations.

As data costs can be anywhere between fifty and seventy-five per cent of GIS development expenditure, the substantial resource over the last year involved in creating this database has been an essential prerequisite if GIS in Anglian Region is to succeed in meeting current spatial data handling needs. The availability of a flexible database means that maps can be updated quickly and users requiring specialized maps of subsets of the database can be more easily satisfied. Thus GIS, although not being available on everyone's desk, can at least meet some of the requirements of users.

The database achieves the creation of a mechanism for ensuring data transfer between NRA functions as well as providing a baseline for most of their data requirements. The project has established the principle of user data responsibility and co-ordinators have been identified to ensure consistency between the maps database and other databases in the organization. The project has also saved the region many thousands of pounds in manual map production costs.

GIS as an operational tool

One of the main users of other functions' information is Planning Liaison. The NRA, through Planning Liaison seeks an effective and professional working relationship with planning authorities for the environmental benefit of the community at large. Planning Liaison ensures that in matters relating to Planning and Development, the public interests which are the responsibility of the NRA, are properly looked after.

The NRA achieves this by internal consultation with relevant staff. The choice of consultees is determined by predefined criteria which can be spatially defined or are based on the descriptive attributes of the particular planning application. For example, the District Engineer, Conservation Officer and Principal Hydrologist must be consulted if an application is in, under or over any water course or within nine metres width of a main river.

The system works by allowing the user to interactively register incoming application boundaries against raster scanned Ordnance Survey 1:10 000 sheets for the area of interest. From the database, using the planning application as a search area, a report of previous applications in the area and the list of consultees can be automatically generated. A listing is then sent to each of the selected consultants.

Spatial information is also required by the consultees in their decision-making. Over 50 items of information have been identified as being of use by consultees, approximately 21 were considered to be essential by over 50 per cent of staff involved in planning liaison. The majority of these have now been made available in a digital form and allow most criteria to be adequately represented within the GIS. The provision of consistent base data by Planning Section staff to consultees will save their time and avoid the need for individual map sets. The eventual aim is to provide each consultee with graphical access to the database which will further enhance the work of Planning Liaison.

The implementation of GIS in Planning Liaison will eventually save the Region approximately three to four man years per annum. The whole project, while seen as a means of improving efficiency and reducing bottlenecks, also improves the public image of the NRA as a professional organization committed to effective decision making based on the highest quality of information. Public interests are being looked after coherently.

GIS as a strategic tool

GIS in the environmental area has long been seen as a valuable tool for exploring spatial relationships and integrating data from a variety of sources and, as shown for Planning Liaison, can easily handle the overlay and geographic searches required for many of our applications. However, we often need to move beyond mapping and simple scenario development. One such application has been the development of the Shoreline Management System.

NRA Anglian Region has just completed the three year Sea Defence Management Study into the provision of flood defences along the Anglian coastline. In the region 1 500 kms of defences protect three quarters of a million people and substantial investment in infrastructure and land from tidal flooding.

Given the magnitude of the task of maintaining these defences, the NRA recognized the need to move from piecemeal approaches to decision making to one based on strategic options. The study increased the NRA's regional understanding of coastal processes and hence allows the development of regional strategies. One of the options available to engineers is to use "soft engineering" as a viable alternative to traditional "hard" concrete defences. This is only possible with acceptable risk, however, if backed by a full understanding of coastal processes and a comprehensive monitoring programme, which is most efficiently and cost effectively handled on a GIS. Through this approach the NRA is seeking to work with nature rather than confronting it and thereby ameliorate the environmental impact of flood defence schemes.

The study in seeking to understand coastal processes brought together information and data from over 40 organizations. The data has been structured into 53 broad categories such as coastal works and administrative boundaries with 325 features, which include protection zones, walls and embankments.

Most of the features have tabular attribute information associated with them. There are 230 tables with over 2 000 columns of information. The data is now in a manageable structure, logically organized, which enables effective administration to occur. A data dictionary keeps track of what is in the database holding information such as source, update requirements and the date of last revision. A monitoring programme will keep key datasets up to date and enable time series analysis.

Anglian Region were advised by their consultants, Sir William Halcrow and Partners, that GIS was the most appropriate tool for handling the data and providing the spatial analysis tools for the development of a management framework. The resultant Shoreline Management System builds on the functionality provided within Intergraph's modular GIS Environment. Additional functions include:

- Customized geographic windows;
- Search area definition;
- X-Y graphs;
- Chainage assignment;
- Photo display;
- Wave rose display generation;
- Statistics;
- Coastal retreat model;
- Defence overtopping model;
- Current vector display generation;
- Data model documentation.

The availability of this functionality enables the users to utilize the information and knowledge acquired as part of the study in order to model and assess the impact of changes in coastal processes both regionally and in relation to a particular defence scheme.

Already "soft engineering" schemes have been started on both the Lincolnshire coastline between Mablethorpe and Skegness and on the Norfolk coast between Snettisham and Hunstanton. In the latter case 40 000m^3 of sand and shingle will be pumped from offshore onto the shore to build up the existing beach so that it will become more effective in reducing wave attack on the defences.

This study, as with the Regional Maps Project, has created a data resource which is being used by other applications. Likewise as other departments increase their use of GIS the engineering function will have easier access to other databases held within functions such as conservation and fisheries.

Conclusions

The NRA Anglian Region is a map-based, multifunctional organization which needs to integrate and share its data resources across functions to be able to better understand how to balance functional and local interests. GIS is therefore a natural device for the NRA to structure and manipulate its data resources.

A strategic view of GIS development is required to be able to implement effectively individual applications. Such an approach enables expensive data capture costs to be seen against an overall framework while enabling individual applications to be developed incrementally on their own merits.

Data resources are of critical importance and much of the success so far of GIS within Anglian Region has been in the drawing together of data into a unified structure with common definitions. Up to date data is the life blood for the continued use of GIS. The existence of GIS has focused the need for monitoring and the necessity of information flow between the GIS environment and other organizational databases.

The desire to work with nature and local interests requires a higher degree of understanding and information provision within the NRA. GIS is already enabling Anglian Region to have a greater insight into the issues facing it, it enables it to develop strategies for the functions of the Region, and allows it to better utilize its capital investment and staff resources.

12

The use of maps to aid the management of the Dartmoor Commons for nature conservation

Jonathan T.C. Budd
English Nature, Peterborough

Background

English Nature advises the Government on nature conservation in England. It promotes, directly and through others, the conservation of England's wildlife and natural features. English Nature is responsible for the establishment and management of National Nature Reserves, and the identification and notification of Sites of Special Scientific Interest (SSSI). The organization also provides advice and information about nature conservation and supports and conducts research relevant to these functions.

GIS has been used within our organization since the early 1980s, though it was not until 1989 that we acquired a fully integrated system. After exhaustive cost benefit analysis and bench marking we purchased an Intergraph system to cover digital mapping, photogrammetry, image analysis, land survey, and spatial analysis. This chapter describes just one of the many projects using the system.

Dartmoor

The warm and wet climate found on Dartmoor has led to the development of a unique flora which are extremely valuable in conservation terms. Some of these are internationally important, for example the blanket bogs, which are the most southerly in Great Britain. The large number of valley bogs present are some of the finest in the country, while the heaths and grassland are some of the best examples of their types. The moor is the only regular breeding site for golden plover and dunlin in southern England. There are also large breeding populations of ring ouzel and wheatear.

The Commons of Dartmoor cover some 90 per cent of the unenclosed moorland found within the National Park, and are divided into 90 Common Land Units covering a total of 35 990 hectares. The traditional use of the Commons for depasturing livestock is still carried out today as this is an essential part of the economy of many of the moorland farms.

Heather is fundamental to the ecology of Dartmoor; it provides food and in particular shelter for moorland birds, insects, and other animals. Not surprisingly because so many animals are reliant on heather, it is a key indicator species for the health of the moorland communities as a whole.

There have been significant changes in the vegetation of the moor over the last decade. These in turn have led to the deterioration of its natural beauty, its conservation interest and in some places the quality of the pasture. The two most serious causes of damage to heather on Dartmoor are over-stocking and too frequent burning. It is clear that overgrazing will stunt and suppress heather and further damage may be caused by trampling. In areas where there has been over-frequent burning species like purple-moor grass regenerate quickly, thus preventing the re-growth of heather. These are not the only causes of heather damage on the Commons resulting from human use. Recreational activities such as walking and pony-trekking can lead to soil erosion and trampling. The use of the area for military training also has an impact. There is general agreement that remedial action must be taken to prevent further destruction of the moorland ecosystem. The Commoner's Council was created in 1985 as part of the Dartmoor Commons act of that year. One of the tasks set this group in the legislation was the creation of a set of regulations that would ensure the commons are managed in a way that will benefit all that use them. This included detailed rules on heather burning and overgrazing. These guide-lines were set up following consultation with various bodies including the Nature Conservancy Council (now English Nature). The Commoner's Council now have in their power the ability to preserve and even enhance the natural beauty and nature conservation interest of the Commons. These regulations were approved by the Secretary of State for the Environment in 1990.

As a result of these regulations the then NCC was asked by the Commoner's Council to produce a document entitled, "Management guidelines for the Dartmoor Commons" (Nature Conservancy Council, 1990). This report concludes that traditional extensive farming rarely conflicts with nature conservation or the natural beauty of the moors.

The management guide-lines outline the different habitat types found on the commons and makes specific recommendations as to how they can be managed. It is not always easy for the farmer to identify these habitat types. For example it can be difficult to assess the extent of heather shortly after a burn, or in an area of overgrazing.

The maps

To try and overcome this problem it was decided to produce a series of management maps for the Commoners and Landowners to use. These would be clear and straightforward maps which would allow them to put the management guidelines and indeed the regulations into practice. In addition to the above it was hoped that the maps would show various additional information. This included the recording of areas of damaged heather, in particular those caused by overwintering cattle. It was also hoped that the maps could be used to identify areas where future heather regeneration schemes could be implemented.

There has been no previous work specifically to study heather condition or to map habitat management requirement for the whole of the Dartmoor Commons. There have been numerous vegetation surveys over the last 30 years, but unfortunately none of these are directly comparable, so it is not possible to make any assessment of the changes in the character and condition of the moorland. There have been more detailed surveys but these are not sufficiently extensive to be of much use. More recently the Countryside Commission commissioned a landscape survey of the National Park (Countryside Commission, 1991) though like the previous surveys, it does not provide sufficient detail on the extent and health of heather.

Since no suitable map data source already existed it was necessary to carry out a detailed survey of heather within the common land. This took place between August 1989 and March 1990. Two classifications were used, habitat management units and heather condition. A more detailed breakdown of these is given in Table 1. The information was mapped onto 1:25 000 OS Outdoor Leisure Map (No. 28). The minimum mappable area was set at approximately one hectare as it was not possible to map accurately smaller areas onto this scale of map. Wherever possible the boundaries followed topographic features, such as streams, walls or breaks of slope. This would hopefully help future users of the maps identify in the field the areas on the maps and also provide practical management units. It was not always possible to do this as certain areas were featureless and also a compromise had to be struck between the accuracy of mapping different vegetation types and the practicalities of management. Fair copies of field maps were drawn up as soon after surveying as possible, one map being prepared for each Common Land Unit.

The use of GIS

The primary role of these maps is to provide the link between what the user sees on their land and the Commoner's regulations. It was thus felt important that considerable effort should be put into the production of clear and helpful maps. It was evident early on that it would not be possible to represent the 14 different feature types clearly using only black and white; the use of colour was essential. Colour maps would also be more attractive to the users and thus in turn encourage their use.

The next question was what was the best way to produce these maps? Traditional manual cartographic techniques could be used to produce a high quality map product. This approach would have involved the use of colour litho printing, an expensive process. It would also mean future updates of the maps would be costly and time consuming. The use of GIS provided a more cost effective approach. Our colour electrostatic plotter could be used for map output far more cheaply than litho printing with minimum loss of quality. The fact that the maps were held digitally would mean that there would be far greater flexibility in the production of future maps. The paper maps only provide a visual image, a considerable amount of information remains locked up and unavailable to the user. The GIS allows the maps to be manipulated and analysed in ways not readily possible with paper maps.

There were three main data sets needing to be input into the system, the Heather Management Unit maps, the Heather Condition maps and the Common Land Unit boundaries; the latter were supplied by Dartmoor National Park authorities. It was decided to use a bureau for the data capture as this was more cost effective than using highly skilled in-house staff. The bureau were asked to follow stringent guidelines for data capture, even so we had problems with data quality. The problem seemed to stem from the fact that the bureau were not clear on how we intended to use the data. It was essential to us that the data was clean, i.e. no duplicate lines or loose ends as we needed to build topology into the data. As a result a considerable amount of staff time had to be spent correcting and checking the data. It

Table 12.1. Classification of habitat management units and heather condition

	Habitat management units (with burning recommendations):	Applicable heather condition Scores
H*	Heathland that should be burnt on a rotation of not less than 12 years	1,2,3
U	Heathland that should be left unburnt, to allow it to develop fully	1,2,3
B	Blanket and valley bog—this should never be burnt	1,2,3,4,5
M	Purple moor grassland—this should not be burnt more often than every three or four years	4
W	Land with a good cover of whortleberry (bilberry) but very little heather—this habitat cannot readily be burnt	4,5
F	Bracken (fern) covered grassland, where the bracken cover is greater than 50% but there is very little heather—burning here would be counter-productive and is not recommended	4,5
G	Gorse covered grassland, where there is very little heather—this should preferably not be burnt more often than once every 12 years	4,5
O	Other grassland, where there is little or no heather, whortleberry, bracken or gorse—these grasslands cannot readily be burnt	4,5
S	Scrubland or woodland with a greater than 50% cover of trees or tall shrubs	1,2,3,4,5

* Note that this category may be applied to areas where the heather has been so severely suppressed by heavy grazing that it cannot be burnt pending the relaxation of grazing pressure and recovery of the plants.

Table 12.1. (cont.)

Heather condition scores (with stocking recommendations):	
1	Good cover of heather, with no signs of decline due to over-grazing or too-frequent burning. The heather cover may be as low as 30% or even less if recently burnt, or in areas of bog.
	Existing stocking in these areas may be maintained but should not be increased without consultation with the Commoners Council.
2	Heather cover remains good, but is suppressed with plants showing evidence of clumping and twisting due to heavy grazing by sheep and perhaps ponies:
	(a) As above, but heather also showing evidence of significant damage due to trampling and grazing by cattle.
	Stocking levels in these areas should be carefully monitored to ensure that the quality of grazing and the ecological importance of the sward is maintained. Where cattle damage is evident this almost invariably arises from the outwintering of stock, so that measures should be taken to reduce the number of cattle present between 15 November and 15 April.
3	Heather cover severely reduced and certainly now less than 30% due to heavy grazing by sheep and/or too frequent burning in the past, but still readily recoverable if the grazing pressure and/or burning regime is relaxed.
	(a) As above, but plants also showing severe damage by cattle.
	Stocking levels should be reduced as a priority, particular attention being paid to any damage caused by outwintered cattle.
4	Soil and/or existing vegetation suggests heather should naturally be present at a high cover but very little or none now exists. The heather may have been lost recently or a long time ago, but is in any event not now readily recoverable.
	Existing stocking levels in these areas may be maintained, pending action being taken on areas classified as 2 or 3 above.
5	Soil or water table not thought suitable for heather (although it may be for whortleberry).
	Existing stocking levels may be maintained.

is only too easy to underestimate the time needed for this, whether there are problems or not and it should be allowed for when using a bureau for data capture. In addition to digitizing their polygon boundaries, the bureau was asked to encode each polygon, and again checking the data proved to be essential. The code described the habitat, heather condition, and damage present in that polygon.

Once the data was checked it was then given a topological structure using the Intergraph Microstation Graphics Environment (MGE) product. The polygons were then linked to the Informix database using the polygon codes. It was then possible to calculate the total area of each feature type for the whole of the Dartmoor Common Land and for the four ''Quarters'' using Microstation Graphics Analyst (MGA). The ''Quarters'' are recognized subdivisions of the Dartmoor Common Lands. A set of three maps were produced showing the Habitat Management Units, Heather Condition, and Cause of Damage. Scanned 1:50 000 scale Ordnance Survey maps were used as backdrops to the above information. Because the contours were not on the above maps these had to be added as vector information supplied by the Ordnance Survey. The final maps were then output to the colour electrostatic plotter at a scale of 1:50 000 for the full maps and 1:25 000 for the ''Quarters''. In addition to the maps area estimates were calculated for each feature, examples of these are shown in Figure 1 and Table 2.

The production of the above maps and area estimates demonstrated a fairly straightforward use of GIS. This is, however, only the beginning. Now that the data are in a digital form it can be used in a wide variety of ways. For example the contour data can be used to produce a map of slope angles, this could then be compared with the areas of greatest erosion to see if there was any correlation. Alternatively comparisons of damage could be made between areas falling in or outside SSSIs. This information could then be used as a baseline to see if heather outside SSSIs were more vulnerable damage than that found within the sites.

Conclusion

It is hoped that the management maps will be of practical help to those responsible for managing the commons of Dartmoor. These maps should provide guidance for many years to come in the same way that the vegetation maps produced in 1969 by the Nature Conservancy continue to be used today.

It is likely that the heather condition maps will become dated fairly rapidly. Hopefully this will be due to the various regeneration programmes now taking place rather than further damage.

The GIS has clearly demonstrated its power in presenting heather management information. The real power of the system has still to be fully realized as a tool for monitoring the future development of heather within the Commons of Dartmoor.

Acknowledgements

I am only reporting here work that has been carried out by other staff within English Nature. Rob Wolton, the project co-ordinator, has kindly provided the background information for this paper. Steve Edge and John Riggall were primarily responsible for the work carried out on the GIS.

Table 12.2. Analysis by area of heather conditions within each quarter, and for all the commonland surveyed

Code	Heather condition	Area (ha) and, in parentheses, percentage coverage				
		North	South	East	West	All commons surveyed
1	Good cover of healthy plants	3654 (36.3)	2571 (30.2)	2418 (34.8)	1695 (20.4)	10 337 (30.5)
2	Good cover but plants heavily suppressed by overgrazing/burning	2801 (27.8)	1872 (22.0)	1635 (23.6)	1755 (21.2)	8062 (23.8)
3	Cover severely reduced by overgrazing/burning	1417 (14.1)	644 (7.6)	748 (10.8)	864 (10.4)	3674 (10.9)
4	Heather absent although soil suitable or typically associated plants present	1716 (17.1)	2931 (34.4)	1702 (24.5)	3416 (41.2)	9765 (28.9)
5	Soil or water table not suitable for heather	473 (4.7)	503 (6.0)	439 (6.3)	581 (7.0)	2001 (5.9)
	Total areas:	10 061 (100)	8526 (100)	6942 (100)	8293 (100)	33 839 (100)

Fig 12.1. The relative extent of habitat management units within each quarter

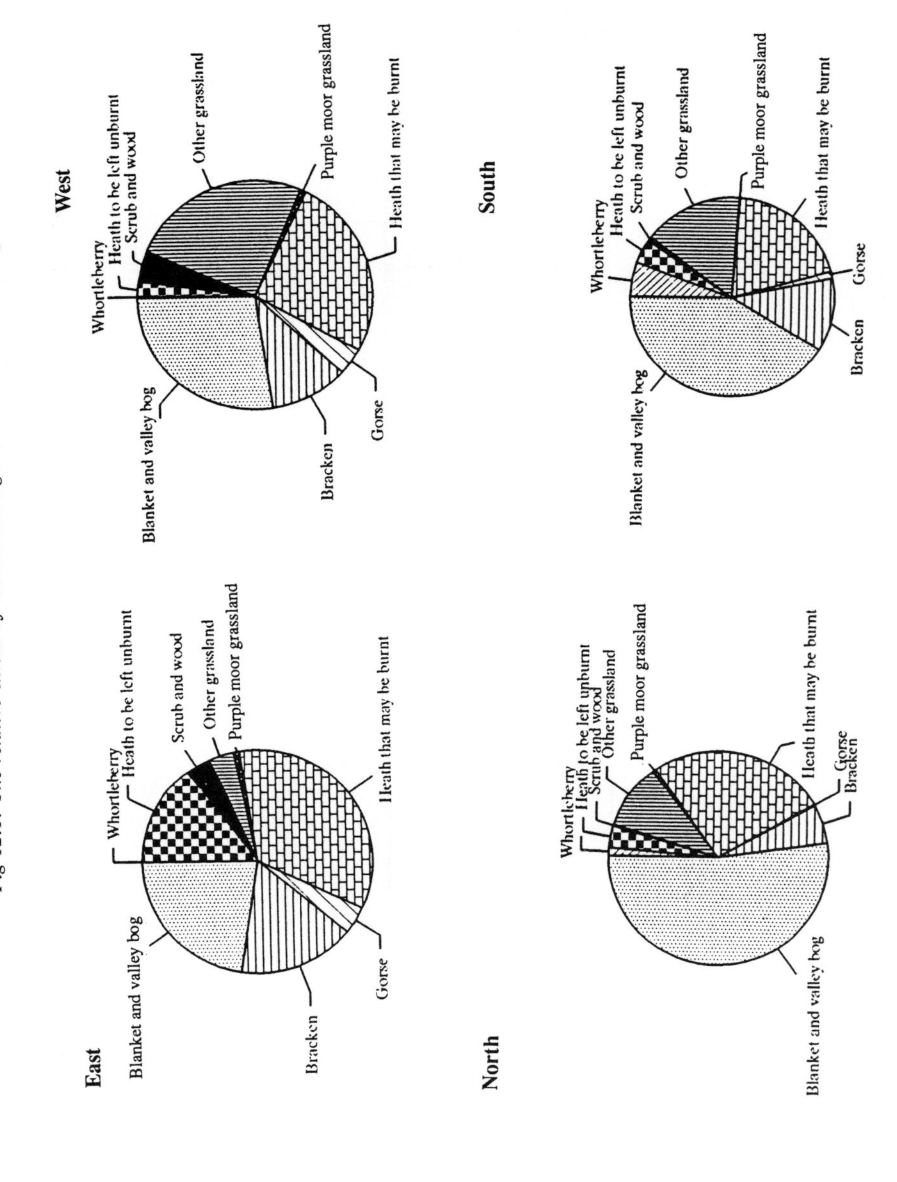

References

Countryside Commission and Countryside Council for Wales, 1991, Landscape change in the National Parks, Manchester: Countryside Commission Publications.

Nature Conservancy Council, 1990, Management Guidelines for the Commons of Dartmoor, Taunton: English Nature.

13

GIS in estuary management: The need, the reality and the scope

Simon Zisman
Posford Duvivier

Introduction

Britain holds greater estuarine habitat than any other country in Europe. The impact of human activities on the estuarine environment is increasing. However, effective solutions to these pressures are hard to find. Posford Duvivier Environment (PDE), leading environmental consultants specializing in water's edge environmental issues, have been active in helping to formulate effective approaches for estuary management. They have been retained to undertake two major estuary management studies in the past year, for the Plymouth Sound, and on the Exe Estuary. The complexity of issues involved in the estuarine environment has highlighted the importance of information management. GIS can fulfil this role, and given appropriate support could provide an essential tool for effective estuary management.

This chapter outlines some aspects where GIS can play a significant part. It then moves on to describe some of the management data required, along with their inherent difficulties. In conclusion the potential for GIS is confirmed. However, because of a lack of co-ordinating, authoratitive body, the major constraint on the use for GIS in estuary management is a shortage of funding and long-term commitment to information management.

The conservation context

Estuaries featured high on the conservation agenda of 1991, having been shown to be under threat from a wide variety of activities and developments. Two studies in particular, brought to light new information on their extent, conservation status, and future (NCC, 1991, RSPB, 1990). Both set estuaries in their national context, highlighting the fact that Britain holds greater estuarine habitat than any other country in Europe. It therefore has a particular international responsibility for their conservation.

Estuary management issues and GIS

The estuary management studies undertaken by PDE identified many issues which could benefit from the use of GIS, in particular, the rationalization of baseline environmental information, analysis of estuary activities, monitoring, and publicity.

Information rationalisation

The lack of useable information is a major obstacle to implementing estuary management. For terrestrial nature conservation (birds especially), water quality, and salmonid fisheries, a great deal of data are available, but not necessarily in a form of practical use to estuary managers who, by the nature of their job, are likely to be generalists and therefore less familiar with detailed technical data. For other activities such as waterborne or water's edge recreation, shellfisheries and commercial operations, there is a shortage of information altogether. Equally serious is the lack of data on the impacts of one activity on another, for example water quality on species of particular conservation interest.

GIS has two roles in resolving this state of affairs. The first is an organizational tool, to collate issue-related data into a suitable common format. This allows data to be clearly presented, and allied to wider regional or national datasets so that local conditions can be evaluated without parochial bias. The second role of GIS is to integrate existing data. In order to achieve a balanced use of an estuary, the integration of information on different interests and activities is essential.

Activity management

With the intensifying use of estuaries, it is inevitable that interactions between different users will lead to more frequent conflicts of interest. As an example, various estuarine activities cause disturbance to birds. Recreation, shellfishing, bait digging, wildfowling, and low flying light aircraft are all potentially damaging examples highlighted during our studies on the Exe. Conservationists object to birds frequently being put to flight from feeding or roosting sites, while others claim that disturbance to the birds is only temporary and does no visible or permanent harm. Estuary management cannot be based on opinions, and clearly a better understanding of the interaction between birds and other users is needed. This requires field research and the integration of bird, habitat, and recreation data. The information and spatial analysis capabilities that GIS can provide can help in developing a better understanding of this problem.

Control of moorings is another frequently highlighted issue in estuary management. The mooring of vessels over traditional salmon netting sites, and the encroachment of moorings into navigation channels are two frequently encountered problems. There are moorings management software packages available, but these lack mapping capabilities or the ability to overlay other administrative or environmental features.

Pollution from sewage outfalls is also of particular concern in estuaries because of their use for contact water-sports, such as sub-aqua and jet skiing, and for shellfish production. GIS can provide storage and mapping facilities to analyse pollution inputs and in combination with numerical modelling output, can provide a valuable insight into pollution patterns. In our work on the Exe, the mapping of water samples points clearly highlighted a mis-match between sample sites and areas where water quality data were needed to monitor suitability for water-sports.

Monitoring

In the Tamar Basin and the Exe Estuary, a range of research into various environmental characteristics is being carried out by statutory bodies, developers, establishments of further education, and amateur enthusiasts. This effort is however, remarkably unco-ordinated because of the vast cultural differences dividing these groups. This inhibits the sharing of data, encourages esoteric ad hoc research, non-compatible data sets, and leads to unnecessary duplication of effort.

A GIS can support a wide range of monitoring requirements, from water quality to recreational use. It provides a framework for monitoring, encouraging standardization, information exchange, and the focusing of research towards real-world management needs. Monitoring is an essential part of management. It is a prerequisite for understanding the environment in question, and is the mechanism for determining the effectiveness of management in reaching pre-agreed objectives. However, it is also expensive, and a low priority because so many institutions work in a reactive short-term framework.

GIS can, however, to some extent overcome these problems. It provides impetus for better monitoring by making the most of existing information, by identifying significant gaps in knowledge, and by providing a co-ordinated framework for data collection.

Data display

Although GIS specialists may resent it, to the majority of people the value of GIS is in presentation of results in an understandable and authoritative-looking format. The people who you are trying to manage will want to see maps with their activities clearly marked. The over-laying facilities of GIS can help highlight the logic behind any management zoning. Sophisticated output is also a powerful tool in planning inquiries. Public consultation is an increasingly widespread philosophy in decision making, and clear and understandable figures play an important role.

Appropriate spatial units for estuary management

Defining the spatial extent of the estuarine entity is extremely difficult. Coupled with this challenge is the problem of the diversity of information, and hence the diversity of scales required to effectively present the information.

To the landward side, all points of land/water interaction need to be considered, including any associated wetlands, high tide roosting sites, outfalls, drainage ditches, coastal footpaths, public slipways, etc. Towards the sea, management needs to consider navigational approaches, associated channel dredging if any is carried out, recreational activities, and areas of conservation interest such as roosting wildfowl sites.

For the studies on the Tamar Basin and the Exe Estuary, defining the boundaries for the estuarine zone have led to protracted discussions. Concern is expressed that including areas of land in a water-orientated management will lead to expensive, unnecessary data collection, and conflict with land-based interests. Others see the broad-based, more holistic approach as essential to resolving the real causes of conflict on the estuary.

Data sources for estuaries management

The breadth of interests that need to be considered in estuary management is reflected in the wide range of data sources that can be used. The breadth of interests also illustrates the lack of co-ordination, the diversity of data scales and quality, and the variability of data availability.

For nature conservation, habitat maps are available from the Joint Nature Conservation Committee. Their coastwatch project has annotated 1:10 000 maps for the whole coast with habitat and adjacent landuse information. The distribution of species of special interest can be derived from reports, where in our experience, sketch maps of unspecified scales annotated with various keys or notes are the norm. Transferring these onto GIS can therefore cause problems, and careful reinterpretation is necessary to ensure the appropriate level of accuracy results. It may even be necessary to contact the original authors to restructure the mapped and annotated data in a more GIS-friendly way.

Species records can also be derived from Biological Record Counties which operate out of most county museums. Records are allied to grid references, and can be mapped relatively easily. Coverage may not be comprehensive however, and is likely to be restricted to certain species groups, depending on the availability of local expertise.

Birds have received most conservation attention because estuaries are essential for their survival. The British Trust for Ornithology (BTO) and the Wildfowl and Wetlands Trust (WWT) provide wader and wildfowl data respectively. This data has been collected regularly over a significant time period but gaps in coverage and changes in the count areas need to be dealt with. As a result, although the data can be obtained by interested parties, to utilize it to the full requires considerable effort and knowledge.

Allied to species and habitat information is site protection data. Areas of special conservation interest within estuaries often receive various types of protection, either statutory or voluntary, or through private land ownership. Sites of Special Scientific Interest (SSSIs) are the most frequently encountered statutory designation. Boundaries are available in digital form from English Nature who administer these reserves, or at 1:10 000 scale in paper map form. Other statutory designations such as National Nature Reserves are available only as paper maps, generally at 1:10 000. International designations such as Special Protection Areas have to be notified as SSSIs first, and consequently their boundaries are available in digital or paper form, although they may cover more than one SSSI. Older types of reserve such as Bird Sanctuary Orders have boundaries which are usually only described in legislation, for example by distances and bearings or between various fixed points. Non-statutory sites of conservation interest are protected through Local or Structure Plans, mapped at a variety of scales in these documents.

Recreation information is less abundant, and consequently has formed a major component of data collection in our estuary management studies. Questionnaires and maps are supplied to recreational interests such as yacht clubs, for annotation with activity areas and details on membership, times of the year most heavily used etc. Mapping of these features in combination with data base storage of information from questionnaire returns allows point-and-shoot interrogation for recreation activities. Which yacht clubs use one particular slipway, or the months during which certain activities operate in particular areas are types of output useful in management.

Other sources of recreation information have been surveys for or by local authorities from earlier decades. Changes in survey techniques, aggregation of data with no indication of the location of original details of survey results and differing styles of presentation all inhibit the optimal use of these reports in analysing patterns of change. These difficulties underline the need for co-ordinated and standardized monitoring, a framework that GIS could readily support.

Estuarine fisheries will usually involve a number or all of the following activities: salmonid or eel netting, shellfish collection or cultivation, bait collection for both commercial and recreation fisheries and netting or lining for seafish. Only data on commercial salmonid fisheries is widely collected at an estuary level and made available to the public. The holder of a licence for commercial salmonid fishing is obliged to provide returns, enabling the NRA to produce monthly statistics for the nature and weight of fish landed. The bulk of other information, including maps of fisheries distribution, is obtained from consultation with the fishermen involved. Location of shellfish grounds, pots, etc. can be derived in this way with sufficient accuracy, although changes in these necessitate annual revision.

Commercial information is perhaps least readily available of any information category in estuaries management. Consultation with businesses lining the water's edge is time consuming, but necessary to locate the features of importance to them, for example for access, shipment of materials, boat handling, waste discharge etc. Annotation of basemaps with symbols or polygons backed up by database information of type of business, access requirements etc. is one way forward where commercial issues are an important consideration.

Information on utilities is available on request from the agencies involved, such as NRA and local authorities on flood defence and coast protection structures. Maps can be sent to these bodies for annotation or are supplied through a routine enquiry procedure.

Other data sources required in estuaries management studies include, water quality, landscape protection designations, local planning policies, sites of archaeological significance and land ownership. As with the above examples, these too suffer from being held in a wide variety of formats and at varying scales. A particularly difficult data type is that of ownership which is often poorly recorded, or hotly contested, yet is a crucial element in estuary management.

Constraints on the use of GIS

A powerful GIS package offers a range of facilities that can be used in assessment of the issues highlighted above. Predictably enough, however, a number of constraints hinder this process. Familiar ones include lack of understanding of GIS capabilities and therefore a perception of its role by clients as an unnecessary computer gadget. Very little of the information required for management is in computer-compatible form, and even where it is, conversions may not be cost-effective. Data entry is therefore expensive. The ability to use the information technology of the client such as a Local Authority, even when they share the same GIS software, can at first appear to be a potential solution to data entry expenses. However, their GIS team may not have corresponding work priorities and may, therefore be unable to help out within the time frame required.

Conclusion

It is important to reiterate certain key points in conclusions on the use of GIS in estuary management. Estuary management should be implemented in response to certain needs and issues. GIS therefore, as a management support tool, needs to be evaluated in terms of its capabilities in helping to resolve these same issues. In our experience, it is suitable for dealing with the information rationalization, activity analysis, monitoring and display issues we have encountered.

The constraints on this are three-fold: lack of digitally compatible data to reduce time and costs of data entry, the severe economic constraints facing potential estuary managers, especially local authorities, and the lack of tried and tested working systems that could demonstrate the value of GIS in estuary management.

Overcoming these difficulties will take time, but estuary management will falter without better mechanisms for dealing with the necessary information.

References

Davidson, N.C. *et al.,* 1991, Nature Conservancy Council (NCC), *Nature Conservation and Estuaries in Great Britain,* ISBN 0 86139 708 8.

Rothwell, P. and Hovisden, S., 1990, Royal Society for the Protection of Birds (RSPB), *Turning the Tide: A Future for Estuaries.*

14

Recreation potential and management in the Cairngorms: use of GIS for analysis of landscape in an area of high scenic value

D. R. Miller, R. J. Aspinall and J. G. Morrice
Macaulay Land Use Research Institute

Abstract

This chapter describes the use of GIS for analysis of scenery in Badenoch and Strathspey District. This is an area of recognized landscape importance and where there are significant land management issues associated with recreational activity, tourism, and land use changes. We present three methods of characterizing visual quality of scenery in an area and their use for assessing impact of land cover change. We also identify some applications of these approaches for integrating recreation and tourism into land management planning. The land use considered is forestry which produces a land cover of high scenic impact.

Introduction

The landscapes and scenery of the mountain areas of Scotland are internationally acknowledged and recognized as among the principal resources attracting tourists and other visitors (CCS, 1990). Badenoch and Strathspey District in the Highland Region is particularly important since it includes the Cairngorm Mountains, an area generally agreed to be one of the most important mountain areas in Britain. Recreation and tourism are important components of the local economy (Payne, 1972; Watson, 1988); in 1989 visitors spent some 1.9 million bed nights in the Spey Valley and many more visitors come for day trips, the area being within two to three hours driving time of Glasgow, Edinburgh, Dundee and Aberdeen.

Three studies emphasize the importance of the scenery and landscape of Badenoch and Strathspey for recreation and tourism. Sixty-seven per cent of a sample of visitors to the District rated scenery as the most attractive/enjoyable visitor attraction (Watson, 1988) while "high quality" scenery was rated as "important/ very important" by 95 per cent of respondents in a questionnaire survey of visitors to the Cairngorms (MacKay Consultants, 1988). A third study identified "drives, picnics and outings" to be the recreational activity with the largest participation (59 per cent of visitors; Survey Research Associates, 1981). In addition, the Cairngorms are a popular area for a wide range of recreational activities, including hill walking, nature study, downhill skiing and mountain biking (Watson, 1988; CCS, 1990). The potential and value of scenic and nature conservation interest of the Cairngorm area is evidenced by the wide variety of designation it receives (for example National Scenic Area, National Nature Reserve and SSSI).

The qualities of the scenic and environmental heritage in Speyside and the Cairngorms which encourage tourism and recreation are, however, highly sensitive to many of the tourist activities they promote. A report by the Countryside Commission for Scotland (1990) identified a number of key issues of significance in the Cairngorms area and, more recently, the Cairngorms Working Party has been considering management of the area and similarly identifies a number of issues which have direct relevance for recreation. These include lack of a coherent strategy for visitor provision, polarization of opinion over the development of skiing, an increase in footpath erosion in some areas, and attrition of the remote and wild mountain areas (which are slow to recover from damage).

In the context of complex issues associated with recreation planning and management and with a need to promote integrated management of resources based on principles which take account of both the sensitivity of resources to their use and the needs of local communities and visitors, GIS offers a powerful tool for providing information in support of decision-making. In this chapter we have chosen the assessment of scenery and identification of "visual impact" to describe the use of GIS for assessment of recreational resources. Other information suited to production with current GIS capabilities includes analysis of the sensitivity of footpaths to erosion, design of footpath networks (for example planning routes which avoid or visit particular features), monitoring and management of land use change (e.g. managing the harvest of forest resources to minimize visual impacts), and production and evaluation of scenarios for provision of visitor centres and other facilities.

An extra dimension to the methods presented here is related to legislation associated with European Community Directive 85/337 (European Communities, 1985). This requires that an environmental assessment be undertaken as part of a development plan or proposal. The aim is a systematic approach to provision of "relevant information on the environment concerned and the likely impact" (European Communities, 1989), and, although aspects of the environment requiring study are specified (these include visual impact), assessment protocols are not. The approaches here provide some protocols for assessing visual impact.

Badenoch and Strathspey District GIS

A GIS has been developed for Badenoch and Strathspey District and incorporates data describing environmental and cultural resources (e.g. soils, land cover, land capability assessments for forestry and agriculture, nature conservation and landscape designations, visitor centres, roads, built-up areas). The database was compiled from maps and air photos and also includes a digital terrain model (from the 1:250 000 scale, 100 metre resolution data produced by the Directorate of Military Survey; Smith *et al.*, 1989). Land cover is available from satellite imagery (LANDSAT Thematic Mapper imagery), air photo interpretation (The Land Cover of Scotland Project; Aspinall *et al.*, 1991) and the Cairngorms "Look-back" Study (Gauld, *et al.*, 1991; Miller, *et al.*, 1991) and video photography.

The "Cairngorm Visitor Survey" (Mackay Consultants, 1988) found that 53 per cent of visitors had consulted Ordnance Survey maps as a source of information and 42 per cent had consulted maps from another source; viewpoints designated on Ordnance Survey 1:50 000 Landranger maps are also included in this study and provide the focus for one element of analysis. Both raster and vector formats are used for storage and manipulation of the datasets. Resolution of the raster data is between 25 metres (for land cover) and 100 metres.

Since the area has a considerable forestry interest and tree type has an influence on visual assessment of landscape (Crowe, 1978) the nature of the forest/woodland was incorporated into the GIS database. Analysis can consider the relative contribution and distribution of coniferous, broadleaf and mixed forests.

The range in altitude contributes, in part, to the appeal of the scenery. The highest point is Braeriach (1 296 metres) and over 50 per cent of the District lies above 450 metres, with a further 37 per cent between 250 and 420 metres above sea level. Rice (1988) divides the District into seven landscape tracts. Five of these occupy about 50 per cent of the District and comprise the higher land around the Spey catchment—the Cromdale Hills, Cairngorms, Badenoch (Dalwhinnie, Loch Ericht and south to Drumochter Pass), the Monadhliaths and a northern moorland area from the Slochd to Lochindorb. These areas have peat, peaty gley, alpine and sub-alpine soils (Walker *et al.,* 1982) and associated moorland and blanket bog communities; mountain heath communities occur on the highest ground (Towers, 1988). These open vegetation communities contribute both to the high nature conservation value of the area and to the scenic appeal through their "wild land" qualities (CCS, 1990). The lower elevation land, described by Rice (1988) as the middle valley landscape tracts, between Glen Truim in the south and Grantown-on-Spey in the north also has a high scenic value (Rice, 1988).

Analysis of scenery using GIS

In the same manner that locations may be classified by their altitude, land cover or land use, so they may be classified by the amount of land visible from them; that is, areas can be compared on the basis of the relative area of land visible from each location. This allows changes in land cover/use visible from different locations to be assessed in the context of the land cover/use visible from all locations. Conversely, the visual impact of any land cover/use change at a location can be measured in terms of the area from which it will be visible. An extension of this assessment is measurement of the features within a scene visible to an observer. Three types of analysis are presented. Two relate to analysis of a large area (the whole of Badenoch and Strathspey District in this example) while one is more local and focuses on visual impact at a particular viewpoint in a particular direction.

Wide area analysis

The digital terrain model is used for identifying land visible from particular locations using viewshed and intervisibility calculations. This analysis is used as the basis of the wide area assessment of potential visual impact. Two approaches are used:

1 A census of the total area visible from all locations (where each location is a pixel in a raster dataset) within the District; this describes the relative (and

absolute) intervisibility of land within the District. An intervisibility calculation was made for every pixel in the digital terrain model, recording the total number of pixels visible from each. The application of this approach in architectural design is described by Benedikt (1979). The set of all locations visible from any specified location has been termed the isovist by Benedikt who describes the importance of the observer's location as central because it is "representing the position of the observer whose spatial experience we are trying to explore". The objective identification of the boundaries of this "spatial experience" is the basis for this analysis of the digital terrain data.

2 Calculation of viewsheds for selected visitor viewpoints to identify priority zones of particular visual importance to tourists; this is calculated in a similar way to the census approach but the locations used as centres for the calculation of intervisibility are the viewpoints and the map output is recoded to describe the number of viewpoints which can "see" each location in the District.

Figure 14.1. Visibility census of Badenoch and Strathspey District.

The terrain data was the only dataset analysed at this stage and an assessment of the level of land cover visibility can be made independent of current land cover. Figure 1 shows the visibility census output from the GIS. Vertical angles are calculated making corrections for earth curvature and atmospheric refraction, to determine whether a point is obscured from the observer's view. This analysis considers a complete 360 degree rotation (equation 1) around each location for a radius of up to 100 kilometres. The analysis is carried out on the 100m resolution raster DEM.

$$Nv = \sum_{1^\circ}^{360^\circ} \left(\sum_{0.1km}^{100km} (x_k > mx_{k-1} \right) \quad (1)$$

- Nv = number of visible pixels (visible area to observer)
- x_k = Height_k + Earth Curvature Correction + Atmospheric Refraction Correction
- mx_{k-1} = Maximum Height of $x_1 \rightarrow x_{k-1}$

The calculation produces the total area of land visible from every pixel, the maximum area visible from any location in this example being 1 555.5km^2. The analysis includes land outside the District which is visible from within it to ensure that results are not influenced by the administrative boundary of the District; forest cover is not included although this may obscure views and impact on intervisibility.

Figure 2a illustrates the results of this analysis. For diagrammatic purposes the histogram of visible areas (ranging from 610ha to 1 555.5 km^2) has been used to present the results as two equal area groups representing "high" and "low" visibility. The higher visibility land is along the bottom and sides of the Spey valley while, in general, the more mountainous regions of the District have lower values. This is due to their relatively enclosed nature. In Figure 2a the visibility census dataset is presented in units of 610ha to illustrate the pattern of relative area of visibility throughout the District. The modal area of visible land is 4 270ha. The five upland landscape tracts identified by Rice (1988) broadly coincide with the low visibility land presented in Figure 2a although there are exceptions, notably over the high visibility of the Slochd, mountain summits such as Cairngorm and certain hill and valley sides.

This visibility census demonstrates one approach to objective analysis of elements of landscape with respect to impacts of land use on the observer. In particular, this assessment provides an audit of the intervisibility in the District against which other resources may be compared; this complies with one objective of the Environmental Assessment Directive.

Analysis for viewpoints

Targeting analysis of scenery with respect to visitors can be done through considering the most popular recreational activity, "drives, picnics and outings", and the viewpoints to which these visitors may be attracted. A measure of scenic potential is assessed by determining the number of viewpoints from which any terrain location is visible. Figure 2b illustrates the District zoned according to the visibility of land from five viewpoints. Forest cover, which may intrude on the scene or obscure the

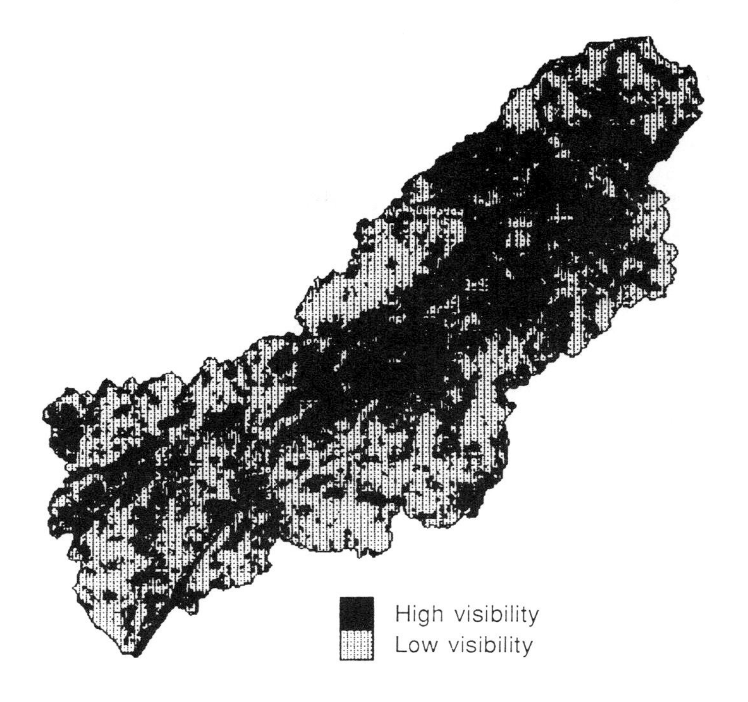

(a) *Terrain visibility census (divided into High and Low visibility by equal areas of visible land)*

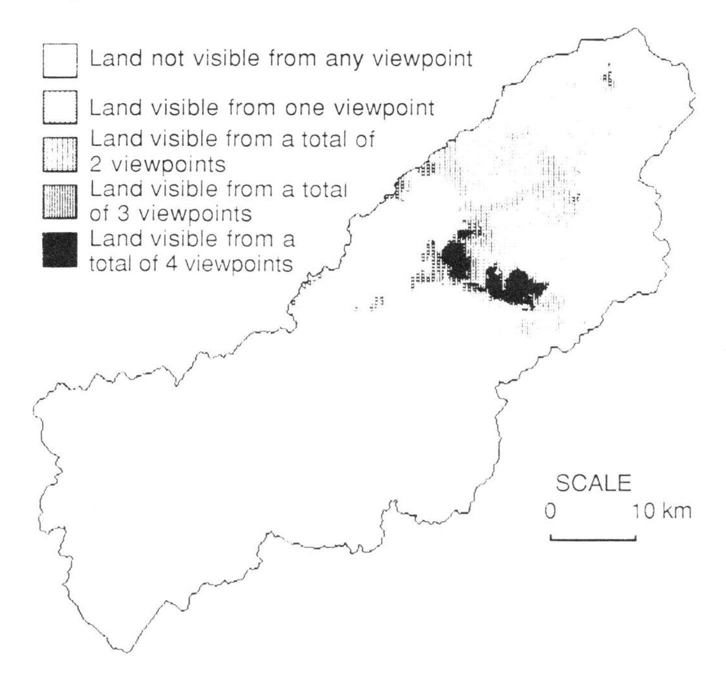

(b) *Terrain ranked by visibility from selected viewpoints*

Figure 14.2. Land visibility maps for Badenoch and Strathspey District.

view is included in this analysis. Scores range from "0" (land not visible from any of the viewpoints) to "4" (land visible from any four viewpoints); none of the District is visible from all five viewpoints.

These two maps (Figure 2a and b) can be used to assess the relative impact of different land cover/uses on scenery. This GIS output delimits land which is most sensitive to land cover change in terms of potential visual impact and provides a framework within which to assess the impact of any proposed changes in land use. The analysis can therefore be used to help plan the location, planting and harvesting programmes for forestry in the area while considering the contribution to the "scenery" of the District. It can also be used to assess the contribution and importance of existing land cover features. For example, Figure 3 shows the area of land under different forest cover types visible from viewpoints (Figure 2b). Forest occupies over 60 per cent of the land visible from four viewpoints, and of this 60 per cent forestry, over 90 per cent is coniferous woodland. This proportion of forest in the area visible from the viewpoints is ten times that in the land most widely visible land within the District and also almost twice the proportion of visible forestry in the District. This comparison between forestry as a percentage of land with different visibility using the two different visibility assessments illustrates the impact that forestry might have on the visual impression of landscape for a visitor at specific viewpoints.

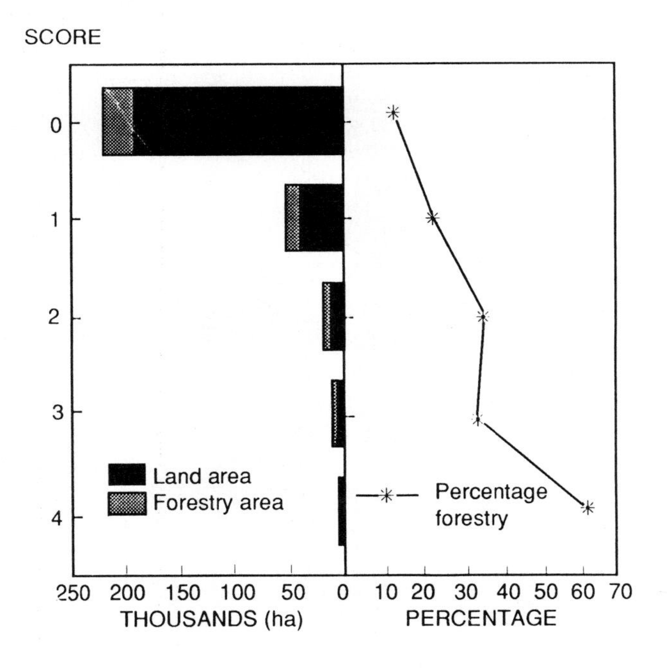

Figure 14.3. Land visibility and forest cover: Forestry area by the number of viewpoints from which it is visible.

Local analysis of scenery

Wide area assessments of visibility in the landscape of the forms described above highlight areas of greatest importance with respect to visitors across the District and are of use for strategic planning and management. A GIS framework also allows more detailed studies of visual impacts at particular locations and for particular scenes.

The scene from a location can be segmented into zones ranked according to predictions of visual impact on an observer at that location. The environmental features which contribute within a scene are topography, landform and land cover. By applying some of the principles of scene assessment (Land Use Consultants, 1990) and knowledge of factors influencing object clarity, a model can be constructed, calibrated and applied to describe the view from a specified location. McLaren and Kennie (1989) describe the advantages of visualization techniques in impact analysis. In improving the quality of landscape visualization, image rendering, with attenuation by atmospheric effects, is used to assess the visual clarity of objects in a landscape. Three further factors are considered in the model used here:

- Distance depth cueing and size perspective. A distance decay function is employed in calculation of visual clarity as distance between an observer and object increases. A projective transformation is also employed to calculate a score which incorporates the effect of an object's physical dimensions.
- Angle of intersection of the view with the terrain. As the angle of view increases with respect to the normal of the terrain, the proportion of terrain visible at that location decreases. However, if land cover at that point has definite height (for example forestry or buildings) the angle will decrease and visual impact increase. This may be most apparent on the visible horizon.
- Land cover. The discrimination of objects and their contrast against view background diminish with distance. Thus, land cover types within a scene provide the context against which to analyse and model visual impact and significant visual impacts may occur where contrasts in hue or brightness between background land cover and the land cover change is greatest. For example, even if a land cover change occurs and is visible from a viewpoint it need not be discernible from the background and the visual impact of that change should be identified as low by the model. Land cover provides texture, colour, hue and brightness, although these are also influenced by distance and atmospheric effects. Land cover (and associated texture, colour and intensity) are derived from LANDSAT Thematic Mapper satellite imagery in the GIS and from videos of scenes in the study area. These sources allow scores for cover type, texture and intensity to be calculated. These factors have been combined in a model for scoring the scene from a viewpoint and applied in the GIS. The model produces a simple index (Vi):

$$Vi = \frac{1}{D_2} + bH + Lf - Av \qquad (2)$$

- D = distance from viewpoint
- H = distance below horizon
- Lf = weighting according to landform type
- $-Av$ = verticle angle to feature

The viewpoint at the car park at the Cairngorm Ski Centre is presented as an example of this local analysis of scenery. Forest land cover is recoded according to average heights for tree type and forest stand. Adding this dataset to the topographic background provides the framework against which to measure the impact of forest land use change on the view from the viewpoint.

A true perspective view from the viewpoint, along a bearing of 270 degrees (toward Aviemore) is derived from the digital terrain data and rendered with the output from the model of scenic impact (Figure 4). Using this output to measure the impact of forest land use change in the example scene shows that the visual impact of afforestation has been low, although the Aviemore centre (a large white building!) has a much higher visual impact in the scene.

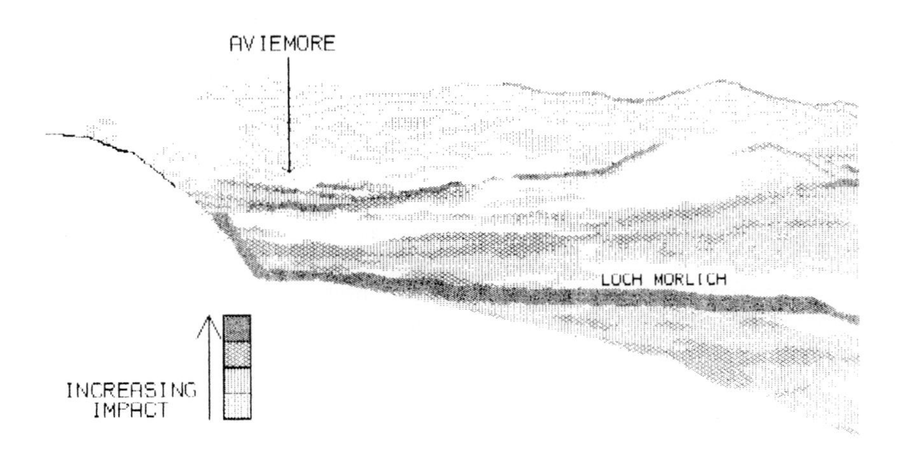

Figure 14.4. Impact model of forest land use for a view from the viewpoint in the Cairngorm Ski Centre cark park.

Conclusions

The methods presented here allow visual impact assessments to be made which are sensitive to the geographic context within which they occur and provide a role for GIS in data synthesis, analysis and evaluation as part of visual impact assessment. This type of analysis has several applications, including providing some basic tools which can contribute to Environmental Assessment.

Quantifying the area visible from any location provides a framework for measuring the extent to which a change in land cover will be visible to an observer.

It also allows for comparison between different locations and creates a framework within which to assess relative visibility in an area. In Badenoch and Strathspey District, the measures of visual impact described illustrate forest land cover having a high visual profile. This has implications for forest management and also for the management of landscapes which are of high scenic interest and which are an important element in the local economy as visitors are attracted to the scenic amenity.

The methods presented here can be extended to other recreational groups. For example, views from hill footpaths, minor roads or laybys can all be generated using GIS and integrated to provide a more wide ranging description of scenic interest and sensitivity of visitor pressure and land cover changes. The approach can also be linked to traditional methods of landscape analysis using questionnaires which typically incorporate some assessment of "place" through analysis of what may be termed "non-georeferenced locational data". Such a study has recently been initiated in the Cairngorms area as a joint programme between the Macaulay Institute and Landscape Architecture at Heriot-Watt University. This approach has potential for analysis of the visual environment as perceived by different groups (e.g. tourists, residents) and offers possibilities for multiple representation of the same scene under a range of scenarios related to both observer and land use.

At present these tools are utilized for analysis of land use change either historically (Gauld *et al.,* 1991; Miller, *et al.,* 1991) or for predicting future changes (Aspinall, 1990). Development of the methods within GIS coupled with their application to regional and local land use and land management planning will help develop useful information which assesses landscape for its amenity potential and which supports decision making as well as land and resource management.

Acknowledgements

The authors would like to thank Paula Horne for her assistance in the project in which the methods were developed, Alfred ODell at the Royal Aircraft Establishment (RAE) Farnborough for development of software, and the Directorate of Military Survey for experimental use of their 1:250 000 digital terrain data. Thanks also go to Calder Miller of the Forestry Commission for information relating to Environmental Assessments and William Towers for helpful comments on drafts of an earlier paper. Sarah Cornelius deserves special thanks for her patience and encouragement while we prepared this manuscript.

References

Aspinall, R. J., 1990, An integrated approach to land evaluation: Grampian Region. In Bibby, J. S. and Thomas, M. F. (Eds) *The Evaluation of Land Resources in Scotland,* Aberdeen: MLURI.

Aspinall, R. J., Miller, D. R. and Birnie, R. V., 1991, From data source to database: Acquisition of land cover information for Scotland. In *Remote Sensing of the Environment. Proceedings of Image Processing 91,* Birmingham, pp 131-52.

Benedikt, M. L., 1979, To take hold of space: Isovists and isovist fields, *Environment and Planning B,* **6**, pp. 47-65.

CCS, 1990, *The Mountain Areas of Scotland: Conservation and Management,* Battleby: Countryside Commission for Scotland.

Crowe, S., 1978, *The landscape of forests and woods,* Forestry Commission Booklet No. 44, London: HMSO.

European Communities, 1985, The assessment of the effects of certain public and private projects on the environment, *Official Journal of the European Communities* (85/337/EEC).

European Communities, 1989, *22nd General Report on the Activities of the European Communities 1988,* para 550, Luxembourg.

Gauld, J. H., Bell, J. S., Towers, W. and Miller, D. R., 1991, *The measurement and analysis of land cover changes in the Cairngorms,* Report to the Scottish Office Environment Department and the Scottish Office Agriculture and Fisheries Department, Macaulay Land Use Research Institute, Aberdeen.

Land Use Consultants, 1990, *Landscape Assessment: Principles and practice.* A report to the Countryside Commission for Scotland, Energy House, Lichfield, pp. 28.

MacKay Consultants, 1988, *Cairngorm Visitor Survey: Summer 1987,* A report for the Countryside Commission for Scotland, Highlands and Islands Development board, Highland Regional Council and the Nature Conservancy Council, Perth: Countryside Commission for Scotland.

McLaren, R. A. and Kennie, T. J. M., 1989, Visualisation of digital terrain models: Techniques and applications. In Raper, J. (Ed.), *Three dimensional applications in geographic information systems,* pp. 79–98, London: Taylor & Francis.

Miller, D. R., Gauld, J. H., Bell, J. S., and Towers, W., 1991, *Land Cover Change in the Cairngorms,* Proceedings of the AGI 1991, pp. 2.13.1–2.13.7.

Payne, M. A., 1972, *Speyside project. Second interim report.* Perth: Countryside Commission for Scotland.

Rice, D. E. A., 1988, The landscape of the Spey catchment. In Jenkins, D. (Ed.), *Land use in the River Spey catchment,* ACLU Symposium No.1, Aberdeen Centre for Land Use.

Smith, J. M., Miller, D. R. and Morrice, J. G., 1989, An evaluation of a low resolution Digital Terrain Model with satellite imagery for environmental mapping and analysis. In *Remote Sensing for Operational Applications,* Remote Sensing Society Annual Conference, Bristol, September.

Survey Research Associates, 1981, *Market survey of holidaymaking and countryside recreation in Scotland,* Edinburgh: Scottish Tourist Board.

Towers, W., 1988, Agricultural Opportunities and Limitations in the Spey catchment in relation to soil, climate and landform. In Jenkins, D. (Ed), *Land use in the River Spey Catchment,* Aberdeen Centre for Land Use (ACLU) Symposium No. 1, 6–8 November 1987, Granton-on-Spey: ACLU, pp. 100–105.

Walker, A. D., Campbell, C. G. B., Heslop, R. E. F., Gauld, J. H., Laing, D., Shipley, B. M. and Wright, G. G., 1982, *Eastern Scotland.* (Soil and land capability handbook No. 5) Aberdeen: Macaulay Institute for Soil Research.

Watson, R. D., 1988, Amenity and informal recreation on and around the River Spey: A consideration of conflicts, issues and possible solutions. In: Jenkins, D. (Ed.), *Land use in the River Spey catchment,* Aberdeen Centre for Land Use (ACLU) Symposium No.1, Granton-on-Spey: ACLU.

15

GIS support for conservation and development planning in Belize

Peter A. Furley and Neil Stuart
University of Edinburgh

Introduction

Confront most people with a map of the world and ask them to point to the location of Belize and you find that eight out of ten times their finger roams hesitantly around the west coast of Africa. In fact, the ex-colony of British Honduras lies on the Caribbean coast of Central America, with Mexico to the north and Guatemala to the south and west. When one hears that this English, Spanish and Maya-speaking country contains Indians, Creoles, Caribs, descendants of pirates and refugees from the American and Central American Civil Wars, within an environmental stage containing ancient mountains, lush rainforests with the only jaguar reserve in the world, mangrove lagoons, beaches, and one of the largest barrier reefs in the world, Belize might seem to be the best kept secret of the Caribbean. The truth is that it has only just discovered its selling potential and is facing up to the unpalatable conflict between development and preservation. It is not surprising that Belize is rapidly becoming the focus for many tour companies offering holidays under the newly unfurled banner of "eco-tourism"; it is equally apparent that unplanned development could quickly spoil the assets that tourists come to see.

With a surface area roughly that of Wales, mostly covered with tropical forest, and a population of under 200 000, Belize may not seem to be suffering the pressures on forest clearance that are alarming features of other Central American countries. However, localized population growth, combined with large tracts of land unsuitable for agriculture, has increased pressure on critical habitats. Increasingly, agricultural clearance occurs on marginal land, leading to a rapid erosion of soil and decline in fertility. The growing rural population, plus a large migrant group from strife-torn neighbouring countries, practice a shifting "milpa" cultivation that is difficult to monitor and almost impossible to control. Pressure on land is leading to the deregulation of protected areas and illegal occupation of private and national lands. In coastal areas, the need for building land as towns expand means that extensive strips of mangrove are being removed entirely. This destruction is despite legislation to prevent its removal because of its known role in defending the coast against hurricanes and in supporting the rich marine resources.

Against this backdrop of the sometimes conflicting needs of conservation, recreation and development, this paper reviews how GIS has been applied in both national and local projects that aim to support Government policies to provide an equitable and sustainable balance between these different demands. The Department of Geography at Edinburgh University has been working in Belize for some 25 years (Furley, 1987), and the remainder of the paper shows how developments in the technology and methods of GIS have been adopted and applied as part of the work on conservation and development projects over the last ten years.

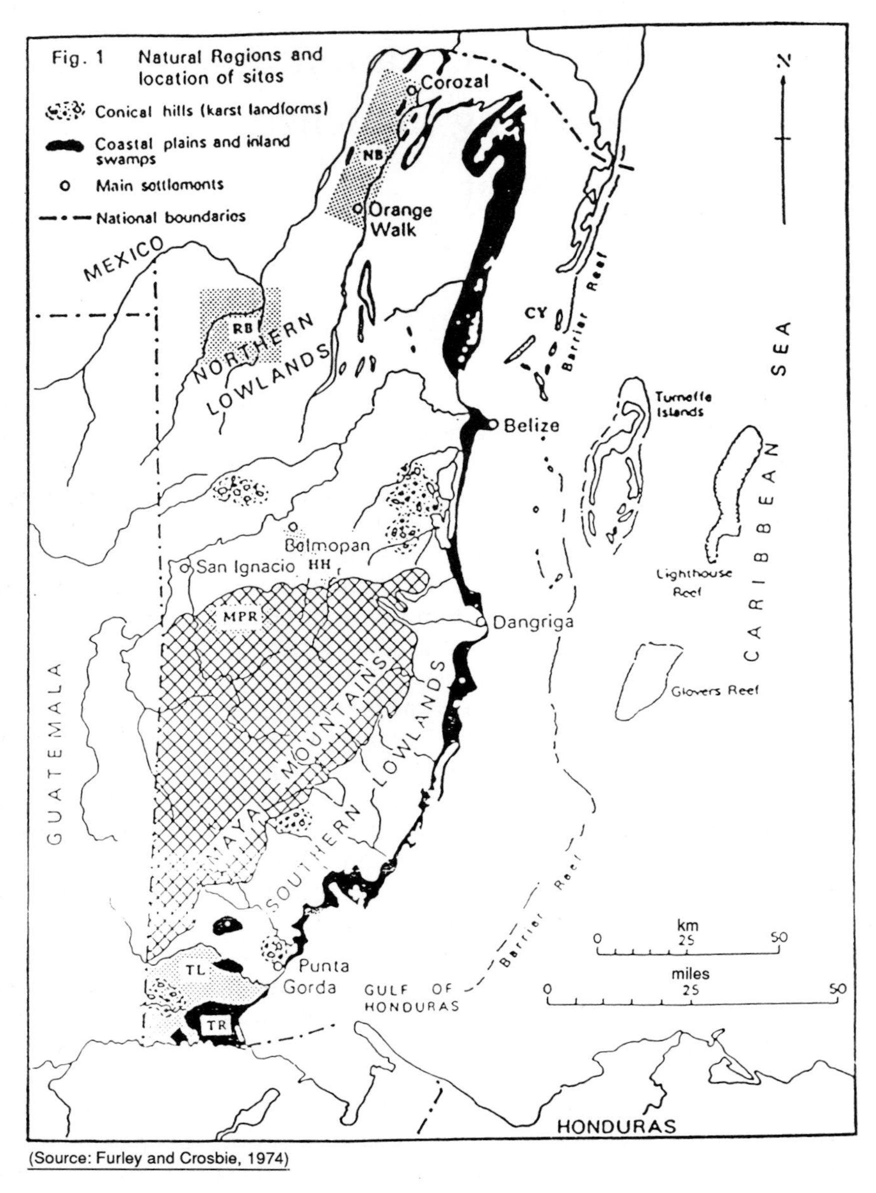

TR	– Temash River	RB	– Rio Bravo	TL	– Toledo Lowland
CY	– Sand Cayes	HH	– Hummingbird Highway	NB	– Northern Belize
MPR	– Mountain Pine Ridge				

Figure 15.1. Natural regions and location of sites.

Developing digital geographic data sets for national development planning

Belize offers great potential for the development of a national GIS capable of storing comprehensive digital data sets and subsequently analysing these for development planning. Some of the advantages which favoured the early implementation of GIS techniques included a manageable area, a relatively small but diverse population, clearly established enumeration districts and above all, data that was reasonably reliable despite many of the usual problems of quality that beset data acquisition in developing countries. Belize possesses well surveyed and mapped terrain, with fairly reliable information on demography and settlement, land use and a small manufacturing and commercial sector. This information has been collected consistently over the colonial and independence periods.

The earliest use of GIS was in connection with the pilot agricultural census (Robinson *et al.,* 1989). This was started in 1983 (Robinson, 1983) and acted as a test survey before the full census, with FAO assistance, which took place between 1984 and 1986. The pilot survey produced a registration of farmers and, while this missed a great number of smallholders, it paved the way for the larger and more accurate census that followed (Furley *et al.,* 1986). The pilot survey illustrated the versatility and usefulness of conventional DBMS, since the initial problem was to construct a consistent, geographically referenced data base of agricultural land ownership. This involved a great deal of tedious checking and cross referencing of, for example, smallholders with place of residence and the proportion of time spent in agriculture. Anyone with a purely European or American background would despair of making any sense out of the apparently conflicting and unfamiliar evidence! Once the data was screened, the next stage used the database to selectively derive new information such as the labour inputs required for the different farming systems.

Agricultural land registration is one part of the larger problem faced by the government in maintaining records of land tenure and of establishing title. Maintaining up-to-date land information is not as chronic a problem in Belize as elsewhere in Central and South America, since historically land related records were well maintained. However, there are presently long delays in establishing and transferring titles, and recent estimates suggest that perhaps 80 000 land parcels are not registered at all, representing a major loss of tax revenue to government. The Ministry of Natural Resources in the capital city, Belmopan, are about to install a land information system (LIS) specifically to address these problems (Hyde, 1991).

Work has been underway in Edinburgh since 1989 to develop a comprehensive GIS to support a strong interest from Belize to have a management and planning system for natural resources. This originally used the database of agricultural land and a digital version of the land use and land capability maps recently produced by the Overseas Development Administration (ODA) for Toledo and Stann Creek Districts (King *et al.,* 1986, 1989).

Using proven geo-relational technology, it has been possible to interrogate a data base and select particular combinations of, for example, soils, topography and vegetation to produce comprehensive land suitability assessments. A series of 1:50 000 digital maps have been produced and widely distributed to field officers. Efforts are underway to ensure that digital data of common interest can be shared between the GIS and LIS databases (Gray, 1989).

A further development has involved incorporating the benefits of remote sensing (RS) with GIS data handling. Until the 1970s, the only remote sensing was aerial photography carried out by the RAF for the Directorate of Overseas Surveys and for military defence planning. This photography is still of great use since the scale (1:40 000 standard) permits better discrimination of certain land and offshore features than can be obtained from commercially available satellite imagery. However, the multispectral capabilities of LANDSAT TM data have been used successfully to identify sugar cane areas and to discriminate different growth stages and varieties (Gray, pers. comm., 1989). Geometrically correcting the classified satellite images to a UTM base created a remarkably useful and comprehensible map, showing for the first time the full extent and distribution of the cane cover (previously estimated) and containing qualitative information that might indicate crop stress or disease.

This ability to undertake comprehensive monitoring of vegetative resources using a combined RS/GIS methodology has now been extended to a reconnaissance survey of mangrove distribution along the entire coastline and the offshore "cayes" (Gray, Zisman and Corves, 1990). This survey was later extended to the inland mangroves and wetlands, particularly in northern Belize (Furley and Ratter, 1990). Difficulties of access, combined with the locally complex pattern of mangrove communities and the recent rapid clearances have made it impossible to produce an up-to-date inventory of mangrove resources by field survey. The satellite image classification provided the first reasonably accurate picture of the national distribution of these important ecological zones. This imagery has now been converted into a more conventional map format, to be used for conservation and tourist planning and for a monitoring capability.

Through developments in both methodology and technology, we have now reached the stage where national natural resource data sets can be automated, maintained in a digital database and used in the development planning process. By combining RS and GIS, environmental information which was previously unavailable can be derived and integrated into land management decisions. While further national and local data will still need to be collected and updated, we are now experimenting with GIS techniques for testing various strategies for land use (either by visually comparing outputs or by modelling) and monitoring consequent decisions. GIS has come of age in Belize—but its real value is just about to be realized.

GIS support for tropical forest conservation programmes

One of the most promising uses of GIS lies in the field of land planning and conservation. This is currently being applied to two areas in Belize, the mainly forested tracts run by the Programme for Belize and the coastal strip and offshore coral and mangrove cayes which form part of the coastal management plan.

The Programme for Belize is a Belizean charity with fund raising centres in the UK and USA. Its objective is the conservation of strategic areas of tropical forest in the northern third of the country. Two initial studies have recently established a framework and provided basic resource information for a local GIS for this area. The first study used stereo aerial photography to derive base mapping for the Forest Reserve. Interpreted land use/land cover data and essential reference data such as roads, rivers, buildings and field boundaries were digitized from stereo models and downloaded into a common geo-referenced mapping grid of a GIS. Stereo photography also permitted the creation of digital terrain models to help in the visualization of relief controls over vegetation and drainage in the Rio Bravo part of the reserve in the far north west of the country (Lee, 1991).

The second study used LANDSAT TM as a basis for the generation of land cover maps and it examined in particular the impact of squatter invasions (Brown, 1991). This is one area where the monitoring capability of RS linked to a GIS can provide an immediate assessment of forest clearance and occupation. In the first instance, the objective of these GIS applications is to delineate boundaries in what is a remote and mostly unsettled part of the country. This will enable the establishment of a grid across the reserve so that the charities can allocate "ownership" of plots to corporate and private donors. This is partly for publicity purposes but also for the very practical need to control the deed allocation for many hundreds of small plots. It also allows for the detailed allocation of different tracts of the reserve to different conservation, experimental or fund-raising purposes. GIS will permit a degree of data manipulation and control essential for the scientific management and overall policy planning of the Programme lands. It is hoped that the system will develop further in a number of ways:

1. Acting as a spatial data base for all conservation and scientific work, including the numerous and important Mayan archaeological sites;

2. Nesting different scales of research from micro-plot and permanent plot studies to overviews of the local and regional distributions of ecological and environmental variables;

3. Planning logging activities and monitoring experimental and developmental aspects of the Programme's work in the area;

4. Using the data base to help formulate and evaluate different strategies for developing and/or conserving different sectors of the total area.

The current objectives of this work are to develop this experimental system into an operational GIS and to improve access to this data by local workers in Belize. Presently, we are providing a bureau service, with analysis run and reports/maps produced in response to specific requests that reflect strategic goals of the Programme. As the users of this data become more aware of what GIS can achieve, it seems likely they will wish to have closer access to the database and to explore "ad hoc" enquiries. The longer term aim is therefore to transfer base mapping and important, current data to a PC-based system, which should allow browsing, exploration and presentation of data in the headquarters of Programme for Belize in Belize City and at Rio Bravo (the field station). Opening up access to the data and technology raises several questions on the need for training non-specialists in GIS: where the master database should reside; how this should be updated and what levels of access different individuals, from field collectors to the Board of Directors should be permitted. While critical to success, these important considerations are beyond the scope of this paper.

GIS support for conservation zoning in the coastal management plan and barrier reef

The barrier reef off the Belizean coast is subject to increasing pressure from marine recreation as each year more people stay on the cayes and take diving, snorkelling and fishing trips to the reef. Ecotourism is being actively promoted, for example people are encouraged to visit the Hol Chan Marine Reserve. Here fishing is prohibited and there is a great abundance of marine fauna and flora, including many large fish. However, the increasing concentration of people in this area is causing its own problems, with anchors and divers causing damage to the living coral. Clearly, recreation and conservation do not necessarily go hand-in-hand.

The Government department responsible for balancing the demands of conservation and recreation offshore is the Department of Fisheries. They are drafting a Coastal Zone Management Plan (CZMP) which incorporates the idea of defining protected areas where the reef is particularly fragile or where the flora and fauna are of particularly high quality. It is hoped that recreational activity could be directed away from these areas.

As a precursor to designating conservation areas, the quality of marine habitats must be assessed and some form of inventory produced. The Department of Fisheries has been assisted in collecting this information by diving groups and marine biologists such as Coral Cay Conservation Ltd, a private company which funds a continual programme of diving and reporting of coral, fish and general marine conditions and who will use their resources and expertise to assist in drafting the CZMP.

A considerable archive of data describing marine conditions has been amassed for selected dive locations. However, for this to be collated, mapped and analysed spatially the data needs to be organized according to a geographical reference. A project has recently begun to create a prototype information system to support marine conservation around certain Belizean cayes. This uses a combination of GPS technology for locating dives and GIS technology to convert this into usable map based information. The project aims first to develop a geographically referenced database which permits the analysis and monitoring of "spot dive" data and second, to incorporate the dive data into a more comprehensive marine GIS. This will provide more powerful analysis and referencing to other environmental data sets, including aerial photography and satellite imagery of the reef.

The results from this work will be useful for several purposes, including the maintenance of a pin map showing the current status of underwater surveying; the identification of areas of missing or redundant data and the analysis of patterns of species inter-association and diversity. By processing dive data using the spatial analytic functionality of GIS, we hope it will be possible to delineate marine habitats according to specified and repeatable criteria; to explore relationships in space and time between point observations of coral or fish and their relations with continuous environmental parameters such as bathymetry, general bed cover or water quality and to produce maps to support a continuing programme of marine conservation. GIS should aid in explaining and justifying the selection of specific sites for protection, while showing other areas more able to support the demands of ecotourism.

Implications and conclusions

From these national and local projects that our institution has worked on over the last ten or so years, a number of conclusions can be drawn regarding the use of GIS technology for conservation and development planning. While some are specific to the country of Belize, we believe many of the points have more general implications for the use of GIS in developing countries.

Technical implications

Over the last ten years, the use of GIS has developed from simply automating a manual process (database searching for census mapping), to overcoming the difficulties in managing and maintaining large volumes of data (national land evaluation) towards making new, previously unavailable products through an integration of RS and GIS (mangrove mapping and agro-ecological zoning). Using the GIS concept of geo-referencing to a common base, multiple applications have been able to use the same basic data. Now, combining some existing digital data with more detailed mapping, remote sensing and field surveys we are investigating specific regional conservation issues and alternative management strategies.

In the same way that RS permitted great leaps forward in providing comprehensive and timely mapping of resources, GPS technology is about to have a major impact in providing quicker and more accurate location of specific features used for reference, such as road junctions in densely vegetated areas. Perhaps of even greater importance, developments in data viewing software and the declining costs of sufficiently powerful PCs to run such software will significantly improve access to and understanding of GIS capabilities by non-specialists. We believe we are finally about to produce appropriate technology GIS for developing nations.

Institutional implications

We are witnessing an increasing acceptance of GIS products by the institutions of Belize. Maps derived from GIS are now used alongside traditional mapping to perform resource assessments. GIS seems to have been accepted by being introduced initially for tasks where the technology has shown benefits before in other countries —notably for automated land inventory and evaluation; also the information is produced in a traditional form (land suitability map) and according to an understood methodology (the FAO framework) and interpretation is supported by the expert local knowledge of scientists and field workers. In this way GIS is seen to aid in the management and analysis of a large volume of environmental data and to provide a more effective method of map production.

More recently, innovative and less conventional products have been produced, such as mapping from satellite imagery. Experience shows that these products will be adopted if they provide information that cannot be obtained from other sources and if guidance in their interpretation can be given. A typical example of guidance is to produce map-like products, with legends and reference information such as a grid and cultural data overlaid. While we are aware of the limitations of some of these products, the practicality is that for some resources, such as mangrove, approximate mapping is satisfactory if it reliably conveys the general situation.

The government of Belize wishes to have a GIS capability in the country for more conventional applications. Belize is similar to many developing countries in that the first government installation will be an LIS for land registration, where obvious fiscal benefits can be identified from improved tax collection. One would hope that the surveyors, who typically gain early access to the technology in these situations will promote the spin-off development of GIS for environmental needs, where financial returns are less clear. We must hope that the "spread effects" of the technology are not withheld, as in some cases where officials see career advancement opportunities through the new technology. Information is certainly power in these countries and GIS means lots of information.

Balancing these factors, conservation agencies with outside funding may be the most likely catalysts for environmental GIS, but perhaps the greatest challenge to institutionalizing GIS in Belize will be the need to impart the necessary skills to create a trained base of local personnel. The present shortage of local skills means that GIS installation is founded on an insecure base not only of short term foreign funding, but more importantly, the knowledge of short-term foreign consultants.

Environmental implications

The GIS methods outlined above provide a far better means than previously possible to monitor and analyse changes in the extent and quality of Belize's environmental resources. Depending on whether local Belizean environmentalists are able to use these GIS products to motivate government to a stronger commitment and enforcement of environmental conservation will ultimately determine whether GIS provides the basis for rational policy making or simply remains a means to map the decline of Belize's natural environment in ever better detail.

References

Brown, M.T., 1991, The feasibility of a multi-sensor approach to change detection for monitoring forest clearance in the vicinity of the Rio Bravo Resource and Conservation Area, Belize, unpublished MSc thesis, University of Edinburgh.

Furley, P.A., 1987, Research in Belize: twenty one years of Expeditions 1966–1987, Edinburgh: University of Edinburgh.

Furley, P.A., Robinson, G.M. and Healey, R.G., 1986, The registration of farmers: summary and data. In Furley, P.A and Robinson, G.M. (Eds), *The Agricultural Census of Belize, Part 3,* Edinburgh: University of Edinburgh Occasional Publication. No. 4, pp. 1–275.

Furley, P.A and Ratter, J.A., 1990, *Mangrove distribution, vulnerability and management in Central America,* Project outline R4736. Unpublished. Oxford Forestry Institute/ODA.

Gray, D., 1987, Report to the Belize Sugar Industries Ltd. Unpublished.

Gray, D., 1989, A preliminary report on the feasibility of establishing a geographical information system (GIS) in Belize. Report to the Government of Belize, London: ODNRI.

Gray, D., Zisman, S.A. and Corves, C., 1990, *Mapping the mangroves of Belize,* Technical Report, Edinburgh: University of Edinburgh.

Hyde, R.F., 1991, The feasibility of a land information system for Belize. Central America, *International Journal of Geographical Information Systems,* **5**(1), 99–110.

King, R.B., Ballie, I. C., Bissett, P.G., Grimble, R.J., Johnson, M.S., and G.L. Silva, 1986, Land Resource Survey of Toledo District, Belize, Land Resources Development Centre, London: Tolworth.

King, R.B., Ballie, I. C., Dunsmore, G.R., Grimble, R.J., Johnson, M.S., Williams, J.B. and Wright, A.C.S., 1989, *Land Resource Assessment of Stann Creek District, Belize,* Report, Overseas Development Natural Resources Institute, Bulletin OB-19, Surbiton, Surrey.

Lee, Y., 1991, Use of photogrammetric data capture and GIS technology for multiple use resource management in the Rio Bravo Conservation Reserve, Belize, unpublished MSc thesis, University of Edinburgh.

Robinson, G.M., 1983, The registration of farmers. in Furley, P.A. and Robinson, G.M. (Eds), *The agricultural census of Belize, Part 1,* Edinburgh Department of Geography Publication No 2, pp. 1–54.

Robinson, G.M., Gray, D.A., Healey, R.G. and Furley, P.A., 1989, Developing a geographical information system (GIS) for agricultural development in Belize, Central America, *Applied Geography,* **9**, 81–94.

Part IV

GIS and Society

16

GIS in Society

Anna Cross
University of Newcastle upon Tyne

Introduction

This part offers both a snapshot of GIS in 1992, as well as emphasizing the way in which GIS should be progressing, with particular reference to the experiences of four application end-users. The issues that are tackled however are not restricted to the areas covered by this part.

First of all it is necessary to give the chapter some focus—especially with an all-embracing title such as "GIS in Society". What does this actually mean? One definition found in a trusty geography textbook was:

> Human beings create society at the same time as they are created by it . . . Society is both a cluster of socially constructed institutions, relationships and forms of conduct that are reproduced across time and space (Johnston, 1981).

This definition incorporates all the key GIS buzz words, such as time, space, clustering and relationships. Thus, this may suggest that GIS is "the" technology for describing and analysing aspects that affect society. As a result though it is probably true to say that most of the subjects targeted in this book, including environmental issues, infrastructure, and commercial users could quite logically be found in this section, since they all have the potential to affect society in some way or another and vice versa.

The papers in this part were chosen therefore to reflect those organizations who have a major interest in GIS but do not necessarily fall into any other subheading found in this book. The result is four very important examples chosen from areas where the implementation of GIS could and does have a direct impact up on society i.e. they are all involved with the provision of a service. In other words they are responsible in some way or another for changes in our urban surroundings, the safety, well-being and general quality of life.

Building on "GIS in society" 1991

Previous GIS literature, and the last edition of the AGI Yearbook is no exception, tend to emphasize the overall euphoria of GIS and its benefits. This culminates in a series of papers which essentially reiterate what GIS stands for, what it could do and basically the advantages that the technology offers to various organizations. While this is a reflection of GIS as it stood in the application world in 1990-1, the situation is somewhat different now. The role of today's AGI contributors should be one of evaluating where GIS has arrived, what problems are starting to raise their ugly heads and essentially focusing on the future course of GIS. A major annual publication such as the AGI Yearbook therefore is an effective communication medium which can pass on valuable knowledge to actual and/or potential GIS end-users, which themselves form one part of the "society" with which this chapter is concerned.

In response to the latter, the authors contributing to this section were asked to provide an objective view of what they saw or expected from their GIS, giving a combination of "this is the wonder of GIS" with that of the possible pitfalls which they realistically expected or had discovered. In addition some personal opinions and a vision of the long-term projection of GIS were also expressed. It was considered that this more balanced and rational approach to documenting GIS would provide an agenda for its development over the forthcoming year(s), with the hope that this would breed success and dictate GIS's continual prominence within the market place.

As a result this part offers a melee of views which reflects the tumultuous and exciting situation in which GIS finds itself at the beginning of the 1990s. It highlights those GIS users who can afford to follow the bandwagon effect and invest in GIS, while others have the vision and the need for such a system but do not have the finances to bring this into fruition. So . . . .

What do the papers say?

The first chapter was contributed by a member of the Northumbria Police Authority, Chief Inspector George Campbell. The viewpoint that he adopts is one of a prominent member of a team involved in the first crucial stages in evaluating the possible implementation of a Police GIS. It serves to emphasize the expectations and desires of potential GIS end-users who are having to decipher what GIS can actually do for them, based upon the confused and conflicting messages that are being received via the vendors. It demonstrates ideally how GIS's technology is assessed by essentially inexperienced users. The number of advantages noted are varied, but the key factors affecting operational requirements for the set up seem secondary in comparison. Yet it is equally important that an understanding of the impact of GIS on the organization as a whole, as well as the "we can do this" syndrome must filter through into the important decision making stages.

The next two chapters describe applications which are being developed within a university environment. Andrew Lovett at the School of Environmental Sciences, University of East Anglia, looks at the role of GIS in the Health Service, while Craig Whitehead and Guy Dominy at the Geographical Information Research Laboratory, London School of Economics, focuses on the impact on Social Services, in particular the potential of GIS in managing data for child protection with an operational case study for Southwark Social Services. Both of these chapters serve to highlight what could be done with a GIS in these key service areas and the benefits that could be accrued operationally from implementation and then passed on to the public.

In addition, the academic environment allows them to adopt a far more objective framework. They can jump ahead to offer an outlook of the realistic potential and effectiveness of GIS which could be realized with further development. Factors concerning basic data requirements and the need for improved understanding of GIS in such service areas are also reviewed.

The final chapter in this part is presented by Roy Laming, the GIS Support Team Leader at Kent County Council. Essentially this provides an example of how Kent County Council has successfully completed and overcome the GIS up-take barrier. In addition it highlights their ability to manage local authority services efficiently. It is, in short, a vendor's dream. It describes the development and operational use of a GIS that has been built into a multi-agency framework. This emphasizes the benefits that can be accrued from sharing and working together with people and organizations that use mutually compatible geographical data, such as the Local Authority, Police Authority, Planning Departments and so on. This is both constructive and cost effective, as Roy points out.

This introductory chapter is used to pull out and expand upon some of the points raised in these chapters. It should be noted at this point though that any pitfalls which are referred to in this part are not aimed at signalling the demise of GIS, but are to be construed positively as a means of preventing others from experiencing similar problems. In turn they offer a strategy for future research which will hopefully offer fresh thinking to any problems and will help to alleviate technological foibles and extend GIS functionality.

The overall message that is received from the papers though, is one of GIS is here to stay. It is also becoming increasingly apparent that the end-user does not have much choice any longer as to whether or not they partake in this Information Technology revolution. Yet GIS is still in its infancy and as a highly attractive technology suffers from many of the problems that affect all new technologies, i.e. that of being wrongly sold as the solution to all the spatial data handling needs of the end-user, the product of which is to lead the end-user into a false sense of security. This false security is heightened by the reality that the novice end-user invariably does not know what he/she wants from a GIS in the first place, or more specifically how its capabilities and limitations will affect their particular application area. Thus these chapters should be accepted as important examples of operational GISs which provide some valuable pointers for others in the process of, or contemplating, implementation.

The latter suggests that GIS is not simply a question of relational databases, the ability to manipulate, store and present data and cost analysis. It is also about people and their response to Information Technology (IT) and their ability to get the best out of a GIS. So the title of this part, ''GIS in Society'', may be more aptly phrased as ''GIS with Society'' (i.e. those in control of the implementation and direction of GIS once taken on board) and in turn, ''GIS for Society'', as demonstrated by the chapters here, which serve to emphasize how their adoption of GIS technology and the subsequent benefits have been or could be passed on to the general public one way or another.

The immediate future therefore must be to ensure that GIS comes down from ''cloud nine'' to establish itself as an everyday nationwide tool which is taken for granted. This will involve a major educational plan, not only in software use, but also in the understanding of what GIS is there for and how it will have an impact on any particular organization. GIS therefore is more than hardware and software; as this quotation by Campbell (1991) emphasizes, it involves:

> the concept of computer technology as a package which includes not only hardware and software but also people, personal skills, operational practices and corporate expectations.

People and GIS

For GIS to succeed on a widespread basis, the general perceptions of the technology must be fashioned in such a way that users understand and accept the advances in the "state-of-the-art" technology and its potential. A couple of the chapters in this part demonstrate that some organizations have succeeded in overcoming the barriers to something new, alternatively the others show that there is a whole range of potential end-users out there that are adverse to the same changes. The reason for this may be that they see it as a threat to their comfortable and already well-established manual and/or computer based systems which they understand and may wish to retain.

In addition, the potential end-user is not always sure what they want from GIS and as a result they cannot necessarily convey their needs to the vendor. In the past this has led to inadequate customization (Openshaw *et al.*, 1990b), which only served to complicate the GIS, rather than make it easier. Vendors must avoid this situation through establishing a continual dialogue between themselves and their clients—such bouncing of ideas is compulsory for success. Thus, the ultimate goal should be to build user-friendly tolerant interfaces which extend the functionality and data access to a wide audience of interested parties.

The latter would not simply involve a six month period of selling, setting up and answering queries of a technical nature. For example, in order to satisfy the requirements of a Police GIS, as listed in Chief Inspector Campbell's "wish list", a far deeper understanding of the relationship between those demanding and those supplying the system was required. The aspect of Crime Pattern Analysis has been under discussion at the University of Newcastle upon Tyne for over six years and still there is a long way to go in terms of meeting all the needs of a police environment—so is a superficial six months really enough for successful GIS implementation?

Another barrier often attributed as an inertia factor to GIS uptake is that of finances, or, more particularly, the lack of them. A typical cynical response for this excuse is usually that the organization already has databases (OK so they may not be linked!), they have graphics packages and maps with pins in that equally demonstrate the geographical element of their data (be it time consuming, laborious and messy!), and they have statistical packages for churning out numbers and statistics for end of year reports (even if this is a lengthy job in itself and continual evaluation, monitoring and updates are restricted to an annual review). In other words the counter-argument for GIS in this scenario is that everything can be done outside a GIS environment, so why invest in this new system at all? This viewpoint is rather short-term and narrow minded, and only seems to focus on the immediate cost-benefits of such technology. It is accepted that the initial set-up costs required to obtain digital databases, convert existing records to a GIS format and other necessary infrastructures are extremely expensive, but the benefits for the future and the time and effort that can be saved from the resultant changes through the introduction of GIS, if employed correctly, would be far greater than these seeming financial disadvantages.

It would seem therefore that while the vendors are multiplying at every GIS conference that is held, each with the aim of convincing people that their system is the best for them and exuding enthusiasm about their nicely customized interfaces and databases, it is all-important to remember that this enthusiasm has to be sufficient to allow GIS to thrive outside the comforts of their exhibition stands. It is essential that the knowledge of, and desire for, GIS must be cultivated at both the strategic and the nuts and bolts end of the hierarchical structure of any organization. This includes those that hold the purse strings and make the organizational decisions right down to those who will be directly affected by the implementation of the new technology, i.e. the operators.

Thus, GIS and its associated software/hardware must also be accompanied by developed channels of communication, data sharing, corporate collaboration and mechanisms to ensure that the results are used and can be used routinely in the key areas where they were intended to contribute. The message here is that it is extremely important not to under-estimate the psychological aspects of misplaced knowledge and expectations which can accompany any advances in technology and may act ultimately as a failure mechanism for GIS development.

The key to overcoming many of these people-orientated problems, as illustrated in the applications to follow, has been to develop a well thought out strategy and decision-making structure with which all parties were involved. This is particularly emphasized in Roy Laming's chapter. This approach is extremely useful during those periods when, as with every application, there will be a setback or two, both technologically and mentally. It is at this point that the researcher/co-ordinator must look to his/her goals as an incentive, because these provide a means of rationalizing the problem in order to find an alternative way of tackling it. Documentation of GIS limitations and other end-user experiences such as those contained in this part, and the book as a whole, therefore are very important in offering alternative methodologies. They also serve to reassure other end-users that they are not alone with their frustrations and that there is a light at the end of the GIS implementation tunnel.

Providing the information for success

The AGI, through its conferences, special interest groups and literature is an ideal communication medium for satisfying many of the issues raised thus far in this section, and those to follow. It can offer credibility to GIS through the experiences of others. This in turn may suitably minimize the hand over process from one IT system to another, in this case GIS. This role also applies to academic institutions where cost benefits are not the main priority and hierarchical barriers are few. Thus they too can aim to enhance and relay the crucial messages concerning GIS and its potential.

Both the AGI and academia should continue to adopt an objective and informative stance. We know all about the applications and research that are working well, the benefits, and implementation requirements. Now we should be looking to express our views and solutions on GIS inadequacies, a theme which is not over-saturating the present GIS literature. AGI has this vital position of disseminator, i.e. relaying both the negative and positive sides of GIS, because the vendors will not!

After a decade of GIS hype therefore it would seem that a new phase in GIS has begun where users, in particular, academics, are not afraid to say that the reality of GIS is not always meeting their original expectations. This situation provides an excellent position for launching alternative ideas and functionality. The following section will now highlight another important aspect that affects the future development of GIS and that is the need to extend GIS functionality itself in order to meet the specific needs of end-users and the society they serve.

Technology

It is hypothesized that there will become a time, in the not so distant future, when the GIS end-user will sit back having acquired his/her GIS, input masses of data, and found that they can manipulate and present their data in a variety of interesting and flexible ways. They may then find themselves expecting more, probably as a result of questions stimulated by the new and innovative system itself, but without the necessary tools to proceed along these new avenues of research.

For example, returning to the dialogue that has been struck up with Northumbria Police and the Centre for Research into Crime Policing and the Community (CRCPC), at the University of Newcastle upon Tyne. A number of additional questions concerning police needs have been revealed, such as what are the relationships between crime and the housing/social profiles of a particular area? Why do there tend to be more traffic accidents here than there? Where should the highways division plan to increase lighting facilities and openness to ensure that the area is much safer for pedestrians? and so on. These are very important organizational and management questions with which, given the masses of data in the GIS, any user may expect the system to aid them. However this is not the case yet!

The latter debate is mainly focused around the need for more sophisticated spatial analysis functionality in order to answer pertinent end-user questions of what, where and why. The vendors may argue that the academics are being greedy. In fact their aim is to anticipate the future needs of those GIS end-users entering the market now, as it is assumed that they too will probably reach a stage when the excitement of GIS has worn off and they will want more from the standard toolbox. This cannot be a bad attitude to have, because ultimately the solutions that are adopted will filter through to the end-user audience as a whole and then again through to the society that is being affected by the services provided.

Last year's article for the AGI Yearbook, entitled "Working for a safer society" (Cross, 1991) illustrated this need to develop the technological aspects of GIS, with particular reference to a health and environment GIS. The article basically argued that there was a whole set of tools lacking from the GIS toolbox. It then offered an end-user experience and response which demonstrated that with the right data and knowledge one could begin to push GIS to its limits and beyond, particularly through the development of spatial analysis functionality.

Over the last 12 months several debates have followed with numerous articles being devoted to the subject of "Spatial analysis and GIS". Initially these papers focused purely on the problem rather than offering constructive and improved techniques to overcome this. Reassuringly though this situation is beginning to change and more is now being offered in terms of solutions.

The new techniques which have been proposed tend to be written as separate programs sitting alongside the GIS in a complementary role. There is no particular software firm collaring the market; in fact they seem very reluctant to take on board these new ideas. Maybe they are waiting until there is sufficient demand from the market they will listen to, i.e. the buyers. Again this emphasizes the need for end-users not to be afraid to ask for more, because they are all-important in stimulating the development of the relevant tools to satisfy their end-user needs.

Last year's article highlighted two forms of alternative descriptive analysis, including a Geographical Analysis Machine (Openshaw *et al.,* 1987) and a Geographical Correlates Exploration Machine (Openshaw *et al.,* 1990a), which were developed as a response to spatial epidemiological questions concerning disease patterns and relationships. In the meantime, recent work has been involved in applying a similar approach to the needs of a police environment, Northumbria Police in fact. The aim is to help reduce the onus upon the Crime Pattern Analyst (CPA) to find patterns in data. This is especially important under the present circumstances of ever-increasing crime rates and accompanying data, as well as the general computerization of incident logging systems which only serve to alienate key personnel from the data. The result is that any possible hunches or "feel" for related incidents and possible patterns have become virtually impossible. So while GIS allows the police to store the data in a flexible system, and present them beautifully, it does not really solve the Crime Pattern Analyst's dilemma. Chief Inspector Campbell's paper does suggest, however, a number of areas within police operations where GIS implementation would have immediate benefits for the police.

The response to the latter CPA problem was for the CRCPC to develop analytical tools that would flag areas of high risk, using a GIS framework to allow the end-user to be as general or specific as they liked in terms of the search procedure. They also offered programs that took into account the historical aspect of the data by flagging areas of unusual activity for any given street, beat and/or division. This was seen as potentially very useful for highlighting the escalation of crime over short periods of time. Patterns according to the time of day or particular shifts were also seen as a potentially useful tool for debriefing or resource management. Further details on the development of these tools can be gathered from articles which document the stand-alone and GIS version of the Crime Pattern Analysis System (Cross and Openshaw, 1991; Openshaw *et al.,* 1991).

In fact the resultant system is now being prototyped in a live operational environment within a sub-division in Northumbria, demonstrating that a lot can happen in six months—from the time that Chief Inspector Campbell wrote up his experiences to the completion of this chapter. So positive changes are happening! In addition, its success is not only in Crime Pattern Analysis, but the fact that the policemen on the beat, Neighbourhood Watch and Community Officers are also using the system. This example therefore serves to demonstrate that with good customization and a clear view of what is needed, the data stored in a GIS can affect a far wider audience of expertise and interested parties than initially intended. This example is only the tip of a very big iceberg and demonstrates how with right tools and education GIS can effectively turn data into valuable information.

The reason for emphasizing this type of development in technology is not to advertise Newcastle University, but to illustrate that the initial discussions which surrounded "spatial analysis and GIS" have now progressed to a more profitable platform of "doing". It also shows that academia can have an impact on real and much needed areas of research that affect society. Academics are not all involved in blue skies research or secluded in their ivory towers as some people would like to believe. They provide valuable information which can and must be effectively communicated to the public and private GIS end-users. The role of conferences, and in particular the AGI, must be to draw on a much wider audience in order to get this message across. In turn it is hoped that the issues raised in this year's part will provide the basis for future work and development over the next 12 months. Perhaps some of the problems that are noted here will have been alleviated by the time the next edition is published.

This section also demonstrates that essentially anything you need, however extreme, could be possible. The Police GIS example is encouraging in that it shows that small beginnings can provide the nucleus for a snowballing effect. This is accompanied by more confidence to ask the ''right'' questions which puts you well on the road to establishing a system that does adequately suit your needs.

GIS vendors therefore should avoid becoming complacent. Their software may be meeting the needs of the present end-user audience. Their flashy graphics and flexibility may entice users to part with their money, but there is still a lot of educating to be done in terms of GIS. Also it must be stressed again that it is still a very young technology and any future maturing stage also requires refinements of the general GIS tools presently available.

Looking to the future

The future of GIS is, to some extent, in the hands of those who have already acquired the necessary expertise and experiences. They have a responsibility to ensure that GIS can, is, and will be extremely useful in every area in which it was originally intended. The net result will hopefully be to the benefit of society, be it directly or indirectly. The aspects raised in this summary are not necessarily immediate concerns, but should be seen as an integral part of the overall evolution of GIS. They should be rectified in the course of its development, in other words over the next five to ten years.

This part also emphasizes that the response to any problems that GIS throws up must not be that of passive acceptance of the ''state-of-the-art'', or the moving of research and application goal posts to suit the technology. This fiddling will not push GIS forward! Flagging problems and barriers to GIS should be perceived as a positive action, because by overcoming them GIS can only get better and more useful. This applies to both the technological and human element which forms the overall GIS package.

Two views on the way in which GIS may go in the future can be hypothesized. The first of these is negative and suggests that GIS will fragment and disappear, being nothing more than a memory at the end of the century. This would occur if the number of disillusioned end-users increases, i.e. those that find that the technology does not always answer some of their more fundamental questions, or they fail to fully understand the system in order to make it a success. This will only happen, however, if the software manufacturers cannot or choose not to acknowledge the limitations of their packages and thus fail to develop the technology to higher levels, which is very unlikely.

A more positive view would be to see researchers and other end-users emerging from pilot studies with suggestions and answers, where possible, for better tools which can be employed to extend GIS's capabilities. This should include an increase in GIS awareness, improvement in data standards (which the AGI are actively trying to establish), increased information on data quality, training and education. The development of co-operation between vendors, academics and the wider end-user community therefore should lead to the emergence of a strong set of core concepts and modules (Goodchild, 1991), forcing the pace of GIS development and providing the basis for an exciting and fruitful future in GIS.

The AGI therefore is a major player in the overall GIS environment, because it can help the wider end-user market with respect to awareness, education and the stimulation of the future direction of GIS. It should be emphasizing that implementing GIS is realistically just another cycle in the general evolution of Information Technology and not the ominous sounding ''revolution'', as it is so often described.

This introductory chapter has only expanded on a couple of the topics raised in the forthcoming papers. Obviously the following chapters will offer far more, be it only briefly, on the implications of present GIS and its future. They serve to provide a very strong message as to the "state-of-the-art" GIS and the ways in which we now should be moving forward. They satisfy Rees' (1990) statement that, "the free circulation of ideas is, I believe, essential to planning a better future for the human environment."

Finally, the four following chapters will further reiterate the general theme of this section which essentially builds upon the vague title "GIS in Society". They emphasize that GIS is in fact far wider reaching than this phrase suggests, in that it is more a case of "GIS with Society", while in the long run we should hope this means "GIS for (the benefit of) Society".

References

Campbell, H., 1991, "The Impact of Geographic Information Systems on British Local Government", presented at the UDMS Conference.

Cross, A.E., 1991a, Working for a Safer Society, in Cadoux-Hudson, J. and Heywood, D.I. (eds) *The Yearbook of the Association for Geographic Information 1991,* London: Taylor & Francis, pp.123–5.

Cross, A. and Openshaw, S., 1991, "Crime Pattern Analysis: The development of ARC/CRIME" presented at the Association of Geographic Information Conference, Birmingham, UK, November.

Goodchild, M., 1991, Towards a science of Geographic Information, in Cadoux-Hudson, J. and Heywood, D.I. (eds) *The Yearbook of the Association for Geographic Information 1991,* London: Taylor & Francis, pp. 212–8.

Johnston, R., 1981, *Dictionary of Human Geography,* New York: Free Press.

Openshaw, S., Charlton, M. and Wymer, C., 1987, Mark I Geographical Analysis Machine for the automated analysis of point pattern data, *International Journal of Geographical Information Systems*, **1**(4), pp. 335–58.

Openshaw, S., Cross, A. and Charlton, M., 1990a, Building a prototype Geographical Correlates Exploration Machine, *International Journal of Geographical Information Systems*, **4**(3), pp. 297–311.

Openshaw, S., Cross, A., Charlton, M. and Brunsdon, C., 1990b, "Lessons learnt from a Post Mortem of a failed GIS" presented at the Association of Geographic Information Conference, Brighton, UK, October.

Openshaw, S., Waugh, D., Cross, A., Brunsdon, C. and Insp. Lillie, 1991, A Crime Pattern Analysis System for subdivisional use, *Police Requirement Support Unit*, Bulletin 41.

Rees, P.H., 1990, Environment and Planning: What's in a name?, *Environment and Planning* A 22, pp. 1–2.

17

GIS in the police environment

Chief Inspector George Campbell
Northumbria Police

The perspective

It is stressed from the outset of this chapter that the uses of GIS in policing are expressed from a highly personal viewpoint. The perceptions outlined should not be regarded as direct police policy, or indeed an indication that Northumbria Police will eventually head along this GIS route. It may even be the case that some of the aspects noted are already on the agenda in other police forces in the UK, but unfortunately such information is not readily disseminated.

The expectations, which are noted in this article, for employing GIS technology were mainly formulated during the research stage leading up to June 1991. This constituted the all important period for evaluating the possibility and benefits for taking GIS on board in a highly complicated operational policing environment. This project is now in the hands of others in the department. They are continuing to assess and develop pilot GIS studies to demonstrate its potential. Thus there is likely to be a follow up to this chapter in the very near future which it is hoped will document the progress of Police GIS with the possibility that some of the ideas expressed in this chapter will have then been realized.

Why GIS?

The concept of GIS evolved within Northumbria Police over two years ago when early attempts were made to digitize beat boundaries using geographic format. This exercise in itself made it apparent that a computer-aided system for capturing, storing, retrieving, analysing and presenting spatial data would be extremely useful, especially in the presentation of facts based on digitized maps, which could facilitate in the identification of patterns and possible relationships between the activity being mapped and the area concerned.

It was believed that as the force moved into the computerization of crime recording in 1992 then a GIS, possessing an ability to display and analyse the data integrated by the Crime Recording System (CRS) would allow the force to identify much more quickly patterns of crimes. They could be related by type of crime, area, means of commission, time and date, offender type, property type and so on. The list is almost endless in terms of the flexibility that GIS will allow to do the necessary searches on the databases in a bid to isolate patterns of interest. This is both useful in terms of crime trend prediction and management.

The greatest benefit therefore that is immediately perceived has to be the ability to make such enquiries without the restriction of traditional boundaries, such as the beat, sub-division and division. GIS would allow the demarcation lines to be chosen in an instant and the request for information to be at one's fingertips, with the desire being that the predictions and patterns that are observed take on a new and more meaningful dimension.

It is also anticipated that GIS technology with the capability to build up user-friendly and robust interfaces will allow this range of enquiries to be made from a varied audience of users. From a local nature, where the question is posed by the Police Constable on the beat wanting to know what has been happening on his or her "patch" since he/she was last on duty, to information that may be sought by the Chief Constable to assist in the future planning and strategies of resources.

Benefits to society

The police service as a whole, and Northumbria Police is no exception, is continually striving to provide its public with a better service. Public expectations are capable of being raised by effective police action in reducing crime. A sophisticated and versatile crime pattern analysis system therefore is considered to be of great benefit in this regard.

Keeping in sight our goal of improving the quality of our service to the public, the GIS objectives might be summarized as:

a) to provide all officers with easy access to information which will assist them in the better performance of their duties;

b) to provide a management tool to aid with better decision-making and the best possible deployment of resources;

c) to provide a cost-efficient means of integrating databases.

To pick up on the last point, the integration to other operational and administrative database systems would obviously ensure that cost benefits are maximized. In turn there would be considerable advantages from combining relevant databases, i.e. establishing additional factors concerning an incident such as domestic violence could include being able to determine whether drink and/or firearms may be involved.

Areas of impact

GIS is envisaged to work at both the strategic resource allocation, database management level and that of hard core crime pattern analysis. The following section highlights some of the examples of its potential use, these are in no particular order of priority, and only constitute a fraction of a long list of desirable requirements.

Data access and beat boundary changes

As briefly highlighted earlier, the ability to access a variety of databases and the flexibility to interrogate the data in whatever manner is relevant is the key to improving the efficiency and management of data on a regular basis. This would be very important when carrying out manpower use and deployment studies. Again the indirect effect of this would be to further the police's bid to improve its quality of service. An illustration may clarify this point; deployment studies establish key factors concerning the policing of areas, i.e. the needs of an area at night may be completely different, in terms of policing, to the needs of the same area during the day.

Another area in which it is felt that GIS will be of great benefit is that of evaluation, and if necessary the implementation of beat boundary changes. This is a particularly valid issue at the moment, as the local authorities are considering changing their own administrative areas. At present the divisional boundaries within Northumbria Police follow more or less the local authority boundaries of Newcastle, North Tyneside, Gateshead, South Tyneside and Sunderland, and these divisions are subsequently split into a total of 22 sub-divisions. As it can be envisaged, any changes in boundaries that may or may not be enforced from other authorities would be an extremely time-consuming and laborious task, which it is hoped that GIS technology could go some way towards alleviating.

Even if the latter is not dictated by outside bodies, a GIS may allow the police to carry out their own beat boundary changes in the interest of better policing for the benefit of the community. In other words changing beat, car areas or even sub-divisional boundaries may be necessary to offset changes in crime loads. GIS could facilitate this type of monitoring by allowing the summary of crime loads, incidents attended, officer levels and the demographic profiles of an area to be calculated on a regular basis.

Once the optimum boundary has been established, via whatever model is initiated to compare and analyse the necessary statistics, then changes should be a simple task. The latter must also serve to change all other databases associated with these new areas, i.e. that of the transfer of property data, streets and premises. It should also be a prerequisite that the police can feed upon Local Authority databases of interest, and that any changes which they may also undergo should be immediately up dated in their system. The latter would require a concept of multi-agency GIS, which could complicate the development of an integrated GIS, however on the plus side it would be important for future developments and increased cost efficiency.

The initial links which should be established are those within the organization, however. This envisages GIS in the long term integrating not only into the Crime Recording System (CRS) and the Streets and Premises databases, but also to Computer Aided Despatch (CAD), Criminal Information System (CIS), Vehicle Information System (VIS), Firearms Records and Accident Records. It must therefore have the flexibility to move with the police in order to respond to their operational and everyday requirements. For example rapid access to a variety of data may be vital. Take the domestic siege situation again. In a single search the GIS should allow an officer to become aware, before even going to the scene, of the number of calls to that address in the recent past, the nature of the calls i.e. history of domestic violence, occupants, possession of firearms and so on. In short this information could help to save someone's life—a rather dramatic statement but true!

Crime Pattern Analysis (CPA)

In addition, GIS should also affect the area of Crime Pattern Analysis (CPA). Again the list of functional requirements is rather long. In fact, it is probably the full range of capabilities and potential influence of GIS on the organization as whole that is making the task of deciding actually to acquire a GIS so ominous.

In terms of CPA though it is assumed that GIS can pictorially depict possible clusters of incidents, or via the CIS look for clusters of criminals. Together these two sources may serve to demonstrate a possible relationship between crimes committed in an area and known criminals living, or active, in the area. This will not, of course, detect crime but will provide officers with the pertinent information to formulate ideas themselves from data that was not previously so readily to hand. This efficient and rapid interrogation of data can be employed in several areas of expertise and the latter form of analysis could be converted into a resource allocation tool. For instance, being able to view all the incidents where a car was broken into or damaged during the last home football match, may be sufficient to infer areas that should be considered for extra manpower deployment in future events, thus preventing a recurrence. This role of GIS serves to take policing from a situation of data saturation and over-stretched policing which at best is reactive, to one which is working towards proactive policing. This can only be for the best in securing a safer society.

Police planning strategies

An ability from the outset to produce from the main maps a series of one-off, specialized, maps may also be necessary. These can then be used to add and remove data/information as and when it is appropriate. This function would be particularly advantageous in house-to-house enquiries, or as a base at football matches to monitor crowd movement and pre-match incidents. Indeed any operation which may require some form of scene containment. For instance, a digitized map on the screen could isolate the area of concern, register the officers in attendance and from this base, swift continual incident monitoring and response would be possible.

In addition contingency plans may need to be drawn up rapidly in the event of, say, a city centre explosion, or a major disaster at a chemical works. This would require instant knowledge of wind speeds and direction, as well as criteria for accessibility which would affect any possible evacuation. Again this is looking to more wide-spread linkages with other associated emergency services. However these service areas are also in a similar position to the police in that their interest, in this fairly young technology, is still in its embryo stage. Other immediate forms of planning that spring to mind include those events that are known about well in advance, such as royal and parliamentary visits. These involve detailed route planning, which can be monitored on the screen along with the manpower deployment allowing the police to keep abreast of every movement and stop incurred on the itinerary.

Implementing GIS?

In general the use of GIS is really all about manpower being used as effectively as possible to serve the needs of the community. This is a recurring theme throughout this chapter and it forms the essence of the benefits that can be accrued from the implementation of GIS technology.

This chapter outlined a number of areas in which it is personally perceived to be of relevance to policing. The examples offered will hopefully emphasize this further, but it should be noted that the list is by no means exhaustive. Obviously as the system is accepted and developed the uses and potential will be extended. However, before the grandeur of full GIS development can be realized, it is imperative to get the basic and fundamental areas right first. Thus meeting the primary objectives of data access, interrogation, manipulation and boundary changes should be the initial objectives of any Police GIS implementation agenda. The next logical step would then be one of full utilization to develop crime pattern analysis and integration into outside agency databases. So while it may be tempting to try out all the ideas in this paper immediately, it is envisaged that this would only result in a system that does a number of things badly. An AGI Annual Conference paper entitled "Lessons learnt from a post-mortem of a failed GIS", written by Openshaw *et al.* (1990, of the Centre for Research into Crime, Policing and the Community, at the University of Newcastle upon Tyne), clearly demonstrates this point and outlines the pitfalls that can be incurred. It is hoped that a well-informed and more gradual and cautious approach to GIS up take will prove a better route for affording long term and successful development of GIS technology in a police environment.

Bearing in mind this idea of walking before running, it may be realistic to regard many of the aspects listed in this paper as forming a police IT "wish list". It is accepted that GIS is not going to be the sovereign remedy to all police data problems. It is however a launch pad, from which careful nurturing and development could see the most exciting innovation in handling police information, which in turn could set the strategy for, and beyond, the 1990s.

The handling of data via a GIS therefore, by skilled and dedicated members, may soon be a reality within police environments, although not necessarily in Northumbria. Presently one of the major constraints for this time-lag in up take can be said to be that of finances. If this problem were to be alleviated then there is probably no saying what the full potential of police GIS could be! The ultimate objective is that the improved efficiency and flexibility of police data handling would be passed on to the community that the police serve with the benefits being to provide more readily the quality of service which that community desires.

References

Openshaw, S., Cross, A., Charlton, M. and Brunsdon, C., 1990, "Lessons learnt from a Post Mortem of a failed GIS", presented at the Association of Geographic Information Conference, Brighton, UK, November.

18

GIS and health services

Andrew A. Lovett
University of East Anglia

Developments in the National Health Service

In the past four years there have been two policy initiatives which have done much to increase interest in the use of GIS within the authorities responsible for hospital and family practitioner services. These have been the renaissance of public health functions following the Acheson report (DHSS, 1988) and the introduction of an internal market in health care under the provisions of the NHS and Community Care Act 1990. The Acheson report recommended increased monitoring of public health through such means as an annual report from each district and greater evaluation of preventative or treatment measures in terms of their impacts on the health status of populations. Under the 1990 Act, the purchaser and provider roles of health authorities were separated, cost-efficient execution of the former requiring ready access to information on the current and future health needs of the population to be served (Wrigley, 1991). Both initiatives, therefore, have generated needs to handle large amounts of spatially-referenced information (e.g. addresses of people registered with a particular general practice or attending a hospital clinic) and GIS has been seen as a means of interrogating such databases and providing output in a easily understandable form.

Benefits of using GIS

A continuing issue within the NHS has been how large amounts of raw data can be converted into useful information for the purposes of policy formulation and management. Following on from the recommendations of the Korner report (DHSS, 1984), the vast majority of health service records are now postcoded and consequently an important benefit provided by GIS is the scope for data integration. A number of applications have illustrated this point (e.g. Curtis and Taket, 1989; Haynes *et al.*, 1991; Kivell and Mason, 1992), often focusing on the derivation of health indicators for small areas which are of relevance for policy or resource allocation purposes (e.g. neighbourhoods and general practice catchments). Another benefit has been to stimulate collaboration between the authorities responsible for hospital and general practitioner services who, as Twigg (1990) notes, have sometimes been reluctant to exchange information.

The use of GIS as a monitoring device is a second area of application where benefits can be observed. Mapping the spatial distribution of disease has long been recognized as a valuable means of epidemiological monitoring (e.g. Lloyd and MacDonald, 1984), but the facilities of a GIS greatly increase the range of questions that can be addressed and the speed with which answers can be obtained. This point is well illustrated by the work of Openshaw *et al.* (1987, 1990) on childhood leukaemia in northern England. There is also scope for using GIS to evaluate the uptake of services, recent examples including studies of child immunization (Gould, 1991) and cervical cytology screening (Bentham *et al.,* 1991). The latter example highlighted differences in the impact of different call/recall procedures and was able to identify several variables which were significant predictors of variations in uptake rates between practices.

A third category of application concerns the use of GIS for scenario modelling. Accessibility measures are readily calculated within a GIS and, for instance, have been used by Gatrell and Naumann (1992) to evaluate the impacts of altering the distribution of accident and emergency facilities in a pilot study for South East Thames Regional Health Authority. Hale (1991) also describes some relevant examples, particularly in the context of analysing patient flows to permit informed decisions when negotiating contracts for hospital services.

Problems in implementing GIS

One feature of the examples listed above is that none of them was undertaken by people employed by the NHS. Many projects have involved collaboration between health authorities and academic institutions or consultancy companies, but it remains the case that GIS expertise within health authorities is limited. This reflects the general shortage of qualified GIS personnel, but the situation in the NHS has probably been exacerbated by the difficulties of competing with the salaries available in the private sector. Staff turnover is one of the problems noted in the AGI-sponsored survey of GIS use within the NHS (Cummins and Rathwell, 1991), but even more striking is the limited range of software with few authorities having systems with capabilities beyond basic mapping. There is also evidence of situations where equipment has been purchased under wider regional or Department of Health initiatives, but subsequently has hardly been used.

Even when suitable equipment and personnel exist, data issues can be a significant barrier to the operationalization of GIS applications (Matthews, 1990; Twigg, 1990). Relevant information may well exist within the health service, but obtaining direct physical access to it can be a lengthy process, even when the project is actually health authority funded! Similarly, all patient records may be postcoded, but the accuracy with which this is done can be questionable (Cummins and Rathwell, 1991) and further errors are likely when using details from the Central Postcode Directory (CPD) to link such information with census sources (see Lovett *et al.,* 1991). Gatrell *et al.* (1991) have demonstrated how the accuracy of the CPD can be improved by using point-in-polygon procedures, but few health authorities have the software to undertake such operations or the facilities to generate boundary files for non-standard spatial units (Cummins and Rathwell, 1991). A realistic view, therefore, might be that the benefits gained from an initial GIS project are likely to be more in the realm of basic data audit than from the results of the analysis attempted.

Future prospects

The main conclusions of this brief review can be simply stated. There seems to be a clear role for GIS within the NHS and a range of benefits have been demonstrated by pilot projects, but the infrastructure for the day-to-day utilization of such systems is largely absent. Awareness of the contribution which GIS can make certainly exists, but needs to be extended, particularly to those in managerial positions. More importantly still, interest has to be converted into practical skills so that existing computing resources are used more effectively. It may not be appropriate or financially possible for each district health authority to invest in extensive GIS capabilities, but it is surely important for them to have basic facilities for data analysis and mapping, along with knowledge of where to obtain additional expertise when necessary and the relevant questions to ask such organizations.

How might such a situation be achieved? There is really no simple answer, but ultimately much must depend on the ability of those involved in GIS applications within the health service to demonstrate tangible and sustained benefits from their work. If GIS can be shown to increase the cost-effectiveness of services, improve standards of health care and reduce morbidity or mortality, then gaining the attention of NHS managers should not be difficult and word-of-mouth will do far more than glossy publicity or expensive conferences to increase awareness. There should also be scope for persuading the higher tiers of the NHS (such as regional health authorities or the Department of Health) to do more in terms of negotiating discounts on equipment purchases and arranging training programmes. It might, of course, be argued that it is difficult to demonstrate benefits without the initial investment in necessary resources, but there are already established collaborative projects around the country and it is to be hoped that more will be initiated under the new NHS research strategy (Department of Health, 1991). A challenge for the 1990s, then, is to translate demonstrator projects into routine applications. If this can be achieved, the benefits for health and health care in the United Kingdom should be substantial.

References

Bentham, C.G., Hinton, J.C., Lovett, A.A., Haynes, R.M. and Bestwick, C., 1991, *Factors Affecting Non-Response to Cervical Cytology Screening in Norfolk: An Analysis Using a Geographical Information System,* Centre for Health Policy Research Working Paper 4, Norwich: UEA.

Cummins, P. and Rathwell, T., 1991, *Geographical Information Systems and the National Health Service: A Report of a Survey,* AGI Education, Training and Research Publication No. 3, London: AGI.

Curtis, S.E. and Taket, A.R., 1989, The development of geographical information systems for locality planning in health care, *Area,* **21**, pp. 391–9.

Department of Health and Social Security, 1984, *Fifth Report to the Secretary of State, Steering Group on Health Service Information* (Korner Report), London: HMSO.

Department of Health and Social Security, 1988, *Public Health in England: Report of the Committee of Inquiry into the Future Development of the Public Health Function* (Acheson Report), London: HMSO.

Department of Health, 1991, *Research for Health: A Research and Development Strategy for the NHS,* London: HMSO.

Gatrell, A.C., Dunn, C.E. and Boyle, P.J., 1991, The relative utility of the Central Postcode Directory and Pinpoint Address Code in applications of geographical information systems, *Environment and Planning A,* **23**, pp. 1447–58.

Gatrell, A.C. and Naumann, I., 1992, *Hospital Location Planning: A Pilot GIS Study,* Paper presented at the Mapping Awareness '92 Conference, London.

Gould, M., 1991, *Health Service Restructuring and Geographically Based Analysis: With Reference to the Multi-Level Modelling of Immunisation Uptake,* Wales and South West RRL Technical Report No. 33, Cardiff: UWCC.

Hale, D., 1991, The healthcare industry and geographic information systems, *Mapping Awareness,* **5**, No. 8, pp. 36–39.

Haynes, R.M., Hinton, J.C., Lovett, A.A., Bentham, C.G. and Bestwick, C., 1991, *Measurement of Health Needs in Small Areas of Norfolk using a Geographic Information System,* Centre for Health Policy Research Working Paper 3, Norwich: UEA.

Kivell, P.T. and Mason, K.T., 1992, Monitoring the healthy city: GIS and health care management in North Staffordshire. In Rideout, T.W. (Ed.), *Geographical Information Systems and Urban and Rural Planning,* Planning & Environment Study Group, London: Institute of British Geographers, pp. 69–79.

Lloyd, O.L. and MacDonald, J., 1984, Continuous epidemiological mapping: A needed public health watchdog, *Public Health,* **98**, pp. 321–6.

Lovett, A.A., Hinton, J.C., Haynes, R.M., Bentham, C.G. and Bestwick, C., 1991, *The FHSA Patient Register: A Problematical Source for Health Service Planning,* Paper presented at the "Disasters in GIS" conference, University of Salford, April.

Matthews, S.A., 1990, Epidemiolgy using a GIS: the need for caution, *Computers, Environment and Urban Systems,* **14**, pp. 213–21.

Openshaw, S., Charlton, M. and Wymer, C., 1987, A Mark I Geographical Analysis Machine for the automated analysis of point data sets, *International Journal of Geographical Information Systems,* **1**, pp. 335–58.

Openshaw, S., Cross, A. and Charlton, M., 1990, Building a prototype Geographical Correlates Exploration Machine, *International Journal of Geographical Information Systems,* **4**, pp. 297–311.

Twigg, L., 1990, Health based geographical information systems: their potential examined in the light of existing data sources, *Social Science and Medicine,* **30**, pp. 143–55.

Wrigley, N., 1991, Market-based systems of health-care provision, the NHS bill, and geographical information systems, *Environment and Planning A,* **23**, pp. 5–8.

19

Managing change in care in the community: The case of geographical information and child protection

D. Craig Whitehead and Guy R. Dominy
London School of Economics

Introduction: Changes in local authority service provision

Over the last decade local government in Britain has been transformed, with over 50 pieces of legislation affecting local government services and finances being enacted. Among these have been legislation introducing major changes in several areas. Local Management of Schools, Compulsory Competitive Tendering, Community Charge and Right to Buy have all introduced fundamental change in both the functioning and role of local government in Great Britain. Change has not all derived from legislation however. A major move towards the decentralization of services into area based units has also taken place during the 1980s (Hoggett and Hambleton, 1987). These changes in local government can be broadly categorized as belonging, essentially, to one of three dimensions of change, being from "finance to service led" and associated with this "from provider led to client led" and third, "away from direct provision", i.e. the Enabling Authority. Overall this involves a major structural shift to smaller business units and an integral part of this shift has been the spatial decentralization of service delivery and service management.

One area that has particularly been affected by these changes is social services. Many social services departments (SSDs) have pursued decentralization during the 1980s—in particular geographical decentralization to area or neighbourhood offices and patch teams (see Elcock, 1986 and Flynn, 1987 for example). In the 1990s "Care in the Community" and the shift towards market models of health care are going to further transform the nature of social services in Great Britain.

Requirements: "Information from data for different users"

Management information

This more complex, and ideally more flexible and responsive, structure has very different information requirements from the traditional, hierarchical, town hall-based authority. Additional and more accurate information is required and access to this information is also required throughout the organization—information systems can no longer be designed purely to funnel reports to senior management and instructions to front-line staff. The last decade has seen a huge increase in the amount of money local government invests in Information Technology (IT). An increasing proportion of this investment has been in the communications, departmental and personal systems necessary to support these changing information needs (ICL, 1988).

Increasingly SSDs are going to have to turn to IT to enable them to cope with the changes outlined above. The introduction of quasi-markets for healthcare and the need for a much greater degree of co-operation between different agencies (public, private and voluntary) will transform the requirements both for internal management information services and for inter-agency collaboration and partnership in the development and operation of information services. As David Streatfield states, "the traditional approach of managing without adequate information systems will not work" (Streatfield, 1992).

The new requirements for inter-agency collaboration, embodied in the Children's Act 1989 and the Community Care Act, mean that not only do SSDs need to ensure that the necessary information is available to the right people within the department, but that information from all relevant agencies is available to those who need it. In order for the new system to work, information must be a vital component integrating users and caring agencies.

Peter Whittingham of KPMG (Whittingham, 1992) has identified three types of data that are becoming important in what he calls the "exchange network":

- socio-economic and demographic aspects of the population;
- attributes and factors associated with particular user groups; and
- data concerning individual users of services.

Alongside issues concerning the technical design of such shared systems and information, potential ethical problems should not be ignored. Much client data is extremely sensitive; if clients feared that the data one agency collected might be used by another they might avoid coming forward at all.

Within this "exchange network" SSDs are in a key position. They are going to be the linchpin of the new system; if Care in the Community, and the new model of social services more generally, is to work then the information systems of the SSD must be right.

Within the decentralized SSD Flynn identifies the team manager as a key figure. Team managers, he argues, need to know:

- the performance of the units, including costs, volume of care provided, quality of care both over time and in comparison with other units in the department;
- the general performance of the team, over time and in comparison with others;
- the relative costs of different packages of care; and
- intelligence about the area for which they are responsible: demographic trends, indicators of needs and alternate resources available (Flynn, 1987).

Senior Managers, Flynn states, need information for resource allocation, "steady stat" monitoring of the department and the evaluation of experiments and new approaches adopted by teams. For resource allocation they require aggregations of the "patch" intelligence that team leaders need and, besides these demographic data and needs indicators they must have an idea of workloads and throughputs so that mismatches of resources can be smoothed across the authority (Flynn, 1987).

Geographical elements of management information

The primary value of geographical information in this context lies in its ability to place client and user group data in context for service planning, demand modelling and resource allocation. Both Flynn and Whittingham, for example, identify the importance of socio-economic and demographic data for the area served for these purposes (Flynn, 1987, Whittingham, 1992). With effective use of information from these geographic data a greater degree of spatial equity and efficiency can be achieved by optimizing the resources of service delivery.

Case study

The child protection register in Southwark's social services department provides an example of how the management of changes in the local authority system is assisted by using innovative information technology solutions. Maintenance of a Child Protection Register is a statutory requirement of Area Child Protection Committees. These are multi-agency organizations drawn from social services, health authorities, education departments, police and probation services. All children deemed to be at risk are entered onto a register and remain upon it until they are no longer at risk. A large amount of data is held for each case, including domestic arrangements, details of siblings and regular visitors to the child's domicile.

The social services of Southwark maintains the register for children within the borough. The traditional method of maintaining this data was in paper form with access being strictly controlled. With the distribution of the service centres as Social Services Area Offices, access to the data by case workers and area managers became difficult. To reduce the impact of this problem a solution was sought through the use of information technology.

The initial requirement was to provide distributed access to a database that replaced the existing Child Protection Register while replicating and enhancing its functionality. An important requirement was that the data should be provided in a manner that allowed staff to elicit the maximum amount of relevant information from the database with the minimum of computing expertise. Besides providing effective database query mechanisms it was thought that statistical and geographical analysis software could enhance the amount of information extracted from the data held in the register. The technological solution selected for implementing the database (see below) allowed the supporting software to be provided in the form of a Federated GIS (Hershey and Whitehead, 1991).

IT solution

Given the range of skill levels and work loads of the individuals required to access information, the traditional local authority IT solution of a centralized database on a Mainframe or a Mini computer with a synchronous terminal access was rejected. Rather, a network of "easy to use" microcomputers accessing a database server was chosen. Apple Macintosh technology with its intuitive interface and corresponding advantages of convenience for occasional users and reduced learning periods for new users was adopted. The database server, itself a Macintosh with a WORM media back-up, is connected to the remote sites (area offices and hospitals) using AppleTalk over an Ethernet system that uses bridges to connect to the British Telecom Kilstream service, which provides secure communication between sites that are up to three miles apart. The central machine is a true database server using the software package Double Helix. This system was chosen to reduce the amount of data traffic on the network with only the result of a query being transmitted and the local machine handling screen displays, validation, etc.

Spatial data components and analysis

The requirements of the system included a wide range of display and analysis functions from the detailed mapping of individual cases by their place or residence to borough-wide analysis of aggregated data about demographic characteristics. The software solution adopted was MapGrafix, a database independent mapping and spatial analysis package.

While some digital information was already available from other sources, such as census and health boundaries, it was necessary to capture many spatial data by digitizing from existing paper maps. As a large amount of digitizing was to be carried out, it was also decided to digitize the street layout rather than use Ordnance Survey digital data. This decision was not taken lightly, but the OS data would have required extensive editing, ultimately resulting in a higher cost; it was with much regret that this approach was taken.

The fixed spatial elements, held in layers, include: street plans with names; local authority housing; open spaces; administrative areas of police; social services and health organizations; location of social service's offices and care centres; bus routes; census areas and a range of landmarks.

The spatial elements that vary in time is the location of the child's home. One approach evaluated for geo-referencing individuals was to use the postcode element of the address. This mechanism is currently in use but is flawed by two factors: first, incorrect and unknown postcodes; and second the coarse resolution of postcodes for this scale of study and in areas of high density housing. Solutions being investigated include manual grid referencing when entering a record, and address matching.

In many GIS applications with high volumes of data the latter would be the immediate choice. In this instance the number of individual cases entered into the system averages one per day and data entry is by a restricted number of individuals who, given the legal nature of these data, are required to validate all components. The additional requirement to grid reference an entry would not be an onerous task and would remove the requirements to validate the postcode. Error checking could be done using the database to hold grid references of sections of streets.

Use of the system can range from displaying the location of case for a particular case worker (or those cases not yet allocated to a case worker), to analysing the distribution of active and archived cases with other socio-demographic indicators in relation to the reduced number of area offices to ensure spatial efficiently and equity in the arrangements of boundaries. The system could also be used to allocate cases across area boundaries where overloading was taking place. This could be monitored and the data fed back into the allocation of resources between areas and the drafting of area boundaries, possibly even obviating the need for area boundaries in the long term.

This example illustrates the nature of GIS adoption within an organisation whose primary requirements is not for geographic information but where an effective Management Information System can, at various levels, be enhanced the provision of Desktop Mapping and Geographic Analysis Software.

Conclusions

The locational information required by SSDs is neither complex in form nor difficult to capture. However, a key element for SSDs remains the client. It is spatial referencing of the client's residence which is most problematic in a metropolitan setting. The unit postcode, much favoured in aggregate spatial analysis (for example direct marketing) is inadequate in this context. Its resolution and client error make it unsuitable for locating individual dwellings, especially in the case of medium and high-rise accommodation. For large numbers of clients an alternative method of geo-referencing needs to be developed. The use of manual digitizing rather than OS data raises the question also of the availability of digital data. If the OS is not meeting local authority requirements for digital data who is? Given the significant proportion of GIS costs derived from data capture, this lack of appropriate data might be an important factor preventing the more widespread adoption of GIS in local authorities.

The adoption of IT in general inevitably implies a modification of working practices. However, the transformation of SSDs and local authorities as a whole resulting from other causes dwarf these changes in working practice. Nonetheless, it is important that IT solutions are relevant to the user, particularly front-line staff, and provide appropriate, timely information through "easy to use" systems. The implementation of IT systems benefit from an evolutionary approach, being structured initially around the primary functions with subsidiary features introduced once the primary functions have been fully assimilated. In the example, itself representative of most SSD functions, the client data was clearly the priority. The spatial analysis component, as a supplementary capability, needs to operate in harmony with existing software and organizational solutions. This implies using a database independent GIS in a federated approach, as advocated by Hershey and Whitehead (1991).

This begs the question what is a GIS? The term GIS is to many synonymous with difficult to use systems on expensive hardware platforms. It is also evocative of stand-alone, operating within a single subject domain. Perhaps GIS has come of age and is now an important component of Management Information Systems (MIS) or in this case Social or Community Information Systems (SIS or CIS)—rather than a tool for those specialists whose primary requirements is spatial information e.g. planning departments. This is a significant development and the example shows how the MIS/SIS/CIS can be enhanced by the addition of geographical analysis software and desktop mapping capability that any end-user can exploit, when necessary.

More flexible IT solutions have enabled the transformation of local government. They have underwritten the implementation of Compulsory Competitive Tendering, Local Management of Schools, the Children's Act and Community Care, for example. The essential importance of these systems lies in their ability to provide fine grain management information about levels of demand and the costs of service delivery. Geographical information is now beginning to be incorporated into these systems and playing its part in facilitating this new era of urban governance. If GIS has come of age then perhaps the terms and its existence as a specialized art shrouded in mystique will disappear as users of information systems come to take for granted the ability to analyse and display the spatial information inherent in their data.

References

Elcock, H., 1986, Decentralisation as a Tool for Social Services Management, *Local Government Studies,* July/August, pp. 35–49.

Flynn, N., 1987, Delegating Financial Responsibility and Policy Making within Social Services Departments, *Public Money,* **6**, March pp. 41–44.

Hershey, R.R. and Whitehead, D.C., 1991, The Federated GIS: An accessible geographic information system, LSE: GIRL Working Paper Number 3.

Hoggett, P. and Hambleton, R. (eds), 1987, Decentralisation and democracy, Occasional Paper 28, Bristol School for Advanced Urban Studies (SAUS), University of Bristol.

ICL, 1988, Local Government in Britain. An ICL Report on the Impact of Information Technology.

Streatfield, D., 1992, The Computer Minefield, *Community Care,* January.

Whittingham, P., 1992, Integrating Care, *Community Care,* January.

20

County Council GIS—benefits and barriers

Roy Laming
Kent County Council

One of the main areas of activity in which Geographical Information Systems are starting to be of benefit to society is in the increasing adoption of this technology by Local Government County Councils. These systems offer a wide variety of tools which can greatly assist such large organizations, who employ a large number of staff, hold large volumes of information and deliver a wide variety of services to many people. Kent County Council is one of those authorities which is now starting to realize benefits from the implementation of GIS applications, and pass those benefits on to the people of Kent. The Council delivers a wide range of functions including Education, Social Services, Highways and Transportation, Strategic Planning, Police, Fire Brigade, Arts and Libraries, Trading Standards and Magistrates Courts. It serves a population of some 1.5 million in an area which contains both high density population and rural environment.

Information systems play a very important part in the delivery of these services and are the key in maintaining quality services with finite resources. The previous concept of centrally based mainframe computer facilities and staff in Kent County Council has now evolved into department information systems and staff units. This has given departments the ability to choose the technological solutions which best satisfy their own strategies and business needs while operating within an overall corporate strategy and framework. This has been made possible by open systems standards and the improving price/performance of UNIX processors.

It was important that the Council's approach to GIS was complementary to this distributed, multi-disciplinary and multi-vendor computing environment and would also allow for maximum benefit from the use of such systems to be obtained. At the outset it seemed that a GIS "toolbox" would be most appropriate, i.e. a product with high flexibility from which a very wide variety of different applications could be built to satisfy specific business needs in departments. The promotion and adoption of a common "tool" would also ensure that the wider benefits, such as information sharing and exchange, could be obtained. Therefore, Arc/Info was chosen as the preferred GIS software product.

In the long-term, most benefit from GIS is likely to be obtained from those wider and often intangible activities, but it is these that are most difficult to measure. Kent County Council's approach is proving successful because applications can be cost-justified and implemented primarily according to immediate, specific and individual needs and benefits, within a longer term strategy and vision. With the existence of previously implemented applications and datasets each subsequent application then becomes easier to cost-justify, develop and implement and the benefit per unit cost increases, resulting in an exponential curve.

The benefits that the County Council will receive from the use of GIS in service planning, provision and management are as many and varied as the number of applications themselves. Obviously not all County Councils will be organized and operate in the same way, and GIS use and benefit will be different from county to county. In general, the main benefits being gained in Kent are as follows:

- better service to the customer;
- better quality information—correct, complete and current;
- easier access to information, authority wide view;
- linking and adding value to existing data and systems;
- better communication to members, colleagues, public;
- greater productivity, better use of resources;
- links to external agencies.

In Kent, most applications now operating within the County Council have been implemented to obtain improved productivity, efficiency and quality. This immediately gives the people of Kent better "value for money" for the community services they receive, for which they are paying in some way or another.

Other societal benefits can be directly related to the departments in which the GIS tools are now being used. For example, the capture and analysis of wildlife habitat and constraints information by the Planning Department will help protect the environment of the "Garden of England". Linking of GIS to employer, training and development opportunities databases in Economic Development will be a useful tool in the work of promoting and maintaining a healthy economy for the county. The analysis of the 1991 Census statistics by GIS will have benefit in many departments, such as Social Services and Education, helping to ensure that services are provided according to different needs in different geographic areas. There are also many GIS applications in use in the Highways department which help to improve road conditions, infrastructure and reduce traffic congestion. Highways are also using GIS techniques to input and analyse accident data which could help improve road safety, possibly saving lives by reducing the number of road accidents. It has been estimated that a serious accident on a motorway requiring use of emergency services, hospitalization of the injured, causing delay to other road users and so on can cost society £500 000!

It will not be too long before the public themselves will be able to make direct use of GIS, for example in schools and public libraries, which both currently fall within the County Council's remit. Indeed, several schools already have satellite imagery systems and other such systems with GIS functionality. The possibilities for providing GIS "screens" in libraries are enormous, with the general public using such facilities to resolve queries for themselves such as, identifying roadworks and traffic flows along alternative holiday routes, planning the location of a new small business, or even looking at the facilities or development plans around a house that might be purchased.

There are, of course, many barriers which currently may prevent GIS benefits being obtained on a very large scale by both County Councils and the people they serve. Most GIS products are still quite complex and require specialist skills and knowledge. Even ArcView, the new sister product to Arc/Info, with its highly innovative and easy to use multi-window graphical interface may be beyond the initial capabilities of many computer users who are just coming to terms with using a single "window" on a screen. There is increasing use of computers in schools and the home but it may be another generation or two before large numbers of staff or customers are equipped to take full advantage of this technology.

GIS hardware and software is still comparatively expensive and this can also prevent the large scale uptake of GIS. Cost justification can still be difficult, particularly where benefits will not be realized in the short term or are intangible and difficult to cost. However, the cost of the technology is often insignificant to the cost of data capture or acquisition. Digitization is still the only really effective way to get paper map-based information into a system and this is a time consuming and costly process. Many existing computer-based datasets will probably not be in suitable form or to a suitable accuracy to be readily used in GIS and this may take much time and effort to resolve. The purchase of commercially available data, such as Ordnance Survey large scale maps, also requires significant funding to get complete coverage for the geographic area required.

Within a large, multi-discipline organization with several independent business units it can be difficult to establish and get agreement on the value of a corporate approach. Kent County Council has been reasonably successful in achieving this and nearly half of the departments are now realizing the benefits of sharing not only data, but skills, experience and applications. The Council has also been able to procure complete Ordnance Survey large scale digital map coverage of Kent. This is delivering significant benefits in both the provision of up-to-date mapping and as the basis for linking much geographically related information.

Undoubtedly, technology, data working practices and skills will continue to evolve, develop and change. This may solve some problems, introduce new opportunities and benefits, and probably introduce a whole new set of problems. However, the use of GIS by County Councils is likely to continue to increase and assist in the delivery of more efficient and effective services to the communities they serve.

Part V

GIS and Infrastructure

21

GIS in the infrastructure

Dr Sara Finch
Logica Industry Limited

Introduction

This part looks at the use of GIS by companies that have to manage assets in the public domain—most usually in the road. Specifically, it concerns the suppliers of water, gas, electricity, telecommunications, transport and cable TV and covers the major issues currently being faced by them as a result of new legislation and changes in their business needs.

These organizations own many millions of pounds' worth of equipment such as tracks, pipes, cables and pylons, which they have to manage efficiently. They therefore need to understand as much as possible about them in terms of installation date, fault and maintenance history, and exact position. GIS is the perfect tool for asset management and it is no surprise that it is predominantly in this area that the technology is being applied.

Levels of GIS investment

There are identifiable benefits to be gained by the use of GIS in all the infrastructure organizations, but the degree to which investment is being made varies considerably. The recently privatized water, gas and electricity companies understand the significant benefits offered by GIS technology and are consequently making considerable investments in such systems. There is no doubt that they are currently the most significant users of GIS, even though the utilities have been working with GIS far longer than most other organizations. They have come to recognize that, when integrated into an overall information technology strategy, GIS can bring enormous and far reaching benefits.

The fact that GIS systems provide a spatial reference to data means that previously incompatible sets of data can be integrated. It is evident that the initial cost of setting up a GIS is high, predominantly because of the huge task of data capture that has to take place. However, once the main database has been established it can be used to integrate other sets of corporate data, enabling disparate datasets to be reviewed and analysed in ways that have previously been impractical.

Several of the utility companies, particularly those supplying water and gas, are now seriously investigating ways to exploit the significant benefits of having an accurate record of their assets. The electricity and telecommunications companies are not so advanced in the use of GIS, largely due to the size of their infrastructure networks. For them the task of data capture is considerably more daunting.

The electricity companies are increasingly beginning to evaluate and invest in GIS technology. A number of successful pilot projects have been run throughout the industry, and many of these have shown tangible business benefits which justify the case for further investment. The telecommunications companies are still exploring the costs and benefits associated with investment in corporate GIS. These organizations are keenly exploring the potential GIS might offer their businesses, and in doing so are considering some of the diverse value-added facilities which can be provided once a basic asset management system has been established.

The transport sector is still in the early stages of applying GIS technology to its business requirements. It is easy to see how transport companies could apply GIS—in controlling vehicles and planning the position of stations, airports and roads, for instance—and, indeed, examples of such systems are beginning to proliferate. One of the best examples to date has been with the planning of a high speed rail link from London to the Channel Tunnel, where GIS has been used to identify and monitor property and the environmental impact.

We know that British Rail has made some use of GIS. It has looked at how the systems could help in the identification and condition monitoring of infrastructure and associated engineering activities. Clearly, the scope for further applications is manifold but, as with many large organizations, there is a fundamental need for corporate recognition of the benefits of GIS to enable the large investments necessary to be made.

In the UK the benefits of GIS are now being exploited by some of the cable television companies, although they are concentrating investment on laying cables in order to increase their customer base. Unlike the old established utilities which have a very difficult task in gathering historic data, the cable television companies have the opportunity to create a complete and accurate database which incorporates the plant data as it is generated.

New legislation

The single most important change in the GIS world since the last AGI Year Book was published, is the introduction of the New Roads and Street Works act of 1991. It replaces the Public Utility Street Works Act of 1950 and makes some dramatic changes to the way in which water, gas, electricity and telecommunications companies have to operate. It is a far reaching act, covering many different areas of operation, including reinstating roads, marking diversions, training people undertaking work in the streets and roads, inspecting the work, and—most interestingly from a GIS point of view— keeping records of all activities undertaken.

The objective of the Act is to create an efficient system that will enable utilities and local authorities to exchange information on all planned street works more quickly and easily, so that activities can be better co-ordinated by the organizations and disruption to the public can be minimized.

A major new element of the new Act is the compilation of a Computerized Street Works Register. Work has started on this already and the first phase of the system is planned to be operational by 1994. This will then enable information to be exchanged between utilities and local authorities in digital form, as an alternative to current methods which are based on post and facsimile.

It is important for the success of the Act that the utility companies can define exactly where they plan to carry out street works. Up to now this has been difficult for there has been no formal agreement on the definition of a street. Recently, however, a specification has been established for a national street gazetteer which will provide a definitive list of all streets as compiled by a local government led working party. At the time of writing, the draft specification has been proposed as a British Standard.

Competition and security

There seem to be two other driving forces towards investment in GIS by the utilities and telecommunications companies in particular. The first is the need to improve customer service, a need which, in some cases, has been laid down by independent regulatory bodies who will impose fines if agreed service levels are not met. By using GIS as an integration tool, utility companies will be able to promote higher levels of service by, for example, managing fault repair more efficiently and thus responding to clients more rapidly.

The recognition of the vulnerability inherent in storing valuable data in paper based form, all under one roof, in another incentive for these organizations to invest in GIS. In the unfortunate event of fire or flood, thousands of original maps and engineering records could be destroyed. However, when the information has been digitized it can more easily be copied and stored in different locations.

The five chapters included in this part concentrate for the most part on the topical developments in the application of GIS within the utilities. The huge impact of the New Roads and Street Works Act is reflected from different perspectives in the first three articles. Subsequent articles look at specific applications of GIS in the water industry and at London Transport. The first chapter comes from the Department of Transport itself, and gives a factual, chronological review of the events and issues that have led to the new legislation.

The National Joint Utilities Group (NJUG) is an association of gas, water and electricity companies, in addition to British Telecom and Mercury. NJUG was set up in 1978 to provide a voluntary discussion forum for the utilities to consider problems of mutual interest, particularly with regard to street works. The chapter from NJUG describes, in a practical sense, the issues involved in digging ''holes in the road'' and focuses on the need for utilities to collaborate, so that disruption to the public and costs are minimized.

A joint chapter by a former county surveyor and the electricity company, Seeboard, takes an historical look at the situation, pointing out some of the anomalies and inadequacies of the old Public Utility Street Works Act of 1950. It is fascinating to find out, for example, that British Gas, unlike the electricity suppliers, has until very recently been under no obligation to keep detailed records of the location of their assets. The chapter underlines the need for the new act, and highlights benefits to be gained.

In addition to chapters concerned with the street works legislation, this part also includes a comprehensive review of the application of GIS in the water industry, written by two of my colleagues from Logica. It explains how water utilities currently use GIS, and then looks ahead at how GIS could be used as an integrated tool to help meet diverse business requirements. The article also covers the impact of the new legislation and discusses some of the practical issues involved in implementing a GIS.

The final chapter in this part describes a project undertaken by London Transport to compile a register of its property assets. All the main steps that had to be taken to develop the system are discussed, including the tasks associated with cost/benefit analysis, bench marking, data capture and data conversion. The experiences of London Transport are likely to prove useful to other organizations starting out on a GIS project.

22

The New Road and Street Works Act 1991 and the Computerized Street Works Register

P. Wallwork
Department of Transport

The New Roads and Street Works Act (HMSO, 1991) entered the statute book in June 1991, heralding the largest change to the way utility street works are carried out in Great Britain since 1950.

Background

The utilities (gas, electricity, water, telecommunications and sewerage) in Great Britain are obliged to provide and maintain a service to the public. When the utilities were established, by Act of Parliament in the nineteenth century, they were given the right to break open the street to lay and maintain their equipment. In the twentieth century as motorized traffic grew, and with it the complexity of road pavement structures, the activities and demands on utilities increased and the original legislation became inadequate. To overcome this the Carnock Committee was set up in 1938 and their suggestions became the basis for the Public Utility Street Works Act 1950, known to most of its users as PUSWA.

The Horne Report

Pressure on the road network continued to grow as traffic rapidly increased in volume and weight and the number of utility excavations multiplied. By the early 1980s PUSWA was on the verge of breakdown and the government set up a committee under Professor Horne to review all aspects of PUSWA in the light of present day circumstances. After taking evidence from all interested parties the committee published its report in 1985, the Horne Report, that contained 73 recommendations for change.

Some of the report's general conclusions were that:

- responsibility for utility openings was confused;
- there was no common standard specification for reinstatements;

- training of operatives and supervision was poor;
- the notice system was cumbersome, unsatisfactory and inappropriate to the high volume of activity;
- the public perceive that there is a lack of co-ordination between the utilities and the highway authority;
- the requirements for recording underground plant are inconsistent and inadequate.

The New Roads and Street Works Act

The Government accepted most of the recommendations made in the Horne Report and the New Roads and Street Works Act was born. The street works provisions in the New Roads and Street Works Act will, in particular:

- transfer responsibility for reinstatements to the utility;
- introduce a new system of training, inspection;
- introduce a common reinstatement specification and performance standards for reinstatements;
- require the street authority to keep a register of all works (including its own) in streets for which they are responsible;
- require those undertaking street works to give appropriate notice in advance, or immediately after completion in certain circumstances;
- place on street authorities the duty to co-ordinate their own and undertakers' works to minimize the disruption they cause;
- place on undertakers a duty to cooperate with the street authority to enable them to minimize the disruption caused by their works;
- introduce common standards and requirements for recording plant in the highway.

To implement the recommendations made by Professor Horne, the highway authorities and the utilities realized that a large amount of work was required to produce a workable system. To this end they established the Highway Authorities and Utilities Committee (HAUC). HAUC set up several working parties, with representatives from the highway authorities and all the major utilities, to prepare the codes of practice and specifications required.

The Computerized Street Works Register

The key to implementing these new requirements is the method by which notice is given and the way in which the notices are used. The Horne Report described the notice system used by PUSWA as "an inadequate, inefficient and expensive system of communication, which is probably acting to the detriment of organizational and individual relationships rather than progressing utility roadworks". Horne proposed that a Computerized Street Works Register (CSWR) should replace the existing notice serving procedure. The CSWR would allow more rapid passing of information and allow the system to operate on a "need to know" and "protection of self interest" basis. The register should:

- replace the existing, unmanageable flows of paperwork, fax and telephone traffic;
- improve the information flow such that local Highway Authorities might be better placed to co-ordinate street works;
- create a diary of events during the interest life of all road openings;
- provide a facility that all interested parties can interrogate for street works information.

In 1987 HAUC established a working party to examine the proposal for the CSWR. The first stage was to undertake a feasibility study and KPMG Peat Marwick McLintock were commissioned to undertake this in 1988 using the Government's Structured System Analysis and Design Method (SSADM). For the feasibility study several options were developed with users and one chosen to test the CSWR's viability. The study (Department of Transport, 1988) concluded that the CSWR was feasible and the tangible financial benefits were likely to be equal to or greater than the cost of operating the CSWR.

The development of the CSWR then slowed while discussions took place over the future development. In 1990 the Department of Transport agreed to fund the design of the Central System software and with the New Roads and Street Works bill before Parliament development started again. The development is following the SSADM path, concluding with the production of the Logical Design. This contract was awarded to KPMG Management Consultants in April 1991 with the Logical Design due in July 1992. The system will then be set up starting with basic operations in April 1994, with the full functionality required by users being progressively introduced up to April 1996.

Location referencing and the CSWR

One major problem with PUSWA is the difficulty that many recipients have in deciding where an excavation is going to take place or where it did take place. This problem causes many hours of wasted effort by the highway authority both in the office, clarifying the location with the utility and on site, trying to locate an excavation. The definition of a street contained in the NRSWA does not help. This definition is very wide ranging, covering any highway, road, land, footway, alley, passage, square, court or any land laid out as a way whether it is for the time being formed as a way or not! Therefore many pieces of land that do not, to a lay person, resemble a street are classified as such for utility street works purposes. This problem will not be solved by the introduction of the CSWR on its own.

A simple solution to the problem of locating works would appear to be for every notice to be accompanied by a map. This has many disadvantages, it increases the amount of data that has to be transferred and usually it is not required, as a description is more than adequate. Users have shown, with the existing paper system, that a vague description and an A4 copy of a piece of map, but without any features or grid references by which the works can be located, can confuse all recipients.

The way that was seen to overcome this was to use National Grid References (NGRs) for each job. NGRs are unique and widely used both by the utilities and highway authorities. The problems are:

1 How to produce them for every job no matter how small?
2 What accuracy is required?
3 Does the accuracy needed vary as the job progresses from coarse, at initial notification, to ± 1 metre when the works are complete?
4 If the accuracy is recorded to ± 1 metre does the boundary of the excavation need to be recorded?
5 Can the cost of producing NGRs be justified?

This is now seen as probably an expensive and unnecessarily complex way of working for most organizations at the present time. Therefore other alternatives have to be sought until all users have digital maps and records.

Alternative solutions that can be considered are:

1 To use the highway authorities street referencing system. The problems here are that:

- many highway authorities do not have robust systems;
- there is no standard system in use;
- they would be cumbersome for utilities to use, if based on section and chainage from nodes that are not physically marked;
- each utility would have to know how to use many different systems; and
- most referencing systems only include streets that the highway authority has an interest in, which can be only 60 per cent of the streets in which a utility has an interest.

2 To use a test based system, as used now. This would be simple and flexible to use but has the same problem of imprecise location as PUSWA.

3 To use a system that holds a map base, like the Dudley Digitial Records Trial CSWR (Yarrow, 1987) that uses an Ordnance Survey digital map base. This would provide both area and precise point information as well as a range of different types of data, such as proximity of two affected streets. This would require the CSWR to have a map base covering the whole of Great Britain which would be very expensive. It would also be unnecessary for those users who had digital map systems and it would be unlikely to be compatible with their system so data transfer could be difficult.

Users generally see a text based system as most desirable until digital mapping becomes widely used. The simple solution of a free text facility would not solve the problem with the existing PUSWA system. To overcome the problem of being able to identify which street is involved, an agreed list of streets could be used. This may not solve the problem of imprecise location but it does allow for searching on geographic areas, therefore it has been proposed in the Government's Consultation Document on the Options for the CSWR (Department of Transport, 1991) to use the National Street Gazetteer (NSGS, 1991) and a detailed description of the location of the works in the street. The street list or the gazetteer could be used as either a pick list or to check notice locations for accuracy. The National Street Gazetteer would be produced for each District Council area in a standard format and contain every street. For the CSWR it is envisaged that the gazetteer would initially be based on the highway authorities' list of adopted streets with further streets added when an organization noted an interest in it. An advantage of the street list or gazetteer approach is that additional information about the street could be added, e.g. the street manager, the reinstatement specification required, whether it was traffic sensitive or contained any features requiring special protection.

Co-ordination of street works

A major benefit of introduction of the CSWR is that it will give highway authorities the ability to co-ordinate the activities of undertakers carrying out works in the highway and so reduce the delays suffered by the travelling public. To do this the highway authority needs to know the proximity of works to other works in both space and time. Initially this will be done by checking the entry of a work against other works registered in that street in a particular period. The problem of works in another street but very close to that street can only be partly addressed by the CSWR as now proposed. Users will need to use either the spatial searching possible within the constraints imposed by the National Street Gazetteer or to build networks of streets using another means, possibly their highway management system.

The co-ordination of street works would benefit from the ability to display the location of the proposed works on a map so that their proximity both on the map and in time could be displayed. This is where having a map-based GIS would be beneficial. It is unlikely that highway authority users with no map-based GIS would be able to justify purchasing this on the strength of the CSWR but it is another source of benefits for a GIS.

Record keeping

The NRSWA also amends the requirements undertakers have in relation to records of the location of apparatus. Many undertakers have been reconsidering their needs and methods of keeping records for some time because of the high cost of operating their existing paper-based systems. The possible new standards to be introduced may provide the opportunity for a simple and quick system of plant records to be introduced. It is also proposed for the first time that highway authorities keep records of their plant and plant they allow under licence.

This system will for the first time enable users to obtain a complete view of all plant in a street. If the new record systems are map-based GISs this could provide the link to the CSWR.

Conclusions

The New Roads and Street Works Act 1991 introduces the most radical change to the way that works are carried out in a street since 1950. It will allow the quality of utility works to be improved at stages of their life. A major element in delivering this improvement to the public will be the Computerized Street Works Register. To be of value the information in the CSWR must be accurately located. The long term use of map-based GIS is seen as the best way to achieve this. Until users are all using compatible GIS the use of the National Street Gazetteer is seen as the way forward. The new record system will improve the quality of the records of the plant in a street and could in the future provide the link to the CSWR.

References

Computerized Street Works Register—Options in Detail, 1991, A consultation paper from the Department of Transport, the Scottish Office and the Welsh Office, 12 August.

National Street Gazetteer Specification, 1991, Luton: Local Government Management Board.

New Roads and Street Works Act, 1991, London: HMSO

Peat Marwick McLintock, 1988, Feasibility Study for a Computerized Street Works Register, Department of Transport, London (unpublished).

Roads and the Utilities—Review of the Public Utilities Street Works Act, 1950, London: HMSO.

Yarrow, G., 1987, Joint Utility Mapping, Paper to NJUG 87 Conference, NEC, Birmingham, 10 June.

23

The quiet revolution

R. Guinn
National Joint Utilities Group

It will happen this year, during July if all goes to plan, and will affect every one of us. The plans for it were made more than five years ago and, once it happens, things will never be the same again. Yet few people appear to have noticed the preparations for this revolution which will occur in an area upon which everyone has a forthright opinion. I am talking about street works—or ''holes in the road'' to you and me!

Contrary to the popular view, both utilities and Highway Authorities have been talking to each other. Not only that, but they have been making plans to change the way in which they operate and the government has at last given its seal of approval by changing the law of the land to allow them to do so. The results should be enjoyed almost immediately by every traveller on the highway and GIS has not only had a hand in the preparations but will play an increasingly important part in the future management of ''holes in the road''.

In order to appreciate the magnitude of the changes which are about to occur, it is necessary to understand the real story behind ''holes in the road''; why they happen, why they seem to happen more frequently these days and why they cause such disruption when they do. Once a few of the myths which abound are laid to rest, the impact of the changes should become apparent.

Utilities' services have been around a long time. Water and sewer pipes are the oldest, in excess of 200 years in many cases; gas pipes were in many of our cities more than 150 years ago and electricity cables upwards of 120 years ago. Even telecoms, the youngest of the services, were around more than a century ago. As the networks developed, they were installed in public streets; they had to be, since, when you think about it, there really wasn't anywhere else for them to go! Indeed, each utility was given and still has statutory powers to use public streets for this purpose.

Consequently it was very much a case of ''first come, first served'' and the new services were laid down alongside those already there, so that, in many instances, all the available underground space was quickly taken up. In addition, when supply pipes and cables came to the end of their useful life they were often abandoned *in situ* after continuing to supply customers right up to the minute that the new pipes or cables took over from them, so they too were laid in the next available space. Apart from the provision of duct lines for some cables, replacement using the original route is a technique which has only been perfected in the last 20 years or so. It can easily be seen that before this, further digging to remove an old pipe or cable was of no benefit to the original supplier, who would have merely provided space for a potential competitor, and so was never an attractive prospect.

Of course, traffic levels in those days bear no comparison with today. However, by the 1930s they had reached a point where they were coming into conflict with utilities maintenance work, so a new law governing the way in which street works were carried out was prepared. Unfortunately the second world war prevented the law coming to the statute book until 1950, by which time matters had deteriorated still further. The law was entitled "Public Utilities Street Works Act 1950", known affectionately as PUSWA—about the only affectionate part of it these days!

PUSWA requires utilities to seek approval from Highway Authorities before carrying out works in the highway by providing a notice in advance, together with a drawing in some cases. The Act carefully specifies the way in which the many different forms of notice must be sent and all involve a paper notice, which can either be posted or delivered by hand. Electronic means was not an option at the time, so is not permissible under the Act. PUSWA also allows the Highway Authority, if it wishes, to itself carry out the permanent repairs to the highway, at the utility's expense, once their work is completed. Therefore a utility is required to make a "temporary reinstatement" of the highway surface which was intended to last for up to six months or so, until the Highway Authority could complete the "final reinstatement".

Despite the changes which occurred before PUSWA could be enacted, traffic levels since 1950 have increased by more than seven times, while the available road has increased by just 20 per cent. On the other hand demand for utilities' services, in an increasingly affluent society, has mushroomed by an incalculable amount in the interim period and this is further enhanced by a new utility, cable TV, entering on the scene. In short, therefore, more traffic is being disrupted more often by more street works. Highway Authorities are generally unable to cope with the number of paper notices which arrive on a daily basis and temporary reinstatements frequently have to last much longer than the originally intended six months. When this is sometimes further exacerbated by poor quality of temporary reinstatement by the utility in the first place, it doesn't take much imagination to see that the situation has been getting out of hand almost since the day PUSWA was enacted, 42 years ago. Clearly something had to be done, but the problem could not be solved by disadvantaging either the travelling public or utilities customers, who are very often one and the same.

Back in the late 1970s the utilities recognized that, though they might be in competition for customers, their practical difficulties of operating in the highway were the same and there were many advantages in co-operating in this area. This is how the National Joint Utilities Group (NJUG) came into being and it very soon produced a number of initiatives and agreements which resulted in a significant degree of improvement between themselves, most of which may not have been immediately apparent to the travelling public. Perhaps the most significant initiative related to the way in which plant details are exchanged so as to avoid damage during excavations. A number of methods were explored, some of which continue to operate very successfully, but the two most important were "Susiephone" in the Lothian Region of Scotland and the Dudley Digital Records Trial in the West Midlands.

Susiephone (Scottish Underground Services Information for Excavators) provides a "one-call" system whereby anyone digging in the street can be provided quickly with details of underground plant. The system is computerized and has been expanded to respond to the notice exchange arrangements under PUSWA. This system works very well and is being continually refined. It is currently being expanded to cover the whole of Scotland.

The Dudley Digital Records Trial was commenced in the early 1980s. It too was initially concerned with the exchange of underground plant data, this time using a common Ordnance Survey digital map base, but in recent years it has also included a computerized street works register which allows PUSWA notices to be exchanged electronically. It remains, therefore, as one of the earliest working GIS systems.

These two initiatives proved significant in terms of an improvement in the levels of plant damage alone, but they also benefited from a willingness to co-operate, both between utilities and with Highway Authorities. From this stemmed the idea that more could be done in this way and the systems expanded into the field of notice exchange. In Dudley's case, agreement was actually reached with the Department of Transport that electronic exchange of notices would be sanctioned—if only on a temporary basis.

In 1983, after a crescendo of public complaints about "holes in the road", the government appointed Professor Michael Horne to head an enquiry into the street works regime. His three-person committee examined every aspect of current practice and took evidence from all interested parties, including the travelling public. They visited both the Dudley Trial and Susiephone and saw first hand the way in which matters could be improved by better co-operation at local level combined with the use of modern technology.

"The Horne Review" as it became known, was published in November 1985 and was immediately accepted by the government, following with very few reservations by the utilities and the Highway Authorities themselves. It made 73 specific recommendations which centred around the concept of a Computerized Street Works Register (CSWR) which would facilitate the electronic exchange of notices and provide a "whole life" record of all works in the highway. This CSWR would then enable the Highway Authority to co-ordinate all intended works and provide a suitable records system to enable utilities to take responsibility for permanently reinstating the highway themselves, after completion of their work. However, in order to accommodate these changes a change in the law was required.

The spirit of co-operation which had been displayed to Professor Horne during his enquiry then took hold on a national basis and the Highway Authorities and Utilities Committee (HAUC) was formed to turn his recommendations into practice. Joint Working Parties of experts were set up to look at all 73 recommendations and provide a balanced approach for the future. This work was largely complete by 1990 and a new bill entered Parliament in November of that year, emerging as the "New Roads and Street Works Act 1991" in June 1991.

As I write, work is continuing on the drafting of secondary Regulations under the Act—the real meat of its provisions and this, together with final "tweaking" of the HAUC Codes of Practice produced by the Working Parties, will be completed in time for implementation of the Act during July. Meanwhile HAUC is spreading the message to its constituents from John-o-Groat's to Land's End.

The only temporary disappointment is the CSWR. It has not developed as quickly as the parties would have wished, mainly due to a reticence by the government to finance its development. However, these difficulties have now been largely overcome and the basic system, which will be operable in 1994, will be expanded to accommodate all the required enhancements by 1996. Meanwhile a much simpler post and fax facility will be employed in the interim. The enhancements to the final system will include a facility to exchange digital maps, thereby opening up the possibility for the CSWR to become a full GIS in due course.

So, as you go about your daily travels, keep an eye on ''holes in the road'' and the street works scene generally. I hope you can see from my explanation the reasons why they have to occur and what is being done to improve the way they are managed in future. This ''revolution'' has been initiated to improve matters for the travelling public, so unless you and I are able to see an improvement as we travel about each day, we the initiators cannot claim the success we so confidently expect.

24

GIS down the hole in the road: The impact of the New Roads and Street Works Act 1991

F.D.J. Johnson
Former County Surveyor, Somerset

and

K.J. Crawley
General Manager, Seeboard

In 1983, in response to growing criticism and a general concern among Highway Authorities and Public Utility Operators about "holes in the road", Professor Horne was commissioned by the Secretary of State for Transport to review the working of the Public Utilities Street Works Act 1950, commonly known in the "industry" as PUSWA. This Act was conceived in the mid-1930s, when the services such as water, sewerage, gas and electricity were provided by City or Borough Councils and a mixture of large and small public and private companies. PUSWA was enacted in the post war years and was designed to regulate the way the "utilities", as the services had become, went about their statutory duties in the highway and how such works interfaced with the responsibilities of the Highway Authorities. At that time the Highway Authorities were either the same County, or City or Borough Councils who had run many of these "new" utilities.

In 1950 the extent of utility mains penetration into towns and villages and the volume of traffic carried on the roads was nothing like that which it is today. For example, in the 40-plus years, the volume of traffic has increased by over 500 per cent with only a 20 per cent increase in total road length. In the same period it has become the norm to provide the full range of utility services to virtually all properties, with the result that all five service networks have approximately doubled in size. Taken together there are well over 120 million direct connections of services to customers and the length of buried mains is over five times the total road length. This growth in services and traffic is heavily concentrated on urban roads with the consequent dramatic increase in the potential (and actual) conflict between "travellers on" and "travellers in" the highway.

The old PUSWA required that each time it was proposed to open a road or a footpath, the initiating utility was required to notify the others and the Highway Authority, in writing ("the only legal way to communicate"). The result was, with today's rate of work, over 100 000 pieces of paper changing hands every working day! This figure does not include maps or plans required to define or elaborate further the scope of the intended works. The efficacy of the PUSWA system had undoubtedly diminished with the growth in activity and practices were tending to vary, from one area to another, at an accelerating pace with the development of information technology. The PUSWA legislation did not require any authority, highway or utility, to co-ordinate work or ensure works by one were phased with those of others. In practice, however, because of the potential for traffic disruption where major activity was concerned, efforts were made by Highway Authorities to exercise a degree of control and co-ordination.

It might be thought that those who lay pipes, cables and services, dig sewers or tunnels, or bury plant of any nature, whether they be utilities or Highway Authorities would have been made to keep detailed records of the location of their equipment by the requirements of some legislation or other. However, this is not the case. For example, until the ''Street Works Act'' is commenced, British Gas is not required to keep a record of its mains, (other than H.P. Transmission). On the other hand, electricity, since 1899, has been required to keep a map of all their apparatus, but only to update it once a year.

None of the legislation that exists specifies any standard for the quality or accuracy of any records that are kept. One statutory requirement is that ''a map shall be kept''; to meet this requirement all that is required is a line on a road on a plan of any scale. In one piece of primary legislation a power was given to the appropriate ''Secretary of State'' to define the map scales and the detail required, but this was never taken up. Similarly, no common map base is required, so any base will do to meet requirements. It is only the Telecom Act which requires the use of Ordnance Maps as a base, but even then there is no definition of quality or accuracy for operators to use.

The situation with the Highway Authorities is just as bad, with no legal requirements to record and therefore no accuracy standards. Often, like the utilities, more reliance has been placed on the local knowledge of the local foreman, than on any requirements to keep detailed records of drains or lighting and traffic signal cables. However, all organizations have already reached the point where these people have, or are about to, retire.

So how will the ''New Roads and Street Works Act 1991'' help improve these situations and where does GIS fit in? To answer this question there is a need to examine briefly developments in the wider context.

First, dealing with highways: in the past, data on the use, life, interruption of highway service and the control of traffic, were, if collected, heavily demanding of manual resources. Today, extensive systems are in existence or in the course of development which enable the Highway Authority to manage the highway asset and to determine the most cost effective way to maintain the network, taking into account both public and authority costs. The management of highways, however, cannot be treated in isolation from the utilities services and this is fully taken into account in the new Act.

Second, utilities have, with the exception of works positively requiring co-ordination, gone about their own business seeking to expedite their own work at minimum costs. The management of these works, however large or small, cannot be treated in isolation from the management of the highways, and this too is taken into account in the new Act.

Third, the work of any others affecting the highway, together with the requirements of standards of service performance by the service provider, all interplay, and therefore the Act imposes duties of co-ordination and co-operation.

The works of all parties, what is to be done and what is there already, are linked by a common attribute—their geographical location. In the legislation there are two principle areas where GIS addresses these requirements directly.

1 Section 53—The Street Works Register
2 Section 79—The Record of Location of Apparatus.

The concept of the Street Works Register is to broaden the availability of information on proposed works, those in progress and the ownership of reinstatements, to any interested person or organization. Thus a fundamental need for accurate and readily available data should be satisfied. Such information is the basis of all highway works planning, traffic management and preparation for improvements. Under the Act the Highway Authorities have a clear duty to co-ordinate work carried out by the utilities within the context of a highway system managed for all users. For this to be effective, however, early and regular information on the intentions of all utilities and developers is essential. As the duty is imposed on the Highway Authority to co-operate in this co-ordination, the key to the collaboration being the provision of geographically referenced data.

Although the form of the Register is not decided, at the time of writing, its design will draw heavily on the experience gained in the development of the Susiephone System in Lothian and Strathclyde, and the mapping and PUSWA Notice exchange work developed in Dudley. Both of these systems, sponsored by the utilities and the Local Highway Authorities, have demonstrated convincingly that such information systems work and give positive benefits. These include faster, more accurate, easier and more comprehensive information exchange plus management control information for current and historical work as well as providing statistical information and the ability to track the progress of works.

A valuable result of this control, is that damage to buried plant is significantly reduced, with the consequential result that operating costs, the time of loss of service to consumers and the time of occupation of road space are all reduced. The opportunity to obtain statistical information also enables management to plan a variety of improvements, for example, in working practices or liaison.

In any form the Street Works Register can only operate from a Street Gazetteer and even this has limitations unless accurate locations can be given. It is here that the link to a full GIS system pays off in that the notification of works, monitoring over the whole life and reinstatement guarantee period, linked to the plant data, street traffic information and other works, can all be brought together for the co-ordination process. It is worth mentioning that such a gazetteer has many other uses, particularly in town and country planning, pollution control of contaminated land, as well as the Land Registries.

The requirement under Section 79 to keep records is the first real attempt to standardize the requirements for record keeping. As has been said earlier, the existing provisions of legislation on record keeping relate to specific utilities and services and date back to the mid-1800s! The most recent updating is in the "Electricity Supply Regulations 1988", but even this document is ambiguous and does not specify levels of detail or accuracy but, for the first time, it does provide for records to be held electronically.

The new Act still does not state what level of detail of apparatus is to be recorded, or made available, nor does it specify levels of accuracy. However, because it has at last been recognized that interchangeability and availability of information is paramount if co-ordination is to be achieved effectively, the Secretary of State has taken powers to make regulations on recording apparatus. A group has already been established to draft a Code of Practice for this purpose and by the time this paper is published the Code of Practice should be in place.

Because of the unregulated state of the current records of buried apparatus, all the responsible bodies have developed records of one sort or another, but in a manner to meet their own needs. This means that levels of accuracy, intelligibility and reference points vary enormously. For example, some records are held on strip plans whose locational information is referenced only to nearby fixtures, and many records are still held on maps based on the old "County" series. The new Act requires that apparatus found in the ground unrecorded or recorded in the wrong place, shall be re-recorded. With this as a requirement, exchangeability as a prerequisite and understandability a demand from the non specific expert, a "public view" presentation of data for exchange will be required. These factors, with the pressure on co-operation and co-ordination will drive the requirements to GIS systems with a common base.

The objective of the New Roads and Street Works Act is to minimize the inconvenience and hence cost to road users, vehicles and pedestrians, and in the interests of time, safety and cost, to protect the structure of the highway and the apparatus in it. Legislation is now in place to enforce the provision and exchange of information and co-ordination of works. Geographic Information Systems provide the means to make it work.

25

Use of GIS in the water industry

John Martin and Dave Taskis
Logica Industry Ltd

Introduction

In England and Wales, the water supply industry consists of private water companies which supply water only and were in existence before 1 November 1989 as well as the newly privatized water services companies. These water services companies, brought into being from the Water Authorities by the 1989 Water Act, supply water (although not necessarily for all of their customers) and remove and treat sewerage. The arrangements for Scotland and Northern Ireland are different and are not discussed further in this chapter.

The water industry in England and Wales is in a period of consolidation, settling down from the major privatizations of 1989 and the establishment of regulating bodies, such as the National Rivers Authority (NRA) for water resource planning, coastal defence, quality and pollution issues and the Office of Water Services (OFWAT) for consumer and commercial affairs. The legal framework has also been reviewed by the government and the 1991 Water Industry Act (Anon., 1991b) has consolidated much of the existing legislation into one, albeit thick, document. The use of GIS by the NRA and OFWAT is not addressed in this chapter.

This chapter provides a brief overview, drawn from Logica's experience of the use of GIS in the water industry, as it approaches the next century and examines where some of the opportunities exist to derive benefit from the application of this technology over the next few years.

Overview of water utilities GIS at present

Water services companies in England and Wales are concentrating mainly on the Digital Mapping aspects of GIS, as these focus on potential benefits in areas where they are most readily realized, notably asset management and facilities maintenance (AM/FM) activities. The benefits of GIS applications have been recognized but companies have in general adopted a cautious approach to their implementation.

Introducing GIS into the business

Digital Mapping technology is a useful tool, as it provides companies with a means to capture and maintain information accurately about the state and location of their assets (underground and overground pipeworks, furniture and buildings). By committing to a water (and in many cases, sewerage) records capture programme, followed by ensuring that data maintenance activities are carried out, business benefits may be identified:

- production of paper plans from electronically held information is a relatively easy task, allowing for more up to date and accurate information to reach operations staff more readily (e.g. valves and manholes in the right places);
- records may be more easily updated where new information becomes available; for operations following field work;
- as data is held electronically it is far more durable and is a common information source to all potential users—all users get the same information from a corporate data source;
- more accurate valuation and depreciation statements of assets which form a major part of the paper value of water services organizations.

All these points are made easier by access to a GIS but the organization of staff and their training has to make it happen. The introduction of a GIS into a company has usually called for a realignment of working practices and the adoption of a revised approach to records management. Some organizations have committed at least as much to supporting their systems within the business as they have spent on the initial procurement (Sage, 1992) and this is a key success factor.

Many water services companies have gained experience of these systems through a trial of limited geographic coverage within their businesses. Some companies are now well advanced into the programme of capturing their underground pipework and pipe furniture. This has not been possible without some pain. Four main areas of difficulty have had to be overcome:

- convincing the company's board that the cost benefit calculations really do show a benefit in a sensible time frame;
- defining a practical implementation plan that allows experience gained from each stage to be built on—in some cases this had led to the re-evaluation of objectives and/or the pilot system being replaced by one from a different vendor;
- ensuring that the data capture objectives are driven by the business needs and not by map availability from the Ordnance Survey (OS), where necessary influencing OS and its priorities;
- ensuring that the data capture staff are given adequate training and realistic targets which do not neglect verification of the quality of the data being captured.

Data capture from paper maps and registers

Capture of asset location and condition information is usually carried out on large scales, typically against 1:1250 or 1:2500 OS background geography. These are standard OS mapping products in digital form, the paper version of which will already be familiar to drawing office, records and field maintenance staff. Logica's experience has shown that this data capture is a major activity requiring the expenditure of many man years and perhaps millions of pounds. Usual estimates put the cost of the data capture programme at between 50 per cent and 70 per cent of the total system.

The scale of the challenge is clearly dependent on the size of the company but a typical starting point might have been:

- vast asset record base, often tens of thousands of km of mains and sewers;
- asset information held in a wide number of locations, sources and media;
- source data quality variable from excellent to poor;
- statutory obligation to meet interim and final time scales.

Records maintenance activities initially puts water services companies into the camp of Digital Mapping users rather than GIS users. However, applications to provide more business benefits are available and as asset bases become adequately mapped, both in terms of accuracy and coverage, fuller advantage of them can be taken. GIS is thus of relatively little use early on until the data capture activity has progressed. If a sensible data take-on plan is followed, it should be possible to migrate an entire operating area (e.g. a sewerage inspector's patch) to digital records and introduce GIS applications early there. Apart from shift working (which can have an impact on system management), at least two other methods of data capture might be available to speed up this process and bring GIS benefits nearer.

Data capture from existing systems

Data capture programmes are being supplemented by validation and conversion of geographically referenced asset data, captured prior to the advent of Digital Mapping, in asset survey systems (Aybet *et al.*, 1991). This approach can significantly reduce the time required to populate the database. It should also be a source of high quality asset data, since it will have been in day to day use on the existing system and any errors should have been gradually removed. Transfer of data from such systems allows the business to take maximum benefit from previous investment.

Data capture through external contracts

Where a company's record maps are suitable (usually complete and of high quality, reducing the need for local knowledge to resolve ambiguities), it is possible to set up digitization contracts with a data capture bureau or other contractor (Woodcock and Clennell, 1991). These contracts typically define the conventions used on the paper maps and registers, and specify the data structure, coding standards and quality threshold for the digital data to meet on delivery.

These contracts can produce good results, especially after the two parties have conducted a trial to explore what is achievable and to understand what type of data is most suitable. It can be advantageous for the client to agree to a flowline system with the contractor, rather than capture a fixed number of records and registers. However, the management effort needed for such contracts from the client's side must not be underestimated.

A significant degree of trust between the digitizing contractor and the water services company must be established since the original operational records usually have to leave the records office for the digitizing bureau. This can be a culture shock for many water services companies. More reassurance than just a provision for safe-keeping in the contract Terms and Conditions is needed to satisfy many records office supervisors.

From capture to operational use and reporting

Through Logica's activities it is evident that the majority of water services companies are in the data capture phase of their programme for GIS implementation, although a few of the small ones have moved into full operational use and the data maintenance and reporting phase. Much effort is going into capturing the data; more probably needs to be done to ensure its correctness, and prototype reporting applications are in use.

Data capture and the improved production of water and sewer record maps may satisfy the legislators but it is not an end in itself as far as the business is concerned. As the data capture programme proceeds, it is possible to report on the assets across a wider area, more effectively than before. It is likely that when data capture is complete, many water services companies will need to re-evaluate their market value, given precise figures on their underground assets. The impact that this may have in the City is unclear.

The digital record base with appropriate reporting software can provide figures such as, location of pressure reducing valves inspected in *w* district last month, or length of unlined, pre-1950 iron pipes in *x* operating area. These answers, which form part of the decision process when allocating resources for a mains rehabilitation programme, had previously been estimated with a planimeter pushed across the original paper maps. As well as providing figures for local managers, the data also has more strategic uses.

Management and executive reporting systems generally distil detailed figures down to major exceptions, trends and statistics, often quoted by operational areas. Asset information and related non-asset information may be usefully captured at a smaller scale or better, derived from large scale data, to enable trends to be established and reported on a geographic basis. For instance, location and other information about the pipework discharging effluent from a sewage treatment works might be combined with the non-asset information defining the permitted effluent quality (the consent) and a bacteriological analysis, for all works discharging into a named watercourse.

This style of reporting has traditionally been done through text only reports but the application of graphics, either as the main report or as a map illustrating the distribution of the data reported, has great power and allows trends to be visualized that might otherwise be missed. This style of presentation is also extremely useful for reporting under the current regulatory regime.

Director General statutory reporting requirements

There is a variety of legislation against which water services companies monitor themselves for compliance. This section considers the impact of some of these on the operational aspects of the business, rather than issues such as employment or health and safety regulations.

The Director General of Water Services expects water services companies to have put in place information systems which provide quality data to satisfy reporting requirements. It is a condition of the operation of these systems that the data contained within them can withstand an independent audit. As operational use continues it is expected that the confidence is discussed later.

The 1989 Water Act established the Office of Water Services to regulate the water services industry. Water services companies are now obliged to report each July to the Director General of Water Services by a number of different criteria. This legislation has been subsumed by the 1991 Water Industry Act (Anon., 1991b) but the provisions are essentially unchanged. Currently, there are four major categories:

- the investment programme,
- output and performance measures (including Levels of Service),
- the drinking water and bathing water compliance programme,
- risk management, disaster planning and sewerage agency policy.

These are in additional to a company's own routine monitoring of the levels of service it provides to its customers.

GIS applications have concentrated on some of the statutory Levels of Service reporting (known as the Condition J indicators), together with Levels of Activity, Asset Condition Indicators and a company's Underground Asset Management Plan (see Table 1).

Reports under DG1 to DG5 all have a geographic component and can benefit from a map-based representation of the detailed figures, at various levels of aggregation. The report will then be much more readily understood both by the regulator and the company's directors who may need to take action to rectify shortcomings or plan preventative investment to avoid operating difficulties.

Reports under DG6 and DG7 are less likely to benefit from a GIS treatment and, apart from indicating the precise sources of complaint or enquiry in relation to major assets or customer service centres, may not need amplification by graphic representations.

Digital Mapping in meeting business requirements

For the majority of water services organizations, the initial requirement of a Digital Mapping system is to provide and maintain the water and sewerage records, building an asset register which identifies the location of each plant item and a range of non-geographic information about it. Through this data capture phase, a (usually large) database of information, referenced on an individual asset basis, is created and then used as a source of information for a range of reporting functions outlined below. The asset database creation and attribute and topological validation is a fascinating subject in itself, beyond the scope of this paper.

Large scale data applications

The majority of large scale Digital Mapping applications are linked to the use of the digital mapping dataset as a master register of overground and underground asset information. This data store is traditionally populated by digitizing from paper records although some organizations are now making use of existing non-graphic data capture front ends which operate very effectively in the hands of experienced users.

Table 25.1. Statutory reporting criteria and content for the Water Industry in England and Wales.

Condition J Category	Report content
DG1	Raw water availability: How reliable are supplies when reservoir capacities and aquifer yields are compared with customer demand?
DG2	Pressure of mains water: How many domestic properties are at risk of not receiving adequate water pressure (and where are they)?
DG3	Interruptions to supply: How many domestic properties have received unplanned interruptions of >12 hours (and where are they)?
DG4	Water usage restrictions: How many domestic properties were subject to hosepipe bans, drought orders, voluntary reductions, standpipes etc. (and where are they)?
DG5	Flooding incidents from sewers: How many domestic properties are at risk from foul flooding (and where are they)?
DG6	Response to billing queries: How quickly did you answer?
DG7	Replies to written complaints: How quickly did you answer?
DG8[1]	Replies to development control consultation: How quickly did you answer?

[1] Not reported in July 1991 returns

As the location of each asset and its relationship to other assets and background geography can be maintained with high precision, this dataset forms an ideal source from which to produce consistent hard copy output for use within the organization and to generate statutory and other reports.

Typical uses of large scale datasets include:

- production of record drawings within the business;
- production of statutory maps for external use (e.g. members of the public, solicitors);
- generation of and reply to PUSWA notices;
- construction of work plans.

Use of Digital Mapping makes the production of large numbers of high quality plots possible. Thus, one of the company's operating divisions could have its old paper records replaced by frequently updated hard copy plots. Another benefit of a single data source is that the information shown on all plots is consistent and that the digital data once updated may be readily viewed by the whole organization to indicate the revised situation.

It is important that operating procedures are well understood. A plotter capable of high quality output must maintain this quality throughout its working life. If high volume plotting is to be carried out overnight, it will be worth investing in extras which allow the large number of completed plots to be managed effectively. Operators quickly become disenchanted when the technology that allows them to report on their fire hydrants in a way they only ever dreamed about, leaves 240 beautifully coloured, A1 sized plots all over the floor and another ten jammed inside the plotter each morning.

Small scale data

In contrast to the large scale datasets, Logica's work with water services companies has shown that some have seized the opportunity to use Digital Mapping in their establishment of the strategic overviews. These are typically required for maintenance of Underground Asset Management Plans (UAMP) as required by the legislation introduced in the run-up to privatization. These plans identify fixed (overground) assets, the major pipework between them and forecast financial provisions for rehabilitation over the next 20 years or so. These UAMPs are now approaching formal review by regulators and as such, digital versions are not only easily audited but may also be readily updated.

Small scale datasets also provide a useful means of reporting statistics on the basis of organizational or operational sections of the business. Through the maintenance of boundaries relating to these divisions, the Digital Mapping system may be used to present sets of statistics against these boundaries, producing thematic style mapping as an added benefit. This provides a powerful presentation tool for regulatory reporting and allows previously unidentified trends or relationships to be located and investigated. If marketed successfully within the company, it can also be used to demonstrate to managers the progress being made by their teams capturing the water and/or sewerage networks.

Data sources and mixing scales

In general, water services companies require geographically based information at two broad scales reflecting the operational and strategic nature of their management: the large scale (typically captured at 1:1250 and 1:2500) for accurate detail reporting and the small scale (captured 1:25 000 or coarser) for production of strategic overviews of the complete organization. While these two information sources will contain common data and may be supported by common references through both datasets, they are also likely to include datasets which may only be used effectively at one range of scales. For example, the OS 1:625 000 vector digital map is a digitized equivalent of the Routeplanner map sheet and not a digitally generalized composite from large scale datasets. Thus the route of a major watercourse at 1:1250 is unlikely to be coincident with the same feature at 1:625 000.

Ideally, GIS data will only be held (and thus updated) once; however the extremes of scale can make this mode of operation impractical. Maintenance of information captured at widely differing scales presents challenges in keeping common information consistent within small and large scale datasets during data maintenance activities. Since sophisticated regeneration software does not yet exist, this may be partially resolved by manually generating small scale features from those held at larger scale. Maintenance procedures must then also be imposed to ensure that such regeneration is performed as necessary to retain data integrity.

Data confidence, precision and accuracy

The level of confidence that may be placed in the large scale Digital Mapping dataset and the reports derived from it are central to its ability to form a sound basis for any of the range of uses outlined above. This confidence may be split into several contributing factors:

- the precision with which the Digital Mapping system holds the location of each asset (e.g. an asset's location may be quoted to 10 cm precision but this may not be correct);
- the accuracy with which the information has been captured and maintained;
- the level of confidence in the data which has been presented as source information (e.g. the data may be the best that is available, or it may be absolutely correct, having been surveyed last Tuesday).

The level of precision associated with the dataset should be linked to the future use of the information (see small and large scale datasets above) and the nature of the data capture process adopted should be sensitive to this requirement.

Value may be added to the dataset by inclusion of confidence indicators for each asset. How far this is taken needs to be justified by each business. Some companies have a confidence level against every attribute in the database, others have none. This can support an auditable dataset which also meets the Director General's requirements and will assist the company in activities such as establishing where mains relining would be most effective (since the probable condition of each section of the main is indicated by an attached confidence level).

Future opportunities for Digital Mapping and GIS

A fully functional GIS is a complex and expensive investment. The widest benefits must be sought to justify such a level of expenditure. In a mature Information Technology (IT) environment, this can be achieved through sensitive integration of the GIS within the framework of a corporate data strategy. This ensures that applications may be integrated across business data structures and more than one department can benefit.

Logica's work has identified that Digital Mapping and GIS are at greater risk if they are conceived and implemented in isolation from IT across the rest of the company. Not all departments will be able to realize gains from the GIS at once, but the integration of the GIS into the corporate strategy should provide the framework for such benefits to be achieved through a phased implementation.

Once a corporate ownership for GIS has been established, the data integration possibilities of the technology can be exploited through a multitude of applications. In a typical business, profitability can be enhanced by increasing revenue and minimizing costs and wastage. These are key areas where GIS applications can deliver, given a quality assured base of asset data.

GIS and revenue recovery

All new properties must be fitted with a water meter and most industrial water users are already metered. The Director General of Water Services committed considerable effort to canvass domestic consumers in England and Wales about the best way forward for the basis of bill payment, with a questionnaire enclosed in their recent water and sewerage bills. Revenue recovery is thus an important issue.

Meter installation at new properties, sewage catchment studies and by-law inspections of domestic properties can all be linked by GIS—if the data model exists—into the company's revenue and billing system. This can relate meter serial number to property and hence to customer, ensuring that only people connected to services receive a bill and, perhaps more importantly, that all properties connected are sent a bill.

Such a water revenue application could trace along a distribution main, up the service connection to the meter, from the meter to the stop tap and arrive at the property. This can be reported and then inspected, perhaps not interactively, in the billing system to ensure a bill is being sent. A similar application could be devised for sewerage, indeed it can be argued that sewerage billing is a less certain area of the company's revenue collection and this application may provide the greater benefit. A sewerage catchment can be investigated by tracing upstream from the treatment works, following sewers and investigating house connections and then checking with the billing system that the customer is being sent a bill.

To be successful, both of these applications require a robust network tracing algorithm, a connected data model (preferably through non-graphic linkage), sound coverage of data that has a known level of confidence and that the results of the rolling by-law inspections programme have reached the GIS and been entered by the records staff. The sewerage application may recover enough in the analysis of the first few catchments to pay for its development.

There are by-law provisions for five year maintenance inspections of valves and ten year inspections of water stop taps and domestic service pipes. These statutory visits provide an opportunity to record and capture both location and condition information on these classes of asset. They also enable data to be entered into the underground asset database at a pace aligned with the relatively long period of revisit. It should be possible to enter this data without disrupting the high intensity data capture effort which is required in order to digitize the water trunk and distribution network quickly.

These data are important to the companies' ability to meet statutory requirements, since they must be able to demonstrate that the visits have been carried out. It also appears to be a requirement of the New Roads and Streetworks Act 1991 (Anon., 1991a). At the time of writing, the industry is unclear as to whether the location of service pipes must be recorded under this Act and there is a Working Party to resolve the issue by establishing a Code of Practice for record keeping. Whatever the results of the deliberations, the new regulations come into effect on 1 July 1992.

The impact of the regulations may be profound on database sizing and the performance of the GIS host. The pipe lengths involved may be at least as much as the rest of the distribution network and the numbers of individual pipes and stop taps could run into millions for a large water company.

GIS and the reticulation networks

With an effective tracing algorithm and a robustly connected data model, new uses of the GIS become attractive. Applications such as network analysis, maintenance of supply to sensitive customers (such as hospitals, home dialysis patients, laundries), notification of interruption to supply, emergency decision support (in order to determine which valves should be closed) can be built if a well understood, carefully captured dataset exists.

As effective PC-based software is already extensively used throughout the industry Logica's experience indicates that it may not be effective to undertake hydraulic performance modelling within the GIS. However, it should be possible for software to be implemented which can generalize and simplify portions of a network for export and offline analysis by industry standard analysis products such as WATNET, WATCHMAN, WASSP and WALRUS. This also removes the need for a separate (and probably unmaintained) digital version of the pipe network.

GIS and the New Roads and Streetworks Act

The influence of this Act on the water industry is not well understood at the moment. The Act implies a major impact on working practices of field teams but on GIS the situation is less distinct. The need for a link with the planned Computerized Street Works Register (CSWR) is evident. The costs of establishing and using the CSWR in its various stages currently exist only in outline and while some Highway Authorities have published their commitments (e.g. £50,000 for the first year in the draft 1992/93 corporate plan of the London Borough of Havering), it is not clear whether budgetary provisions are yet being addressed by the water industry.

If a company's GIS is involved in its work planning or PUSWA issuing operations, then a notification application will be required, at the latest by 1996. Codes of practice are being drawn up at present to support the operation of the Act and it may be that electronic data interchange will be authorized as part of the notification procedure.

The Act defines a duty to record position of plant if it is exposed by works, whoever is undertaking the work. If a water services company exposes another organization's plant then that plant has to be recorded and the owners notified. Finding the owner may be a diverting activity but can the company afford for its GIS to gradually be filled with short lengths of asset record about another company's plant? It seems sensible to maintain this data on paper, since it is unlikely to get updated, merely passed across once the owner is found.

If the owner is quickly identified, it opens the possibility of renting the hole to them so that they may come and inspect the condition of their plant. We have not investigated the value of such an activity to the market but the technology exists to manage such transactions.

Other areas of benefit

On 1 January 1993, the single European market will have been established. One of the topical issues relating to this is that of Environmental Impact Assessment. Most water companies have an environmental department and some are already experimenting with PC-based polygon handling systems which analyse the intersections of planned capital schemes with environmental boundaries, such as Sites of Scientific

Interest, Areas of Outstanding Natural Beauty and others. It seems sensible that as part of the corporate nature of GIS, such applications should be supported, where a suitable alternative does not already exist.

There are other areas of benefit such as minor capital works, sludge to land, bursts and leakage applications, sewer rehabilitation, mains extensions, tracing watercourses to identify potential upstream areas of concern from a sampling point, integration of legal, estates and property terrier records with the pipe and building records. It is Logica's experience that in these areas projects can get defocused and lose track of the primary objective. Even if a figure can be placed on the apparent intangible benefits of such system integration, can the business afford it today? Judging by the changes of the past few years, the shape of any business may have changed by the time it could afford to carry out integration and, therefore, it may no longer be desirable.

Conclusion

GIS applications allow companies to produce clear, well illustrated and supported reports for their own managers, boards and for the water industry regulators.

Asset valuations determined from the digital asset register, customer connections through connectivity tracing and links to the billing system provide key opportunities for GIS applications to be used. Such applications can allow water services companies to understand their proper market values and realize their full potential revenues. This use must lead to greater operating effectiveness.

GIS as an integrating technology is at its best when implemented in a mature IT environment. In such an environment, the corporate view of information will have been understood. The value of a business data model, applied across the whole organization is key to effective use of GIS.

It is important to know how GIS is going to be used, as well as what it is going to be used for within the organization. Identifying functionality, applications and wide ranging issues such as how working practices may have to change with the introduction of GIS, should not lead to specific operating details being overlooked. It can often be the small issues that affect the acceptance of the system by its users.

A company should not wait until its whole operating area is covered with data before it attempts to implement GIS applications. It should capture data and deliver benefit incrementally by concentrating on a small operating unit and convert it to digital operations. Each application should be carefully chosen and then implemented well.

Data quality is crucial to the delivery of accurate reporting, on which strategic business decisions are made. This principle does not just apply to GIS but when viewed against the implementation and data capture costs associated with such technology, the contrast is greater. Effort spent capturing data which is not subsequently validated, is effort misdirected. Poor quality data in a high quality reporting environment results in a reduction in the value of the GIS and its data and could ultimately lead to misdirected capital infrastructure investment.

References

Anon., 1991a, New Roads and Streetworks Act 1991, London: HMSO.

Anon., 1991b, Water Industry Act 1991, London: HMSO.

Aybet, J., Martin J. and Taskis D.M., 1991, Protecting investment by capturing asset data from existing systems. Proceedings of the Association for Geographic Information 1991 Conference, Birmingham, UK, paper 3-2.

Sage, R.A., 1992, Anglian Water's GIS in operation—A case study, Water Supply, 10(3), Proceedings of the IWSA Specialized Conference on Geographic Information Systems "Mapping the Future", Lyon, 1992.

Woodcock, M. and Clennell, R., 1991, Can I use a Digitizing Bureau? Proceedings of the Association for Geographic Information 1991 conference, Birmingham, UK, paper 3-3.

26

How would you like to compile an asset register?

R.F. Ashby and I. Bush
London Transport

Abstract

The title of this paper is taken from the immortal words spoken in October 1987 by the London Transport (LT) Group Property Manager, Colin Smith, FRICS, to Michael Guerin, FRICS, former Commercial Property Manager of LRT and now retained as a consultant. The question arose following identification of the need for an accurate record of land ownership for the LT Group of Companies created out of the London Regional Transport Act 1984, which split the former London Transport Executive into separate operating businesses.

Aims

The record should hold sufficient information to fulfil financial and accounting requirements together with cross-references to detailed plans and deeds locations relative to the efficient management of the newly formed companies. This chapter considers the compilation of a comprehensive paper-based property asset register and its subsequent development into a geographic information system (GIS). The chapter considers the justification of the investment in a GIS, selection procedure, installation, data capture, benefits gained to date and looks into the future to consider some of the further developments being considered.

Introduction

London Transport is a major property owner, by value, in London. Its present ownership is in excess of £300 million and includes investment properties as well as those required for the operation of a large transportation organization. Tenures include Freehold, Leasehold, Licences, occupations by subsidiary companies under the LRT Act 1984, occupations under Parliamentary Powers, Informal Agreements and Running Rights agreed with BR.

In October 1987 London Transport Group Audit asked that a comprehensive asset register be compiled to meet their changing requirements. These requirements had been fully endorsed by the Board of London Transport.

Setting up the initial asset register

The initial ownership list was compiled between October 1987 and March 1988, from Property Department records, consultation with other departments in the group and other research.

During 1988 information concerning each property on the list was collected and entered onto data input sheets by a team of temporary staff. This data was verified for correctness and completeness. Following verification the data was entered into a custom built database residing in London Transport Property's IBM System 36 computer. Data entry was performed by in-house computer staff and cross-checked.

Ordnance Survey (OS) paper sheets were purchased for each area containing a property or land parcel in which LT had an interest. These were sent in batches to the LT Muniments Department (responsible for the company's deeds records) for the ownership boundaries to be extracted from the legal documents and marked onto the OS map sheets.

On their return to the Property Department the boundaries were colour-filled to indicate the company having an interest and also the asset's tenure. Only surface features were marked. The sub-surface detail such as tunnel locations, are not part of this map register (although included in the associated textual database) and the maps for these are under the stewardship of London Underground Limited.

The coloured sheets were returned to the Muniments Department for final verification.

This system, OS sheets and IBM System 36, was released for general use in March 1989.

Version two of the property asset register

The initial asset register was used successfully from March 1989 and underwent continual updating to reflect purchases, sales and to correct data inaccuracies. It was a valuable aid for the efficient running of the Property Department and for the associated company accounting functions.

In 1989 the recently appointed Property Asset Register Manager was asked to investigate the feasibility of upgrading the existing asset register to a fully computerized system holding the ownership extents graphically on a computer and the textual information in a relational database, in other words, a geographic information system. This upgrade was requested by Group Audit to meet changed audit requirements and also because of the proposed disposal of the IBM System 36.

The project was initiated by analysing the perceived user requirements of the GIS through discussion with Managing Surveyors and Officers from Group Audit. These requirements included all those for the existing system plus others such as the use of Apple Macintosh computers and the production of customized plots at various scales and orientation. At the same time an assessment of the likely on-line disk storage, networking and interface requirements was made. This assessment was conducted between October 1989 and December 1989.

In January 1990 an initial costing exercise was performed to give an indication of the expected costs. Cost information was obtained by informal discussion with GIS vendors. These figures were used in a budget approval submission to LT Property and approval was given in March 1990 to select actively a suitable system.

A visit to a working asset register at Milton Keynes Development Corporation was undertaken and following this a firm of independent consultants were retained to advise on system selection, procurement and implementation of the system.

The invitation to tender

The perceived user requirements document was produced in December 1989. Interviews with key users of the existing system and officers from Group Audit were conducted and a detailed Statement of User Requirements document (SOUR) was produced and submitted to London Transport Property for consideration and approval.

After approval of the SOUR an Invitation to Tender (ITT) document was produced. This was based on all the requirements contained in the SOUR and distributed to vendors who, it was felt, could supply a solution in the timescale expected by LT Property. This timescale was for a system to be fully operational by October 1991.

Cost benefit analysis

Having analysed the tenders the expected cost for the computerization of the Asset Register was assessed. The expected benefits and cost savings from installing the system were analysed and these included:

- The ease of maintaining an up-to-date map which in the paper based system involved, on average, two hours labour to transcribe the ownership extents onto the new OS map sheet. By using the GIS this task was expected to be achieved in a negligible time.

- The increased efficiency in performing ownership enquiries received from both LT companies and external agencies. These enquiries were running at 1300 per year.

- The ease in production of plans to a professional standard for legal planning, development and operational purposes at any scale and orientation. This was performed using a photocopy, cut and paste methodology in the original system.

- The increased security from the computerized asset register. The existing system was a one-off i.e. no duplicate copies of the paper maps containing ownership extents were kept thus making the system very exposed to fire and flood. The GIS would have all its data backed up on magnetic media and stored in fireproof safes in Townsend House and off-site.

- Greater flexibility in the production of textual reports of property asset data to satisfy LT Group, Internal and External Audit requirements. This was impossible with the original system without the assistance of the system designers.

It was estimated that the cost savings from implementing the GIS would be in excess of £50 000 per year.

Benchmarking

While waiting for the deadline for return of completed tenders, information and data for a typical area of Central London was collated. This data was to be used in a benchmark of prototype systems of the solutions tendered by vendors.

The area around Baker Street Station was chosen because of the complex ownership in that region and the fact that the area lay at the corner of four OS 1:1250 map sheets. The data collated included:

- The digital data for the sixteen OS 1:1250 map sheets centred on Baker Street Station;

- Copies of the four OS paper map sheets covering Baker Street which showed the ownerships in that area. These overlays were colour coded to indicate the owner and tenure of the asset. Vendors were asked to enter the ownership data into their prototype system;

- A key diagram which described the colour coding of the above map sheets;

- An extract from the IBM System 36 database for each asset in the Baker Street area. This was supplied as a computer listing. This textual data was to be entered into the proposed relational database and attached to the centroids of the relevant ownership polygons;

- Plans of sub-surface detail in the area. These plans were of varying quality and the requirement was to scan them and display them on the system. The best quality plan was to be warped to fit the building footprint digitized from the ownership extent plans.

- The digital data of London Underground Limited (LUL) tunnel detail for the four 1:2500 map sheets covering the area;

- The definition of the schema for the existing IBM database.

During each benchmark the critical aspects of the functionality of the system were rated. These aspects included:

- display speed,
- ability to perform benchmark tests,
- ease of use,
- ability and ease to system customization,
- system security,
- quality of documentation,
- ease of database use,
- ability to export database,
- ability to change database structure,
- speed of plot production.

These scores were used to compute a weighted sum designed to represent the ability of each system to meet the requirements of LT Property.

Data capture

In order to achieve the earliest commissioning date for the system it was decided that the capture of the graphical ownership extents from the OS map sheets should be carried out by an outside agency. However, as it was not considered desirable to allow the unique map records to be removed from the premises a home-based solution was adopted whereby the data capture was carried out at Townsend House, using loaned equipment and started while the order for hardware and software was being processed. This method not only minimized the disruption to the existing system but enabled prompt answering of queries as they arose.

The analysis of the OS paper map sheets started on 2 April 1991 and the data capture started in earnest on 22 April 1991 and was completed on 19 July 1991. The process highlighted several discrepancies in the existing system and some of these have still to be resolved because they entail detailed investigation of the deeds.

Textual data conversion

The existing database on the IBM was downloaded to floppy disks on 19 March 1991 and translated into a flat ASCII format suitable for bulk loading into a relational database on the GIS workstation.

This data conversion exercise took five weeks, during which time the database on the IBM was kept up-to-date in the normal manner and all alterations were logged in a ledger. The database on the GIS workstation was updated, to reflect the changes logged in the ledger, after successful loading. The accuracy of the database was confirmed by random inspection of 5 per cent of the records by two independent people.

System delivery and installation

After the hardware and software had been delivered and installed, the property extents were captured, the relational database built and populated and the 1500 OS map sheets loaded onto the installed system. The system user interface was customized by consultants to enable easy production of plots and database reports by casual users of the system. The system went live to general users on Monday, 15 November 1991.

Updating

LT Property are introducing new systems to facilitate continual updating of the register allowing for new purchases, sales and developments.

The future

The future will see the expansion of the system to include other graphics data sets including that for the extent of the tunnels. These data have not been included since they have not been available in sufficient accuracy.

Other data sets which may be incorporated include:

- vector data for the extents of commercial units at booking hall level which are leased to private individuals and companies;
- vector data of the detailed station fire plans;
- raster data produced from scanning photographs of properties.

It is also possible that the local Ethernet network between the server and the two workstations be connected to similar networks in other sections of LT Property and to networks in other LT departments.

Although there are many possibilities for expansion of this system, none will be embarked upon until the system is accepted by the user base. This acceptance will be indicated by the decline in the use of map cabinets and the production of all plots and database information from the workstations.

The system quickly received favourable responses from everybody who saw it in operation. As has been the case with many similar systems the first output was the easiest to achieve. This was a scaled map of an extract of several OS map sheets. Although easy to produce the surveyor was suitably impressed.

One of the first requests for output from the GIS came from the Director of Property and was for a plot of a development site which was to be used in a proposal to the LT Main Property Board. This proved an ideal opportunity to show the capabilities of the system at the highest level. The plot was produced and drew such a favourable sign of approval from the Board that a directive was issued by the Director that all plans for inclusion in proposals for Board consideration must be produced using the capabilities of the GIS. Additionally, all future plans for inclusion in legal documents are, where possible, to be produced by the GIS system to alleviate problems of site identification encountered due to the use of inferior quality plans produced by photocopying. It is hoped that these procedures will greatly enhance the company profile in dealings with outside property agencies.

In conclusion there are many benefits to be gained from putting a property asset register onto a GIS. They are all achievable with dedication, professionalism and total commitment of everybody concerned with the project.

Part VI

GIS and Remote Sensing

27

GIS and remote sensing

Matthew Stuttard
Earth Observation Sciences Ltd

Introduction

Earth observation (EO), or remote sensing, is the technique of deriving information about the Earth by recording the interaction of electromagnetic radiation with its surface or atmosphere. The results—mainly images and point measurements—are an important source of geographic information. This chapter briefly introduces the subject, describes what is currently happening in remote sensing, explains what the practical difficulties are in using the data and how some of them are now being overcome. It concentrates especially on the benefits which are possible by integrating image processing and GIS technology. Specific examples are provided by the case studies which are referred to throughout this chapter. They have been chosen to show practical, operational uses of remotely sensed data in conjunction with other types of geographic information.

Aerial photography is the most widely used form of remote sensing, and the techniques for measuring from and interpreting stereo photographs are maturely established. Even so, major changes are currently occurring in the aerial mapping field, a few of which are mentioned here, and Cassettari gives a fuller treatment. Observing the Earth from satellites is newer, but is set for a data explosion in the next five years. The potential use of data from satellites is enormous but in practice it has been fraught with so many difficulties that it has remained a specialist preserve, and the information derived from the data often falls short of what is really required. Evidence is presented here to show that this situation is changing so that EO data or derived products are becoming more commonplace in GIS installations.

This part emphasizes land because it is the subject of this book. It is not possible to discuss ocean, ice and atmosphere as well, but it is worth pointing out that the physical interaction of radiation with ocean and atmosphere is more tractable than it is with the land surface. For example sensors on ERS-1 (orbiting 780 km above the Earth) can measure sea-surface temperature to within 0.5°C and the elevation of large water bodies to within 50 cm. Such accurate measurements, in physical units, cannot often be attained by remote sensing of land because it is more lumpy and less homogeneous than air or sea (Barnes, in the next chapter, mentions use of land temperatures accurate to within 3°C). For this reason, obtaining useful information by remote sensing of the terrestrial environment often requires an interpretative or empirical approach which combines many types of data from a variety of sources. It is within this context that the integration of image processing and GIS technology is especially beneficial.

Remote sensing products

Satellite data as supplied

Unlike the air photo products outlined by Cassettari (in this part of the volume), satellite derived data is available at a range of levels. The following definitions are the data processing levels recognized for EOSDIS, the planned Earth Observing System Data and Information System (NASA 1991):

level 0: Reconstructed, unprocessed instrument data at full space-time resolution;

level 1a: Level 0 data annotated with ancillary information (radiometric and geometric calibration coefficients and georeferencing parameters) computed, but not applied to the level 0 data;

level 1b: Level 1a data processed to sensor units (e.g. a digital number representing voltage);

level 2: Derived environmental variables at the same resolution as level 1 source data e.g. radiance corrected;

level 3: Level 1b or 2 data which has been spatially and/or temporally resampled to a uniform scale, including averaging/compositing;

level 4: Model outputs and results from the analysis of lower level data.

The above definitions are necessarily abstract. Putting this in practical terms: the majority of Earth observation data is currently supplied at level 1b, and therein lies one of the main problems. To use the data at all in a GIS requires that it is processed to at least level 3 (e.g. an image geometrically warped to fit the same map grid as is used in the host GIS) and more often, level 4. Even then the data might comprise a "potentially useful" image rather than the actual information desired (e.g. thematic maps of deciduous woodland in 1982 and in 1992). The consequence has largely been rejection by potential users without the knowledge or facilities to transform data from level 1b to level 4. One solution, adopted by Steve Fletcher (in this volume), is to let a contract for provision of data at the appropriate level.

Value added products

Another solution is for the data suppliers to produce higher level products automatically as part of their standard data processing chain, thereby adding value during mass-production. This raises two problems: a) that if all data is processed at higher levels there will be a processing bottleneck (see chapter 5) and b) that a high level product suitable for all potential users may be impossible to define. These problems will be solved because system engineers are beginning to realize that GIS compatible products are necessary in order to justify project funding. For example the OMI system planned for operation in 1998, has a planned routine product range which includes DEMs and image maps (orthoimages with vector data overlays) in a standard projection, such as UTM, delivered within days (not months) of acquisition.

For now, value added products are created outside the standard processing chain, for example the majority of the chapters in this part emphasize land cover classifications made by, or for, a single end-user from low level remote sensing products. What is interesting here is that the purposes to which the results are put are so varied—from site selection for radioactive waste disposal (Beaumont) and estimating nitrates in groundwater (Fletcher) to identifying conservation targets (Mortimer). This begs the question—why not create a single national map of land cover and sell it as an off-the-shelf product? Metcalfe *et al.* explains that the Institute of Terrestrial Ecology (ITE) is engaged in doing just that. In the light of Fletcher's chapter one wonders if the NRAs will be early customers for the ITE product. If marketed properly and priced sensibly, derived products such as this one will bring the benefits of remotely sensed data to GIS users who lack the knowledge, resources or inclination to start with a conventional level 1b EO product.

It must be stated, however, that the precise definition of the classes in a land cover map and the methods by which it is made are often application specific as shown by the case studies in Mortimer's chapter. The lesson here is that while thematic maps (such as land cover, biomass or temperature) can be produced from remotely sensed data for general purpose use, there is also a place for special purpose thematic map production to specifications and standards which depend on the particular application —these will often be made from lower level EO products. This is the approach adopted by the MARS project to meet an operational need (Barnes).

The benefits of remote sensing

This section sketches the application benefits which, potentially, stem from the intrinsic properties of remotely sensed information.

Image maps

The immediate appeal of remote sensing comes from the beauty of the images, as Fletcher comments in his chapter—you can put them up on the wall. Though it is fashionable to disparage colour composite images as expensive pretty pictures, they are easily produced and can have immense value, for example petroleum geologists continue to use satellite image prints to identify oil bearing structures and Mortimer recommends the use of prints during field survey.

In digital form, geometrically corrected images provide a context for other geographic data—for example both Mortimer and Beaumont overlay satellite images and classifications with vector thematic map data from other sources for on-screen comparison, editing and analysis. Cassettari also emphasizes the value of scanned air-photos as a backdrop for topographic map compilation and revision.

Synoptic view

Remotely sensed data provides a synoptic view, the extent of which can range from a few hectares for a single air photo, through the 60 km $\times$ 60 km coverage of a single SPOT image, to the whole Earth disk imaged every 30 minutes by the Meteosat satellite. There are no other practical methods of obtaining these views, and they have all found uses in mapping, monitoring and forecasting. The benefit of a synoptic view is well illustrated by Metcalfe *et al.*'s chapter, which shows how national land cover mapping is possible at a field scale.

A synoptic view can also yield unbiased coverage of all regions irrespective of physical ruggedness or administrative/political differences, this is a critical advantage in activities such as monitoring European agriculture (Barnes) or mapping remote areas (the Antarctica project launched in 1991).

Detailed mapping

For some applications the amount of detail available from satellites is far greater than that available from other data sources. For example, Fletcher's alternative information source for estimating nitrate leaching was the MAFF statistics based on parishes; these were insufficiently detailed for his purpose. Similar inadequacies in existing data, and remote sensing solutions to them, are referred to by Beaumont and Metcalfe *et al.*

Repetitive coverage

Remote sensing produces geographical snapshots which are invaluable for change detection, and a GIS is the natural tool for the inevitable database construction and analysis tasks. As the remote sensing archives age, the potential for usefully comparing old and new data increases.

An example of short term monitoring is the 1992 set-aside monitoring trial. This used high resolution SPOT and TM data to confirm that land declared to be taken out of arable production for 1992 was actually productive in 1991 and remained out of production throughout 1992. The method adopted involved merging digitized field polygons with results from periodic interpretation of the latest satellite imagery. In the UK this trial is administered by MAFF, but the methodology was developed by the EC within Action 4 of the MARS project (see Barnes) and may now be adopted throughout Europe at government level.

In the UK, two projects engaged in longer term monitoring are: a) monitoring landscape change in the national parks (Bird, 1991) and b) land cover of Scotland project (Miller *et al.,* 1991). Both of these substantial projects have involved interpretation from aerial photography with repetitive coverage spanning more than 10 years. Interpreted land cover overlays for each "snapshot" were digitized into a GIS for analysis. The combination of analytical GIS and remote sensing in both projects has allowed detailed tabulations and visualizations of change to be created over large areas. The result is an effective planning tool and a sound basis for continued monitoring.

Timely results

Getting high quality remote sensing coverage of the desired area at the desired time is a well known problem (see Beaumont or Fletcher). The potential, however, for using remote sensing to obtain timely information is very great. Barnes shows that the MARS project provides continual forecasts of crop yield and final estimates within six months of harvest. This information is of immense economic importance because conventional crop returns for Europe cannot be collated until 18 months after harvest, which is too late for setting support prices or planning for surplus production.

Super-human data source

Some remote sensing instruments provide us with images at wavelengths which correspond to human vision. Parallax measurements on stereo images allow us to determine surface elevation. Optical sensors which respond to wavelengths outside the visible range provide super-human capabilities, for example the water vapour channel of Meteosat, the near infra-red sensitivity of special aerial films, SPOT, TM and AVHRR, or the thermal infra-red bands of many airborne imagers, AVHRR and TM. Active and passive microwave instruments expand the range of sensing capabilities even further to include derivatives of surface roughness, dielectric and distance. These, and many other sensors, provide us with information which we could not otherwise obtain.

Digital data

Though satellite remote sensing products are inherently digital, hardcopy prints have often been used as basic data from satellites because digital facilities for GIS integration were unavailable. There are still cases of organizations which use their digital image processing system to enhance satellite images for producing prints. Manual interpretations of these prints on tracing overlays are then digitized to get the results back into digital form in a GIS. Not only is this inefficient, but it also risks introduction of errors which are almost impossible to detect and difficult to correct. Seamless integration of GIS and image processing software reduces the risk of errors, permits automatic quantification of errors and facilitates the process of error location and correction, as shown by Mortimer. Integration also unlocks the potential to derive improved high level remote sensing products by using additional geographic information to help create the product—see Barnes' discussion of knowledge based systems (KBS).

Air-photos suffer severe geometric disadvantages for making maps which have been fully overcome by well-established photogrammetric techniques. Cassettari refers to the digitization and rectification of air-photos in the developing generation of digital photogrammetric workstations (DPWS). Trotter (1991) expects that airborne scanners currently under commercial development will supplant the large format aerial camera, eliminating some of the geometric distortions in the raw image and providing digital input direct to DPWS for final rectification and on-screen mapping. The aerial photograph is not yet obsolete, but a strong challenger will emerge from the combination of digital aerial survey data with DPWS.

For remotely sensed data from both satellites and aircraft, the advantages of end-to-end digital processing and integration in GIS are enormous, but are only just starting to be realized. Some users (e.g. Mortimer or Barnes) already benefit from these advantages but the next two years will see an increasingly widespread adoption of the fully digital approach through the convergence of software and data and the gradual replacement of obsolete technology.

Economy and efficiency

Remote sensing often does not give sufficiently detailed information—however, several of the case studies show how it can provide a first approximation to identify areas for more detailed study. Beaumont shows how target areas for site investigation were speedily and economically identified in the office by intentionally adopting "quick and dirty" short-cuts during analysis of satellite images.

Limitations of remotely sensed data—current status

The perception that more accurate and appropriate information can be obtained more easily and cheaply from other sources is perhaps the greatest blockage to the use of remote sensing technology. In some cases the traditional method employs aerial photographs and the current view is that satellite derived data could be used, but is not as suitable. Examples include large scale topographic mapping (Hartley, 1991) and the detailed land cover classification of the National Parks. The main limiting factors are evaluated below.

Cost

A detailed discussion of costs is not possible here, however, some broad indicators of current costs are possible. The cost of commissioning 1:10 000 scale air photo coverage for an English county, 400 sq. km in area, was approximately £20 000 in 1991. The cost of purchasing uncorrected digital multispectral SPOT images for the same area could vary from £2400 for two scenes up to £4800 for four scenes. Geometric correction of SPOT images costs about £700 per scene. The cost of data is comparatively small—and to some organizations could be seen as a distinct advantage. For both air photos and digital imagery the main investment is in equipment and effort to obtain the desired information from the data. Many organizations which could potentially use remotely sensed data have been kept from doing so by the capital and skills investment required. An image processing system (and the staff to run it) is, to many, a costly and overly specialized item with limited and unproven benefits.

Availability

As Fletcher, Beaumont and Mortimer all mention, acquiring good data is difficult—the primary problem being weather conditions. There is an inverse relation between spatial resolution and the frequency of satellite overpass which means that there are fewer chances in the year of getting cloud free satellite coverage for high resolution imagery such as SPOT than for low resolution data (e.g. AVHRR). Beaumont indicates that the all-weather and night-time capability of radar instruments (such as ERS-1 SAR) might overcome this problem. However, the jury is still out on whether SAR can provide the same type of information as the more established optical instruments, and there is considerable scope for scepticism—when compared to visible light, a reflected microwave signal is a response to completely different physical properties of the reflecting surface.

Data management

Once digital imagery is purchased a host of data management problems arise. The first one is reading the media and format. Half-inch tape is still the standard distribution medium because it is cheap, well proven and can handle the large data volume—however, many PC or workstation users would prefer smaller data volumes on floppy disk or QIC tape rather than investing in a costly half-inch tape drive. Data providers have responded to this problem; Beaumont mentions purchase of 15 km × 15 km mini scenes on floppy disk. Raster image data formats are generally less of a problem than for vector data and are usually not difficult to read into alien systems.

Having read the data, the second data management problem is data volume: a seven band TM scene covering 185 km × 185 km occupies about 260 Mb. Raster images are inevitably disk hungry and their storage volume cannot be reduced significantly using lossless compression methods. Fletcher's solutions were to use first the classified thematic map (not the original image data) and compress it into a quadtree structure. Second, a smaller map file was made at coarser resolution, but the eventual solution was to purchase a larger disk. The potential reduction in data volume for a thematic map compared to the original image is worth noting. Thematic maps do respond well to raster compression and can also be vectorized, so GIS applications employing thematic map information may not need bulky images (or image processing facilities) at all, if a derived digital map product is purchased as the basic input.

Data integration

Rectification of image data takes substantial effort and judgement, it may involve radiometric and geometric correction and mosaicking. In some cases short cuts can lead to an effective end-result, as shown by Beaumont. Preparing data for extraction of the desired information can be an expensive task—Barnes mentions that 80 per cent of effort is spent on data preparation and quality checks.

Information extraction

A lot of skill and imagination is required to extract good information from remotely sensed data as shown by Barnes, Mortimer and Metcalfe *et al.,* this is the domain of the remote sensing specialist and will remain so even though many of the procedures are being simplified or enhanced by closer integration of GIS with image processing.

Inadequate classification accuracy is a problem with all thematic maps. It should be emphasized that some measure of classification accuracy is always necessary—Barnes' comment on the seductive credibility of thematic maps is pertinent. Barnes states that automatic classifiers which look at each pixel in isolation can produce 60-70 per cent accuracy. Humans are currently more accurate than these algorithms, but unlike computers they get tired, make mistakes and do not apply the same rules consistently, they are also slower. Improvements with KBS of 10–15 per cent are possible (Barnes), and Mortimer mentions the need for this type of contextual, probability-based classification algorithm in his applications at English Nature. Barnes' system was specially developed for the MARS project, Mortimer's is a commercial packaged system which does not yet have these features. There is a message here for system vendors: pixel-based classifiers are no longer adequate, remote sensing data analysts will want to seize the data fusion opportunities offered by integrated GIS and image processing systems to perform contextual and knowledge-based classification, and to perform accuracy assessment against ground data at the click of a mouse—built in functions to do this will be a selling point.

GIS for Earth observation data

This chapter has concentrated on using remotely sensed data with a GIS, but the converse is also important. An enormous quantity of geographic data is used in generating EO data products and the header or metadata for the products, even at level 1b. Land/sea flags are required to adjust parameters in the processing chain, administrative boundaries and terrain characterizations are needed for each product granule (i.e. one image), and so on. GIS have made the task of developing and maintaining these ancillary data sets easier—though, as Barnes states, in order to gain maximum throughput, operational systems to perform the data processing tasks will tend to be highly integrated developments rather than GIS sub-systems bolted together from commercial packages. Low level libraries of spatial algorithms would be of great benefit in building high performance data processing facilities but there is currently a notable absence of suitable spatial algorithm libraries in the commercial GIS market.

Geographic interfaces to EO data archives

The user interface to the vast archives of EO data scattered around the world is always a catalogue of some sort. The problems of using these catalogues have been a major impediment to adoption of EO data products. GIS technology, coupled with rapid online retrieval of quicklook data products has the potential to open up catalogues so that users can define in geographic and other terms precisely the data they want, and then see a graphic summary of what is available (Stuttard and Cleden, 1991). Geographic interfaces for the next generation of interoperable catalogues are now moving from prototype to operational status. It is worth noting that much of this technology could be applied to other catalogues dealing with spatial data such as maps or air-photos.

EO data systems under development

New EO data systems under development include EOSDIS, the Data and Information System for NASA's Earth Observing System. This will have to handle 720 terabytes a year by the year 2000 (compare this with the currently substantial pre-processing requirement of 0.2 tb per year for the MARS project, given by Barnes). An important initiative in 1992 was the approval by the European Council of Ministers of the Centre for Earth Observation (CEO) which will be a user oriented data centre. A rough tally of the data output from all EO systems planned for the year 2000 yields a total of 7000 tb/year. The software infrastructure of EOSDIS, the CEO and other data centres will make extensive use of highly optimized GIS technology in order to facilitate access this awesome quantity of data.

Trends

Remote sensing systems

In the 1992-93 period ERS-1 and J-ERS-1 are providing novel microwave sensing capabilities, while Landsat 6 and SPOT-3 will be launched to perpetuate delivery of well established products. Radarsat will add a further microwave system. Towards the end of the decade, the massive growth in the variety and volume of EO data indicated in section 5.2 will occur as a result of the POEM, J-POP and EOS programmes of ESA, Japan and NASA.

Data

GIS compatibility is becoming apparent in remote sensing products in response to demand; for example it is now possible to purchase geometrically corrected and radiometrically adjusted image data by the map sheet from Satellitbild, SPOT Image, NRSCL and other suppliers, on distribution media including QIC tape, floppy disk and CCT. The number (and visibility) of specialist remote sensing bureaux able to provide bespoke thematic mapping will continue to grow as demand from non-specialist GIS users increases. As shown by the example of OMI, given earlier, these trends will eventually trickle down to the satellite ground segment—so that higher level products become available at an earlier stage in the processing chain.

Hardware and software trends—Fusion tools

The most important trend in software is the convergence of image processing and GIS capability. Closer integration of raster, vector and database functions is essential for remote sensing products to become more widely used. Products from Intergraph and Genasys have possessed this integration for some time, ERDAS Imagine has improving GIS function while Arc/info version 6.00 introduced raster capabilities in 1992. The effect of this trend is that an expensive and dedicated image processing system will no longer be required to use remote sensing derived data with other geographic information—the capability will become available on existing systems resulting in more widespread use of EO data products. It is not necessary for all GIS to have specialist image processing tools such as geometric correction or spectral classification, but a basic raster module has become essential for use of images and terrain data.

Among high-end systems it is possible to foresee more workstations seamlessly incorporating GIS, satellite image processing and photogrammetry. The savings in file transfer and ease of data fusion will lead to better image understanding and higher quality derived products. The potential for building DEMs from satellite and airborne digital data will stimulate production of better visualization software for draping images on terrain models. Fusion of the human brain with geographic information will increasingly be achieved using polarizing glasses synchronized to the display to create a virtual reality from real data (Ahac *et al.,* 1992).

Conclusion

In the last two years many articles and symposia have taken the theme that GIS and remote sensing have a great future together. The other chapters in this part show that the future has, in fact, arrived—that it is possible to overcome the technical and practical difficulties involved in using remote sensing and engage in useful work which would not otherwise be possible. The key word upon which to conclude is integration; there is always room for improvement but more appropriate products, standard approaches, improving software and faster hardware are allowing integration of remote sensing technology within the realm of GIS. Specialist skills will still be needed to exploit low level remote sensing products but the derived results are increasingly available for non-specialists to use in existing GIS installations. In the absence of specialist staff the high level products can be bought "off-the-shelf" or purchased to specification through service companies. With the advances in sensors, data products and software which are indicated here, and in other studies (Euroconsult, 1991), many remaining difficulties will be removed so that remote sensing technology can be widely used as just another tool in the spatial analyst's kit bag.

Further information

To find out more about remote sensing the following sources may prove useful:

The Remote Sensing Yearbook contains much useful information including comprehensive international contacts and data suppliers. The latest edition is edited by Arthur Cracknell, Ladson Hayes and Huang Wei Gen, and published by Taylor & Francis, London, 1990, 330 pages, ISBN-0-85066-808-5.

The Remote Sensing Society has an international membership from both academia and industry. Enquiries on remote sensing can be made through the secretary Karen Korzeniewski at The Remote Sensing Society (0602 587611).

Information on British companies specializing in remote sensing can be obtained from BARSC, the British Association of Remote Sensing Companies, by calling the secretary, Christian Ripley (0252 541464).

Acknowledgement

Thanks to my colleagues at EOS, Graham Read and Pam Vass, for suggestions and reviews and to my wife, Mandy, for her forbearance.

References

Ahac, A.A., Defoe, R. and van Wijk, M.C., 1992, Considerations in the design of a system for the rapid acquisition of geographic information, *Photogrammetric Engineering and Remote Sensing,* **58** (1), 95–100.

Bird, A.C., 1991, GIS based data on land cover change in the national parks of England and Wales. *AGI '91 Conference Proceedings,* paper 2.12.

Euroconsult, 1991, Earth observation data processing and interpretation services: Analysis of the sector and the conditions for its development. Executive sumary. Report for CEC, DG XII.

Hartley, W.S., 1991, Topographic mapping and satellite remote sensing, *International Journal of Remote Sensing,* **12** (9), 1799–810.

Miller, D.R., Gauld, J.H., Bell, J.S. and Towers, W., 1991, Land cover change in the Cairngorms. *AGI '91 Conference Proceedings,* paper 2.13

NASA, 1991, *EOS Reference Handbook,* USA: Goddard Space Flight Center.

Stuttard, M.J. and Cleden, D., 1991, Using GIS to find spatial data sets, *Mapping Awareness,* **5** (2), 20–23.

Trotter, C.M., 1991, Remotely sensed data as an information source for geographical information systems in natural resource management: A review, *International Journal of GIS,* **5** (2), 241–51.

28

Operational systems for monitoring of agriculture in Europe using GIS and image processing systems

Ian R. Barnes
CEC—Joint Research Centre

Introduction

The Common Agricultural Policy depends on a regular supply of reliable data on agricultural production within the European Community. The member nations provide statistics which are not necessarily compatible. The Council of Ministers therefore established the "Pilot Project for Remote Sensing Applied to Agricultural Statistics" to test the feasibility of monitoring agriculture using remote sensing. The project, now known as the MARS project (Monitoring Agriculture from Remote Sensing), is now in its fifth year and has moved from feasibility studies to operational systems. This transition has been greatly aided by the integration of GIS tools into the domain of image analysis and product generation. In this chapter, Section 1 outlines the MARS project and Section 2 highlights some of the ways in which integration of GIS and remote sensing is beneficial to the project.

Section I: Monitoring agriculture in Europe

Agricultural production in Europe

Some 127 million hectares (1.25 million km^2) of Europe is classified as agricultural. Of this, 68 million hectares are arable. The percentage that agriculture contributes to the GNP varies greatly from country to country; in Greece it reaches nearly 20 per cent, while in the UK it is 2 per cent. In absolute terms, Italy and France depend on agriculture for some 25 000 million ECU per year. Even in smaller countries such as Portugal, Eire and Belgium, agriculture is worth 2000 million ECU per year. Given the financial implications, for example, with farm subsidies, statistics about agriculture are a sensitive matter.

Present status of agricultural statistics in Europe

Conventional data collection for agricultural statistics requires the area and yield of crops to be obtained and depends, very broadly, on four techniques:

1 Using expert estimates, in selected areas,

2 Legally requiring a crop return from farmers,

3 Conducting an agronomic survey on a national sample of fields,

4 Collecting returns from merchants.

Most statistical services in Europe use a combination of two or more of these methods. However, that is where the similarity stops. There are no common strategies for collecting, analysing and presenting agricultural statistics but all methods are subject to accidental or intentional error.

Problems associated with unknown accuracies, incompatible methodologies and missing or late data delivery provide a very difficult framework within which agricultural statistics can be collated at a European level.

Monitoring with remote sensing

Remote sensing has several advantages for agricultural monitoring:

1 It is objective, if one accepts that the signal is subject to interference, and is not restricted by national boundaries or administrative/political bias.

2 It can provide information at different scales, ranging from the single field to the continent, depending on the resolution of the sensing instrument.

3 Temporal comparisons can be made both between and within years.

4 Data can be processed in near-real time provided the hardware/software infrastructure is suitable.

The MARS project

The Institute for Remote Sensing Applications (IRSA), in particular the MARS project, is required to provide operational systems for "monitoring European agriculture" to the CEC. It achieves this through the operations of four main "actions". These are given in Table 1.

Table 28.1. MARS Project Overview

	Action	Method/activity	Geographic area covered	Main Input Data
1	**Regional inventories**	Regression analysis applied to image classifications. GIS support to "expert system" approach to classification and stratification.	Selected NUTS II administrative regions within Europe	High resolution satellite data (SPOT, TM)
2	**Vegetation conditions and yield indicators**	Spatial or temporal comparison of "vegetation indexes" and "surface temperatures". GIS support for product generation.	All Europe	Low resolution satellite data, i.e. NOAA-AVHRR
3	**Models of yield prediction**	(i) improvement of existing agromet. models (ii) integration of satellite data with agromet. models (iii) deriving agromet. parameters (RET)., GIS support for data management, analysis and product generation.	All Europe	Meteorological daily data. High resolution satellite data (SPOT, TM) Low resoloution satellite data, i.e. NOAA-AVHRR
4	**Rapid estimates of crop areas and potential yield**	Computer assisted photo-interpretation. GIS support for ground survey data management, analysis and product generation.	53 selected sample sites within Europe.	High resolution satellite data (SPOT, TM)

Currently actions 1 and 4 are fully operational, with actions 2 and 3 due to be complete by 1993. Both actions 2 and 3 have achieved pre-operational status, meaning that prototype products are available and being validated.

The MARS approach to image data and GIS

Though founded within a remote sensing institute the MARS project has regarded the "image" as just one of a number of data sets to be utilized. Emphasis is placed upon the combination of geographic and image data to achieve reliable statistical products. All the above actions utilize a GIS to a greater or lesser extent with the principal benefits being;

1. Spatial data management: (e.g. data storage, reprojection etc.);

2. Data analysis and enhancement: deriving application specific data from existing general data (e.g. derivation of an "Available Water Capacity" map from the "European Soil" map);

3. Map design and operational product generation (e.g. generation of summary A4 maps and tables for the CEC). Each action has its own specific data and processing requirements.

Image data

Project MARS normally receives 100 NOAA AVHRR scenes and five high resolution scenes (either LANDSAT or SPOT) per week. This approaches 200 gigabytes of image data to be preprocessed every year[1]. Without the considerable effort of reducing the data volume by applying sampling techniques, these data would be unmanageable. The use of GIS tools have assisted in this process of data reduction.

Geographic data

The provision of agricultural statistics to the CEC is required at any resolution between the levels of NUTS I and NUTS III (Nomenclature Unites Territoriales Statistiques)[2]. This requires a corresponding level of accuracy in the core geographic data sets. These include:

1. Soil map of Europe (JRC/INRA, 1990);
2. European Administrative Boundaries (CORINE, 1990);
3. World Database II;
4. US Navy 5 minute Digital Elevation Model (USN, 1990);
5. European Available Water Capacity Map (JRC/INRA, 1991).

The problems of data

The word "operational" is used in this paper to describe the current status or eventual aim of the various MARS actions. In this context "operational" implies that the output data represents the truth. The problem is that environmental data can be sufficiently inaccurate as to mislead and confuse systems or people that just want a simple answer.

The MARS project has spent 80 per cent of its time and money investigating the quality of data it uses. This has been necessary because all data is received without any quality assurance information. As a consequence, much of the "operationality" of the actions is dedicated either to cleaning up the data. In the context of remote sensing, this involves calibrating and atmospherically and geometrically correcting images. For geographic data it has been necessary to evaluate both the spatial accuracy of features (e.g. coastlines) and investigate the consistency and "truth" of geographic attributes (e.g. soil types). GIS tools have been used extensively for these tasks, though it is easy to be led into a false sense of security when data can be presented so convincingly by such graphical systems.

Section II: Operational crop monitoring

Examples of techniques using GIS and image processing systems

Both the following techniques have used GIS extensively and are committed to the use of GIS tools for their operation, though future developments will use, independently, only the necessary GIS functionality, without including all the functions available in commercial systems.

GIS as a tool for image classification

MARS Action 1 (crop inventories, see Table 1) requires the classification of high resolution images (Gallego, 1991). The methodology to date has employed classic pixel-based image processing techniques in association with a regression estimator (Alonso *et al.,* 1991, Gallego *et al.,* 1991). The image classification accuracies range between 50–80 per cent with values most common between 60–70 per cent. These results so far have proven statistically adequate for crop area estimation, though it is clear that more accurate image classification techniques are desirable.

To improve upon these accuracies the MARS project has used geographic data in a Knowledge Based System (KBS) for Image Analysis (Kontoes *et al.,* 1991, 1992). It has been shown that improvements of between 10–15 per cent are possible over classic "parametric" methods. To achieve this, ancillary digital geographic information, residing in the ERDAS raster GIS, has been integrated into an image interpretation process. The geographic data provides additional contextual information to assist the classification process. The objective is to develop "image understanding" methods which automatically select "optimal classifiers" exploiting background geographic information relevant to the imagery.

The role of GIS

The KBS approach requires inputs from auxiliary data, referred to here as "geographic data". GIS are used in this technique for the following reasons:

- data may be stored and retrieved easily;
- data may be created and modified easily;

— analytical functionality may be applied on an operational basis. In this case the GIS can support the use of ''context'' rules applied to data as discussed below.

In this project the original data sets were created and enhanced in ARC/INFO. However, to enable the application to run with the ERDAS image processing system on an operational basis, it was necessary to convert them to ERDAS raster GIS files.

The Knowledge Based System Method

This method uses geographic data to ''refine'' or ''re-define'' classes which have already been derived purely from image information. This is achieved by combining the classified image output with new class evidence, coming from RULES which are ''triggered'' by CONTEXT information. This context information may be derived from both the imagery itself and ancillary geographic data sets. A ''forward chaining control strategy'' is used for triggering rules which can eventually effect the assignment of a class from the image. The method is illustrated in Figure 1.

The initial process in the system uses a statistical classifier (maximum likelihood) to generate two products:

1 'Super-Class'' classification based on image texture. It has been found that texture information based on computed statistical methods (Haralick, *et al.,* 1973) has some potential to discriminate land cover classes, even those with overlapping spectral characteristics (Kontoes *et al.,* 1991); and

2 ''Land Cover Class'' based on just the spectral data. This leaves each pixel in the image with two classes based purely on image data.

To each pixel is applied a ''reasoning'' process based on both image and geographically derived rules. Each rule is comprised of a CONDITION and ACTION. For image based rules, the condition is applied to the image and for the geographic rules, the condition is applied to the ancillary data in the GIS. Both types of rule will generate a ''support value'' for a particular class or super-class for each pixel. Tables 2 and 3 show an example of a geographic rule and an image rule.

The support values generated by the rules indicate the strength of evidence for accepting or rejecting the class of any particular pixel. The execution of geographic rules is realized through the analytical functions of the GIS and the image rules through the image processing system. After all the rules have been checked the evidence is ''weighed up'' based on the Dempster-Shafer theory of evidence. The procedure (Lee, Richards and Swain 1987), based on this theory involves the computation of ''belief'' values representing the combined degree of confidence in each class as a result of the combination of evidence from the various sources.

Ancillary data

The system has been tested in the Region Centre of France employing only two ancillary data sets, which are as follows:

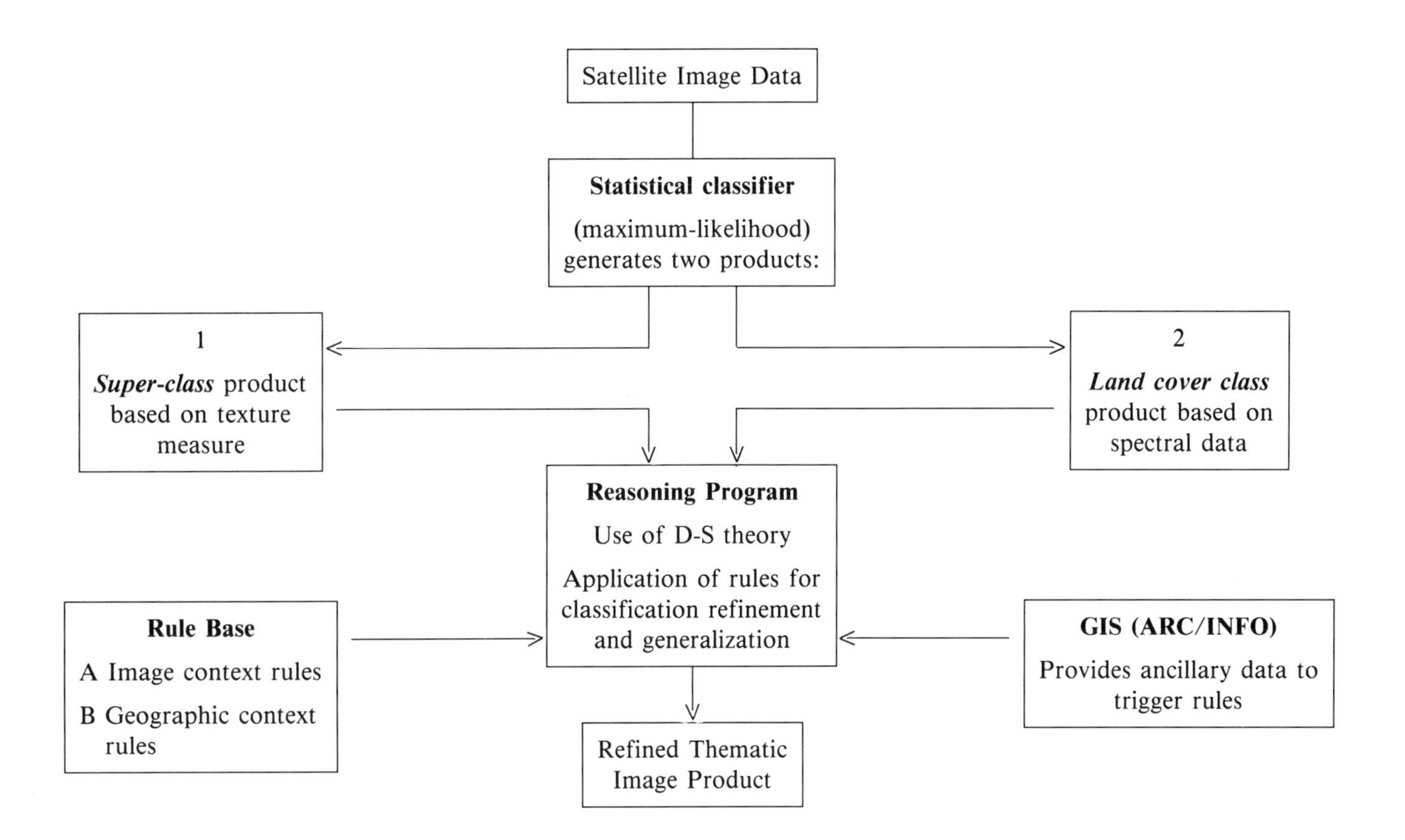

Figure 28.1. An expert system approach to image classification

Table 28.2. Example of a geographic context rule

Rule 391: Soil Type

CONDITIONS:

- The dominant soil class within the immediate 4-connected window centred on this pixel as "calcareous clay".
- The initial spectral class assignment for this pixel is "maize".
- The dominant initial spectral classification for a 9 × 9 pixel window centred on this pixel is "sunflower".

ACTIONS:

- disconfirm the initial class assignment of this pixel with support value 0.9 [Very Strong Disconfirming Evidence]
- Confirm the class "sunflower" as a condidate class for this pixel with support value 0.9 [Very Strong confirming Evidence]

EXPLANATIONS:

- There is considerable spectral confusion between "maize" and "sunflower" early in the growing cycle owing to background soil reflectance especially on calcareous clay soil with low water retention capacity. The rule aims to connect isolated misclassified "sunflower" pixels which have been initially classified as "maize" where they exist in a "sunflower" dominated neighbourhood. The initial "maize" class is thus strongly disconfirmed and the "sunflower" class is strongly supported.

Soil Type: Soil type has a significant influence on land use. Characteristics such as depth, available water capacity and organic content provide strong indicators of suitability for different use. The data used in testing the system was digitized from the Esquisse de Carateres Hydriques des Sols de la Region Centre from the Chambre Regionale d'Agriculture du Centre (1:250 000).

Accessibility: Land use is affected by access to transport links. This is particularly apparent in remote agricultural areas, where land use is characteristically low in agricultural value and production. The Institut Geographique National (IGN) 1:25 000 topographic maps were digitized for this purpose.

Conclusion

The task of implementing a KBS for image classification is not trivial. Rules have to be applied based on sound evidence. This can only come from expertise in image interpretation. Deciding upon rules is one thing, applying a probability to the evidence is quite another. In this sense, it is an inexact science.

Furthermore, for operational purposes, the system must rely heavily on extensive geographic data, in this case on a European scale. The functionality provided by GIS is essential for storing, quality checking and preparing the data for the system. In the future, the necessary tasks presently performed by ARC/INFO and ERDAS will be integrated into a single software system, to improve its portability and operating speed.

Table 28.3. Example of an image context rule

Rule 290: Image Context Rule

CONDITIONS:

- Three of the immediate 4-connected neighbours of this pixel have been same initial spectral class assignment as the pixel.
- The spectral class assignments of this pixel and of its three matching 4-connected neighbours are NOT all supported by the super-class assignments derived from texture.
- The super-class assignments of two out of these four pixels are consistent with their initial spectral class assignment.

ACTION:

- Commit confirming evidence to the initial spectral class assignment for this pixel with support value 0.5 [Medium Confirming Evidence].

EXPLANATION:

- If three of the immediate 4-connected neighbours of this pixel have the same initial spectral class assignment and the super-class information derived from texture is also consistent for two of the four ''matching'' pixels, then the initial class assigned to this pixel is likely to be correct and should be supported, though its support should only be with medium strength because the texture based evidence is not highly supportive in this region of the image.

Agrometeorological models for yield forecasts

The MARS project has also engaged in monitoring European agriculture without the direct use of satellite imagery. Action 3, entitled ''Agrometeorological Models for Yield Forecasts'' (Vossen, 1991), as yet, does not utilize remotely sensed data. Its aim is to monitor the state of crops and ultimately to forecast yield. However, the use of satellite imagery is expected to be a feature of a fully operational system. The integration of image data into what is a wholly vector GIS based approach will form the second part of this section. To begin with, however, a description of the action's methods follows.

The methods

Action 3 draws upon the fact that the development of any crop is primarily influenced by a combination of weather and soil conditions. The action presumes that crop growth can be satisfactorily modeled by using meteorological data, soil information, regional cropping characteristics and management inputs. This requires synthesis of supporting data and the models themselves. This synthesis involves (i) data management, (ii) data analysis and (iii) product generation. It is for this functionality that a GIS has been chosen, not just as a tool for development, but as a host for full operation.

The methods, though complex in detail, are relatively simple to summarize. A network of reference points has been established throughout the EC countries, based on a 50×50 km grid. For each reference point, meteorological data is interpolated from both an historical archive and real time data. Every 10 days, crop models are run for each point, (of which there are approximately 1500) drawing upon these data and the underlying soil information. The model outputs, measured in terms such as "stage of development", "water stress" and 'forecast yield", are compared with the "average" and "previous" years progress, and mapped on a 10 and 30 day basis.

The data

Meteorological data

The meteorological data is divided into real-time data and historical data. The real-time data is received on a daily basis from over 600 stations within the EC countries and their immediate borders. Readings from each station are processed every ten days to calculate parameters such as mean rainfall and temperature which are then used directly by the crop models. The historical archive of data goes back, on average, some 30 years for approximately 350 stations. This large data set is processed every year to regenerate the short term (10 day), medium term (monthly) and long term (yearly) averages. During operations only these averages and the real-time data reside on-line.

Ancillary data

Available Water Capacity map (AWC). The crop models require information on available soil moisture. This information is not easily available at European scales. In 1989 Action 3 sponsored the creation of a European Available Water Capacity (AWC) map based on the information supplied in the European digital soil map (version 1, JRC/INRA). This work required a revision of this version of the soil map, effectively doubling the amount of data within it (version 2, JRC/INRA). The generation and modification of both the soil map and the AWC map have been managed entirely within a GIS environment.

US Navy DTM. Altitude information is taken into account for the purposes of interpolating meteorological data to the 50 km reference grid. The five minute European DTM of the US Navy has proved to be the only reliable source of elevation information at European scales. This data has been degraded to provide information at the resolution of the 50 km grid.

Integration of remote sensing data

The integration of remotely sensed data into agro-meteorological models exists only as a theoretical concept. However, the future activities of Action 3 plan to incorporate the use of the NOAA/AVHRR and METEOSAT meteorological satellites (SuGrAm, 1991). The overall approach is summarized as:

$$y = f(\text{agromet}) + f(\text{metsat}) + c$$

where y = yield
agromet = agrometeorological model
metsat = meteorological satellite spectral models
c = constant

The more traditional approach tries to incorporate image derived data directly into the agrometeorological model (e.g. including satellite derived surface temperatures directly within evapotranspiration estimators). By contrast, this approach accepts there is a useful statistical relationship between yield and meteorological satellite data. The following section defines the three satellite derived inputs proposed for the suite of MARS agrometeorological models; namely (i) biomass indicators, (ii) surface temperatures and (iii) global radiation estimates. These inputs would reside as digital maps in the GIS covering the EC member states.

Biomass indicators

The celebrated Normalized Difference Vegetation Index (NDVI) which utilizes the red and near-infrared wavebands of Earth observation sensors is known to have a strong correlation with vegetation biomass. Though this index has no physical theory supporting it, it has been used successfully in small scale vegetation monitoring. The MARS project has tested the NDVI for quantifying biomass and even crop yield. As a result, the use of NDVI for Action 3 will be limited to the following:

1 qualitative validation of spatial variability in biomass production, by EC regions; and

2 quantitative evaluation of "season beginning" and "season end", by EC regions (through human interpretation of NDVI time series). These data, in the form of digital maps, will be made available, on a daily basis if necessary from the AVHRR processing system of Action 2 (See Table 1).

Surface temperatures

Similarly, the "surface temperatures" derived from the thermal channels of some satellites (in particular NOAA/AVHRR channels 4 and 5) can be used, but with caution. With an accuracy of no greater than 3 degrees Celsius over land, these data can at present only be used as qualitative support. As with the biomass indicators, these data, in the form of digital maps, will be made available by Action 2, and will be used primarily for validation purposes.

Global radiation estimates

Global radiation is a very important component of the agrometeorological model as it defines the total incoming solar radiation available to the plant. Though it is best measured directly with solarimeters, estimates over large areas use cloud cover and day length. With almost continual monitoring of cloud cover available from the METEOSAT satellites it is possible to provide accurate and automatic estimates of cloud. Given the subjectivity of cloud cover estimation (using "look and see" methods which are still common), this source of data may be of greater value than that currently available.

Conclusion

Action 3 both manages its data and issues products based on a 50 by 50 km grid cell. Satellite information is available at far finer resolutions than this. Though image information would be both misleading and difficult to handle at the pixel level the provision of classified images or statistical summaries will provide additional information which will be valuable for product validation. The AVHRR image products required in Action 3 will be supplied operationally from action 2 as digital thematic maps.

Overall conclusion

This paper has described two activities of the MARS project which use, or plan to use, both GIS and image processing systems together on an operational basis. However, operationality is compromised by the need to integrate one or more commercial software systems, i.e. Arc/info (GIS) and IIS (image processing). To be able to export the methods and systems to other non EC countries, (which is an aim of the MARS project), it is preferable to remove both software and machine dependency. In making this decision it is necessary to define whether operations will be biased to a vector or raster system. Whatever is chosen, systems will need to handle a variety of image and geographic data sets.

Operationality is also compromised by the lack of complete and reliable data. Most of the system development time and system run-time is dedicated to checking and enhancing data before use. This situation is not expected to change, as environmental data is notoriously prone to error or misinterpretation. The advantage of the project is that common methodologies for data gathering, preparation and analysis now exist for monitoring agriculture at a European level. This makes the system auditable, giving some accountability (i.e. estimate of quality) to the final products.

Notes

1 Preprocessing involves correcting image data for atmospheric and geometric distortions.

2 NUTS I is approximately at the regional level, i.e. NW England. NUTS II is approximately at the county level.

References

Alonso, F.G., Soria, S.L., Gozalo, J.M.C., 1991, Comparing Two Methodologies for Crop Area Estimation in Spain Using Landsat TM Images and Ground-Gathered Data, *Remote Sensing and Environment,* **35**, pp. 29–35.

Gallego, F.G., 1991, Remote Sensing and Agriculture: Image Classification and Area Estimation, Second International Meeting on Agriculture and Weather Modification, Leon, Spain, pp. 173–184.

Gallego, F.J., Delince, J., Rueda, C., 1991, Crop Acreage Estimates Through Remote Sensing: Stability of the Regression Correction, *International Journal of Remote Sensing*. In press.

Haralick, R.M., Shanmugam, K. and Dinstein, I.H., 1973, Textural Features for Image Classification, *IEEE Transactions on Systems, Man and Cybernetics*, **SMC-3**, No 6, pp. 610–21.

Kontoes, C., Rokos, D., Wilkinson, G.G. and Mgier, J., 1991, The Use of Expert System and Supervised Relaxation Techniques to Improve SPOT Image Classification Using Spatial Context, Proceedings IGARSS, Helsinki. IEEE Session WP-13, Knowledge Based Systems, pp. 1855–8.

Kontoes, C., Wilkinson, G.G., Burrill, A., Goffredo, S., and Mgier, J., 1992, Integration of GIS Data in a Knowledge-Based Image Analysis System for Rapid Estimates of Crop Acreages by Remote Sensing, *International Journal of Remote Sensing*. In Press.

Lee, T., Richards, J.A. and Swain, P.H., 1987, Probabilistic and Evidential Approaches for Multisource Data Analysis, *IEEE Transactions on Geoscience and Remote Sensing*, GE-25, **3**, pp. 283–93.

SuGrAm, 1991, EC Support Group for Agrometeorology. Literature available from the MARS Project, Joint Research Centre, Italy.

Vossen, P., 1991, Forecasting National Crop Yields of EC Countries: The Approach Developed by the Agriculture Project. Proceedings of the conference on The Application of Remote Sensing to Agricultural Statistics, Belgirate, Italy, November 25–26.

29

The integration of GIS and remote sensing within an environmental siting project

Peter Beaumont
Sir Alexander Gibb & Partners Ltd

Introduction

Pennsylvania has formed a compact with West Virginia, Maryland and Delaware to manage and store their low-level radioactive waste. Since Pennsylvania is the major producer of such waste it has decided to assume the responsibility to site, develop and operate a Low-Level Radioactive Waste Disposal Facility (DER 236, 1989a) for 30 years. In assuming this responsibility, the Commonwealth of Pennsylvania, with public involvement, developed regulations to site such a facility. An important siting criterion established that the facility could not be closer than half a mile from an important wetland. Thus, with changing regulatory requirements and the large area of land that must be assessed for suitability, any effort that may reduce the potential significant field work is necessary. Pennsylvania has an area of approximately 46 000 sq. km within which the three most favourable 500 acre sites must be located for future detailed site characterization. Law Environmental Inc and Sir Alexander Gibb & Partners Ltd are involved in the siting process which is being carried using GIS.

Protection of wetlands has become an important issue in the United States, and wetlands delineation has correspondingly become an important application for Geographic Information Systems with remotely sensed data being an important data source. Many individual states have regulations restricting development on wetlands. The Commonwealth of Pennsylvania currently has such legislation in place, based on a regulated definition of wetland. However, proposed changes in the legislation, requiring mapping of wetlands on a county-by-county basis, are currently being considered.

The US Fish and Wildlife Service has performed wetlands mapping across the United States, as part of the National Wetlands Inventory (NWI) Project. The Service is now in the process of producing a georeferenced digital wetlands database for the entire state of Pennsylvania. This effort has resulted in digital NWI maps available for approximately 17 per cent of Pennsylvania. However, these maps have limitations which prevent their use for regulatory wetland delineations.

In order to assess the effort involved in mapping wetlands for the entire state, an attempt was made, in a selected area, to use several other data sources to refine the NWI digital mapping data. Hydric soils were identified from County Soils Surveys, and their locations obtained from digital soils survey maps, and Thematic Mapper image data was used for classification of soils and vegetation, based on wetlands with known locations and characteristics. The digital NWI map/database was re-examined in light of our analysis and any re-delineation of the wetlands will be performed if required. An attempt will also be made to define the level of effort required on a statewide basis to perform wetlands mapping.

The US EPA, the US Army Corps of Engineers, the US Soil Conservation Service, the US Fish and Wildlife Service (Federal Manual, 1989), and the Pennsylvania Department of Environmental Resources (DER Manual, 1990)—agencies responsible for wetland management—all now use the same method of delineating wetlands. However, each agency still use individual definitions that define wetlands somewhat differently. Wetlands defined by all these agencies have a common thread that includes hydrophytic vegetation, hydric soils, and wetland hydrology.

A common start for any wetland delineation is the use of the wetlands maps produced as part of the National Wetlands Inventory Maps (NWI) project (Tiner and Pywell, 1983 p. 103). In order to assess the effort involved in delineating wetlands for the entire state of Pennsylvania, an attempt was made, in a select area, to use several other data sources to refine the NWI mapping data.

Hydrological flow patterns in the area have been greatly affected by glaciation. The soils in the area are formed from glacial till and have generally low to moderate infiltration rates, and in many places are poorly drained. This is especially obvious in the end moraine from the Wisconsinian glacial advance which crosses the study area.

Data availability

The preliminary wetlands assessment reviewed both digital and non-digital data. The primary digital data sets included the US Fish and Wildlife's NWI digital maps, Landsat Thematic Mapper satellite imagery, Landsat Multispectral Scanner satellite imagery and the US Soil Conservation Service digital soils data set. The non-digital data included aerial photographs (both black and white and false colour), US Geological Survey 7.5 minute quad maps, hardcopy NWI maps, Pennsylvania Geological Survey geology maps and results of field reconnaissance. It is important to note that the data, both digital and hardcopy, were created over a period of 15 years. While this may seem to present, at first glance, the potential for serious error in the assessments, the reader must remember that the study is a preliminary investigation and more importantly, the data represents what is currently available off the shelf.

The US Fish and Wildlife NWI digital and hardcopy maps provide basic information used for general land use planning. NWI maps are developed using photo interpretation from aerial photography typically from March to April and obtained from the early 1980s. The scale of older photography (1970s) is 1:80 000 with newer imagery at 1:58 000. Because of the scale difference and changes in technology some maps capture wetlands greater than five acres, while other maps capture wetlands greater than two acres; both omit wetlands below these limits. The study area included NWI maps developed from black and white aerial photography taken in October 1976 at 1:80 000 with the interpretations placed directly on US Geological Survey 7.5 minute quad sheets created in 1965 and photorevised in 1983 using aerial photography taken in 1981.

Landsat Thematic Mapper (TM) 15km × 15 km mini-scenes were purchased with acquisition dates reported as 25 October 1985, 1 January 1986, and 8 August

1986. These mini-scenes were available on floppy disks and were extracted from Landsat WRS path D014 Row 032. Additionally, Landsat Multispectral Scanner (MSS) data acquired on 6 October 1987 Path D014 Row 032 was also available digitally for use during the study. The three TM mini-scenes were selected to allow comparison between the wetland spectral response during the autumn, summer and winter.

Black and white aerial photography was obtained from the Pennsylvania Geological Survey (PAGS). This photography was used by the US Fish and Wildlife Service to develop the NWI maps. False colour photography was also reviewed, but could not be taken out of the PAGS Office so only a table top review was accomplished.

Digital Soils data was obtained from the US Soil Conservation Service (SCS) County Soil Survey Report, published in 1975.

Digital data analysis

All analysis was performed using both the ARC/INFO vector based GIS software package and the ERDAS image processing software package running on SUN Sparcstations. The TM data was received from EOSAT in the Space Oblique Mercator projection. In order to perform overlays using the digital NWI data, the TM data was rectified to the Universal Transverse Mercator co-ordinate system. The original TM scenes were resampled using cubic convolution and nearest neighbour techniques. An initial review of the data revealed that the 26 October 1985 data was offset four pixels in the y-direction and two pixels in the x-direction from the other two coverages. Corrections were made to the October image to allow accurate comparisons between the three images. All three images were rectified within ERDAS using 25 ground control points. Rectification was performed to facilitate the overlaying of the digital NWI ARC/INFO vector data required for evaluation of wetland inventories.

All image enhancement and spectral classification were performed using the raw data as received from EOSAT. Cloud cover and cloud shadow on the January and August scenes coupled with three different sun angles presented significant scene variances. The January and August scenes were rejected for further image processing. The reader is cautioned not to expect that significant effort was expended to perform the analysis. It is important to reiterate that the goal was to perform just enough analysis to identify potential problems in the source data and to identify the effort required when the siting process moves to the local or 7.5 minute quad or sub-quad level.

Processing of the October image was limited to False Colour Composites, Principal Components Analysis and Tasselled Cap Transformations. Prior to any visit in the field unsupervised image classification was carried out. The most successful image for preliminary interpretation of the wetlands was a false colour composite on bands 4, 5, 3 displayed on red, green and blue.

After all the TM image processing was complete, the digital NWI maps were overlayed on the TM scenes using the ERDAS-ARC/INFO LIVE/LINK. Additionally, the digital soils maps, (hydric soils only) were also overlayed in the same manner. These two overlays coupled with the field ground truthing formed the base of information use to assess the use of digital information as our primary site screening tool.

Ground truthing of NWI maps

The ground truthing was conducted from 9 July 1991, through 11 July 1991 by three field personnel experienced in wetlands identification for regulatory permitting. The field survey was primarily conducted from a vehicle to assess generally the accuracy of the classification of selected wetlands and was not to evaluate the exact wetlands' boundaries.

Fifty-two areas from portions of three adjacent NWI 7.5 quad sheets (Blakeslee, Pocono Pines and Mount Pocono) were field-checked. The Blakeslee and Pocono Pines maps were produced from black-and-white aerial photography taken in Oct/Nov 1976, while the Mount Pocono map was produced from colour infra-red photography taken in April 1981. The field surveys were conducted from a vehicle to assess generally the accuracy of the classification of selected wetlands, not to evaluate the exact wetland boundary. This limitation was due to lack of access to private lands.

Field observations indicated that for the most part, NWI-mapped wetland were located accurately, although small wetlands (one acre or less) were observed but not indicated on the NWI maps. In general, wetland habitat classification was also accurate, but minor discrepancies were observed. Certain wetlands mapped as shrub-scrub are now forested, a result of ecological succession occurring during the 10 to 15 years that have elapsed since the photography date. Also, other wetlands mapped as forested now appear to be shrub-scrub wetlands, perhaps due to a change in the local hydrology during that same time. One other common discrepancy was a greater presence of deciduous trees than indicated for certain wetlands on the NWI maps. This discrepancy may be a result of deciduous trees being under-represented in winter aerial photography.

Summary and conclusions

Since the NWI maps were produced from 1976 aerial photography, wetlands populations may change significantly. Thus, one cannot assume that the NWI classifications are consistent with image classification derived from current TM data. However, by incorporating ground truthing information, one can classify TM imagery to correlate with existing NWI wetland location but not specifically with wetland types.

Using image classification, smaller wetlands that are not mapped on the NWI maps may be identified. This is important for the siting project as we are required to identify all wetlands that are potentially impacted from the facility. The creation of the NWI maps imposed limitations during the photo-interpretation process. These limitations were based upon the scale of the aerial photos and limited wetland delineation to certain sizes (five acres or larger for the 1:80 000 photos and two acres or larger for the 1:58 000 photos).

Digital soils data were found to be of limited use in the study area due to the methods of soil classification and the nature of the glaciated soils. Soils identified as having major hydric components correlated well with the NWI wetland delineation. However, soils having inclusions of hydric components were difficult to use in a meaningful way.

Cloud cover will prove to be our biggest enemy in using remote sensing for the operational delineation of the wetlands. In this study two out of the three scenes were of little use due to the cloud cover.

In short, using TM data in combination with the digital NWI maps one can perform a broad based classification that correlates known wetlands with the TM imagery to locate areas for future field investigations. Additionally, we have found that using the broad classification scheme updates to the NWI map may be possible.

From the standpoint of the siting project, we have found that it appears possible to use the TM imagery, when a potential candidate area is located, to identify, in the office, those areas requiring field investigation. This is quite important as resources must be used in a cost effective manner and in a way that moves the siting effort towards facility construction and operation.

The future?

The integration of remote sensing within GIS will be greatly eased in the future by the introduction of:

- User-defined mini-scenes,
- Geo-coded imagery,
- More commercial GISs integrating vector and raster data.

These facilities will encourage the greater use of remote sensing data as a useful and very economical GIS data set.

In Pennsylvania, the satellite imagery will form an important data set within the GIS database and will be used by the client for future monitoring of environmental change. We are investigating the use of both ERS-1 data and airborne radar imagery to aid the discrimination of wetlands and to remove the problems of cloud.

This project was carried out with John Barron, John Dwyer, Dennis Lewis and Sue McCuskey of Law Environmental Inc. Both Law Environmental Inc. and GIBB are members of the LAW Companies Group.

References

Barron, J., Dwyer, J., Beaumont, P.R.K. and McCuskey,S., 1991, Use of Digital Data to Refine Wetland Delineation, Christ Church, Oxford: Spatial Data 2000, September.

Department of Environmental Resources, 1989a, Title 25 Chapter 236, *Low-Level Radioactive Waste Management and Disposal,* Chapter 236, 128, pp. 4676.

Department of Environmental Resources, 1989b, Title 25 Chapter 105, *Dam Safety and Waterway Management Rules and Regulations,* Chapter 105, 17, pp. 11–12.

Department of Environmental Resources, Commonwealth of Pennsylvania, 1990, *Wetlands Protection: A Handbook for Local Officials,* Environmental Planning Information Series Report, No. 7.

Federal Interagency Committee for Wetland Delineation, 1989, *Federal Manual for Identifying and Delineating Jurisdictional Wetlands,* Washington DC: US Army Corps of Engineers, US Environmental Protection Agency, US Fish and Wildlife Service, and USDA Soil Conservation Service, co-operative technical publication.

Tiner, R.W., Jr., Pywell, H.R., 1983, Creating a National Georeferenced Wetland Data Base for Managing Wetlands in the United States, 1983 National Conference on Resources Management Application, Washington: Energy and Environment, pp. 103–115.

30

Towards integrated image-based systems for aerial photographs

Seppe Cassettari
Kingston Polytechnic

Introduction

Aerial photographs are a well established source of geo-referenced information. They are used extensively in map compilation and revision, for the collection of inventory data, morphological and vegetation studies and for monitoring landuse and environmental change. They are also used to provide a pictorial backdrop to more traditional map data.

Traditional photogrammetric methods remain important for deriving highly accurate spatial data in two and three dimensions. The development of analytical plotters and, more recently, digital photogrammetric workstations (Dowman *et al.*, 1992) can be seen as building blocks which introduce photogrammetry into a more integrated GIS framework.

The role of photo interpretation is gaining credence within GIS circles as an alternative approach to certain types of data collection. Typically such information has been manually extracted to meet the particular needs of individual studies, with little or no consideration of the wider use of such data. The introduction of viable digital solutions for photo interpretation studies increases the potential for using data, which might otherwise not be collected. The number and range of such studies is increasing, bringing aerial photographs out of the cupboard where they have traditionally resided, unused except by the specialist.

The trend is for more frequent and extensive aerial surveys. The development of high quality and relatively low cost colour aerial photography has convinced many of the value of such imagery. Photographs are establishing a complementary role as both a source of spatial data and a contextual tool on which to overlay other geo-referenced information. Colour imagery is increasingly used to provide geographic context for GIS display. The developments in GIS technologies and more recently in image-based systems presage a new revolution in the use of aerial photography across the UK in the 1990s.

UK aerial photography in 1991

During a census year there is a laudable desire to link the collected statistics with concurrent imagery thus giving an integrated socio-economic and landuse dataset. This desire in turn leads to an increase in commissioned aerial surveys. Despite prevailing economic conditions, 1991 proved to be no exception.

Large areas of the UK were flown over during the year. GEONEX UK Ltd completed aerial surveys of some 14 counties, all of which were in colour, at scales ranging from 1:5 000 for areas such as West Glamorgan to 1:25 000 for East and West Sussex.

A relatively complete and current aerial survey exists for the whole of the UK at a variety of scales ranging from 1:3 000 to 1:25 000 and flown by a number of aerial survey firms. The ownership and marketing rights to these surveys varies, depending on the commissioning agent and the survey company involved. The time of year at which the surveys were undertaken also varies, and much of the imagery is still black and white. The result is a customer or task specific library collected on an *ad hoc* basis. There is surely a case to be made for a regularly updated national aerial survey to be used as the aerial photographic archive from which the growing body of users could purchase imagery as required.

A recent agreement between the Ordnance Survey and the main UK aerial survey companies has led to the establishment of a national information service on the availability of aerial photography. The National Air Photo Library, NAPLIB was launched in October 1991 as a successor to the long defunct air photo register maintained by the OS. Details on the available photography are supplied to OS as coverage diagrams by the six main UK aerial survey companies and OS operate a query service on aerial photo availability through the Air Photo Sales Office (contact number 0703 792584).

Recent improvements in the photographic films used in aerial cameras has extended the range of flying options, making it possible to conduct surveys in poorer weather conditions and at different times of year while still retaining image quality. Added to this is the introduction of a new generation of aerial survey cameras such as the Carl Zeiss RMK TOP. These have computer-supported control systems with forward motion compensation, high performance lenses and filters and integrated navigation instruments for automatic navigation and overlap control. The result is improved image quality and feature definition even from small scale photography.

Integrated photogrammetric solutions

Survey standard aerial photography flown to a high order of precision is designed for the accurate collection of geo-referenced data using photogrammetric techniques. In particular the three dimensional model created in the photogrammetric plotter provides an alternative method to ground survey for the collection of precise planimetric and heighting data.

Photogrammetry is used not only for traditional map compilation and revision but also for a wide range of specialist application studies. Examples include detailed contouring for coastal defence planning and the heighting of buildings in urban areas for determining telecommunication transmission blind spots. Such projects involve considerable photogrammetric expertise and are expensive to complete. There must be a well justified case for adopting a purely photogrammetric solution to data capture but where high accuracy is critical it is often the preferred solution, for example in creating a digital elevation model for coastal zone planning.

Increasingly aerial photographs are being used in digital form within integrated image management systems. These draw together the hitherto distinct strands of photogrammetric data collection, image analysis and image-based map compilation. In addition, aerial photographs are becoming available as digital orthophotos, or photographs which have been geo-corrected by removing all the inherent distortions due to aircraft movement and ground height variation. The advantage being the ability to overlay photos directly with map data.

Examples of such image management systems are the Carl Zeiss PS1 PhotoScan and PHIPS system and the Intergraph ImageStation 6187 which have a conceptually integrated approach to GIS and photogrammetry. The PS1 PhotoScan scanner for the digitization of photographs is a joint venture between Zeiss and Intergraph. This high resolution scanner has a pixel size of 7.5 microns, a scanning rate of 2 megapixels per second that will capture a 230mm square colour image in about 20 minutes.

The PHIPS system is a Photogrammetric Image Processing System, designed as the data processing kernel of a digital data manipulation suite of products. It will digitally produce orthophotos for use in application studies based on orthophoto mosaics. PHIPS can run in a multi-processor environment in order to satisfy the high computing demands required for the manipulation and management of such large digital images.

The ImageStation 6187 incorporates image processing capabilities. It combines raster and vector data for simultaneous viewing and manipulation on a single screen. In addition, photogrammetric data extraction can be undertaken by viewing overlapping images which are digitially offset through polarized glasses giving the operator a three dimensional model on the computer screen. The acceptance of this approach to large scale photogrammetric projects over more traditional optics on analytical plotters could revolutionize photogrammetric data capture but much depends on the availability of suitable computing platforms and how operators adjust to the glasses. The ImageStation 6187 is a fully compatible solution within the Intergraph RISC Workstation range of GIS products.

Digital photogrammetric workstations require a high level of computing power. The ImageStation 6187 typically might operate with 60Mb RAM running at 120Mhz (although this is not a minimum configuration). Single black and white aerial photographs require in the order of 30 to 40 Mb of storage at 30 micron resolution, while colour images are three times as large. The stereo viewing capability doubles the storage requirement.

This is not an exhaustive review of such systems. In the UK for example, GEC Avionics market a range of digital stereo image and photogrammetric workstations produced by Helava Associates Inc., a subsidary of General Dynamics. It does however serve to demonstrate the recent developments in this type of high cost solution to handling digital aerial photographs.

Low cost photo-interpretation systems

Not all users of aerial photographs can justify the use of such solutions; for many an alternative approach needs to be considered. The needs of the professional photo-interpreter and the more casual user of imagery are much less constrained by high accuracy requirements. Their need is for large area coverages from which data, often only interpretable from aerial photographs, is collected as map overlays with lower orders of accuracy. Typically the importance of the photos is to provide a geographical context for the data being collected, before it is transcribed to map based overlays (Young, 1992).

The technological requirements of such users are contradictory. Large geographical areas are wanted, usually comprising many tens or hundreds of frames, which have to be high quality photographs displayed and stored on low cost computer platforms and associated storage devices.

One approach to this problem has been the GEO-DAS system. Based on Apple Macintosh with an additonal 24 bit colour graphics board, the system displays partially corrected, geo-referenced photo tiles that are still of an acceptably high quality. These tiles are extracted from the original images to form edgematched blocks which create a "seamless" photo world. The images are displayed together with a transparent map layer and user defined layer on which photo extractions may be compiled using a standard range of graphic tools (Cassettari, 1991).

The contrast in this approach to that of the digital image workstation is highlighted by the image file size. A single tile is displayed in the GEO-DAS six-inch image window at a resolution of 72 dpi. Each colour photo is stored as a 24 bit image but only requires of the order of 650k storage, uncompressed, which can be reduced by up to a factor of 10:1 with current compression techniques. Uncompressed the photo, map and user layer require less than 1mb of storage. While the image quality is good, the resolution does not allow image enlargement without the pixel structure becoming apparent.

In order to capture large areas cost effectively the images undergo a simple linear transformation but are not orthophotos. Hence they still contain some of the distortions inherent in any raw aerial photograph.

While the digital photogrammetric workstation may be regarded as the "professionals tool", GEO-DAS is aimed much more at increasing the awareness and widening the use of colour aerial photography through photo interpretation studies. It is not too difficult to conceive of future solutions combining the benefits of the orthophoto and stereo model with large "seamless" photo coverages on relatively low cost platforms. The continuing reduction in computer costs, coupled with increased power and storage capabilities over the next five years are likely to provide the impetus for such developments.

Increasing the use of aerial photographs

The development of GIS concepts is increasing the use of geo-referenced information within a consolidated decison-making structure. Aerial photographs in tandem with maps form a very valuable graphic element to the visualization of geographic data. GIS is only one part of the information revolution currently underway, however. Potentially the development of multi-media technologies and their integration within spatially referenced information systems has enormous implications for the wider user of aerial photographs.

It is already possible to buy atlases and small scale map data for the whole world on CD-ROM for use on the average PC with the appropriate player. Sony have launched their home CD collection for the PC environment and Kodak will be providing CD formatted home photography. Alternatively videodisk technology can be used to store large numbers of still pictures as analogue TV frames, offering enormous capacity and fast retrieval; for example one side of a double sided disk can store 54 000 8-bit digital images of 640 by 480 pixel resolution. Retrieval times would be in the order of three seconds.

Aerial photographs linked to text information, ordinary photographs and sound have enormous potential. Add to this moving images, including video from aircraft, and a wide range of applications become possible, such as the development of local atlases and guides and many image-based inventory systems, all of which use the geo-referenced aerial photograph as the index. However, to be truly effective such integrated datasets will normally require an element of user interaction. Software solutions need to be added which allow the user, in the office or at home, to scroll around the photo world, add his or her own graphics, sound or images and call up information in any form, including video, based on the geo-referenced index.

Assuming that the market predictions made by Sony and Kodak are realised, CD-ROMs will be part of every computer and home TV system in the very near future. Aerial photography is potentially an important component within such systems. The critical elements are the simplicity of the software and the development of more intuitive user interfaces. The World Wide Fund for Nature's SATCOM project is an example of a well structured approach to multi-media technology with a central geographical focus, developed for the purposes of environmental awareness in schools. There are many others.

The wider use of aerial photography has implications for those commissioning and flying the surveys. Wider use of photography increases the copyright fees, giving the commissioning organization a greater return on investment. This may in turn lead to more frequent flying of certain areas and the availability of various scales flown at different times of year, thereby meeting the disparate needs of those interested in leaf cover, soil moisture, crop type and such like. The development of time series coverages is of increasing interest as a method of monitoring change and thus modelling future developments within a GIS environment.

Integrated approach

It is clear that the flying of aerial photography has continued across the UK during the early 1990s, given new impetus by the availability of colour imagery. The development of integrated digital photogrammetric and othophoto systems is part of the move to establish photogrammetry as a key component in data capture processes, linked directly into integrated geo-information systems. Likewise the development of low cost archive and management systems developed for a different range of users is addressing the same issue but from a different perspective. The continuing downward trend in computer cost for increasing capability will draw these two approaches together. At the same time the development of multi-media technology is likely to provide a wider market for map and graphic information. The aerial photograph has a particular role to play in this development.

A clear objective over the next few years is the increased awareness of the potential of aerial photographs as a source of data, both highly accurate and less stringent in positional terms, and as a means for deriving geographic context. The aerial photograph has the potential to complement the traditional map as a means of communicating geographical information.

References

Cassettari, S., 1991, ''Integrating vertical aerial photography with GIS—The problem and a solution'', *Proceedings International Cartographic Association Conference,* Bournemouth, UK, pp. 631–9.

Dowman, I.J., Ebner, H. and Heipke, C., 1992, ''Overview of European developments in digital photogrammetric workstations'', *Photogrammetric Engineering and Remote Sensing,* **58**, 1 pp. 51–6.

Young, R.S. 1992, ''Digital Imaging System Link Airphoto Interpretation to GIS'', *GIS Europe,* **1**, (3), pp. 34–7.

31

Use of remote sensing in the Severn Trent Region of the National River Authority

Steve W. Fletcher
Severn Trent NRA

History

The first steps along the remote sensing trail were taken by the NRA's predecessor organization, the Severn Trent Water Authority. The problems of nitrate in drinking water were becoming appreciated and there was an obvious need to assess the regional implications of fertilizer applications on crops and land use management in general. The initial problem of determining the distribution of crop types was an obvious candidate for remote sensing, so in 1988 a pilot project was undertaken to classify an image into crop types. We proposed acquiring an image of Nottinghamshire during the growing season. As might have been predicted, there were no suitable cloud-free images that year so one from 1984 was used with the attendant difficulty of obtaining ground truth information. The image was processed and a large poster-sized map produced which would look excellent on the office wall but, apart from that, had no real use. We had no way of analysing the data so a successful project produced no useful information other than a very rudimentary count of the areas of each crop on our aquifer outcrop.

Nitrate sensitive areas

The NRA was formed in 1989 at a time when the government was formulating its policy on reducing the amount of nitrate leaching into groundwater. The imposition of an EC maximum limit of 50 mg/l of nitrate had meant that urgent action was necessary in many areas. Staff at the Severn Trent Region of the NRA had already been in the vanguard of nitrate studies and subsequently led the NRA involvement with Nitrate Sensitive Areas (NSAs). The NRA defined 10 areas around major public water supply boreholes which were subject to rising nitrate trends. These areas were those which contributed directly to the abstracted water. Within the areas, farmers were asked to join a scheme where, in return for changing their husbandry to lower leaching patterns, they would receive compensation. Nitrate leaching and groundwater movement are slow and in the five year life of the scheme, few of the boreholes would show any resulting improvement, so it was important to be able to predict the long term effects with some confidence. Staff at the Ministry of Agriculture, Fisheries and Food (MAFF) produced tables of the leaching to be expected in terms of use, fertilizer application, ploughing etc.

For long term trend predictions, we needed to assess the spatial distribution of such leaching as well as its total area and it was decided that a field by field calculation of leaching was required with the results being fed into a finite difference groundwater model. To show the results and to assist in the calculation we saw advantages in having a basic GIS available and purchased SPANS from Tydac Technologies running on a PS2/70. To acquire the data for this we ran a small project to cost remote sensing against field visits. A trial site was flown using an aircraft-mounted video camera with individual frames being analysed in a normal image processing system. While the costs of remote sensing and field visits were comparable, the project became overtaken by events when MAFF decided to visit each farm individually and record not only the present cropping patterns but also previous patterns and exact fertiliser usage as well. At the same time, the tables of leaching were becoming more complex requiring more data than could be determined by remote sensing. Whilst the remote sensing element was discontinued it became apparent that the modelling capabilities of SPANS were far greater than was initially appreciated and the model finally built took into account:

1 crops,
2 fertilizer application,
3 ploughed out grass and its age,
4 the number of years since the grass was ploughed out,
5 soil,
6 type, amount and season of applied farmyard manure.

This data was all included as maps of distribution and associated attribute files. The results of the modelling equation were gridded by SPANS and produced as a file of grid references and leaching amounts for direct inclusion in a finite difference model.

The SPANS software we had purchased included a raster conversion package and it was realised that this could be the vital tool we had been short of two years previously. The digital data for our "crop type wall map" was still available and was loaded onto SPANS and built into its internal quadtree format (with something approaching a seven-fold reduction in data volume at an appropriate quadtree level). Now, not only could we analyse the crop types and varying areas, but we had a powerful regional tool for assessing nitrate leaching. The modelling equations used in the NSAs were simplified and the fertilizer usage estimated to produce a nitrate leaching map of the whole of Nottinghamshire.

On the basis of the success of this exercise, we let a contract for processing the most recent (1990) LANDSAT TM images of the whole of our aquifer. Again the images were not acquired at the ideal moment for crop determination but two frames covered the whole Severn Trent region and both were cloud free apart from the mountains of mid-Wales. As a result of an internal newsletter, colleagues expressed an interest in using the resulting thematic map so eventually the whole area was processed rather than just the aquifer outcrops originally specified. At this point we became victims of our own success and ran out of space on our hard disk! The raster thematic maps for our area, produced from the imagery, occupied two 50Mb files which had to be built up individually and joined. This was only possible by clearing all the other data off the machine which, no matter how sure you are of your backups, still manages to produce heart stopping moments. Moreover even when converted to quadtree format the map of the area was unmanageable, so a reduced resolution (lower quadtree level) map was built which only occupied 5 Mb. All the present work was done on this computer until a machine with a larger hard disk could be purchased.

Present and future work

Groundwater vulnerability

While the nitrate work was progressing, a catalogue of useful maps were being accumulated either by in-house digitizing or small contracts. By using SPANS for analysis rather than cartographic purposes we have largely avoided the problems of major digitizing projects; however we have now digitized groundwater units, surface subcatchments and river networks as well as individual field boundaries within NSAs. As a result of the nitrate awareness three years ago, we produced paper maps of aquifer vulnerability which was a thematic map of soil properties and geology overlain to indicate where groundwater was particularly vulnerable to contamination. An attempt to reproduce this on SPANS was frustrated by the lack of digital geology maps so a contract was let to digitize the thematic maps themselves. When this map was overlain by a contoured map of groundwater nitrate concentration, the high nitrate coincided remarkably closely with the highly vulnerable areas. This match was good for Shropshire with relatively uniform agriculture but less good for Nottinghamshire where there are large areas of low leaching woodland. Work is in hand to improve the relationship by including groundwater vulnerability in the nitrate modelling equations.

Equilibrium nitrate values

The NRA is nationally undertaking a major project to define contributing, areas around 750 major supply boreholes. When it is finished the total nitrate leaching into each borehole area can be calculated from the previous model allowing us to assess the scale of the future drinking water quality problem in the area. Predictions could also be made assuming a changing pattern of cropping.

Acid runoff

This project again arose when a colleague asked if the classified land cover map produced from the Landsat image could quantify the areas of woodland in the mid-Wales headwaters of the River Severn. While the basic task was well within the capability of SPANS, the area in question lay under the only clouds in the Landsat image. However the Institute of Hydrology had been working in the area using remote sensing and had a partially analysed image of the area. We let a small contract to them to complete the area required and loaded it onto SPANS. While the initial brief was merely to calculate the area of woodland, SPANS'' versatility has meant the project has again grown ''like Topsy''. We have now digitized the subcatchments, the topography, soils and effective precipitation and have produced cross correlations of, for example, areas of conifers above 300m. At the same time a basic modelling exercise has been completed using a point scoring system in an effort to predict areas likely to produce acid runoff. Even using such a crude model, we are able to model successfully which rivers have high acid runoff contributions. Work is now in hand to modify the point scoring system to a more representative and quantitative method with the aim eventually of producing a tool whereby we can predict the effects of a major afforestation scheme on river water quality. Such a tool could indicate whether we should or should not object to such a scheme or whether we should request more work to be undertaken using, say, a river water quality model.

Groundwater resources

The basis of groundwater resource management is knowing how much water infiltrates through the soil. The processes involved are infiltration and evapotranspiration.

Infiltration depends on soil, slope, geology and incident rainfall. Evapotranspiration depends on crop type and climate. In predominantly sandstone areas where huge amounts of water stored in the sandstone provide adequate buffering, resources are calculated using long term annual average climatic figures. The Severn Trent Region of the NRA has used MORECS data to allow for evapotranspiration which uses long-term annual potential evaporation and translates it to actual evaporation by assuming the crop type is either short grass or some real crop type averaged over 40km squares. It is proposed to recalculate the actual evaporation using real crop types on the classified Landsat image. All the other factors involved in calculating resources are readily represented by maps in a GIS and this is the route to be followed for general resource calculations.

Different methods have been used in the past when finite difference groundwater models have been constructed. Infiltration is calculated using weekly rainfall and evaporation data and only allowing recharge when a soil moisture deficit has ceased to exist. This calculation relies on knowing the crop types represented at each model node and their associated root constants. For the latest model we are building of the Nottinghamshire Triassic Sandstone Aquifer, the grid references of each model node have been entered into SPANS and Theissen polygons created around them (for the rectangular grid of points each polygon is a square centred on a node). This map has been overlain on the Landsat image and a cross-correlation undertaken to produce a file of nodes with the percentage crop composition of the square kilometre around them. The result will then be used in the model to calculate the recharge with much greater certainty than before.

Pesticides

Pesticide runoff and infiltration are the next problem to be addressed after nitrates. Working in this area is made particularly difficult because of the low concentrations involved, which invariably means that large volumes of samples are taken and then concentrated. Moreover the number and chemical variability of compounds which must be determined are considerable, as is the expense involved. In an effort to target compounds, it is proposed to use the classified Landsat image in conjunction with a map of either surface catchment area or groundwater units to determine the most likely compounds to be found in surface waters or groundwater respectively.

Conclusion

The work we are undertaking is still in its infancy but the acquisition of a complete classified Landsat Image and a larger computer means that we can now push ahead with the projects described. The use of GIS and data derived from remote sensing has opened up completely new avenues in hydrogeology and hydrology which we will be exploring for some time to come.

32

National land cover mapping using LANDSAT TM

Jane C. Metcalfe, R.M. Fuller, A.R. Jones and G. Groom
Institute of Terrestrial Ecology

As pressures on the environment increase, the need for effective land resource management increases in parallel. Remote sensing from aircraft and satellites is used as a source of environmental information. Data on the land surface, such as topography, soils, and land cover are required for land use management and for predicting the consequences of land use change. Scientists at the Environmental Information Centre (EIC), the data centre for the Institute of Terrestrial Ecology (ITE), are at the forefront of applying remote sensing to ecological, environmental and rural land use questions.

The land cover of Britain has not been mapped since schoolchildren undertook the task in the early 1960s. Their output took the form of paper maps, so analysis was a laborious manual task. Relating land cover to other information such as soils, topography, human population or pollution was of limited feasibility. However, satellite images now provide information at a spatial resolution compatible with mapping the complex patterns of the British landscape, and the Remote Sensing Unit at the EIC has developed methods to derive national land cover information from satellite data at the field-by-field scale. The project is now being implemented over the whole country, and is the first of its kind in Great Britain. The results will provide the first digital database of national landcover. These digital data will form part of the ITE's comprehensive geographical information system (GIS) on Britain's environment.

The method uses a multi-temporal classification of Landsat Thematic Mapper digital data (Fuller and Parsell, 1990). For any one area, a summer scene (May to July) and a winter scene (October to March) are geometrically registered to coincide with Ordnance Survey national grid. The data are summarized as the reflectance from the land surface at a resolution of 25 metres. The red, near infrared and middle infrared bands of each scene are combined as six-band datasets. The data are then displayed as images on a visual display unit and photographed for field use.

Figure 32.1. Images produced at different times of the year show different detail. Remote sensing experts at the EIC have developed techniques of using combined data from summer and winter images, which exploit seasonal differences in vegetation.

Figure 32.2. The technique of combining different images (multi-temporal analysis) means it is now possible to provide accurate field-by-field maps, previously only possible using labour-intensive survey methods.

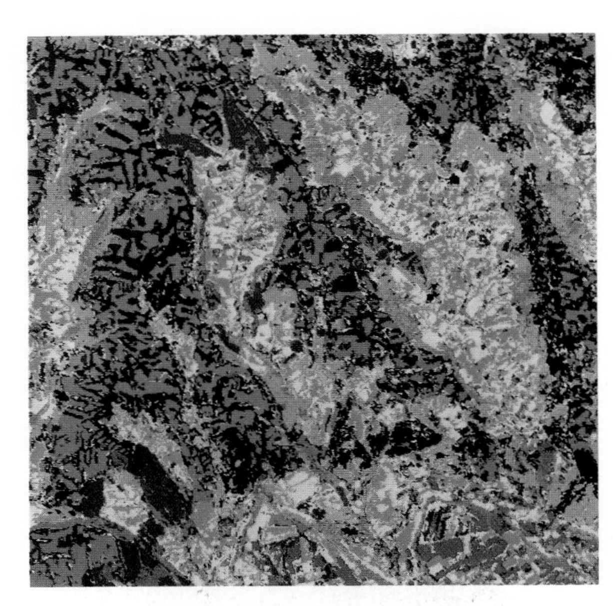

Figure 32.3. A 12.8 km × 12.8 km section of the land cover map for the North Yorkshire Moors; it shows the heather moorlands (mauve), with the regular patterns of burn moorland (dark green) comprising mixed grass and the regenerating heather which is so important to grouse. Note the steep valley sides with bracken (orange), dropping down to the valley floor of pastures and meadows (green).

Figure 32.4. A 1.8 km × 12.8 km section of the land cover map for central London: it is possible to see the urban centre (dark grey) giving way to suburban areas (light grey) and the grass areas (green) of the London parks such as Hyde Park (top left) with the Serpentine (blue); note the fine detail, for example the "herring-bone" of suburban streets or the bridges (grey) across the Thames.

Specific information on actual ground cover is collected during a preliminary field survey. The computer is then "trained" to recognize the different categories of land cover by their particular spectral signatures. This derived knowledge is used to identify the cover types in a test area, and the results are checked for accuracy against the field survey data. The full Landsat scene is scanned and classified by this method. The final maps will show 25 land cover classes for 25m grid cells throughout Britain, with a minumum accurately mappable area of between half to one hectare. The cover-types include:

- sea and inland waters,
- bare, suburban and urban areas,
- arable farmland, pastures and meadows,
- rough grass, grass heaths and grass moors, bracken,
- dwarf shrub heaths and moorland,
- scrub, deciduous and evergreen woodlands,
- upland and lowland bogs.

The land cover maps are produced to an accuracy showing 85 per cent correspondence with the field survey. After this stage they are fully validated by direct comparison with field data recorded in approximately 500 × 1 km squares, taken from the 250 000 in Britain as a whole. These squares were selected from a stratification based upon a statistical evaluation of topographic and thematic map data, which forms the basis of the ITE national Countryside Survey (Barr, 1990). The field records include details of soils, plant species and cover, boundaries, land use and management. Using information collected in this detailed sample survey, based on a full national stratification of all one km squares, inferences can be made about the countryside in the whole of Britain. However, local details may affect such estimates, and the TM-derived land cover information may help improve the accuracy, even of features which are not individually shown on the cover maps. For example, estimates of distributions of unimproved grassland might be modified by a knowledge of overall grassland cover. Crop estimates can be refined by a map of arable land. Estimated density of oak trees can be improved with a knowledge of the extent of deciduous woodlands.

The land cover map will be a vital component in ITE's national geographical information system on the environment. It is planned to analyse patterns within the landscape, such as diversity, fragmentation or continuity of cover, which may determine the distributions of plants and animals. The data will be used in conjunction, for example, with topographic data, soils and geological maps, plus the detailed grid-referenced species data of the Biological Records Centre of the EIC. The combined datasets will be used to develop our understanding of ecological and other environmental processes, and to help model the impacts of environmental change.

The project is due for completion in mid-1993; the sample-based field survey is complete and over half of Britain has been mapped from satellite images. The range of possible uses of the data is impressive, and there are already examples of early experiments which demonstrate potential uses of these land cover data. These include:

- planning of landscape management for conservation,
- detection of changing land cover,
- studying species movements to and from ecological "islands",
- relating bird species' variety to landscape diversity,

- assessment of land use and water quality in river catchments,
- mapping bracken which supports ticks carrying human diseases,
- mapping "critical loads" of pollutants for different ecosystems,
- environmental impact assessments of motorway extensions.

Planned uses include:

- assessment of landscape sensitivity to pollutants,
- the estimation of national and regional carbon budgets for use in global models of climate change,
- the prediction of land use changes in response to potential climate warming.

It is expected that the land cover map will be widely used to monitor, predict, and thereby help manage, our use of the land of Britain.

Acknowledgements

This study is jointly funded by the Department of Environment, the Department of Trade and Industry (through the British National Space Centre) and the Natural Environment Research Council.

References

Barr, C.J., 1990, Mapping the changing face of Britain, *Geographical Magazine,* **62** (9), 44-47.

Fuller, R.M. and Parsell, R.J., 1990, Classification of TM imagery in the study of land use in lowland Britain: Practical considerations for operational use, *International Journal of Remote Sensing,* **11**, 1901–917.

33

Remote sensing in an ecological GIS at English Nature

Graham N. Mortimer
English Nature

Introduction

English Nature (EN) is a statutory organization tasked with setting up National Nature Reserves, Sites of Special Scientific Interest and promoting England's natural heritage. Map based products are used by EN in many of its nature conservation activities, ranging from definitive site boundaries and large scale topographic maps through to detailed thematic maps of vegetation. In many areas of EN's work, remotely sensed data (as both a source of information for a GIS and independently) has a key role; for example in site identification, environmental audit, monitoring and education.

This paper summarizes the developments in the use of remotely sensed data in English Nature's GIS, with reference to specific examples and discusses some of the practical difficulties encountered in their implementation. The future for remote sensing in an ecological GIS is also considered.

Background

The volume and variety of spatial data acquired by Nature Conservancy Council (as it was before April 1991) and EN is very large and includes digital remotely sensed data, aerial photography, textual information relating to species distributions (Biological Records Centre), point data for all major habitats (COREDATA), digital SSSI boundaries, and transparencies and paper maps for various census and site related habitat surveys (e.g. Phase 1 and Phase 2, National Vegetation Classification surveys). In 1988 the difficulty of access to this data, the inability of relating these different products to each other, and a need to improve speed of access to boundary information relating to SSSIs led NCC to install a Intergraph Geographic Information System.

Until 1988 NCC's use of remotely sensed data was piecemeal. The work was typically carried out a) without the integration of other spatial data, b) as pilot studies that were usually site related, c) contracted out to people inexperienced in both ecology or remote sensing, and d) within limited time scales (often budget controlled). The results were indifferent, and it is fair comment that the benefits of remote sensing were not generally accepted by scientists/ecologists within NCC. Equally little or no work concentrated on the potential benefits of updating small scale topographic maps (1:50 000 and smaller) using high resolution satellite imagery.

System environment

In phases 1 and 2 of EN's GIS implementation, all image analysis work was carried out using Tigris Imager software running on a standalone Intergraph 340 workstation with 48Mb of internal RAM and 300mb of hard disk storage. This was networked to the rest of the system via Ethernet. Allowing for baseline UNIX software this system gave an effective online storage of 150Mb (approximately four bands of a full Landsat Thematic Mapper scene). Since Tigris Imager was memory based it was only possible to display and process one band of a full TM scene (34Mb). Most work which required displaying false colour composites or processing many bands was therefore carried out on image extracts. Imager allows both raster and vector processing to occur within the same working file (object space) simultaneously. There are a number of mutual benefits for remote sensing work and for using remotely sensed data in GIS that arise from this capability:

(1) Site boundaries, and detailed habitat surveys could be used to improve the location of training data sets in supervised classification;

(2) Ancillary data, e.g. contour data, soils maps, could be used to improve classification accuracy;

(3) Classification accuracy assessment could be more easily performed;

(4) Remotely sensed data could be used as a backdrop to topographic maps;

(5) Remotely sensed data could be a source of information for map update;

(6) Classified imagery could be vectorized and input into spatial analysis queries.

In late 1991 ISI2 (Image Station Imager 2) software replaced Tigris Imager and many of the problems associated with the latter have been expurgated, while maintaining the advantages of being able to manipulate raster and vector concurrently. ISI2 differs from Tigris Imager in the way that it handles the raster data. Using a tile manager it avoids having to load full images into memory and avoids the common problems associated with UNIX swapping. Elsewhere on the network a 670Mb file server is now used to store image data, access being made via Network File System (NFS). It is now possible to process—but not display—all seven bands of a full TM scene should there be a requirement.

English Nature projects integrating remotely sensed data in GIS

A number of GIS projects involving image data have been undertaken since 1989. These include:

1 National Countryside Monitoring Scheme,

2 Moorland Bird Survey,

3 GIS demonstration project concerned with mapping Calcareous Grasslands in the Derbyshire Dales,

4 National Peatland Resource Inventory,

5 Moorland Mapping pilot project (with MAFF, CC, ITE, DOE and Silsoe College).

The Moorland Bird Survey (MBS) and the National Peatland Resource Inventory projects are good examples of a shift in the use of remote sensing away from "one-off" classifications and towards modelling the relationships between a number of environmental variables and different plant species, birds and mammals and monitoring programs and environmental audit at the national rather than local scale.

Moorland Bird Survey

The Moorland Bird Survey is one example of where spatially related data is being incorporated with satellite imagery for modelling. This work aims to model and predict moorland breeding bird numbers in the uplands of Grampian, Tayside and Central regions. Bird surveys were made on 71 one kilometre squares using a stratified random sample. For each square the information on birds is available as the number of breeding pairs of a number of species including, golden plover, dunlin, curlew, snipe and red grouse. Vegetation data was also collected for each square.

For each square possible variables explaining the bird numbers include: northing and easting of OS grid; maximum, minimum and mean height, aspect and slope derived from digitized OS 1:50 000 contour maps; and vegetation cover information (relating to the per cent of heather and per cent of grass dominated moor) extracted from classified Landsat Thematic Mapper imagery.

In 1989 Macaulay Land Use Research Institute carried out a pilot study (on behalf of NCC) in an upland area of 30km × 50km in Grampian Region (Aspinall *et al.* 1991). Although the pilot study provided a clear method for classifying the satellite data, a number of things conspired to prevent following it to the nearest detail. Not least was the size of the study area and hence the size of the image dataset. A prerequisite was that the model should be developed for Grampian, Tayside and Central as a whole. Effectively this meant processing the two TM scenes (205/20 and 205/21) covering this area as a whole, which because of disk storage and processing capabilities was not possible. The two scenes had to be split into six parts. Because Imager would not allow training datasets—developed from a sample area—to be applied to other scenes, each part was processed separately.

One visible (Band 3) and two infrared (Bands 4 and 5) wavebands were geometrically corrected to the OS national grid; pixels were resampled to 25m. Non-moorland areas were removed from the image using a combination of visual interpretation to mask out agricultural areas and supervised classification to generate masks of forestry, snow and open water. Visual editing was chosen to discriminate between agricultural land and moorland because, first, a false colour composite (Bands 5, 4 and 3 displayed to the red, green and blue colour guns respectively) gives a visually distinct moorland/agricultural boundary and second, supervised classification is not accurate enough to allow all agricultural types to be discriminated from moorland areas. The agricultural land was "masked out" by editing the raster data on screen, the operator making use of context, texture and ancillary information, i.e. 1:25 000

OS maps. The visual editing took approximately 30 man days for the six parts. With hindsight, while visual editing is an acceptable method for an extract—as used in the pilot study—it is not the most efficient method of achieving this over large areas. Although untested, possible ancillary sources of information are digital soil survey information, contour data and land use maps that could provide "a moorland boundary" with which to mask out agriculture.

Unfortunately, a major limitation of ISI2 (and Tigris Imager) is that existing vector polygon information cannot be used automatically in creating a mask, i.e. a new polygon has to be defined while executing the raster editing routines. A work-around has been found at EN using the Microstation Grid Analyst (MGGA) package whereby the vectors are gridded at the same resolution as the satellite data and database entries for each polygon are used to define which parts of the grid are inside or outside the polygon boundary. Grid cells "inside" are assigned a value of 1 and holes or areas outside are given a value of 0. Overlay analysis is then used to multiply the image data by the mask, creating masked imagery. MGGA was, however, not available during the MBS work and for this reason visual editing was used.

The separate masks (agriculture, forestry, snow and water) were then merged to give a "non-moorland" mask. Pre-classification median filtering has been shown to be an effective way of reducing the variability in Landsat TM in order to improve classification accuracy (Williams, 1987). Therefore a 10 × 10 (250m × 250m) median filter was run over the image and mask. Supervised classification—using the field data from the 71 one km squares to drive training data collection—was performed to identify grass dominated and heather dominated moor and masks of these were created. In the pilot study, unsupervised clustering was used to identify 10 grass moor and 10 heather moor sub-classes. As discussed above this was not possible because the two scenes were being processed as six individual images and Tigris Imager is limited by not being able to apply training data to other scenes. Supervised classification methods were therefore adopted in defining classes within the grass/heather moor masks. This required reconciling problems of:

(1) topography,
(2) inconsistency in field survey data between different surveyors,
(3) an ecologically (not spectrally) based field classification scheme.

The effects of illumination on south facing slopes were overcome by identifying two separate training classes for affected cover types, i.e. a north facing class and a south facing class. A noticeable difference existed between statistics for training sets derived from 205/21 to those in 205/20. This was attributable to either inconsistent identification on the ground between surveyors or radiometric differences between the two scenes. Visually it was apparent that both were important. Much effort was accordingly put in to ensure that the training sets created for each part were comparable. This was achieved by reclassifying the field survey data. Spectrally distinct patches were identified and described on the basis of the field survey information. In this way a new classification scheme was derived and meaningful training sets gathered. The classification results were output in the form of a percentage for each land cover type per one km square.

Using this data and contour information—supplied in similar format (i.e. per cent area for ranges of height, aspect and slope), models are presently being derived on a PC-based system. It is hoped that using the grid analyst package the modelling can be done on the GIS to avoid large amounts of data transfer.

National Peatland Resource Inventory

The total extent of the commercially exploitable peat resource in Great Britain—the majority of which is blanket peat—is estimated to be over 1 300 000 hectares (Robertson and Jowsey, 1968); there has, however, never been a census survey to establish the precise figure. The National Peatland Resource Inventory (NPRI) (previously a unit within NCC Chief Scientist Directorate—now within Scottish Natural Heritage) was set up in 1989. One component of the NPRI was a) to provide an initial baseline audit of the blanket peat resource in England, Scotland and Wales to a local, national and international level and b) to maintain an ongoing programme of environmental audit on a permanent basis. Detailed site survey—as used in lowland areas—is not a practicable way of achieving these aims. Therefore Landsat Thematic Mapper imagery, in combination with aerial photography and detailed ground truthing, is being used to complete the work. The main source of information on peat extent is the British Geological Survey 1:50 000 series, which highlight areas of Great Britain with a peat depth of greater than approximately one metre.

Boundary information from these maps have been digitized (using ARC/INFO) and this data is being used by the NPRI to calculate figures on the total amount of blanket peat in Great Britain. However the NPRI requires a more detailed breakdown on the extent of different peat types (for example, *Calluna vulgaris* (heather)—dominated bog with an understorey of moss or Juncus (grass-looking rush)—dominant bog). While Landsat TM has been shown to be a useful data source for mapping peatland in Canada (Grenon, 1989; Palylyk and Crown, 1984), very little work has been carried out in Great Britain on an operational basis; this is somewhat surprising considering the extent of this habitat. Land cover mapping projects (Hunting Technical Services, 1986; NCC, 1987, 1988) where peatland has been a target habitat type, have also been equally interested in other cover types, e.g. forestry, urban. At certain times of the year, this presents remote sensing with problems because very different cover types have the same spectral response. For example, in winter imagery peatland areas have similar reflectance to urban areas. Similarly, different cover types may have the same major vegetation species present and hence the same reflectance. For example lowland heath on mineral soil will in summer imagery have similar reflectance to blanket bog that is heather-dominated in certain wavebands. Aerial figures for peatlands have therefore been variously underestimated or overestimated. In this respect the digital BGS boundaries have been used in two ways to highlight peatland areas: (1) superimposition of the peat boundaries onto a false colour composite, enables polygons defining training areas to be created accurately; and (2) the peat boundaries can be used to physically "mask out" non-peat areas (using the method discussed above). Equally in accuracy assessment and in working out what each spectral class is, the ability to overlay field survey information onto the imagery allows clearer assessments to be made than "eyeballing".

In 1989—before MGGA software was available—an alternative method to highlight peatland areas was devised using Principal Components Analysis. Using the BGS boundaries and field knowledge, pixel values from "regions of interest" i.e. known peat sites, were extracted from wavebands 1 to 5 and 7 of the TM scene being worked upon. This data was used to generate covariance and correlation matrices and using PCA, eigenvectors were applied to the whole scene. This has the effect of improving the spectral variation within the peatland areas and decreasing the variation in non-peat areas. Clustering of the PC images identifies non-peat areas as one cluster, while other clusters relate to variations in the peat. Each cluster (spectral class) relating to the peatland area is then described using the field survey data.

Learning from the Islay project (Belward *et al.,* 1990) and the Moorland Bird Survey work (discussed above), the field data has been collected using the satellite data as the starting point. Homogeneous blocks of tone and colour are identified visually from the imagery and visited in the field—accurate navigation to the centre of a block being achieved using 1:25 000 scale air photographs. Vegetation data in the form of a Domin score of cover for every higher plant species, moss and liverwort (as well as descriptions of microtopography, visual dominant species, management, erosion gully widths, extent of bare peat), slope and aspect were recorded for each block.

Independently phytosociological groups are derived from the vegetation data using TWINSPAN on a PC. A match was then made between the spectral class decriptions and these phytosociological groups. Using a phytosociological key EN regional staff are presently assessing (in the field) the frequency of match between the spectral classes and phytosociological groups to give a "locational" accuracy for each spectral class i.e. the probability that a pixel in the classified image actually represents that category on the ground. Two sample areas covered by TM scenes, 203/23 (Peak District) and 206/18 (Shetland), have been classified and accuracy assessment is currently being made.

Future issues

Future emphasis in EN's activities will be towards conservation in the wider countryside. This presents GIS and remote sensing with many opportunities. It is proposed that England will be divided up into "Ecological Zones" and satellite imagery may be the only consistent method of achieving this stratification. Within each zone, satellite data will be an important source of information for monitoring and change detection over different timescales.

New technology will also have a major impact on the use of digital remote sensing data in GIS work. Workstations with faster processor boards (e.g. Intergraph's C400 workstation) will mean faster turn around in deriving useful map information from raw data. Increased disk storage capabilities will mean larger amounts of data will be available "on-line", reducing the amount of time wasted in archiving and retrieving data. The availability of low-end systems also has implications for the use of remotely sensed data and GIS in the regional offices of English Nature. For example scaled down image processing systems are now on Intergraph's product list (ISI1) to allow users to simply display and enhance imagery, while on PCs, Microstation Review enables non-GIS/remote sensing users to display graphics files. Decentralizing the focus of remote sensing and GIS work means that GIS can be used to solve conservation problems at the front line. However, in itself this has implications for standardizing methodologies, cartographic standards and data integrity throughout the organization.

Research work into problems such as topographic and atmospheric effects on radiance should produce new algorithms that correct for these problems and that are implemented in packages such as ISI2. This will allow cross scene comparisons and improved use of multitemporal imagery, both of which will enhance the use of satellite data for vegetation mapping. Also, research into alternatives to per-pixel classifiers (e.g. probability mapping, fuzzy logic, mixture modelling) should overcome the problems of trying to classify a continuum, such as semi-natural vegetation. Equally, new sensor systems will give improved spectral and spatial resolution, for example Landsat 6 and SPOT 3, while radar data from ERS-1 has untested but potential uses in coastal and marine conservation.

References

Aspinall, R.J. and Veitch, N., 1991, Modelling the Distribution and Abundance of Breeding Moorland Birds, Nature Conservancy Council, Chief Scientist Directorate, Commissioned Research Report No. 998(2), Aberdeen: Macaulay Land Use Research Institute.

Belward, A.S., Stuttard, M.J., Taylor, J.C., Bignal E., Matthews, J., Curtis, D., 1990, An unsupervised approach to the classification of semi-natural vegetation from Landsat Thematic Mapper data—A pilot study on Islay, *International Journal of Remote Sensing,* **11**(3).

Grenon, A., 1989, Peatlands and Remote Sensing—Inventory of Peatlands in Quebec, Service de la cartographie, Centre Quebec de Coordination de la teledetection, Ministrie de l'energie et des resources secteur tries Sainte-Foy Quebec.

Hunting Technical Services, 1986, Monitoring Landscape Change (10 vols), London: Final Report to the Department of the Environment.

Nature Conservancy Council, 1987, Changes in the Cumbrian countryside, First report of the National Countryside Monitoring Scheme, Scotland: Nature Conservancy Council (Research and survey in nature conservation No. 6), 39.

Nature Conservancy Council and Countryside Commission for Scotland, 1988, National Countryside Monitoring, Scotland: Grampian. Perth: Countryside Commission for Scotland and Nature Conservancy Council.

Palylyk, C.L. and Crown, P.H., 1984, Application of clustering to Landsat MSS digital data for peatland inventory, *Canadian Journal of Remote Sensing,* **10**(2).

Robertson, R.A. and Jowsey, P.C., 1968, Peat resources and development in the UK, *Proceedings of Third International Peat Congress,* Quebec: International Peat Society, edited by C. Larfleur and J. Butler, pp. 13–14.

Williams, J.H., 1987, Upland vegetation classification in Snowdonia using Thematic Mapper data, In Weaver, R. (ed.) *The ecology and management of upland habitats: the role of remote sensing,* Aberdeen: University of Aberdeen Press.

Part VII

Inputs into GIS

34

GIS: The data comes first

David Parker
University of Newcastle upon Tyne

In the past few years many organizations have been attracted by the potential of Geographic Information Systems. A significant number of those early to take positive steps to introduce the GIS approach to the handling and analysis of their spatial data have become somewhat disillusioned. They have become disillusioned not with the concept or the technology but with the problems of suppling their data needs.

To review the problems faced it is worth distinguishing between users of spatial data. At one end of a spectrum there are those who are responsible for collecting and maintaining data sets either purely for their own internal use or as a service to others. At the other end there are those who rely totally on datasets compiled by others. There is obviously a continuum of uses and users within this range; those who, to differing extents, mix their own in house data with that sourced from outside their organization.

Of the organizations who mainly collect and maintain their own data, those who have most readily accepted a GIS approach are those who started with no established dataset, even in paper form, and had no established procedures to maintain records. They faced very few data related problems. Another similar category of organization who have also relatively readily adopted a GIS approach are those charged with maintaining a dataset but one which has fallen into disarray. They look to GIS as a tool to assist them in recreating an up-to-date and hence usable record. The data problems they face are mainly those of data conversion. They have the incentive to overcome any data conversion problems since they have the justification of the need to improve their data stock. Organizations who have well-maintained spatial data records and working procedures to maintain them face the problems of data conversion but with less justification to go digital. Their justification must come from predicted cost savings and improved usability.

At the other end of the spectrum are those organizations reliant on datasets supposedly captured and supported by others. The greatest success appears to be with organizations whose data requirements are limited in quantity and extent and accessible largely in the public domain. The significant factors here are that they have minimal data conversion problems but more importantly avoid the problems of data ownership and all the problems associated with the rights to use data. At this end of the user spectrum, GIS is often being used as a project orientated tool. Little real success has yet been shown in exploiting the potential for combining datasets to create new products and services.

Key issues

The data related issues can be divided into two broad groups. For those who employ GIS techniques to create and maintain their own data sets, the main issues are those of initial capture and continued maintenance. These tend to be the more technological problems. For those reliant on external datasets the issues are more varied, rather more institutional, and hence somewhat harder to resolve: the knowledge of what data are available, from whom, in what state and under what conditions; the confidentiality in which some data is held; indexing data within datasets to enable synthesis; the right to use data; the right to income for data suppliers.

The chapters in this part discuss many aspects of the current problems in the supply of data for GIS applications including the role that leading organizations are taking to help solve the problems. The hope is to give encouragement to those who see the potential for GIS in their organisation but shy away from action having observed others scratching around for appropriate data. First, where should we be headed?

Peter Dale in his paper on Domesday 2000 at AGI '91 (1991) sets out his vision of a National Land Information System for the UK. The system will detail land ownership, land value and land use. The concept is of a computer information network linked at nodes to databases of relevant land information maintained by the specialist agency charged with that task: the Land Registry for example. Enquiries would be routed via a hub which would extract data as required equitably reimbursing data suppliers for information accessed.

There are many bottle-necks in the implementation of such a system but not as many as at first sight and none insuperable. The major hurdle is in achieving critical mass; until sufficient data is in the system there will be little demand for access. The concept, however, is extendible to all data, spatial and non-spatial. With support from all factions Domesday 2000 can become a reality and set the standards for the way we access all our required data. Unfortunately this is a little in the future but a lot can be done immediately to ease the way towards the goal.

Institutional issues

To help alleviate the difficulty in knowing what data is available and from whom, this part includes a directory of data indexes. The ideal would be an index by dataset but experience has shown that it is currently too difficult to compile and maintain a catalogue embracing all data for all application areas. The directory is a compromise, pointing to organizations who maintain indexes of datasets from specific sources or for specialist application areas.

A major stimulus to the airing of the issues in cataloguing and making more accessible spatial data is the activity surrounding the Government's Tradeable Information Initiative. This aims to stimulate the British information industry by creating opportunities for the development of value added information data services and value added networks supplied with raw data from government sources. The chapter in this section by Roy Haines-Young details the aims of the initiative, what has been achieved to date and what more needs to be done.

One of the achievements is the cataloguing of all spatially indexed government held data. The catalogue is held by the Association for Geographic Information and is referenced in the directory of data sources previously referred to. There is still some way to go even for this directory. The completeness of the catalogue is questionable and the arrangements for updates uncertain.

The initiative is highlighting the issues. On the negative side for example, more than 35 per cent of the datasets listed are categorized as confidential and because of the charging regime there is little incentive for government departments to actively market their data. On the positive side though, it is being realized that improving the access, integration and structure of the datasets can make them much more versatile in their use and hence much more valuable.

One example of centrally collected data that can clearly be adapted to create added value is that of the 1991 census. All indications are that the market for this dataset will be much larger and more diverse than for the previous one. The outputs and opportunities for use of the new census are detailed in Keith Dugmore's chapter in this section. The chapter provides an overview of the census content and references to where further information can be found. The awareness of the potential of GIS is clear. Four of the basic six output types are to be provided in computer readable form offering considerable opportunities to GIS users. Concurrent with the work of compiling the census statistics are the initiatives to produce an integrated set of digital administrative boundaries; at last, unique cross-referencing of enumeration districts, wards, parishes, districts, counties and constituencies appears feasible. Creating even greater long term potential are the additional datasets that can be combined with the census statistics. There are four broad types: earlier censuses, administrative data, sample surveys and modelled estimates. Details of some specific types are quoted in the chapter.

The two key issues of the right to use data and the right to income for data suppliers are affected by one and the same problem: that of legal copyright. Data is a commodity and the supplier of data is due a legitimate income for its supply. All this income need not come directly from the recipient of course. The supplier could be subsidised for the service of supplying such data, as is the case with Ordnance Survey and many other government administrative datasets. There are those who argue strongly for the total subsidy of such information, placing the data in effect in the public domain. As we are all aware however, the trend in central funding for data suppliers has for some time been in the opposite direction: the subsidy levels have been reduced rather than increased and suppliers are heading for total cost recovery. There are of course many organizations in the private sector for whom this has always been the case and they must welcome this trend as it will enable them to compete sensibly in the open market.

It is the law of copyright which controls the rights of a data supplier. This law is far from straightforward, as will be appreciated by reading Andrew Larner's contribution, "Digital maps: What you see is not what you get". The law basically protects the supplier's rights, hence permitting them to control by licence the use made of the data and the income received for its authorized use. As the interpretation of the law stands, it is the copying of digital data that is controlled, but what constitutes taking a copy? Is each new screen view of a dataset a copy?

What is required in the end is a mechanism for charging. Andrew Larner's chapter distinguishes between an adaptation, an indirect copy and a direct copy. No matter what type of copy is taken however, if this is of a "material form" and there is no "implied licence" to do so, then a charge is due. The opposite of this situation is that if the copy taken is not of a "material form" or implied in the licence as being allowable then no charge is due.

The cards are not all in the hands of the suppliers. A recent legal precedent established that a supplier's copyright may be infringed where it is restricting to a user's right to "enjoy" data that has been purchased.

Technological issues

Mention the problems of data supply and the majority of those who have recently investigated GIS for an application based on large scale mapping and thoughts turn to the mass conversion to digital form of paper-based map and text records. Whether or not this conversion is still really a problem must depend on the quality of the existing paper-based records, the level of intelligence required of the digital data and how much there is of it to convert. These issues are discussed in Jane Drummond's contribution, "Raster to intelligent vector: Current status". There is little problem if an image of the existing paper records is all that is required. For the majority of GIS applications based on large scale data, however, considerable effort is required to achieve "intelligent vectors": structured, featured-coded and attributed-coded. Over the past decade there has been hope for the automatic creation of intelligent vectors from raster scanned data. This appears still to be a hope rather than a reality. Given a suitable map, for example a river network overlay where the level of detail is limited, then fully automated intelligent vectorization appears possible. Given the more typical map sheet which may well have on it a great wealth of information in a range of line styles, symbology and text and the level of intelligence achieved automatically is limited.

For paper records unsuitable for complete automated processing, a semi-automated or "toolbox" approach has developed. As the name implies, the tools of raster vectorization such as line following and pattern recognition are made available to an operator who controls and assists the process. On-screen digitizing is still possible if the tools prove inadequate.

Jane Drummond concludes that a fully automated approach to intelligent vectorization is probably unnecessary. The techniques currently available and the labour of the digitizing bureaus of Eastern Europe and China may be sufficient for the finite task of data conversion.

The priority currently afforded to the problems of large scale spatial data conversion is overshadowing the long term problem of the checking and maintenance of the prepared dataset. Checking and updating can require field survey. Hugh Buchanan, in his chapter in this section, looks at recent technical developments that will lead to more rapid field data collection in the years to come. He divides the developments into three main categories: theodolite developments, satellite surveying and pen-based computing. He reviews them considering that the task of maintenance can be subdivided into the upkeep of the map base and the upkeep of the associated attribute information while maintaining its spatial reference. The first two developments make it feasible for field surveyors to work alone and still determine their position. Positional accuracies of up to 10mm are achievable with lesser accuracies of course possible. The satellite surveying technique appears to have the greatest long term potential for GIS. It is not without its problems however; there are significant problems in relating positions determined in the satellite co-ordinate reference system to that of Ordnance Survey mapping.

Field surveyors are in an ideal situation to collect and update attribute as well as positional data. They have traditionally collected only minimal descriptive data such as feature codes but with position determination becoming simpler and with recent developments in portable computing there is opportunity for a wide range of attributes to be collected. Pen based computers seem to offer the greatest potential. With all the developments reviewed, the correct positioning and data collection technology, a single surveyor can be charged with the update of both the map base and the associated attribute information. Such an approach has appeal.

What of the update of a digital map base and attribute information from the wide range of imagery available: aerial photographs, both vertical and oblique, paper and digital, and all other remotely sensed imagery? The issues are discussed in a recent article by David Tait (1991). To detect change, delete the old and add the new, there is a need for comparison between existing map data and the imagery or, better still, a stereoscopic model from that imagery. This can be accomplished if the existing map is superimposed on the stereoscopic model. Facilities must exist to 'warp'' the existing mapping to the model or vice versa and for digital output in the map base co-ordinate system and data model. Instruments with this functionality exist but are generally optimized for new map production and not digital map data revision. There is a place for cheaper, less precise but more adaptable equipment designed specifically for the revision task to come.

Key players and data integration

What of the key players, those who hold large quantities of relevant data and whose activity can provide the catalyst for many other organizations? In large scale applications the key players are the Local Authorities and the utilities. Both these groups, but especially the utilities, were instrumental in accelerating the digital large scale conversion programme of the OS. However, activity in the utilities since this time has appeared limited. Many have invested in GIS technology and are converting paper-based records to digital form but the level of activity is often more appropriate for a trial and not a full-blown enterprise.

A stimulus has been given to the GIS activity of the utilities by the recently enacted ''New Roads and Street Works Act of 1991''. Two sections are of particular relevance, those on the Street Works Register and the Records of Location of Apparatus. Full details of the implications of the act for GIS are discussed in Chapter 4 of Part 5 by Fred Johnson and Kevin Crawley.

As a result of the Act, a major step has been taken in improving the ability to integrate information held in all utility and other applications. This is the definition of a standard for a National Street Gazetteer: ''An index of streets, their addresses and geographical locations, created and maintained at District Authority level''. The definition of a street is very broad—from motorways to back lanes and tracks. At the time of writing, the working party is working rapidly on the best approaches for its implementation, financing and policing. The activity in the local authorities was slow to start but the interest in recent conferences and other forums has shown that awareness is now high and activity is predicted to rise rapidly with the ending recession!

A major stimulant to local authority interest in data collection and aggregation will again come with the definition of a standard. This time a standard for a National Land and Property Gazetteer. The basic unit of all local authorities, and many other organizations, is the land parcel. The standard defines the Basic Land and Property Unit (BLPU) and the requirements for a national gazetteer. The two gazetteers are closely linked. The land and property gazetteer very sensibly relies on that of the street gazetteer. Progress with the land and property gazetteer is slightly behind that of the street gazetteer. The specification is about to be piloted while the working party puts together a business case for its implementation. It had been planned to include in this section a full status report on both the gazetteers but the high level of activity and the publication delays made this profitless.

The two specifications are destined to follow the same route: to become British Standards. We must all support their adoption, implementation, policing and use as this will give a great impetus to data collection and integration for the key players and hence many other organizations.

GIS: The data issues come first

Maybe the title of this chapter is not quite correct. "The data issues come first" may have been more appropriate. If all the problems in data sharing can be correctly solved then data will be captured, converted, maintained, traded, aggregated and analysed to the benefit of all suppliers, users and society. GIS will have come of age.

References

Dale, P., 1991, *Domesday 2000, in Proceedings of the Association for Geographic Information Conference,* West Trade, AGI '91, paper 3.21.

Tait, D. A., 1991, Instrumentation requirements for modern map revision, *Photogrammetric Record,* **13**(78), 901–8.

35

The Tradable Information Initiative

Roy Haines-Young
University of Nottingham

The "Information Revolution" is having many consequences. An important one is the way it has turned something as intangible as "information" into a "commodity"—that is, something which can be bought and sold. This chapter will focus on some of the ways in which we, as a society, are coping with this change.

Our particular concern is with data held by central government and its agencies. For, while it is clear that central government is aware of the value of its information, it is by no means certain that the steps it has taken to foster the "market", through its Tradable Information Initiative (TII) are adequate in either their pace or scope. These issues are vital, for as many have argued (e.g. Moore and Steel, 1991), central government attitudes are critical to the success of the information services sector in the UK.

The importance of government policy and attitudes towards its data were also recognized by the Chorley Committee in their review of the barriers to the better use of geographic data in the UK (DoE, 1987). The aim here is to examine how things have changed over the last five years and to focus on what remains to be done.

The Tradable Information Initiative—What is it?

Government policy towards tradable information is described in a Department of Trade and Industry (DTI) publication which sets out some guidelines for government departments in their dealings with the private sector in the area of tradable information (DTI, 1990). The Initiative was already in place at the time of the Chorley Review, having grown out of a 1983 report to the Prime Minister by the Information Technology Advisory Panel (ITAP) called "Making a Business of Information" (Garnsworthy, 1990). Among its recommendations the ITAP Report had argued that:

> Government should recognise the current economic significance of the tradable information sector, and enhance the opportunities for future growth and take its interests into account in policy formulation (para 6.5).

It went on:

> We would give prominence first, to the stimulation of the UK Industry through the release of Government-held information, and secondly to the protection of the commercial value of information (para 6.13).

The TII was the government's response to these recommendations. According to the latest published guidelines, the goal of the policy on tradable information is: to supply Government information to private sector concerns who create public access to information retrieval service as the result. (DTI, 1990, p.6).

In drawing up the guidelines three type of trading activity are identified as relevant for consideration:

1 information processed and handled by one government department for one purpose and resold to the private sector for re-use in the same context;

2 information collected by the government for one purpose but resold to the private sector for a completely different purpose; and

3 information collected by the government which is processed by a private sector organization prior to re-use within the government.

The guidelines suggest that it is with the first two types of activity that the private sector organization is acting as an information marketing concern who can add value to information and sell it on to end-users. It is to these areas that the TII mainly applies. In contrast, the type of relationship with a bureau service envisaged in (3), is not regarded by government as "value-added trading" and these activities lie outside its scope.

Moore and Steel (1991) suggest that the information services sector generally is growing more rapidly than the rest of the UK economy. In a recent survey (*EP Journal,* 1990), the turnover for the nine major European Community partners was estimated to be about ECU 2.5m, with the UK having the largest share. The major market sectors were those relating to finance and business data. The survey also shows the UK to be an important exporter of these service areas. While the turnover of the sector dealing with spatially referenced data is small by comparison, it is likely that it could grow given the right market conditions. Policies affecting this active business area are likely to be vital to our long term interests.

The Tradable Information Initiative—What has been achieved?

Twelve of the 64 recommendations contained in the Chorley Report dealt with the general issue of data access and the Tradable Information Initiative. Since these recommendations are well known, we will confine our comments to the government response in these areas.

To a large extent the government response in the area of data access has been built around the formation of a Tradable Information Inter-departmental Working Group (TIIWG; Garnsworthy, 1990). The Working Group is managed by the DoE in conjunction with the DTI. The Working Group aims to facilitate the wider availability of government-held spatially referenced information to those outside government. The group is meant to reflect the government acceptance of Chorley's recommendations relating to the need for minimally aggregated data to be made available on a standard basis.

To date the major deliverables from TIIWG have been: to agree that common "directories" (such as the Central Postcode Directory) should be developed and used to link data sets with differing minimum levels of release at the "lowest common denominators" of the particular data sets involved, with grid references as an essential geographical element; and, to collect detailed information about data holdings from all government departments. The collation of information about Central Government data holdings is perhaps the most significant result of the work of TIIWG. The "information about information" was collected via questionnaire on which the Association for Geographic Information (AGI) was able to comment. The returns allows over 450 datasets to be described in terms of 22 characteristics ranging from simple information such as its name and owner to more complex attributes such as the purpose underlying collection, geographical coverage, updating arrangements, limitations, locational referencing system used and nature of the basic spatial unit. Details of any conditions attaching to the wider use of the data have also been recorded. Under present arrangements these data will be collated and stored in a database and made available through the AGI. The aim is to promote wider knowledge of what government-held data are potentially available.

Despite progress in generating information about government-held geographical data the initial survey conducted by TIIWG leaves two important questions unanswered:

1. Although the number of datasets recorded was significantly in excess of those in Appendix 3 of the Chorley Report, with some departments describing additional datasets over and above those listed in 1987, returns from other departments were lacking. In order to make the survey as wide ranging as possible, the questionnaire survey conducted by TIIWG took in other public sector organizations such as the Natural Environment Research Council, English Nature and the Countryside Commission. However, since the survey was conducted on the basis of a single request "for information" to each department and agency, we have no way of knowing how complete the listing is. If the survey is eventually to provide the basis of a valuable resource then the coverage of the listing needs to be checked and arrangements made for updates and additions.

2. A large number (37 per cent) of the datasets were reported as being in "category D". That is they are unavailable for reasons of confidentiality. Since the decision was made by the department alone, the status of this category needs to be clarified and a standard approach to its application maintained.

Neither of these questions undermines the value of the survey, however, and to a large extent they can only be answered once the database being constructed by the AGI comes "on-line" and is widely available and used. On its own, however, the database is unlikely to release a flood of government data or stimulate much activity without additional steps being taken. The final part of this review will focus on what more is necessary.

The Tradable Information Initiative—What more needs to be done?

In looking to the future, three issues stand out as being of major importance if further progress is to be made in the area of developing a trade in government-held information. These concern the matter of cost and marketing arrangements, the institutional framework in which we find ourselves, and the levels of education and awareness amongst users and suppliers of data.

Cost recovery and market intervention

At the time the Committee took evidence, Treasury and Central Statistical Office (CSO) rules required government departments to charge what the market will bear for data of commercial value, although from 1986 on departments could charge on a marginal cost basis for a limited period where data have not previously been exploited. As a consequence the Chorley Committee recommended that Treasury restrictions on running costs should be lifted where departments could show a net return, and that charges should be at marginal costs and only at the higher rate if the market will bear it. They went on to recommend that greater use of franchising agreements should be made so that outside bodies could act as distributors of government-held data.

While TIIWG is consulting CSO about the marketing and charging arrangements for government-held statistical data, five years on, it seems that no clear guidelines exist. Perhaps most seriously, the major obstacles to releasing data by government departments, namely, the Treasury rules which require such activities to be contained within the ceilings of staff and gross running costs for each department, remain in place. Departments are still required to pass any income from trading information back to the Treasury even when the extra costs incurred in supplying the information could be covered by income generated. As the Chorley Report noted in 1987, even if Departments could meet the requirements, there is little incentive for them to do so. The same broadly remains true today. As a consequence the pace of the TII has been slow.

Progress in this area of charging is difficult because there is a major ambiguity, it seems, at the heart of government policy (cf. Ricerca Internationale, 1991). On the one hand, the philosophy of government action over the last decade has been strongly against intervention in the free market economy—the market must look to its own needs—and on the other hand, services or giving some public services "agency status" and the duty to operate at a profit. Yet the balance struck between contracting out and operating the agency concept has direct market consequences.

If government agencies seek to reach their customers through a series of private sector intermediaries, for example, then this would clearly stimulate the information services market. If, by contrast, they choose to reach the market by forming their own partnerships in order to maximize their returns and operate efficient cost-recovery procedures, then this could have the very opposite effect. The recent announcement that Companies House in conjunction with Mercury would offer an on-line information service (*Government Information Trader,* 1991) illustrates just how the tension between a government agency and the needs of the private sector can arise. Whether they like it or not governments do intervene in markets, even indirectly—and government policy needs to acknowledge that fact.

The extent to which data generated at the public expense by all should be available to a few at less than its true cost is an issue which will not easily be resolved. The extent to which such questions will affect the development of the market for geographical information remains to be seen as key agencies such as the Ordnance Surveys and Land Registries develop and widen their activities. The role of the AGI in such matters is:

1. to press for a clearer articulation of charging policy in relation to the stated goals of government concerning the stimulation of the information services sector, and argue for greater incentives to be given to government departments and agencies to make data available more rapidly;
2. to ensure that charging policies receive the widest public debate and are appropriate to all sectors of the information market; and, more generally,
3. to monitor the evolving market in government-held information and ensure that conflicts between market requirements and policies towards government agencies do not work against the wider public interest.

Institutional frameworks

Many have argued that many of the problems of access to government-held information could be resolved by some kind of freedom of information (FOI) legislation in the UK. Under such legislation the law would protect only those data where it can be shown that access would breach national security, represent a danger to the public interest or constitute an unjustifiable invasion of an individual's privacy. As we have noted above, a number of developed industrialized countries have such legislation. It is instructive to consider their experience.

Freedom of access under FOI legislation to data does not, of course, mean freedom from charges. Even in the United States, where access arrangements to government-held information are among the most liberal, requests for commercial purposes are charged the full cost of the document review (Birkenshaw, 1988; Rhind, 1991). Although these charges may be less than full cost recovery, they must be taken in conjunction with the other "hidden costs" of any freedom of information legislation which will arise out of the additional staffing and administrative machinery necessary to make it operative.

The extent to which the overheads associated with FOI legislation are worth paying are a major issue. Certainly they go far beyond our immediate concerns with the availability of spatially referenced data. Even if such legislation is eventually enacted, experience in other countries suggests that it is unlikely to provide all that is required. Legislation also has to be accompanied by a wide ranging change in constitutional culture, institutions and practices if data are to become more freely available. In the UK we can see something of what might be required at the level of local government.

Birkenshaw (1988) has argued that the local government (Access to Information Act) 1985, is essentially a freedom of information act at local level. Despite its existence, however, authorities still have to have detailed duties and powers assigned to them by public and private acts of Parliament and statutory instruments to ensure data are available in appropriate forms (Birkenshaw, 1991). In the context of geographical information, recent legislation relating to registers of contaminated land and public streetworks are examples.

There seems little reason not to extend similar rights of public access in a piecemeal fashion to particular datasets held by central government as and when appropriate, even in the absence of FOI legislation. In fact this process of broadening public access to government-held data is already underway in the UK, not withstanding the Tradable Information Initiative. The UK government has, for example, accepted an EC Directive which comes into force in 1992, requiring most environmental information held by public authorities to be made available to the public at reasonable cost (Birtles, 1991). In support of the initiative the recent White Paper on the environment (DoE, 1990) states that the "Government will propose arrangements, including registers, by which environmental information can be made available in Britain". The stated position in the White Paper is that public debate must be based on sound information and a thorough analysis of the risks, costs and benefits of proposals. The document accepts that the government can only help people act in a "balanced way" by promoting good scientific research and monitoring and that this requires, the "clear and wide dissemination of the results of research and other environmental data".

These arguments in the environmental arena have moved on, it seems, from issues of simple data access, to arguments about the scope, the terms and the conditions under which access is allowed. It also suggests an acceptance that there are issues of data availability which go beyond commercial ones, to the very nature of public debate and democratic government itself. What more needs to be done?

1 Organizations such as the AGI should use their expertise to act in a strategic way, to identifying those areas where access to government-held data is both appropriate and necessary, and to suggest how access can most effectively be achieved. In particular the AGI should ensure that access arrangements through the TII are not dictated solely by commercial interests but extend to the provision of the data necessary for informed public debate;

2 Advise and monitor the arrangements made by central government for access to data to ensure that information is available in a timely and useable form; and

3 Develop, monitor and encourage standards of quality assurance in the public and private information service sectors dealing with geographical information.

Education and awareness

The Chorley Committee saw lack of awareness of what data were potentially available as a major barrier to their wider use. Clearly, through the formation of TIIWG we are seeing a basis for the better dissemination of information; this is only the first step. Once again, experience overseas illustrates what more needs to be done, over and above cataloguing and "making available" government-held data. Birkenshaw (1988) describes the situation in Canada, where the Information Commissioner's reports describe some of the realities of the operation of their FOI legislation. The Commissioner criticized the absence of publicity or explanations of the act, and the lack of funds available for these activities, observing "the FOIA cannot be improved upon nor can it serve the country well if its very existence is kept secret". On poor response to the facilities of the act, Birkenshaw goes on to describe the gloomy tone of the 1985-86 Commissioners' Report which noted that the most frequent users were journalists, academics and researchers.

Although the TII and other piecemeal measures which make information more widely available are a long way from an FOI act, arguments about better public education still apply in the UK. The register of spatially referenced data held by central government will never develop unless those in central government are made aware of the value of the data they hold and the impact which access might have for society, nor will the potential users of those data become aware of the opportunities unless the existence of information is distributed widely. Thus, in the context of helping develop the Tradable Information Initiative, perhaps the most important steps which the AGI might take following the completion of the register of data held by central government are to:

1 develop as a matter of priority an education and awareness programme which describes the contents of the register, the importance of the cataloguing data and the uses and opportunities which arise out of better access to information; this material should be targeted both at the potential suppliers and users of data; and

2 identify those areas in which centrally held data are lacking either in content or coverage, and to focus on how these deficiencies can best be remedied most effectively in the public interest.

Conclusions

Few would now dispute the importance of the information revolution which is now upon us, nor can we doubt the importance of the role which government plays as one of the major collectors and providers of data. The government has recognized that information is a tradable commodity and that it can use its information resources jointly with the private sector to stimulate the trade in information products in a positive way, but much remains to be done.

Although the trade in geographical information is at present small in relation to the entire information services sector, many of the issues which apply to spatially referenced data apply elsewhere. Issues of cost, marketing arrangements, appropriate institutional frameworks and education are paramount for future progress in all areas. The AGI is uniquely placed to help focus on what is necessary and to help identify what can be achieved.

References

Birkenshaw, P., 1988, *Freedom of Information,* London: Weidenfeld and Nicolson.

Birkenshaw, P., 1991, *Government and Information,* London: Butterworths.

Birtles, W., 1991, The European Directive of Freedom of Access to Information on the Environment, *Journal of Planning and Environment,* July, 607–10.

DoE, 1987, *Handling Geographic Information Report of the Committee of Enquiry chaired by Lord Chorley,* London: HMSO.

DoE, 1990, *Our Common Inheritance—Britain's Environmental Strategy,* CMD 1200, London: HMSO.

DTI, 1990, Government-Held Tradable Information, London: Department of Trade and Industry, DTI/Pub259/2K/5/90.

EP Journal, 1990, Getting the Measure of the information services industry, *EP Journal,* July, 4–6.

Garnsworthy, J., 1990, The Tradable Information Initiative, in *The Association for Geographic Information Yearbook, 1990,* Foster, M.J. and Shand, P.J. (Eds) Taylor & Francis/Miles Arnold, pp. 106–108.

Government Information Trader, 1991, Service Profile: Companies House, *Government Information Trader,* **1**,1, Summer, p. 3–4.

Moore, N. and Steel, J., 1991, *Information Intensive Britain,* London: Policy Studies Institute.

Rhind, D., 1992, Data access, charging and copyright and their implications for Geographical Information Systems, *International Journal of Geographical Information Systems,* **6**, 1, January–February.

Ricerca Internazionale, 1991, Le Gestione Delle Banche Dati Nella Pubblica Amministrazione, *Seminario conti Dirigenti Generali Della Publica Amministrazione,* Stoccolma, 19-20 Settembre.

36

Raster to intelligent vector: Current status

Jane Drummond
International Institute for Aerospace Survey and Earth Sciences (ITC)

As in many professions, our paper archive is now becoming digital and its analysis computer supported. For us, however, conversion from paper to digital archive is made difficult because our symbolization does not use the limited character sets of other professions and varies from organization to organization.

Raster document scanners converting a map to a matrix of binary, greylevel, or colour-coded pixel values, have been around at least as long as GISs, and their potential for efficiently converting the graphic archive to a digital one early recognized. However the resulting data, although rapidly generated and suitable for graphic redisplay, is not thought suitable for many of the required analytical tasks; or at least not without vectorization. Vectorization starts with automatically acquired pixel matrices and the purchasers of conversion services or software expect it to end with vector representations of real-world entities suitably (i) structured; (ii) feature coded; and, (iii) attribute coded for processing through known and unknown forms of geographic analyses; this end product may be called a set of ''intelligent vectors''.

Because vectorization starts with a batch task (raster scanning) there may be an expectation that all subsequent steps can also be performed as batch tasks. This was so rarely the case, however, that organizations requiring to convert their paper archives to digital archives consisting of ''intelligent vectors'' usually opted for labour intensive manual digitizing and the keyboard insertion of feature and attribute codes. Formerly this decision was made easier by the high price of suitable resolution raster scanners, but now adequate desk-top scanners cost about the same as digitizing tablets. Perhaps the available processing tools can efficiently replace the human operator in creating the required set of ''intelligent vectors''?

As has already been said, an ''intelligent vector'' is structured, feature coded, and attribute coded. In the GIS context ''structured'' means that the digital representation of the spatial archive of concern is that which most efficiently contributes to the analytical procedures of the GIS. For example, if route finding is a necessary procedure, then the road archive should consist of road arcs (segments, edges) having nodes (ends) coinciding with accessible road junctions. On the other hand, if the GIS supports road maintenance planning the road archive also needs nodes where the surface material changes (e.g. asphalt/concrete), where bridges begin and end, etc. The structure required in the first application is symbolized on most countries' medium scale topographic map series—road junctions are graphically discernible so software procedures can be developed to discern them. The structure required in the second application is not symbolized on traditional medium scale topographic map series, although in some countries it is symbolized on larger scale series. What if all the other information found in large scale map series is not needed and anyway the spatial resolution is too high for road maintenance planning? Both spatial resolution and structure needs are related to application, but different map series contain different structural information.

If our archive is structured to support an analytical procedure such as route finding and all road arcs have nodes at accessible junctions, then a further enhancement may be referred to as "topological structuring". This requires that the connections between arcs and nodes are explicitly stated—e.g. that road arc 125 has node 872 as its start-node and 873 as its end-node, and that node 873 is also the start-node for arc 126 and 457, and the end-node for 456. This example situation is implicitly supplied by a traditional topographic map where a minor road connects with a major road. In GIS connectedness is one of the most useful branches of topology because we consider the Earth to be populated by a collection of discrete entities, some of which are connected to each other. This model of the Earth is inherent in topographic, cadastral and utilities mapping. Were we to consider a road to be a concentration zone in an asphalt continuum our model would be different, and our GISs might not need vectors.

Feature code and attribute code can both be considered to be part of the taxonomy used to provide (usually non-spatial) information about real world entities. Feature code assigns a real-world entity to a more or less specific class (e.g. building, or school-building, or primary-school-building), but in its digital representation a real-world entity may have one or more attribute codes which can (individually or in combination) be expected to be unique, such as the number of pupils, the school's postal address, the financial status of the school, the name of its head, or a unique identification number.

Maps traditionally do provide feature information, and the extent to which they do is found in the map legend. Larger scale maps may also show some attribute information, such as a unique identification number, a land-parcel's owner, or a ducting material, but usually attribute information is in other registers or archives which may be linked to the unique identification number.

Thus from the foregoing, if it is expected that through the process of scanning a map and batch vectorizing the resulting pixel matrix a structured, feature-coded, and attribute-coded vector data-base can be created, then disappointment may arise. The map to be scanned may contain no information on the required structure, or if it does then far too much other information; information will have to be added to or deleted from the end product. The map to be scanned will contain only the feature information shown in its legend; this feature information will be depicted graphically (linestyle, colour, weight, area or point symbolization) and the vectorization software will have to recognize these patterns. The map to be scanned will contain very little attribute information. If it contains unique identification numbers which link the map details to other archives, these numbers have to be recognized before any linking takes place. The recognition of numeric and alphabetic characters is quite well developed, but we have a tougher time than other professions because we may scale and rotate our characters to fit the space available—while others people keep them the same size and arrange them in straight lines!

Organizations require their data to be structured, and feature and attribute coded in a way that suits them; this reflects an organization's model of the earth. Without a "standard model" data-bases are custom-built and organizations must expect the conversion from map to data-base to be expensive. Some "standard models" do exist and are incorporated into official spatial data transfer formats. Notable are the ATKIS Landscape Model (Germany), DIGEST (NATO), GDF (the European motor and electronic industry), NTF (UK), and SDTS (USA); associated with these formats are definitions of real-world entities belonging to feature classes (Bruggemann, 1991). Structure exists as far as it is represented by these real-world entities (e.g. one format may treat the stretch of road between two junctions which has a change of surface material as two real-world entities but another format as only one), but only some of these formats support topological structuring. Attribute coding is supported via the option of storing unique identifiers for each recorded entity, providing a potential link to attribute tables. It is reasonable to expect data conversion at competitive prices to the level supported by the effective official transfer format. The cost of anything more "intelligent" is unpredictable, although with map conversion via scanning and vectorizing being from two to seven times as fast as manual digitizing it could be worth it. If we are fortunate we may have a good source map, in that the required structure and feature codes are represented clearly and near each graphically represented real-world entity there is a unique identifier depicted in a standard character font. Although good geometry has not been mentioned as a characteristic of an "intelligent vector", if the source map has lines of uniform width and density, with clean junctions and the scanning resolution is adequate—then the quality of the geometry will be higher than that achieved by manual digitizing, even when only using a PC-supported desk-top scanner. Without a map of suitable quality, effort can be invested in improving the map. Given a suitable map, the established vectorization procedures which follow scanning (i.e. sequentially skeletonizing, line tracking, segment merging, (topologically) structuring, and the associated procedures of varying sequence such as noise removal, node improvement, and area-pattern, line-pattern, and character recognition) may deliver the required "intelligent vectors", in batch mode—at least if we have good enough vectorizing software.

Even the cheapest PC based vectorizing software now available and costing about £300 provides correct skeletonizing, line tracking, and segment merging. Such software satisfactorily supplies input for DTM creation, if a contour separate has been scanned. If a network separate (e.g. rivers) has been prepared with deliberate small gaps at junctions and otherwise good line quality, then the resulting vectors are adequate input for the topological structuring procedures available in standard commercial GISs.

However, the cheapest vectorizing packages do not yet appear to support pattern and character recognition. Pattern and character recognition represent areas of gradual improvement—only being achieved by long established suppliers of cartographic conversion software. Given a suitable map and scanner, and good software we can expect "intelligent vectors" to be generated in near batch mode.

What if we do not have a suitable map or it would be too expensive to prepare one? The document we have can still be raster scanned to give a pixel matrix, and we will have a digital record for our archives even if it lacks "intelligent vectors". Accepting such a pixel matrix as a useful digital record has spawned an approach to vectorizing which began to emerge in the mid-1980s, and continues to improve. Sometimes called the hybrid approach, it is also more descriptively called the "toolbox" approach, the tools in the "toolbox" being those that have been independently used for map conversion in the past, and any new ones that come along.

Currently the environments supporting these "toolboxes" are PC or workstation-based. The scanned map is displayed on a colour monitor, and the "intelligent vectors" are displayed as they are created. Most of these tools can be described as semi-automatic. One such tool is semi-automatic line following, initiated by the operator indicating a start point of the raster depiction of the line to be subsequently automatically followed. Certain parameters of the lines to be followed—such as line-width and line-style, actions to be taken at junctions (straight on, branch left/right, stop) or at the end of the line (stop, search for other similar detail), and adjustments to be made in the shape of the line (e.g. smoothing) must be specified. Another tool ensures snapping of vectors to the co-ordinate of the centre point of a pixel in the pixel matrix permitting the subsequent development of clean nodes. Of course this whole procedure is incremental. The operator may only vectorize that detail needed for the data-base, and such a piecewise approach may meet the requirements of smaller organizations, or overcome the problem of excessive map detail.

When semi-automatic line following is not possible, or the required structure is not evident or cannot be determined automatically from the raster data then "on-screen digitizing" may be used. As with manual digitizing, the operator selects the features to be digitized from the display of the raster scanned map, and does so in point or stream mode. This technique is very similar to manual digitizing, but with the zoom capabilities provided by PC and workstation systems permits greater accuracy. On-screen digitizing is regarded as an improvement ergonomically over manual digitizing, and for this reason is sometimes advocated even when the rest of the tools in the "toolbox" are not available. Recent Dutch tests (Goel, 1992) have shown that "on-screen digitizing" scores over manual digitizing in terms of maximum geometric error, completeness, and speed of data capture (except for point features). Some of the vectors may have been created previously by manual digitizing or using a scanning and batch vectorizing procedure which had been found successful for some of the map detail, with the "toolbox" only being used for completion.

As they are being created the vectors can be processed as they would in any modern CAD or vector mapping system—that is their graphic attributes can be modified, they can be assigned to layers to provide feature coding, they can be linked to the tuples of the relevant data-base relations to provide detailed attribute codes, and these tuples themselves can be modified. This last task cannot be performed in any batch vectorization procedure!

As described above the "toolbox" approach began with a scanned map. Obviously it could have begun with a scanned (ortho)photograph or (ortho)image, and most of the vector data could have already existed. By overlaying new imagery with old vector data the tools in the "toolbox" can be used for database maintenance and revision.

Map conversion may be thought of as a finite task; one day all our archives will be digital and maps will only be one way of displaying spatial information, and no more a source of spatial data. A recent American survey (Montgomery and Schuch, 1991) predicts that by the end of the century conversion will be almost 70 per cent complete. According to this survey most conversion is now being done by manual digitizing, although the expectation is that more will be done using automatic approaches. One suspects that this American survey did not take into account the digitizing bureaux springing up in Eastern Europe and China where for not very much of our hard currency well-educated operators of digitizing tablets can gather very "intelligent vectors". To develop a completely foolproof batch approach which takes a pixel matrix and delivers "intelligent vectors" is attractive but perhaps rather academic. Probably using either low labour cost manual digitizing or the "toolbox" approach in one or two decades, we will have completed our conversion. Thereafter it will be maintainence and revision all the way, and we will be using a lot of the tools already in the "toolbox".

References

Bruggemann, H., 1991, ''Exchange Formats for Topographic-Cartographic Data'', Proceedings of the OEEPE Workshop on Data Quality in Land Information Systems, Apeldoorn, September.

Goel, S.P., 1992, ''Building a National Digital Database through digitization of 1:250 000 Maps: An evaluation of Alternatives'' ITC MSc, Thesis.

Montgomery, G.E. and Schuch, H.C., 1991, ''Technology Drives Future of Conversion Services'', *GIS World,* November.

37

Digital maps: What you see is not what you get

Andrew Larner
Association for Geographic Information

One of the characteristics of digital data is the ease with which it copies. Illicit copying of digital data deprives the data supplier of legitimate income. Suppliers have therefore turned to copyright law to protect their work.

The growth of computer technology has required that copyright law take into account new types of work and new ways in which technology copies them. However, computer technology, as well as making new and easier ways to copy, has made it difficult to distinguish between copying works and using them. Some data suppliers have tried to establish a link between the use of their data and copying. By establishing this link they hope to levy royalty charges for the use of their data in addition to the copying charges for which copyright law was devised. Central to the establishment of the link between use and copying is the belief that the process of viewing the data on the screen constitutes copying.

There are a number of ways that viewing the screen might relate to the legal definition of copying. This chapter explores these relationships in order to determine if viewing a digital map on a screen might constitute copying. It aims to assess the relative strengths and the implications of each interpretation of the act of viewing the digital map on the screen.

Figure 37.1. The view on the screen as a broadcast: charging to view the map

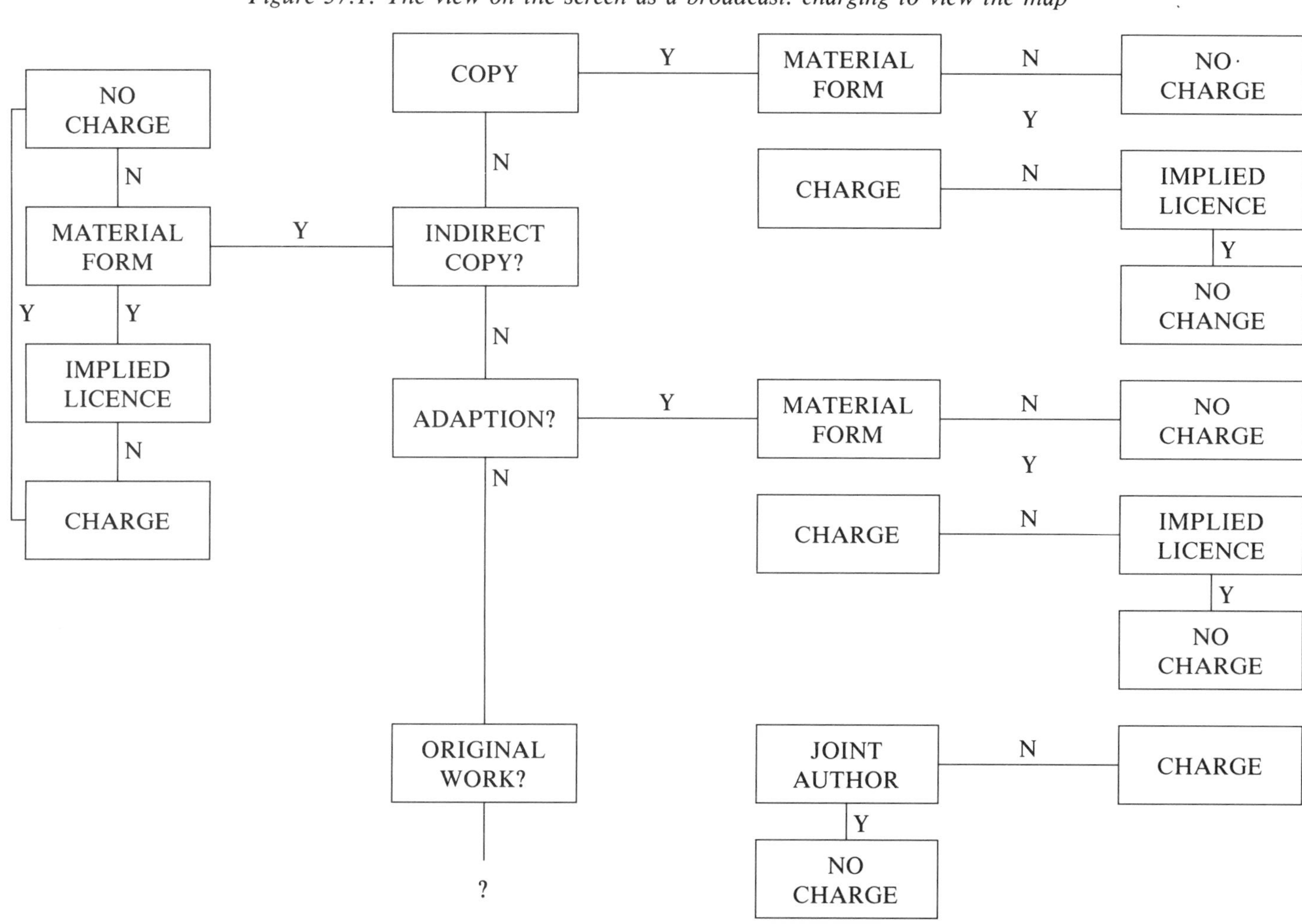

Copyright

Copyright is a property right. The law of copyright is one of a range of legislative provisions relating to intellectual property. Intellectual property has no agreed definition. However, one of its distinguishing characteristics is the fact that the act of copying the property directly deprives the copyright owner of the ability to make a living out of his or her investment. The Copyright Act therefore bestows upon the copyright owner the sole right to perform certain acts, referred to as restricted acts, with respect to the work. The enforcement of restricted acts effectively bars the purchaser from copying the work without permission.

The Whitford committee's report (DTI, 1977) gives an excellent example of the contrast between intellectual and other property. It considers the comparison of the baker of a loaf of bread with the writer of an article about baking bread. In the former case, the baker is protected from theft of the loaves baked. The purchaser is entitled to do with the loaf as he or she wishes. However, the baker does not have a monopoly on the baking of bread. The baker bakes his or her bread without "helping himself [sic] to the product of the skill and labour of his rival".

The author does not hold a monopoly in the writing of articles about the baking of bread. However, he or she does

> have an interest not merely in the manuscript, the words on paper which he [sic] produces, but in the skill and labour in the choice of words and in the exact way in which he expresses his ideas by the words he chooses. If the author sells copies of this article then again a purchaser of a copy can make such personal use of that copy as he pleases. He can read it or sell it second hand. If a reader of the original work is stimulated into writing another article about bread the original author has no reason to complain. But it has long been recognized that the original author ought to have the right to reproduce the original article and sell copies thus reproduced. If other people were free to do this they would be making a profit out of the skill and labour of the original author.

It is important to note that the owner of copyright may not be the author.

The Copyright Designs and Patents Act 1988

The 1988 Copyright Act (The Copyright Amendment Act, 1985) defines those types of work which are capable of protection by copyright. These are divided into different types of work. The paper map is an example of an artistic work.

Chapter II of the 1988 Act deals with "Rights of the copyright owner" [s16–27]. The rights of copyright owners include the sole right to perform "restricted acts" with respect to their work. Section 16 lists the restricted acts as:

(1) to copy the work;
(2) to issue copies of the work to the public;
(3) to perform, show or play the work in public;
(4) to broadcast the work or include it in a cable programme service;
(5) to make an adaption of the work or do any of the above in relation to an adaption.

Any of the above acts would constitute "copying" the work. The restricted acts are not mutually exclusive as indicated by section 16(e). All acts performed upon digital map data will be interpreted in relation to the definitions of these restricted acts.

The "digitized map" as a work protected by copyright

The Whitford committee in reviewing the law of copyright considered the digitized map as a particular example of a "compilation of data". It must be noted that the findings of a committee like Whitford are not binding in the courts. In certain cases recommendations of such committees have been ignored. The reasons given for this have been that the recommendations put forward are not all necessarily taken up in legislation. Also it is not possible to know if the legislators even intended to take up the recommendations of any particular committee. However, Glanville Williams maintains that the use of these recommendations as the basis for an argument in law will increasingly find favour (Williams, 1982).

Discussions held at the European level have come to conclusions similar to those of the Whitford Committee. In 1990 and 1991 the European Commission held hearings in connection with its Green Paper on copyright. These hearings have indicated that digital map data should be protected as a compilation of data. The compilation of data is a particular example of a literary work. Thus digital map data, while in some cases derived from an artistic work, is a literary work. This distinction is important since an alleged infringement of copyright distinguishing between the act of "copying" and "use" could differ according to the description of the work.

The view on the screen as a direct copy

The first of the restricted acts is to copy the work. To infringe copyright with this act the screen view must either be a direct copy of the work or indirectly cause such a copy to occur. Section 17 defines what is meant by "copying the work". It states that: "Copying in relation to a literary, dramatic, musical or artistic work means reproducing the work in any material form. This includes storing the work in any medium by electronic means." [s17(1)]. This section explicitly protects the copying of works into electronic form. Section 17(6) states that: "Copying in relation to any description of work includes the making of copies which are transient or are incidental to some other use of the work." This has been interpreted by some owners of digital copyright works as equivalent to viewing of the "work" on the screen.

It is likely that the view on the screen would be held to be transient, but this does not mean the view is a copy. By section 17(1) the view of the screen must be a reproduction of the digital data and be in material form. There are conflicting opinions within European Commission countries and within the UK itself on whether the view on the screen is in a material form (Metaxas, 1990 and Flint, 1990).

The 1988 Act does not define material form. Despite the *ad hoc* way in which the law relating to copyright has developed, the essence of copyright remains the protection of the right of a copyright owner to profit from his or her investment. This principle forms the basis for understanding what constitutes material form. In the case of digital map data the key consideration is whether the view created on the screen is in a form which will deprive the data provider of the ability to profit from their investment in those data.

Computer technology is now at the stage where it is effectively able to hand out copies. It is as easy to take a digital map on a portable computer into the field as to take a paper map (Cory, 1991). Where the digital map data is held on a network, more than one person may be able to use a single copy of that data at any one time. If the users are able independently to view a map so that they do not have to purchase multiple copies of the data then the transient view on the screen is part of a set of actions that is depriving the data providers of their profit. In the case of the portable computer there is an obvious and deliberate act of copying the original work before use. In the case of a network the copying is "hidden" from the user. The act of copying the data from one computer to another constitutes a restricted act since it reproduces the work by storing it in a medium by electronic means. Even though copying the data from one computer to another constitutes "material form" this does not mean that the map on the screen is a direct copy of the digital map data. The creation of the map on the screen from data does not in itself deprive the copyright owner of sales of that data.

Case precedent and the 1988 Act provide that the making of work in two dimensions, or involving a change from two to three dimensions, is in fact a copy—for example, the making of a statue from a painting. However, there is no provision in the Act which would suggest that the making of an artistic work from a literary work constitutes direct copying. The supplier provides digital map data, not digital maps. If the digital map on the screen is an artistic work and the data are a literary work then the screen view cannot be a reproduction of the data. Whether the internal working of the computer copies the data in order to create the view on the screen will be dealt with later.

The view on the screen as an indirect copy

It is important to consider, quite separately from the issue of direct copy, whether viewing the screen causes an indirect copy to be made. The strength of the argument for an indirect copy would depend upon the source of the data. If the digital data were derived from a paper map then indirect copying might occur. Insofar as the view created on the screen faithfully reproduced the equivalent paper map sheet then there could be said to be an indirect copy of that map. This would be a restricted act in relation to the paper map "work".

The view on the screen created from derived vector data should be readily recognizable as a map. However, it is unlikely to be the same as the map sheet from which the data was digitized. In creating the digital map data all the cartographic input has been stripped away. While the cartographic input needed can be easily combined with the data to create a map, this input is the original work of the software and the user. However, the screen view is closer to a reproduction of the paper map than to the digital data.

A somewhat similar area of copyright is that relating to video games. In this case the screen view and the software are part of the same product. The software contains data which are the parameters necessary to create a view on the screen. Since the game holds all the data and software necessary to create the screen view, the copyright of this view as an artistic work is a part of the copyright for the whole video game. Anyone creating the same image by copying the screen view or studying the software and parameters would therefore be infringing the video game's copyright.

The video game contrasts with the digital vector map. Unlike the digital map the video game does not have user input to create the image. The digital map requires the combined effects of user, software and data to create a view on the screen. The input of the user and software are original. The individual pans and zooms to select a view and can determine all the elements of style, such as line style and colour, that determine the cartographic quality of the map. For this reason it is questionable whether a screen view created from derived vector data constitutes an indirect copy of the original paper map.

Raster map data are derived from two sources, raster conversion of digital vector data and raster scanning of paper maps. In the case of scanned data, the view on the screen will be remarkably similar to the map from which the data was derived. It is probable that the only differences would be that the image on the screen would be in a different scale and be in "black and white". There may also be a change in orientation.

In this case there is a strong argument that the view on the screen is an indirect copy of the paper map. In the case of a map scanned by the user, the scanning forms a restricted act under section 17(2) since the data is stored by electronic means. The paper map supplier would therefore be able to prohibit this act or levy a royalty charge in return for permission to carry it out. With raster data scanned by the user, the map on the screen might easily be used as an excuse not to buy additional copies of the paper map. This would relate to the copying of the data between computers and possibly to each time the raster data was viewed.

It is possible, given the apparent reproduction of the paper map, that the screen view would be held to be an indirect copy. In this case it is necessary to know if the courts would allow the copyright owner to stop the legitimate purchaser of the data from using it for the purpose for which it was purchased. This might equally apply to the whole question of viewing the map on the screen and is dealt with later.

Does the screen view indirectly copy the data

There is the possibility that the computer, in using the data, makes transient copies internally—for example when copying the data from storage into the working memory. This was an issue that was considered by the Whitford committee. The committee was unable to come to a unanimous conclusion from its debate. Whitford did not consider that transfer of a program instruction by instruction constituted reproduction in the legal sense. The majority of the committee felt that this form of copying should be protected for the express purpose of protecting use of software, and by implication the use of data. However, the report also suggests that to follow this logic would mean that it would be an infringement to follow the itinerary in a guide book, or a recipe set out in a cookery book. As the report itself states, protecting use is a conflict with the idea that "... copyright does not protect ideas or schemes, only the forms in which they are expressed".

The United Kingdom is a party to the Berne Convention. This is the international agreement on copyright protection. To change the basis of copyright protection to cover "use" would affect our commitment to this convention and would be in conflict with existing UK law. It is likely that it was for these reasons that the 1988 Act makes a specific provision which allows the copying of the program incidental to its use. This act of transient adaption will be dealt with later.

Whitford suggests that the problems of software and data are the same. However, the 1988 Act, while making specific provisions for the internal copying of software, makes no such provisions for data. Therefore one must ask whether an internal copy of the digital data occurs and if so whether it is in a material form. The internal transfer of the data actually changes it to the extent that is should be held to be an adaption. If it were held to be a copy it would be difficult to argue that the copy was in a material form. Does the internal copying deprive the data provider of income? It is for this purpose that the purchaser buys the data. Rather than depriving the vendor, it actually creates income.

The view on the screen as a broadcast

The precise wording of the Ordnance Survey copyright regulations suggest that the OS might expect the view on the screen to be connected to a transmission. Paragraph 2.2 of the regulations states that:

> No Ordnance Survey (or Ordnance Survey-based) publication or other mapping or survey data may be reproduced, stored by electronic means, transmitted by any means (including being used on a visual terminal screen), or be translated into another form (e.g. hard copy) or language without the permission of the Ordnance Survey. All such acts are Restricted Acts under the Copyright, Designs and Patents Act 1988 and require a licence from the Ordnance Survey (Ordnance Survey, 1991).

By the terms of the 1988 Act the reference to transmission would have no connection with the viewing of the map on the screen. Transmission might, however, have a connection with the exchange of data between computers on a network.

The "transmission" in terms of the 1988 Act is a part of the definition of a broadcast [s15(d)]. The broadcast is defined in section 6 of the 1988 Act as a "...transmission by wireless telegraphy of visual images, sounds or other information...." Wireless telegraphy is defined in section 178 as the "... sending of electro-magnetic energy over paths not provided by a material substance constructed or arranged for the purpose". These definitions would preclude the sending of data down a network cable.

An act associated with broadcasting is the cable programme service. The connection of this act with the sharing of data on a network has been the subject of legal debate (Metaxas, 1990). The definition of a cable programme service [s7 and 7(a)] includes: "A service which consists wholly or mainly in sending visual images, sounds or other information by means of a telecommunications system ... for reception at two or more places (whether for simultaneous reception or at different users)". While this definition would mean that a network might come within the meaning of a cable network service there are exceptions to the definition which would effectively exclude the data network within an organization. This would not necessarily exempt databases being accessed outside the user's organization.

The view on the screen as an adaption of the digital data

It could be held, as described above, that the view on the map is a translation of the literary work into another form, or that the computer internally makes an adaption in creating the view. This would be covered by the Act as an adaption of the digital data (section 21). Section 21(4) of the 1988 Act provides that *"in relation to the computer program a "translation" includes a version of the program in which it is converted into or out of a computer language or code, otherwise than incidentally in the course of running a program".* Given the supposed similarity of software and data one might have expected the same provision to have been made for data.

One can argue that for the internal adaption this section already covers transient adaption data. In running a program the software and the data are mixed in a way that would make them indistinguishable. It was for this reason that Whitford felt that they should receive the same treatment in the eyes of the law. The view on the screen is a tangible product which is distinguishable from the software. In this case the map on the screen might infringe the copyright in the data. The data user would therefore have to argue that the view was an original work, unless the court held that there was an implied licence to use the data.

The view on the screen as an original work

The Whitford committee determined that most computer output would be works of joint authorship. In this case each author as the first owner of the copyright will have the rights to that view on the screen. The Whitford committee considered that ". . . the author of the (Computer) output can be none other than the person, or persons, who devised the instructions and originated the data used to produce the result". However, the "person who originated the instructions" could be the programmer, the person who runs the program or both. The 1986 DTI paper clarified this point by stating that the definition of author includes not only ". . . the creator of the program that controls the computer . . .", and ". . . originator of the data upon which the computer operates to create the new work ..." but also ". . . the person responsible for running the computer to produce the work". But which of the three parties, or combinations thereof, will be the author of the view on the screen?

Michael Flint follows the guidance given by the Whitford Committee. He denies that the programmer has an interest in the output since the software is a "mere tool". He goes on to argue that the balance between the interests of the user and the data originator depends upon the degree of skill and preparation of the data input and the extent to which the output is "computer generated". Computer generated means that there is no human author of the work which does not apply to the map on the screen. The user selects the "map section" its scale and orientation and sets defaults which can radically alter the cartographic style of the map. If this argument were upheld in court it might also undermine the copyright royalty attracted by maps printed from the data. One might even extend the argument to the extent that both the software and the data are "tools" which the purchaser uses in the way that Flint describes.

Case precedent

The edge between what constitutes a copy and what is an original work is very blurred. Copyright holds that there must be sufficient similarity in both the amount copied and the quality of that similarity. The courts rarely rely upon finding this dividing line; rather they refer to the political principles underlying the statute. These considerations would include the amount of effort expended on the new work, the economic return to the copyright owner of the first work, and whether the copyright owner holds a monopoly position. For this reason many cases fail to come to trial.

If, at the hearing to decide if there is a case to be answered, the principle of monopoly were raised it would be clear to the copyright owner that his or her case was unlikely to be successful.

There is no case precedent in the United Kingdom which covers the viewing of a map on a screen created from digital data. The above consideration of the restricted acts suggests that the court might either find that the view on the screen is an original work or an adaption, and in the latter case the screen view might infringe the copyright in the data. It was also stated that the court might hold that the purchaser of the data had an implied licence for its use. This would mean that despite the rights of the copyright the purchaser would be given the right to view the digital map on the screen without any royalty payment.

The decision in the case of BLMC vs. Armstrong Patents Ltd (1986) would suggest that an implied licence would be granted. In this case the House of Lords considered whether Armstrong had the right to reproduce spare parts for British Leyland cars without a licence, and whether BL's copyright in the design drawings was greater than the car purchaser's right to enjoy the use of the car and to ensure this by repairing it at the most economic rate.

The Lords held with a majority of four to one that the BL copyright was infringed but that the purchaser's right to enjoy the use of the car outweighed the consideration of BL's copyright. The second decision is of fundamental importance to the use of digital map data. If the data supplier sells data which the user will clearly use to create a view on a screen then they would have to reverse the principles on which the decisions in this case were made.

The maxim used in deciding that the rights to enjoy use were greater than those of the copyright owner was that also used in property law: "A grantor having given a thing with one hand is not to take away the means of enjoying it with the other": per Bowen L.J. in Birmingham, Dudley and District Banking Co v. Ross (1988). Even if this remark were considered to be obiter (a remark in passing), and not fundamental to the decision, it would have great weight with lesser courts.

The House of Lords' decision in BLMC vs. Armstrong Patents Ltd has far wider implications, relating to other types of use of data, than just the view on the screen. Consideration of use of data other than creating the screen view have been excluded from this chapter. Case precedent relating to the conflict between the free use of intellectual property and the copyright owners' interests will need to be revisited in looking at these other uses of data.

Summary

Where one person is looking at the same map independently of another thus making additional purchases of the map data unnecessary the copyright owner is being deprived of revenue. However, this is due to internal copying between computers. The software industry has taken this into account in their licenses which relate to the number of computers using the software.

A consideration of the 1988 Act alone suggests that there could be varying interpretations of the map on the screen. The most likely interpretations are that the screen view is an adaption of the data or an original work. As an original work the screen map could undermine the data providers' royalty from printed output. As an adaption a link between use and copying would be established and undermine purchasers' right to use the data.

From the users'' perspective the viewing of the map on the screen is no more than ''use''. It is equivalent to taking a paper map out of a cabinet and examining it before placing it back in the cabinet. If data providers force users to defend their position, they risk losing more than they stand to gain.

The precedent created by the case of BLMC vs. Armstrong Patents Ltd (1986) suggests that whatever the outcome of the debate, copyright protection of the view of the screen will not allow the data provider to bar the purchaser from the use of their data. It would be to the benefit of the data user and data provider for the transient adaption of data to be treated in the same way as that of software. This requires existing case precedent to be confirmed or, a change in the 1988 Act.

References

The Copyright (Computer Software) Amendment Act, 1985, London: HMSO.

Cory, M.J. (November 1991) ''Portable Data Capture'', in AGI 91 Conference Papers, The Association for Geographic Information, London.

Department of Trade and Industry, 1977, ''Copyright and Designs Law'', Cmnd 6732, London: HMSO.

Department of Trade and Industry (DTI), 1986, ''Intellectual Property and Innovation'', Cmnd 9712, London: HMSO.

Flint, M.F., 1990, *A User's Guide to Copyright,* third edition, London: Butterworths.

Metaxas, G., 1990, ''Protection of Databases: Quietly Steering in the Wrong Direction'', in *OPINION,* 7 EIPR pp.227-34.

Ordnance Survey, 1991, ''Digital Map Data, Copyright Arrangements—1991'' (Formerly OS leaflet No. 45), Southampton Ordnance Survey Copyright Branch, April (Currently under review).

Williams, G., Q.C., LL.D., F.B.A., 1982, Learning the Law 11th Edition, London: Stevens and Sons.

38

1991 Census: Outputs and opportunities

Keith Dugmore
MVA Systematica

The 1991 Census of Population provides an immensely rich source of demographic information, giving great detail of both geographical areas and subject classifications. Before reviewing the Census and the range of statistical outputs that are being produced, this chapter begins by illustrating the need for and value of geographical information. This is followed by a description of the Census outputs and of other datasets which are often combined with Census statistics to create valuable information, and the ways of linking them. The chapter concludes with some observations about recent past experience and possible future developments.

More and better decisions?

Slowly and erratically many organizations are making increasing use of geographic information to improve their decision making. Public services are using demographic information for small geographical areas to improve the targeting of their services, and private firms are using similar information to pick out their most profitable markets. Both types of organization are assessing the existing and potential demand for their services, and planning supply from point locations such as police stations or petrol stations, and the distribution of staff across territories, whether salespeople or social workers.

The scope for making better decisions is enormous. To take some of the types of questions typically posed:

- How many children are at risk within 10 miles of the nuclear power station?
- Where should we locate our next superstore?
- Which of our jeweller's shops should we close?
- What is the potential market for the leisure centre?
- Why do some local probation services spend more than others?
- How does water consumption vary according to type of neighbourhood?

- How can I target sales of smoked salmon by post?
- What sort of rubbish will the refuse collectors pick up in that neighbourhood?
- Where shall we target the fitted kitchen campaign?
- In what sorts of areas have rural riots occurred?

The last example, rural riots, provides a particularly interesting case. The analysis was carried out following widespread public concern about the riots that broke out in 1987. The Home Office undertook a research project, the findings of which were published in the report "Drinking and Disorder: A Study of Non-Metropolitan Violence" (Tuck, 1989). In brief, police information on incidents of disorder was related to local authority areas which had been classified by the Census Office using information from the 1981 Census (Craig, 1981). Types of area which suffered a high incidence of disorder were identified, and this provided the basis for subsequent survey observation. The study gives a particularly fruitful example of how census statistics can be used with other geographical information to review policy for targeting resources.

These, however, are questions which have actually reached the status of a defined problem. In many other instances the allocation of resources continues to follow precedents set in earlier years, and decisions go by default. The scope for taking a fresh look at resource allocation through the medium of geographical information is almost boundless.

1991 Census

The 1991 Census was carried out on 21 April by the Office of Population Censuses and Surveys (OPCS) in England and Wales, and by the General Register Office (GRO) in Scotland. Although there are some differences of approach between the countries there are many similarities. Geographically, England and Wales was divided into 110 000 enumeration districts, EDs, or for the administration of the census, and these form the basis of the output areas. In Scotland the Census was organized using postcodes, and these have been aggregated to give 35 000 output areas.

The Census form covered the usual wide range of demographic and housing questions. Compared with 1981, five new questions were asked:

- Central heating,
- Term time address of students and school children,
- Ethnic group,
- Limiting long-term illness,
- Hours worked per week.

A general introduction to the 1991 Census is provided by the Market Research Society (Leventhal and Moy, 1991). The Census Offices are producing a comprehensive series of documents including prospectuses and user guides.

Outputs: An overview

Users of census statistics are often initially concerned with two key questions:

- For what geographical areas are the statistics available? Some outputs are at the finest ED level, while others are produced for wards, postal sectors or only for larger areas.

- What subjects do the tables cover and in what detail? For some subjects all (100 per cent) of the responses have been coded; for other questions, which are more difficult to code, statistics are produced for a 10 per cent sample.

Other issues which then arise include timing of delivery, output media (paper? magnetic tape? disc? CD-ROM?), price (a small fraction of true cost) and copyright restrictions/transfer possibilities.

The 1991 Census outputs can be classified into six broad categories:

- Preliminary Report. The first report, published in July 1991, which gives preliminary counts of the numbers of people and households for local authorities and larger areas (OPCS, 1991).

- Local Statistics. These statistics, which include the Small Area Statistics and the Local Base Statistics, provide considerable detail for both small geographical areas and for subjects. Local statistics are produced in both computer readable and paper form. Two other categories of census product are closely related to the local statistics: maps and boundaries, which define and illustrate census areas; and the ED/Postcode directory, which enables users to relate census areas to the Post Office's postcode system.

- Topic Reports. These traditional paper reports on a range of census subjects are produced for large geographical areas such as regions.

- Special Workplace Statistics and Special Migration Statistics. Although categorized by OPCS as topic reports, these statistics are something of a special case, being for small geographical areas and produced on magnetic media.

- Commissioned Tables. The OPCS also produces tables to customers' own specifications of both subjects and areas.

- The Sample of Anonymized Records, which provides a sample of individual records without detailed geographical coding.

Of these outputs, four, which provide information for small areas in computer readable form, offer considerable opportunities to GIS users.

Small area statistics and local base statistics

Many users of the 1981 Census found the Small Area Statistics (SAS) to be the most valuable form of output. The SAS provide statistics about residents of EDs in considerable subject detail for both the 100 per cent counts and the 10 per cent sample. The 1991 SAS provide an unprecedented 9 000 items of information for each area.

In addition to the SAS, the 1991 Census also offers a closely related form of output, the Local Base Statistics (LBS). The LBS differ from SAS only in that they provide an additional 11 000 cells—making *c.* 20 000 in all—and are produced for wards of a minimum size rather than for EDs. The following paragraphs usually link the SAS and LBS together as SAS/LBS, drawing distinctions only where necessary.

Areas. OPCS are initially producing SAS for EDs, and SAS/LBS for the standard higher level of wards, districts and counties. Once the SAS/LBS have been produced for standard areas, OPCS will also offer output for the following areas:

SAS only: Postcode sectors
Civil parishes
Urban and rural areas

SAS/LBS: Sub-divisions of wards
District Health Authorities
Parliamentary constituencies
European constituencies
Standard and statistical Regions.

Tables. SAS/LBS provide an immensely rich set of data, giving numerous tabulations by the full range of census subjects. The LBS provides 99 tables of statistics, while the SAS subset totals 86 tables.

Media. OPCS supply SAS as A4 paper copy for a maximum of only five areas. Otherwise, supply of SAS/LBS is on magnetic tape or cartridge. Supply on other media such as diskette or CD-ROM has been considered but is not being used for the initial output.

Maps and digital boundaries

Users of census statistics for smaller geographical areas often wish to refer to hard copy maps to establish how the areas relate to the underlying pattern of houses and roads. There is also burgeoning interest in using digital boundaries with GIS to display and analyse the statistics.

Maps. OPCS is able to supply paper copies of both ward maps and ED maps. Maps showing ward boundaries are usually drawn on 1:50 000 Scale Ordnance Survey (OS) base maps, with one map for each local authority district. ED boundaries are drawn on OS 1:10 000 scale base maps representing an area of 5 × 5 kilometres. Ward and parish boundaries are also shown and named. The boundary information shown in the 1:10 000 maps is also available at the larger scales of 1:2 500 or 1:1 250 OS base maps in some areas.

Digital Boundaries. With the advent of much improved computing facilities in recent years, census users are increasingly making use of boundaries in digital form for displaying results, searching areas, relating datasets and calculating population densities. In Scotland, the GRO has produced digital boundaries. In England and Wales OPCS does not itself supply digital boundaries, but has encouraged initiatives to produce national boundaries sets. The OS, sponsored by the Department of Environment, has produced administrative area boundaries for England. The dataset includes boundaries for wards, parishes, districts, counties and constituencies, plus coastal high water marks.

There are currently two initiatives to produce national sets of digitized ED boundaries. These represent a great advance on the 1981 Census, when EDs were digitized on a patchy local basis. ED-LINE is produced by partnership between the London Research Centre, MVA Systematica, Ordnance Survey and Taywood Data Graphics. The ED-LINE boundaries are integrated with the OS/DOE ward and parish digital boundaries, and have Ordnance Survey Quality Assurance. ED91 is produced by Graphical Data Capture Ltd. The ED91 boundaries are digitized from the OPCS ED maps independently of the OS/DOE digital ward and parish boundaries.

ED/postcode directory

The OPCS is producing a directory which lists each whole postcode or part postcode in every ED. This is one of the major innovations of the 1991 Census. The directory is of great potential benefit to many census users. It has been produced in response to demands for improved methods of relating SAS statistics to postcoded addresses whether in individual list or aggregate form. The directory is arranged in ED order, and includes the number of usually resident households in each postcode/part postcode, and a grid reference to 100 metre resolution for each postcode taken from the Central Postcode Directory.

Special workplace statistics and special migration statistics

From a user's viewpoint, the Special Workplace Statistics (SWS) and the Special Migration Statistics (SMS), being for small geographical areas and produced on magnetic media, are more akin to SAS than to Topic Reports. Unlike SAS, however, they include information on movements between particular areas: residence/workplace in the case of SWS, and old address/current address in the case of SMS. In contrast to the 1981 Census, uniform sets of statistics are being produced for the whole of Great Britain.

Special Workplace Statistics. The census included questions on address of work and daily means of transport to work to be answered by persons in employment. The questions have been processed for a 10 per cent sample of census forms. The SWS can be a very valuable source of information to all those who are interested in the daytime population of areas—especially urban centres—and in the catchment areas of local labour markets. Apart from those cases where the workplace is at home, not fixed, outside Great Britain or not stated, each address has been postcoded for computer processing. Distance from residence to workplace is a measure of the straight line distance between the postcode of residence and the postcode of workplace, using the grid reference given in the Central Postcode Directory.

The 1991 Census allows flexibility in the specification of customer residence and workplace zones. The standard outputs are for wards or for postal sectors. It is also possible to request EDs for residence zones and postcodes for workplace zones. Production of SWS is scheduled to start in July 1993.

Special Migration Statistics. The Census included a question on usual address one year ago (i.e. 21 April 1990), to be answered by all persons aged 1 year or over. Responses to this question have been used in combination with information on the area of usual residence on census night to form a matrix giving numbers of people in households moving between areas of residential origin and destination. SMS are processed at the 100 per cent level only. OPCS plan to start the production of the standard SMS in March 1993.

1991 Census: Specialist analysis software

The 1991 Census provides a wealth of opportunities to users of geographical information systems ranging from the simplest mapping packages through to GIS software which has a comprehensive range of facilities. For the 1991 Census users will have the benefit of digital boundaries in a wide range of formats. However, for the census statistics themselves, there is the initial task of getting selections into other software systems.

The Census Offices are supplying SAS/LBS on magnetic tape or cartridge county by county, for all counties. Some users may decide to write their own programmes to make their extractions of areas and cells. Others will use the services of census agencies. Many organizations, however, have already decided that they want to have the capacity to load data into their own machines and have the ability to retrieve selections as and when necessary; such organizations have joined the SASPAC consortium.

SASPAC is a software package whose development had been funded by 180 organizations in central and local government, health and water authorities and the academic community. It is built on the success of a similar initiative for the 1981 Census.

MATPAC is the sister software to SASPAC, designed to read and analyse the origin/destination matrices recorded by the Special Workplace Statistics and Special Migration Statistics. While MATPAC enables users to select areas and variables, further analysis and presentation of origin/destination flows will provide considerable scope and challenges for a GIS approach.

Additional datasets

The range of other geographical datasets which can be combined with Census statistics is already wide and is continuously developing. The origins are varied. Some are from government sources, others from the private sector. There are four broad types: earlier censuses; administrative data in many forms; sample surveys; and modelled estimates, where datasets have been manipulated to produce new statistics.

Geographical levels, too, range from the individual person upward. Information may, therefore, be available for the person, household, address, postcode, enumeration district or higher levels such as wards or postal sectors. Point or map data may be at one metre, 10 metre, 100 metre and broader geographical resolutions. The wish for geographical detail often runs counter to the requirements for confidentiality, and in the case of the Census, output at ED level is regarded as the best compromise. It follows that the analyst wishing to combine datasets is naturally aiming to relate other statistics to Census EDs, or if this is not possible, larger units such as wards or postal sectors.

So, what other datasets can fruitfully be linked to Census statistics? The following are most widely used:

Area classifications. Geo-demographic neighbourhood classifications such as ACORN, PIN and MOSAIC—themselves derived from Census statistics—often provide a valuable link between datasets.

Births, Deaths and Electorate. Statistics are produced by OPCS at ward level. Changes in ward boundaries over time, however, can be troublesome.

Updated Estimates and Projections. These typically take the Census as a base and then use the administrative statistics on births, deaths and the electorate to produce modelled estimates and projections. Such statistics are widely produced by local authorities for small areas and are also available commercially.

Unemployment. The Department of Employment regularly publishes its detailed statistics on unemployment. These are available for wards and postal sectors and give a hint of the potential for producing anonymous aggregate statistics that lie within government departments. What chance small areas statistics derived from Inland Revenue or Department of Social Security records?

Lists. Lists of addresses—users of public services or consumers in the market place—can be of immense value when integrated with Census statistics. Among the most obvious examples are patient records in the NHS, customer files in such organizations as banks and building societies, the Postcode Address File and the Electoral Roll.

Surveys. Few surveys have large enough samples to give reliable estimates for small geographical areas. When records are coded with a geo-demographic classification, however, it is possible to analyse respondents according to the type of area in which they live. This opens the way to using major surveys such as the Target Group Index or the General Household Survey to produce modelled estimates for small areas.

Property Locations and Map Data. Again, during the last decade, Census statistics have increasingly been combined with information on property locations such as retail outlets. Ordnance Survey's plans for Address Point hold great prospects for the future.

Linking the 1991 Census with other datasets

There are, indeed, great benefits to be obtained from linking the Census with other datasets, but how do we make the link? Three tools will be commonly used throughout the 1990s.

Directories

Although it might sound dull, the ED/Postcode Directory to be produced by OPCS will be a revolutionary tool for Census analysis. Census geography was related to the postcode system in Scotland as far back as the 1981 Census, but south of the border the 1980s saw various attempts to make approximate linkage between EDs and unit postcodes. The 1991 Census will, however, produce a listing of each whole or part postcode occurring within each ED. This opens the way to taking files of individual addresses and aggregating them to give numbers of occurrences for each ED—this can apply to anything from library users to direct debit payers.

ED centroids

The grid reference for each ED is a population weighted centroid and is expressed to 10 metres or 100 metres resolution depending on the scale of the source maps used for ED planning. This opens up the scope of linking datasets by using the ED's grid reference for spatial searches. The fastidious analyst will, however, wish to bear two factors in mind: the grid reference refers to the south-west corner of the 10 metre or 100 metre square rather than its centre; also, the resolution of the centroid has to be inferred from its value—there is no indicator to state whether the source is 10 metre or 100 metre.

ED boundaries

Digital boundaries, whether used at ED level or aggregated to larger areas, will provide considerable scope for spatial searches to link datasets.

Looking back and looking forward

Although it is easy to point to the continuity between the 1981 and 1991 Censuses, 1991 does offer significant advances. Not only are more statistics being produced but in the field of linking other datasets there will be an official postcode directory and also digital boundaries will be available. These will all present a wealth of opportunities for geographical analysis and targeting.

In the near future the big technical issue will be whether or not the OPCS keeps the directory up to date. As for the next Census, the question will again arise as to whether Scotland has already shown the way. The use of the postcode system for planning both Census enumeration and also output would enable progress in linking datasets to continue apace.

We must not, however, focus purely on the technical aspects. In order to convince decision makers of the value that geographical analyses add, much more effort needs to be put into both evangelizing the potential and quantifying the real benefits.

References

Craig, J.A., 1984, 1981 *Socio-economic Classification of Local Authorities and Health Authorities,* Studies of Medical and Population Subjects No. 48, London: HMSO.

Leventhal, B. and Moy, C., forthcoming, *A Researcher's Guide to the 1991 Census.* The Market Research Society.

Office of Population Censuses and Surveys (OPCS), 1991, *Census Preliminary Report for England and Wales,* London: HMSO.

Tuck, M., 1989, *Drinking and Disorder: A Study of Non-Metropolitan Violence,* Home Office Research Study 108, London: HMSO.

39

One man ground survey

H. J. Buchanan
University of Newcastle upon Tyne

Introduction

GISs are based on data, much of it spatial. The most accurate way to collect this data, and the source that gives the most up-to-date data is ground survey. It is only in this case that we are measuring the object itself, rather than a representation of it.

Data for GIS can be acquired from many sources, and in setting a system up existing paper maps are often used. Setting up a system is only a one-off process. The long-term problem is maintenance of the data. This maintenance task is by definition an infinite one; it needs to continue indefinitely. It can be subdivided into two categories: maintenance of the map base and updating attribute information while maintaining its spatial reference. In practice these two operations will often be carried out simultaneously. As for initial data acquisition, many sources of updated data are available, including Ordnance Survey data, photogrammetric surveys, remotely sensed data and ground surveys.

Of all these, the most accurate and timely will be ground surveys. Even with modern data preparation and publishing techniques there will be a gap of months or years between data collection and map publication. That is not to say that all data needs to be ground surveyed for every use. It will often be more appropriate and cheaper to take data from an existing source. Currency and cost often act in opposing directions.

This chapter looks at recent technical developments that will lead to more rapid field data collection in the years to come. Whether those working with GIS use the equipment themselves, or simply employ a firm to do so on their behalf is a separate issue. The technical developments can be divided into three categories: theodolite developments, satellite surveying and pen based computing.

Theodolite developments

The theodolite for measuring angles, the electromagnetic distance measurement device (EDM) for measuring distances and the level for accurate measurement of height differences have been the surveyor's tools for many years. Since the 1980s the first two have been combined into the clumsily named total station which measures angles and distances, and often performs the simple trigonometry necessary to give co-ordinates and height directly. These instruments still require two people to use them:

one operating the instrument, and the other moving a target around the various points to be measured.

The most significant development in this area in recent years has been Geotronics' Geodimeter 4000 instrument. This instrument allows a single person to carry out a field survey. The instrument has the capability to locate its reflector, to ensure that it is indeed the reflector, and then track the reflector as it moves. Therefore the surveyor needs only to move the reflector between the various points to be measured, collecting attribute information, while the instrument tracks those movements, and records angles and distances. The instrument measures angles and distances to the reflector continuously, but the surveyor controls for which points these quantities are recorded (Fort, 1990). There are several technical features of interest within the instrument.

The instrument must first be able to locate the reflector. This process is accelerated if the instrument knows the approximate location of the reflector before searching for the exact location. This can be achieved if the reflector includes a small sighting telescope with approximate angle measurement devices (Perry, 1989). From the reflector, the surveyor sights the instrument. The resulting vertical angle tells the instrument at what vertical angle to seek the target. This could be extended to horizontal angles if the target device included a magnetic compass, but this is not the case in the current instrument. The surveyor can tell the instrument a particular sector of horizontal angles that is to be surveyed. The instrument then searches throughout that sector.

Once a reflected signal has been detected, the instrument must then be able to ensure that it has found the reflector and not some other reflecting surface. This has been achieved in the Geodimeter 4000 by making the target reflector itself an active unit that sends a signal of a different frequency to the instrument's transmitted frequency. In seeking the target point the instrument seeks both a reflection of its own signal and the presence of the target signal. There are in fact two separate cases of this problem: a surface such as a window that is perpendicular to the instrument, and a surface such as a water surface that is nearly parallel to the line between the instrument and reflector. Different solutions are adopted in each case (Geotronics AB, 1990).

Having located the reflector, the instrument then needs to track the target as it moves from point to point. The instrument is fitted with servo motors that control its movement around horizontal and vertical axes. What is required then is a system to compute the displacement of the instrument from the line to the target. From the literature, there appear to be two ways considered to achieve this. The first (Geotronics AB, 1987) involves modulating the periphery of the signal transmitted from the instrument by varying its intensity. By correlating the intensity of the outgoing and returning signals, the displacement of the instrument can be computed and corrected. The second solution (Geotronics AB, 1990) is simpler and involves the use of a quadrant detector within the instrument to detect the displacement produced by the reflection of the signal around the corner cube reflector.

The advantages of the Geotronics 4000 are several. Since the instrument does not rely on an operator sighting from instrument to target, non-daylight working is possible. This has obvious advantages for surveying on roads, airfields or busy cities.

One key advantage is that the surveyor is always positioned at the reflector, and is recording data such as feature description or attributes. With conventional total stations this has also been possible, and various manufacturers have had telemetric links allowing data recording at the target point. In addition, the ease of use of modern instruments has meant that correct positioning and recording of the target point are more demanding tasks than operating the instrument. However there has still been a tendency for the surveyor to look after the precious instrument, leaving the less qualified assistant with the more important job of correctly locating the points to be surveyed.

The main commercial advantage of this system is that it can be used by one person giving savings in manpower costs.

Satellite surveying

The second major change in positioning techniques for ground surveys has been the development of the satellite based Global Positioning System (GPS). The system will offer (when complete) 24-hour, weather independent positioning in which the surveyor need not actively take any observations.

The technical details of the system will not be discussed here, other than to say that to fix a position a receiver must be placed in a location where it has a direct line of sight to at least four satellites. A technical overview and discussion of many of the issues involved is given in Cross (1991). We can consider the relevance of this for GIS in the two categories mentioned in the introduction: base map maintenance, and attribute updating.

For base map maintenance, where surveying accuracies (better than a metre) are required, specialist equipment and techniques are required. Leica have recently launched their GPS system 200 which is designed for one person use (Schwarz, 1991). It comprises two satellite receivers, one of which can be mounted on a target pole and easily moved between points. The associated data recording unit can be hand held. In use, one of the receivers is placed on a point whose co-ordinates are known, and left there during the survey. The second receiver is then placed successively on each point whose position is required. So long as visibility to each of the satellites is maintained throughout the travel period, sufficient data can be collected in two minutes. If the visibility is temporarily lost through, for example, passing under a bridge or working near to high buildings, then the next point to be measured must be occupied for a longer period (typically five minutes). After all points have been visited, the data is then processed in software supplied with the system, and can result in positioning accuracy of better than a centimetre.

For attribute data collection (where position is only required to several metres), hand-held GPS is well established. Small receivers, comparable in size to a personal organizer or mobile phone give a continuously updated position. Working in isolation (absolute positioning) these give an accuracy of around 100m. Working with a second receiver placed on a point whose position is already known (relative positioning) can improve the accuracy to around two metres.

A point of significance in this area is the mathematical transformation between the co-ordinate system that GPS uses (WGS84), and the co-ordinate system in which Ordnance Survey mapping is published (OSGB36). If this transformation were constant, then its application would be trivial. However for a variety of reasons that are beyond the scope of this paper, this transformation is not constant. Anonymous (1987) portrays these differences graphically, showing that between the extremities of Great Britain, the transformation varies by around 100m in each of latitude and longitude. Therefore, the simplistic approach of applying a single national transformation to GPS co-ordinates to give co-ordinates to be related to Ordnance Survey mapping could give an answer with an error of the order of 100m.

This variation is made up of two components. The first component, which accounts for the largest part of the total, arises from the way in which the co-ordinate systems are defined. Each of the co-ordinate systems is based on an ellipsoid whose shape approximates the shape of the earth. The ellipsoid used by each co-ordinate system is different, and in the area of the United Kingdom the two ellipsoids have significantly different shapes. Since this component of the variation arises from the difference between two mathematically defined surfaces, it is well defined, and can be removed in a straightforward manner.

The second component of the variation arises from distortions within the OSGB36 system, and amounts to some 20m across the whole of the country. These are more difficult to model accurately. However, Anonymous (1987) gives regression formulae which are claimed to remove the effect of these distortions to around one metre.

It is not clear from the manufacturers' literature how individual receivers treat each of these elements. Prospective users should assure themselves that any system they are considering can deliver the accuracy that they require. Part of that decision will clearly be the purpose to which the system is to be put. For vehicle tracking, 100m accuracy may be adequate. For collection of data about individual properties then much higher accuracy will be required to identify the correct property from its position.

The completion of the satellite constellation in 1993 will make this technology much more marketable, since at this point 24 hour coverage will be available. At the time of writing coverage totalling around 16 hours in each day is available in the UK. In the USA, where satellite availability is much better many more applications are already being developed.

Pen based computing

An area of change over recent years has been in the amount of processing and analysis of survey observations carried out in the field. In the early days of total stations, simple recording devices were attached to instruments that logged the measured angles and distances and a simple feature code. These observations were then processed by downloading to an office based computer. An alternative to these simple systems is now provided by systems that carry out calculations in the field to give co-ordinates or other data derived from the observed quantities. These more sophisticated systems are either built into survey instruments or are provided by attached computers. One limitation of such systems is that they have generally been restricted to alphanumeric display, and therefore can only give a very limited view of the survey being carried out. Graphical systems have been difficult to implement because of the obstacle of providing a mouse or some other form of control over a pointing device that is practical for use in a field environment.

The recent advent of pen based computers seems to have taken this process a significant stage forward (Gray, 1991). These computers are characterized by not having a keyboard, but allow control of a graphics pointing device by the application of a pen-like pointer on the screen. In addition they provide a screen display with a resolution sufficient for the display of graphical information. The main disadvantage of these systems would appear to be that alphanumeric entry involves either computer recognition of handwriting, or laborious entry by using the pen on a keyboard displayed on screen. Neither of these options is ideal since computer recognition of handwriting is both prone to error and very computationally intensive, while a screen keyboard reduces the most proficient operator to single finger typing. In situations where the text required can be predicted, then menus can reduce this bottle-neck. One example of such hardware is the Grid Pad from Grid Computer Systems, for which several pieces of software for field survey are currently available.

The application of these devices can again be considered in terms of maintenance of the map base and updating the attribute information. For updating the map base they can be connected directly to the surveying instrument, and programmed to display each point in its surveyed position as it is observed, with appropriate symbols to indicate the feature coding. This allows the surveyor to view the survey as it is carried out. For certain applications it may be important to carry out such a check on the surveyed shape of ground slopes in addition to plan detail. This is an area where an inexperienced surveyor can easily make mistakes since changes of slope are not clearly marked on the ground. The surveyor can therefore be assured by visual inspection that the data collected in the field correctly portrays the ground before leaving the site. Without this technology it would be necessary to visit the site a second time with a plotted version of the surveyed information to ensure its completeness. In addition, a system such as this can be used to allow graphical survey techniques to be used where necessary to complete a survey (Cory, 1991).

For updating attributes, then a copy of the existing map and related attributes can be loaded onto the computer with some simple software to display the two sets of data and their relationship. The update of attributes is then carried out by simple map-reading, visual inspection and recording of the amended attributes. One advantage of the pen based nature of these devices might be that signatures can be collected and immediately integrated with other data. Whether the legal profession will ever be satisfied that this procedure is a legitimate one remains to be seen.

The way forward

Many of the processes and technical developments outlined in this chapter be developed further. It seems very likely that other manufacturers will follow Geotronics' lead into one man total stations as soon as they can see that the market has accepted the technology. There has already been publication of patents to show that other manufacturers are actively considering this way forward (Wild Heerbrugg AG, 1986). Other developments in this area may include the refinement of EDMs that do not require a reflector which will allow measurement to points without visiting the points themselves.

Satellite surveying techniques will also be developed further as satellite availability improves and as software processing techniques simplify the processes required for high accuracy work. For lower accuracies this approach will become greatly simplified if Ordnance Survey take the (very considerable) step of transforming their mapping from the existing OSGB36 reference system to the WGS84 system used by GPS. This change would involve enormous upheaval for users and the Ordnance Survey themselves, but would enormously increase the usability of satellite surveying, especially for attribute collection.

Field computing will benefit from continuing developments in battery technology and low power memory and screen devices. These changes will increase the operating life of pen based computers to allow a full day's work without the need for battery changes. In addition, the development of radio based computer communication will allow such field systems to communicate in real time with central data sources. Many of the other limitations of this technology seem likely to be eroded quickly (Baran, 1992).

This chapter has not addressed any of the techniques of data collection by photography or other remotely sensed images. While these techniques have an important part to play in data collection, ground survey will also continue to be a major source of data. The techniques and technologies discussed here will contribute to the more rapid and accurate collection of data in the field from objects themselves, rather than their representations.

References

Anonymous, 1987, Department of Defense World Geodetic System 1984 Technical Report, Part II—Parameters, Formulas and Graphics for the Practical Application of WGS84. Defense Mapping Agency Technical Report, No DMA TR 8350.2-B, 1 December.

Baran, N., 1992, Rough Gems: First Pen Systems Show Promise, Lack Refinement. *Byte,* April, 212–22.

Cory, M., 1991, Portable Data Capture. *Proceedings of Association for Geographic Information Conference,* West Trade, Birmingham, 20–22 November, Paper 3.23.

Cross, P.A., 1991, GPS for GISs, *Mapping Awareness,* **5**(10), 30–34.

Fort, M., 1990, One man, one system, *Civil Engineering Surveyor,* **XV**(8) p. 19.

Geotronics AB, 1987, Arrangement for holding an instrument in alignment with a moving reflector, United States Patent Application, Patent Number US 4712915, 15 December.

Geotronics AB, 1990, An Arrangement for Performing Position Determination, World Intellectual Property Organization Patent Application, Patent Number WO 90/12284, 18 October.

Gray, T., 1991, Advanced Methods of Capturing and Using GIS Data in the Field, *Proceedings of Association for Geographic Information Conference,* West Trade, Birmingham, 20–22 November, Paper 2.25.

Perry, J., 1989, Position Measuring System, United Kingdom Patent Application. Patent Number GB 2217454, 25 October.

Schwarz, J., 1991, Wild GPS 200—GPS for routine surveys, *Leica Reporter,* **27**, 10.

Wild Heerbrugg AG, 1986, Measuring angular deviation, United Kingdom Patent Application, Patent Number GB 2166920, 14 May.

40

Data indexes and major sources

Introduction

This directory of data indexes and major sources is designed to assist those searching for appropriate datasets to determine what is available and from whom. The ideal would be an index by dataset detailing the contact, content, format and restrictions of use. Unfortunately experience has shown that it is currently too difficult to compile and maintain such a catalogue embracing all data for all application areas.

To assist the user, entries are divided into nine categories. These can only be regarded as pointers to the main type or use of the data. Much of the data can be used in a wide range of applications.

ADM	administrative
CLI	climatic
ENV	environmental
GEO	geological
HYD	hydrological
IND	data index
INF	infrastructure
SOE	socio-economic
TOP	topographic

Every attempt has been made to check the validity of the meta-data provided but complete accuracy is not guaranteed. This is the first year of publication of the directory. Hopefully holders of relevant information will be encouraged to amend and add to it for the next yearbook.

Directory of Government Tradeable Information IND

Organization:	The Association for Geographic Information (AGI)
Contact:	Maxine Allison
Tel. no.:	071 222 7000 ext. 226
Fax no.:	071 222 9430
Other:	915443 RICS G (telex)
Address:	12 Great George Street, London SW1P 3AD

Description: The AGI hold descriptions of government held data and can provide a list of dataset titles or a disk containing details of all the datasets held. The information available includes data content, accuracy, timing, area of coverage, form of spatial referencing used and restrictions as to use. The data have been collected through a survey of all government departments as a result of the government-held tradeable information initiative.

Reference: Government Information Trader from Electronic Publishing Services (Publications) Ltd (071 490 1185). A quarterly publication with articles and descriptions concerning government tradeable information.

GEOQUEST IND

Organization: Department of Surveying
Contact: Sally Smith
Tel. no.: 091 222 6449
Fax no.: 091 222 8691
Address: Department of Surveying, University of Newcastle upon Tyne, Newcastle upon Tyne, NE1 7RU

Description: A provisional directory of sources of data primarily collected to assist civil engineers in locating existing data sets.

ESRC Data Archive Catalogue and BIRON SOE

Organization: ESRC
Contact: Mrs Bridget Winstanley (Assistant Director)
Tel. no.: 0206 872141
Fax no.: 0206 872003
Other: ARCHIVE@UK.AC.ESSEX (email)
98440 (UNILIB G) (telex)
Address: University of Essex, Wivenhoe Park, Colchester CO4 3SQ

Description: The ESRC data archive provides a national data service for the social sciences. The archive acquires, stores, documents and disseminates computer-readable copies of economic and social science data. Data holdings are described in a two volume catalogue but there is also on-line access through BIRON (Bibliographic Information Retrieval ONline) for JANET users or by the Archive on the user's behalf.

Reference: Data Archive Bulletin—contains news on developments, new data acquisitions and research within the Archive. It is published three times a year.

CACI Market Analysis Data SOE/ADM/INF

Organization: CACI Ltd.
Contact: Jill Collins (Manager Public Services Group)
Tel. no.: 071 602 6000
Fax no.: 071 603 5862
Address: CACI House, Kensington Village, Avonmore Road, London W14 8TS

Description: CACI hold a variety of data sets including; demographic, household, geographical boundaries, retail location, infrastructure and market research data.

Eurostat GISCO project TOP/ADM/ENV/SOE/CLI/GEO/INF

Organization: Eurostat
Contact: Mr V. Schreurs (GISCO project leader)
Tel. no.: 010 352 4301 4398
Fax no.: 010 352 4301 7316
Address: Eurostat, Unit F3-GISCO, Batiment, Jean Monet B3/84, L-2920 Luxembourg

Description: The role of GISCO (the Geographical Information System of the COmmission of the European Communities) is to identify the requirements of the Commission Services for basic topographic reference data and subsequently to acquire and maintain this data. The data so far collected includes various coverages at scales 1:3 000 000, 1:1 000 000 and 1:100 000 for the EEC and sometimes adjacent countries. The following data has already been added to make up the basic reference data: coastlines; national, regional and municipal boundaries; water features; road networks; settlements; morphological agglomerations; airports; ports and nuclear power stations. Certain other data, termed ''common data'' has also been collected, including climatic data, land cover, soils, potential natural vegetation and viticulture regions. Furthermore Eurostat maintains two statistical databases which are spatially detailed called ''REGIO'', containing regional statistics on demography, agriculture, industry, transport, energy, employment, etc. and ''LOC'' containing socio-economic small area statistics (municipality level).

References: Eurostat, 1992, Report on the activities of the Spatial Statistics Task Force and GISCO (includes comprehensive database contents and distribution policy guidelines).
Cubitt, R., 1992, The role of Eurostat in GIS, *Proceedings of the Mapping Awareness Conference,* February 25–27 1992, Olympia 2, London.
Eurostat, 1991, REGIO, Regional Data Bank, Description of Contents.

NOMIS (National Online Manpower Information System) SOE

Organization: Mountjoy Research Centre
Contact: Michael Blakemore (Executive Director)
Tel. no.: 091 374 2468 / 2490
Fax no.: 091 384 4971
Other: NOMIS.TEAM@uk.ac.durham
Address: Unit 3P, Mountjoy Research Centre, University of Durham, Durham DH1 3SW

Description: NOMIS supplies the latest and historical national, regional and local information on the UK labour market including government statistics on employment, population, unemployment, migration and job centre vacancies. The system allows for the abstraction of data on flexible geographic areas as small as postcode sectors. The data required are selected by spreadsheet facilities via BT PSS, telephone dial-up and modem or JANET (Joint Academic NETwork). NOMIS also provides facilities for abstacting tables, statistical analyses and producing high quality maps and charts.

Reference: Fact sheets from NOMIS on "What Data are Available", "Geographic Areas", "Analytical and Mapping Facilities" etc.

Social Science Research Data SOE

Organization: Edinburgh University Data Library
Contact: Peter Burnhill (Manager)
Tel. no.: 031 650 3302
Fax no.: 031 662 4809
Other: datlib@uk.ac.edinburgh (email)
Address: Data Library, The University of Edinburgh, Main Library Building, George Square, Edinburgh EH8 9LJ

Description: Edinburgh University Data Library holds a large data collection of interest to academic, policy and market researchers. These data are held in an environment well suited to (remote) multiaccess, interactive computing and which is rich in software for statistical analysis and graphic display (including mapping) display.

Reference: DATALIB is an online information system about the Data Library. To access DATALIB, connect to the Edinburgh castle via the X29 link: *call uk.ac.ed.castle* (this may need to be reversed) or via PSS, code (234)231354354. Type *inform* in response to the login prompt and type *inform* for the password. Menus are provided to access relevant topics.

Scottish Spatial Data IND

Organization: ESRC Regional Research Laboratory for Scotland
Contact: Heather Ewington (Information Officer)
Tel. no.: 031 650 4087
Fax no.: 031 662 4809
Address: 22a Buccleuch Place, Edinburgh EH8 9LN

Description: RRL Scotland provides access to key geo-spatial data sources such as the Agricultural and Population Censuses, geo-directories and digitized map boundaries. It is also a member organization of the Scottish GIS Forum whose aims include increasing awareness of GIS and promoting the compilation of registers of spatially referenced data.

Reference: RRL Scotland Working Paper Series.

Aerial Photography Advisory Service TOP/ENV

Organization: Ordnance Survey
Contact: Aerial Photography Sales
Tel. no.: 0703 792584 (direct)
0703 792000 (switchboard)
Fax no.: 0703 792250
Other: 477843 (telex)
Address: Romsey Road, Maybush, Southampton SO9 4DH

Description: This service is available to assist inquirers in locating recent vertical aerial photography from government and commercial air survey companies for England. Contacts are also available for the location of recent and archival photography for Scotland and Wales.

Directory of Aerial Photographic Collections IND/TOP/ENV

Organization: NAPLIB—National Association of Aerial Photographic Libraries
Contact: Mr D.R. Wilson (Secretary of NAPLIB)
Tel. no.: 0223 334575
Fax no.: 0223 334578
Address: Committee of Aerial Photography, Mond Building, University of Cambridge, Free School Lane, Cambridge CB2 3RF

Description: A list of 200 or more collections of aerial photographs to be published later this year.

Remote Sensing Data TOP/ENV

Organization: National Remote Sensing Centre Ltd
Contact: Derek Gliddon
Tel. no.: 0252 541464
Fax no.: 0252 375016
Other: 859891 (telex)
Address: Delta House, Southwood Crescent, Southwood, Farnborough, Hampshire, GU14 ONL

Description: NRSC Ltd offers a complete commercial remote sensing service supplying data acquired by satellites and aircraft, computer processed imagery and a broad range of consultancy services. The company's large team of qualified applications Specialists are skilled in a wide variety of remote sensing applications, including the integration of remotely sensed data into Geographic Information Systems. NRSC Ltd also operates the United Kingdom facility for processing and archiving data from the European Space Agency's ERS-1 satellite. Data is also available through regional centres in Wales, Northern Ireland and Scotland.

References: A comprehensive range of fact sheets and brochures depicting applications of remote sensing data are available upon request. The company's newsletter, Albedo is published bi-annually.

GENIE (Global Environmental Network for Information Exchange) IND

Organization: The GENIE Project
Contact: Dr Ian Newman
MRRL
Tel. no.: 0509 222687
Fax no.: 0509 211586
Address: The GENIE Project, c/o Department of Computer Studies, Loughborough University of Technology, Loughborough LE11 3TU

Description: The GENIE project has been established to provide a global environmental change data network through a contract from the ESRC awarded to a consortium led by the MRRL, the Computer Centre at Loughborough University of Technology, the Remote Sensing Unit at Nottingham University and Genasys. A system for metadata retrieval and management developed by MRRL will be used to provide information about data holdings relating to environmental change. The system will also provide links to other international data directories. The project started on 1 April 1992 and should be fully operational by the end of 1993.

CORINE (COoRdination of INformation on the Environment) ENV

Organization: European Environment Agency
Contact: Jef Maes
Tel. no.: 010 322 236 8815
Fax no.: 010 322 236 9562
Address: CORINE Program, Rue de la Loi 200, 1049 Brussels, Belgium

Description: The CORINE program has been established to provide a consistent data set describing the land-use cover across the European Community. However, the CORINE program also hold additional environmental data sets. The land use data set is at present complete for Portugal, Luxembourg, Spain and the Netherlands.

References: CORINE database manual, 1989, Directorate General for the Environment, Nuclear Safety and Civil Protection. November.
CORINE—An Information System on the State of the Environment of the European Community, 1992, *Geodetical Info Magazine,* January, 55–57.

Ecological Data ENV

Organization: Institute of Terrestrial Ecology (ITE)
Contact: Dr Jane Metcalf
Tel. no.: 04873 381
Fax no.: 04873 467
Address: Monks Wood, Abbots Ripton, Huntingdon, Cambridgeshire PE17 2LS

Description: ITE is part of the Natural Environment Research Council. With six research stations in Great Britain, ITE undertakes specialist ecological research in all aspects of the terrestrial environment. Research is directed in specific programmes, and detailed information is held on the following disciplines: forest science; land use and the environment; global environmental change; environmental pollution; population and community ecology. Specific databases are held for certain subjects at the Environmental Information Centre: species distributions in GB; critical loads of atmospheric pollutants.

References: Top Level Directory of NERC data holdings. Contact: Angela Morrison, Polaris House, North Star Ave, Swindon SN2 1EU. Tel: 0793 411500.
Species distributions: Biological Records Centre, ITE Monks Wood.
Global environmental change: UK Global Environmental Research Office, Polaris House, North Star Ave., Swindon SN2 1EU.

Geological, Geochemical and Geophysical Data GEO

Organization: British Geological Survey
Contact: Enquiry desk
Tel. no.: 0602 363143
Fax no.: 06077 6391
Other: 378173 BGSKEY G (telex)
Address: British Geological Survey, Keyworth, Nottingham NG12 5GG

Description: The BGS is responsible for research and mapping of the geology of the British Isles, the continental shelf offshore and similar technical aid programmes for countries overseas. It produces several series of geological, geophysical, geochemical, hydrogeological, engineering geology and environmental geology maps at scales ranging from 1:10 000 to 1:2 500 000, some in a digital format. They are accompanied by explanatory memoirs and reports. Spatially-referenced analytical data are held, increasingly in machine-readable form, and the national archive of reference collections holds copies of borehole logs, seismic traces, site investigation reports, mine and quarry plans, photographs etc. Regional data may be accessed through a number of regional offices

Reference: Catalogue of Printed Maps 1992 and Associated Literature, BGS.

National topographic mapping TOP/ADM/INF

Organization: Ordnance Survey
Tel. no.: 0703 792773 (Digital Sales)
0703 792449 (Special Products)
0703 792792 (Information and Public Relations)
Fax no.: 0703 792324
Address: Romsey Road, Maybush, Southampton SO9 4DH

Description: The Ordnance Survey is responsible for producing and maintaining up-to-date surveys of the Nation at scales of 1:1250 for major urban areas and 1:2500 and 1:10 000 for the rest of the country. They also produce maps at other scales, some of which are derived from the basic scales. The 1:1250, 1:2500, 1:10 000 data are largely available in a digital format as are 1:625 000 and 1:250 000 data. Height data are available at scales of 1:50 000 and, for special orders only, 1:10 000. The OS will also sell their most up-to-date unpublished topographic mapping. Some digital data sets of abstracted features can also be purchased, for example administrative boundaries at 1:10 000 and the centre alignment of roads (OSCAR data).

Reference: Ordnance Survey leaflets are obtainable free from Information and Public Relations on the following topics: maps, atlases and guides; large scale maps, services and special products; and digital map data.

MundoCart/CD and World Climate Disc TOP/CLI

Organization: Chadwyck-Healey Ltd.
Contact: Paul Holroyd
Tel. no.: 0223 311479
Fax no.: 0223 66440
Address: Cambridge Place, Cambridge CB2 1NR

Description: MundoCart/CD is the World digitized at 1:1 000 000 on CD-ROM from 275 Operational Navigation Charts produced by the US Defense Mapping Agency. Includes coastlines, international and selected administrative boundaries, major town and city boundaries, rivers, lakes and canals. The data are available to PC users at the scales of 1:1 000 000, 1:5 000 000, 1:25 000 000 and 1:150 000 000 as complete or as eight regional subsets. The global climatic change data combined with a digital map of the world are also provided on CD-ROM and was compiled by the Climatic Research Unit at the University of East Anglia. The data includes monthly mean surface air temperature, gridded monthly mean temperatures, monthly total precipitation, gridded mean sea-level pressure data for the northern hemisphere and gridded mean height data for the whole world.

Digital Chart of the World TOP/ADM/INF

Contact: Barry Thomas
Tel. no.: 081 330 7959
Fax no.: 081 330 7959 ext. 5387
Address: Military Survey, Ministry of Defence, Acquisition and Library Group, Block A Government Buildings, Hook Rise South, Tolworth, Surbiton, Surrey KT6 7NB

Description: The Digital Chart of the World will provide topological vector data at a base scale of 1:1 000 000 for the whole world. The data will include a gazetteer, transport links and contours. It will be provided on four CDROMS and will include VPFVIEW software and documentation.

Reference: Danko, D.M., 1992, Global Data—The Digital Chart of the World, *Geo Info Systems*, January, 29–36.

Topographic Mapping TOP/ADM

Organization: Bartholomew
Contact: Kevin Dickson
Tel. no.: 031 667 9341
Fax no.: 031 662 4282
Other: 728134 BARTS G (telex)
Address: 12 Ducan Street, Edinburgh EH9 1TA

Description: Bartholomew supply small-scale digital data sets of the world and Great Britain for market research and spatially related applications including 1:250 000 data for GB, 1:500 000 data for Ireland, 1:14 000 000 data for the world, GB postcode boundaries and an index of placenames and physical features of the world. Data sets for London and Europe should be available soon. The data sets are also available through CHEST for higher education establishments.

AA datasets/Postal datasets TOP/ADM/INF

Organization: Kingswood Ltd
Contact: Chris Greenwood / Mary Short
Tel. no.: 081 994 5404
Fax no.: 081 747 8047
Address: 19 Kingswood Road, London W4 5EU

Description: Kingswood supply mapping at 1:200 000 for UK and 1:1 000 000 for Europe based on road atlases. They also have Central London Street data and UK and European postal geography. Paper reproductions of the digital data are on sale at most book shops.

Property ADM

Organization: Her Majesty's Land Registry (HMLR)
Contact: John Deas (Customer Service Manager)
Tel. no.: 071 405 3488 ext. 4432
Fax no.: 071 242 0825
Address: HMLR, Lincoln's Inn Fields, London, WC2A 3PH

Description: HMLR, through its regional offices, hold records of about 14 million of the estimated 22 million property titles in England and Wales. For a fee the public may inspect the register of title to property which includes a description of the land, the name of the owner and any legal interests which adversely affect the land. Associated with each register of title is a filed plan showing the property boundary. The property boundary and unique title number is also shown on the Public Index Map. Both the plan and the map are based on large scale OS mapping.

References: Further details as to the information held by HMLR and how to obtain it are contained in the Registry's explanatory leaflet no.15, "The open register". This leaflet may be obtained (free) from the above address.
Smith, P.J., 1988, Developments in the use of digital mapping in HMLR, *Mapping Awareness* **2**(4), 37–41.

Woodlands ENV

Organization: Forestry Commission
Contact: Graham Bull
Tel. no.: 031 334 0303 ext. 2347
Fax no.: 031 334 3047
Other: 727879 FORCOM G (telex)
Address: 231 Corstorphine Road, Edinburgh EH12 7AT

Description: Two main aims of government forestry policy are to provide a steady expansion of tree cover, and the sustainable management of our existing woods and forests to increase the many diverse benefits that forests provide. Data is currently held on OS paper-based maps showing land holdings, compartments which are management units and sub-compartments against which species, planting year and other information is held to assist with the management of the forests. This data is not generally available to the public but, if required, a contact can be provided in a regional office where the data is held. A pilot study is being undertaken to investigate the potential of GIS to aid the commission in its role in managing its own land and its advisory/regulatory role in Private Forestry.

Soil Survey GEO/CLI/TOP/ADM/ENV

Organization: Soil Survey and Land Research Centre
Contact: Ian Bradley
Tel. no.: 0525 860428
Fax no.: 0525 861147
Address: Soil Survey and Land Research Centre, Silsoe Campus, Silsoe, Bedford MK45 4DT

Description: SSLRC operates as a research and professional consultancy organization applying its understanding of soil and soil processes to the management of the environment and agricultural production. The land information system contains data on soil, including physical and chemical analysis and digitized soil maps; climate at five km resolution for England and Wales; topography; administrative boundaries and land use. The data is available for lease.

Reference: Catalogue of printed maps and books; details of data available for lease.

Meteorological Data CLI/ENV/HYD

Organization: Meteorological Office
Contact: Commercial Services Library and Archive Information Officer
Tel. no.: 0344 856207/854841
Fax no.: 0344 854906/854840
Address: London Road, Bracknell, Berkshire RG12 2SZ

Description: The Meteorological Office holds the UK's records of weather. Quality controlled observations of a large range of meteorological variables are kept. These include upper air data from nine sites, hourly synoptic observations from 60 sites, daily climate observations from 600 sites and daily rainfall measurements from 5000 sites. Observations since 1960 are stored on computer datasets. Records prior to this date are in the form of manuscripts or weather charts available from the archives. Derived data from a number of models are also available.

Reference: Details of data availability can be obtained from Met Office Commercial Services. Archive details from the Library.

Hydrological Data HYD/CLI/TOP

Organization: Institute of Hydrology
Contact: David G. Morris
Tel. no.: 0491 38800
Fax no.: 0491 32256
Address: Wallingford, Oxfordshire OX10 8BB

Description: The Institute of Hydrology runs the National Water Archive and supplies many alternative retrieval options for streamflow and catchment water balances. The Institute also holds the UK Acid Waters Monitoring Network archive, a measured soil moisture database, and automatic weather station data from UK experimental catchments. IH both sells and leases spatial datasets, in particular those needed for the UK Flood Studies Report calculations including: 1:50 000 scale UK rivers in vector form and hydrometric area boundaries. The digital terrain model gives drainage directions to the river network and cumulative catchment areas based on a 50m by 50m grid interpolated from 1:50 000 scale contours.

Reference: The Institute's DTM News gives details of progress towards national coverage. Gauged flow retrievals are indexed in the Hydrological Yearbook UK.

River Management Data HYD/SOE

Organization: National Rivers Authority (NRA)
Contact: Public Relations Department
Tel. no.: 0454 624400
Fax no.: 0454 624409
Address: Rivers House, Waterside Drive, Aztec West, Almondsbury, Bristol BS12 4UD

Description: The National Rivers Authority is responsible for all rivers. The Authority is divided into 10 regional offices each of which holds a public register of information. The main information held is consents to discharge into rivers and includes information on the owner of the consent, where the discharge is coming from and where it will go, and details of the consent conditions such as flow and solid content. The results of analyses of all samples are also held. Other details held vary from region to region but can include, for example, registers of prosecutions and water abstraction licenses.

Census Statistics (England and Wales) SOE

Organization: Office of Population Censuses and Surveys (OPCS)
Contact: Mrs Linda Graft
Tel. No.: 0329 42511 ext.3398
Fax no.: 0329 810260
Address: Census Customer and Support Services, OPCS, Segensworth Road, Titchfield, Fareham, Hants PO15 5RR

Description: OPCS provides a service to answer customer enquiries relating to all census statistics and will supply products available or direct customers to appropriate sources.

Reference: Census Newsletter—A link between the Census Offices and census users. The newsletter is published periodically and provides information on many aspects of the 1991 census in Great Britain, details of relevant publications and census-related activities. It is possible to be included on the mailing list by applying to the address given above.

Focus ADM/SOE

Organization: Property Intelligence PLC
Contact: John Taylor (Sales Director)
Tel. no.: 071 839 7684
Fax no.: 071 839 1060
Address: Ingram House, 13-15 John Adam Street, London WC2N 6LD

Description: Focus is designed to provide information on commercial property throughout the UK. Other facilities provided include Property Focus, giving on-line information on ownership, rents, planning etc.; Town Focus for key demographic, socio-economic and property information for towns; Relocation Focus indicating demand for office property; Retail Focus detailing retailer requirements; News Focus giving the latest property and economic news; and special services giving, for example, land title and planning reports. The services are used by the majority of the main commercial property investors and advisers throughout the UK.

Utility Data INF

Description: There is no central co-ordinating body for advice on spatial datasets held by the utilities. Utility companies and authorities maintain records of the location of underground, surface and overhead plant. The local utility should be contacted for information.

Local Government Data TOP/ENV/GEO/CLI/HYD/SOE/ADM/INF

Organization: The Local Government Management Board (LGMB)
Contact: Tony Black (GIS adviser)
Tel. no.: 0582 451166
Fax no.: 0582 412525
Address: Arndale House, The Arndale Centre, Luton, Bedfordshire LU1 2TS

Description: Local Authorities are responsible for recording information on population, economic, land use, environmental, highways, property, client and administration matters. The relevant local authority should be contacted. The LGMB will provide general advice and appropriate contact names.

Reference: GIS News—published quarterly. Contact Rosi Somerville on 0582 451166.

Census Data/Postcode Boundaries (Scotland) SOE/ADM

Organization: General Register Office for Scotland—GRO(S)
Contact: Derek Wilson
Tel. no.: 031 314 4254
Fax no.: 031 314 4344
Address: Census Customer Services, Ladywell House, Ladywell Road, Edinburgh, EH12 7TF

Description: GRO(S) holds population data for each Census from the first in 1801 to the most recent in 1991. The data includes local base statistics and small area statistics from 1971, 1981, 1991 for all regions and island areas as well as statistics on particular census topics. The data can be purchased from Customer Services. 1991 Census local statistics will start to become available from summer 1992 with the publication of Monitors and the availability of Small Area Statistics. Boundaries of the unit postcodes have been digitized and includes 1991 Census output areas, overlays of the same, and index files allowing the user to link the various types of area.

Reference: Local Base and Small Area Statistics Prospectus, Boundary Products Prospectus, Census Newsletter, HMSO bookshops.

Spatial Data: Northern Ireland IND

Organization: Northern Ireland GIS Unit (NIGIS)
Contact: Geoff Mahood
Tel. no.: 0232 611244 ext. 265
Fax no.: 0232 683211
Address: c/o Ordnance Survey of Northern Ireland, Colby House, Stranmillis Court, Belfast BT9 5BJ

Description: Collaboration between Ordnance Survey of Northern Ireland and other major government and public authority partners is establishing the beginnings of an ambitious national integrated network of spatially referenced databases collectively known as the Northern Ireland Geographic Information System—NIGIS—the linkages being achieved through the growing digital topographic database. The NIGIS unit can advise on contacts in all the appropriate organizations, with regard to available digital data etc.

Northern Ireland Topographic Information TOP/INF/ADM

Organization: Ordnance Survey of Northern Ireland (OSNI)
Tel. no.: 0232 661244
Fax no.: 0232 683211
Address: Colby House, Stanmillis Court, Belfast BT9 5BJ

Description: OSNI is responsible for producing, and maintaining an up-to-date survey information archive for Northern Ireland at scales of 1:1250 for major urban areas and 1:2500 and 1:10 000 for the remainder of the country. Other information based at smaller scales, and horizontal and vertical control information is also available. OSNI are currently producing and maintaining a complete edge matched, digital topographical database derived from the current survey archive. Its key component is a logically structured, dynamic graphic database, of approximately 200 structure levels, wherein each polygon, linestring and point feature can be linked by an automatically generated, unique, twelve figure grid reference to the relevant associated textual information. For example, each addressable property is linked, via its grid reference, or geocode, to the full postal address held in the textual database.

Reference: OSNI leaflets and information with regards to availability and charges can be obtained from the Marketing and Publications department.

Part VIII

GIS and Education/Training

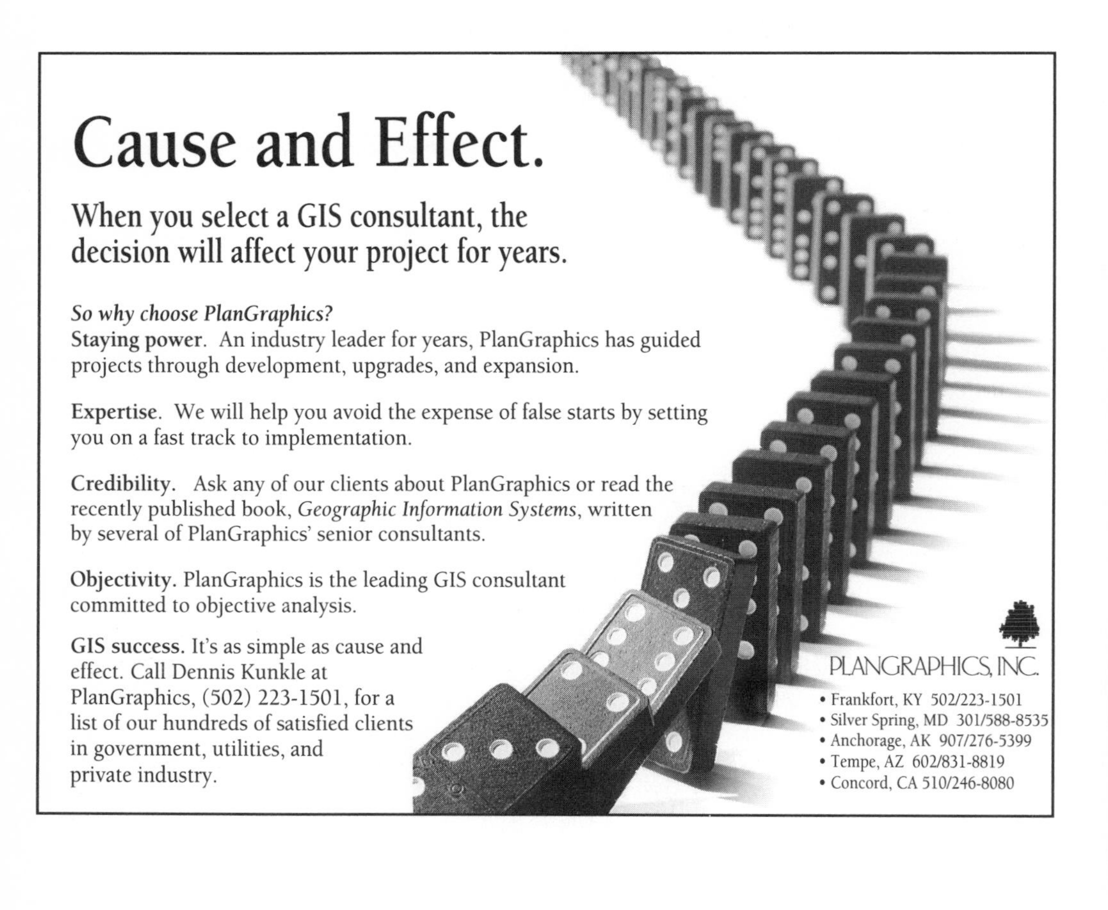

41

GIS education and training: Developing educational progression and continuity for the future

David R. Green
University of Aberdeen

Introduction

GIS education and training have long been topics for discussion in the literature. One notable example is the issue of what differentiates education from training (e.g. Gittings and Mounsey, 1989; Goodbrand, 1991; Green, 1991a; Unwin and Dale, 1989). Others have discussed the need for a GIS syllabus and curriculum for secondary and higher education (e.g. Coggins, 1990; Unwin, 1989a), and GIS as an application of Information Technology (Green and McEwen, 1989). More recently issues such as introducing GIS into school education (Cassettari, 1991; Freeman, 1991; Green, 1991b; Heywood and Petch, 1991), the educational essentials of GIS (Gilmartin and Cowen, 1991), and the educational requirements of the Eastern European countries (NCGIA Special Session—EGIS'91) have become topics for discussion.

With continued growth in the GIS industry the demand for both graduates and staff with a background in GIS for a wide variety of jobs has increased. Not surprisingly the requirements for both GIS education and training have also grown quite rapidly over the last few years—such is the impact this technology is now having on our everyday lives! In response to this demand, new undergraduate and postgraduate GIS degree and diploma programmes, short courses, hardware, software and teaching materials have all been developed to provide the necessary support to enable the education and training required.

GIS education and training are undoubtedly important areas which warrant further development and research in the foreseeable future. It is very likely that the demands for both will also continue to grow over the next few years. Recent developments in the UK education system are now beginning to provide the framework within which it will subsequently be possible to develop an element of educational progression and continuity for GIS education and training. Coupled with the evolution of GIS, the development of new software and hardware targeted at education and training, and new teaching materials, this should serve to provide a sound basis for the future. However, progress will also require co-operation, co-ordination, a curriculum, and a formally recognized body to make it all work properly.

The overall premise of this chapter is that a sound education system is the fundamental basis for future GIS education and training. To achieve this goal it will require provision of a proper educational framework with the scope to introduce educational progression and continuity, thus facilitating a structured approach to both GIS education and training.

The objective of this short introductory chapter is to support the above argument by presenting a brief outline of some of the key areas in GIS education and training where developments have been or are now taking place. A number of the topics highlighted are subsequently explored in greater depth in the chapters in the remainder of the part.

The importance of education

Education is a fundamental right of every person in any country. It is an important stage in anyone's life and, among many other things, offers opportunities to acquire knowledge, to explore and understand the environment in which we live, to mature in thought, and to develop appropriate skills. Moreover education is a vital human resource and the key to our survival in the real and now rapidly changing world.

Education in GIS and about GIS is no less important than for any other subject area. Indeed the impact of GIS on society, on the home, travel and the workplace is now so great that we should, at the very least, be aware of GIS, what it is, what it involves, what it does, and how it is used. Knowledge, understanding and information about this technology are all necessary because GIS will eventually affect us all in one way or another. Education provides the key to the acquisition of such knowledge and understanding. It therefore needs to be supported and carefully fostered.

Changes in the education system

In recent years the entire UK education system (primary, secondary and higher) has been at the centre of a number of radical changes. The structure of the existing education system, the way in which people are taught, what, how and when they are taught, have all been re-examined in detail. The UK National Curriculum, Modularization, Access courses, and Enterprise initiatives are all important examples of these changes. As a direct result many new educational opportunities have opened up for all age groups. Whether it be the school age pupil or someone returning to higher education these changes will all have a significant impact upon learning, the acquisition of knowledge, the development of practical skills, and preparation for future employment.

This new educational framework, or infrastructure, is designed to allow greater choice and flexibility, to prepare people for the real world of today, and to meet the requirements of the modern workplace. Increasingly, therefore, emphasis is being placed upon developing the practical applications in combination with classroom theory. This emphasis is perceived to be the most useful approach for the long term benefit of both the student, and eventually the employer. It is all very well to be a "bookworm", but employers now require evidence of an ability to apply knowledge to a job in a variety of different ways. Information Technology (IT) and computer literacy are becoming integral components of education and the workplace. The communication of information in a variety of different forms is also considered important.

GIS is a technology for today, tomorrow and the future. It is having an almost universal impact on society. The educational changes taking place are enabling GIS education and training for the future. Providing the right educational framework will in turn facilitate a proper means to educate about and use this technology.

A case for developing educational continuity and progression in GIS education

To use GIS properly in the workplace strictly requires a sound and in-depth knowledge of subjects such as Information Technology (IT), mathematics, geography, data analysis, geo-statistics, remote sensing, cartography, environmental issues, error, and many more. However, this expectation may currently be rather idealistic, especially in light of the limited time available to cover all the subject matter in a one year postgraduate degree or diploma programme. A number of problems associated with the relative newness of the technology, incoming student background, and the duration of the current courses still need to be resolved in order to maximize the value and level of GIS education; problems which could be overcome through the development of educational progression and continuity.

For students now enrolling in postgraduate GIS courses, some assumptions still have to be made about their educational background. While many of the current course intake do have a geographical background, some do not. A number of educational prerequisites may therefore need to be imposed, particularly for those students with little background in the so-called GIS component subjects or, for that matter, geography. Unfortunately many geography students entering postgraduate education still have a fairly limited working knowledge of computers, IT, mathematics, statistics, cartography and remote sensing, particularly at a level which could be considered ideal or appropriate background for a postgraduate GIS course. Although this problem can be overcome, it can place some severe constraints on the level of teaching it is possible to achieve in a one year postgraduate course, especially if additional remedial-type modules for each component topic must be included in order to bring students up to a common standard. Furthermore, this situation could worsen in the future as GIS becomes a tool for more and more disciplines, e.g. Archaeology, Zoology, Agriculture, Forestry and the Environmental Sciences.

One way around such problems in the future is to ensure that the essential components e.g. a good working knowledge of the fundamentals of co-ordinate systems, statistics and maps is introduced and taught much earlier on in the educational curriculum. This policy would ensure the necessary foundation for later years. GIS need not be taught as such but it would mean that greater emphasis would be placed upon developing a sound working knowledge of the appropriate components necessary for GIS, together perhaps with some awareness and applications of GIS to place the theory into context.

Although GIS education originated at the postgraduate level, as GIS gradually filters downwards into undergraduate and school education many new opportunities will open up. In the long term they will enable postgraduate GIS courses to build upon the education system as a whole in order to fulfil the ideal prerequisites for good postgraduate GIS education and training. With the introduction of GIS into both the undergraduate and school curriculum, there is now a good opportunity to initiate educational progression and continuity for the future benefit of GIS education as a whole. With the growing multi-disciplinarity of GIS applications there is all the more reason to ensure the development of a sound background in the fundamentals of geography and other related subjects at the school level so that students (who potentially may return to GIS much later on in life) can cope with the geographical and other aspects at the more advanced level.

GIS in the secondary education curriculum

> Some specialized aspects of IT are best taught in geography and are included in the geography Order (NCC, 1991, p. C18);
>
> Geographical Information Systems (GIS) can provide schools with a means of using IT to handle spatially referenced data. (NCC, 1991, p. C18);
>
> Evaluate the effectiveness of a composite thematic map as a Geographical Information System. (DES, 1991, p. 6);
>
> Use transparent overlays or information technology to combine data and show inter-relationships, e.g. use information about relief, geology, soils, land-use, settlements and road systems to select a location for a new local authority rubbish dump; evaluate the resulting composite thematic map as a way of providing spatial information. (DES, 1991, p. 6).

With GIS now written into the National Curriculum for Geography (Cassettari, 1991; Freeman, 1991; Heywood and Petch, 1991; Green, 1991b; Unwin, 1990), attention has turned towards teaching GIS or "about GIS" in the school curriculum. The addition of a GIS component into school geography is an important new development and could have a significant impact upon the overall structure and content of geography and GIS courses throughout the education system in the future.

Reading through the above definitions, however, many school teachers would probably argue that they have already been teaching this subject for many years, though admittedly not under the heading of GIS, e.g. using transparencies and overlay techniques. Although the basics of GIS are not considered new, the placement of such techniques within a computer environment, the potential for new types of enquiry and analysis, and the relevance of GIS to the real world are new for schools. It is awareness of this fact that is perhaps most important and especially when set in the context of the fundamental principles, IT, maps, databases, visualization, and enquiry.

Why is awareness, and elementary knowledge and understanding of GIS at the school level important? First, many of the fundamentals of GIS and some of the basic concepts are in fact already taught, though usually separately, in primary and secondary education in subject areas such as IT, mathematics, and geography e.g. modelling, simulation, remote sensing, maps, and co-ordinates. For example, many mathematics books for the 11-12 year old age group provide examples of questions which use maps, involve measurement of distances, and the calculation of areas. Second, GIS offers a focus, not only for geography but also for many other subjects in the school curriculum. At the very least, GIS offers school age pupils a glimpse of how their knowledge of mathematics, statistics, geography (human and physical), maps, and even remote sensing can be brought together in a practical and relevant everyday application. Indeed, some schools have already been exposed to GIS through the Domesday project which offered school-age pupils some insight into the potential of bringing together lots of different geographical information in many different forms—pictures, images, maps and text—into a single user environment. Overall the advantages are that GIS offers a practical element, has the potential to link disparate areas in the school curriculum, offers new routes of enquiry, visualization, and perhaps most importantly can be related to real world applications.

This potential does not mean, however, that GIS must be taught as a pure subject in schools or that it must replace geography and other subjects. Rather it offers one way to teach about the geography of areas (Unwin, 1990), environmental issues, the fundamentals of geography and other subjects in the National Curriculum.

What still seems to be missing in geography education at the school level is a clearly defined focus for pupils, one that facilitates the integration of knowledge in an applications-type environment. All too often school pupils do not see how knowledge acquired from many apparently disparate subjects comes together. At the end of their school education they do not see where geography and many other subjects can lead to in the future. Surely this is not a desirable endpoint for geography education in secondary schools?

Some teachers have questioned the value of teaching GIS in secondary school. In part this scepticism may stem from a number of misconceptions about what GIS is, and what aspects of GIS (in its most technical sense) actually need to be taught to the school-age pupil. To many teachers GIS seemingly requires the use of both sophisticated and expensive hardware and software and may even involve some retraining. However, teaching GIS or about GIS does not necessarily require the use of computer technology. Schools without computer hardware and software are not therefore excluded. The fundamentals of GIS that are required at an early age. Teaching about components of GIS, together with example applications of GIS using illustrative materials e.g. literature, demonstrations and videos, would therefore be sufficient. On the other hand, the availability of computers could mean that GIS be used as a very specific example application of IT. This method would help to fulfil some of the cross-curricular requirements specified in the National Curriculum (NC) documents.

Other misconceptions have probably resulted from the lack of guidance and provision of resources needed for both teachers and pupils in order to introduce them to GIS as outlined in the NC. While there is now quite a lot of software and teaching materials available for teaching GIS in higher education, most of these are far too advanced for the school-age pupil. There are also few suitable articles available on GIS, and which are jargon free, to provide not only the necessary GIS awareness but also information on how to proceed. Furthermore, the links between GIS and IT, the use of spreadsheets, databases, modelling, cartography (maps), and remote sensing have yet to be described. To date relatively few resource packs have been developed or are widely available. There is also a paucity of suitable computer software.

Clearly, new ways are needed to help raise awareness, provide examples, and speed up the learning curve of teachers and pupils if GIS is to become a reality in schools. However, this implementation will require co-ordination and help from GIS experts in higher education, government and commerce.

GIS in higher education

Undergraduate

In the last few years some institutions of higher education have initiated new GIS undergraduate degree and diploma courses. Many more are offering new GIS course options or modules as components of undergraduate geography degree programmes. In others, GIS applications form a major component of courses on Applied Geomorphology, Hydrology, and Environmental Issues. Where institutions have followed the modularization route, these GIS options are also no longer the domain only of the geography student.

The growth in the number of courses on offer in part reflects the demand from students for awareness of and knowledge about GIS. After all it is where the jobs are at present. There is also possibly a desire by many institutions to jump on the GIS bandwagon at the present time in order to remain institutionally competitive.

With the development of all these different courses there has also been more demand to use computer-based technology for "hands-on" experience to supplement the GIS theory taught in lecture classes. There has been an increasing need for low-cost software to run on basic low-cost hardware for class teaching. Initially many courses have relied upon the standard textbooks, journals and conference literature as sources of information and ideas. However, with growing pressure on staff teaching time there has been demand for "ready-made" course notes and teaching materials. The most widely used of these has been the NCGIA (National Center for Geographical Information and Analysis) Core Curriculum (Goodchild and Kemp, 1990) which has so far proved very popular and extremely good value. These materials have either been used in their entirety, as the basis for a course, or selectively to supplement other materials.

Similarly, tutorials supplied with GIS software have often formed the initial basis for practical teaching. Subsequently, "in-house" exercises and tutorials have been developed to introduce students to specific applications of the software. More recently a number of practical workbooks have been made available by the NCGIA, Clark University, and Leicester. Clark University is also now developing workbooks which cover specific areas of GIS applications e.g. coastal and marine studies, and forestry. In addition, TYDAC, through its Institution for GIS in Education, has developed a range of educational teaching materials for SPANS software (Heywood, 1992).

Online resources for higher education have also grown in popularity. Electronic forums provide a quick and effective means of communicating information for those interested in GIS. The best known of these is the GIS-Listserver from Buffalo. More recently a special IDRISI users' discussion list has been established, providing an opportunity for users of IDRISI software to develop and swap new modules, to discuss problems and different applications.

Increasingly, sophisticated low-cost GIS software is finding its way onto low cost computer platforms. This is ideal for use in a teaching classroom environment where 20 or more machines are required at one sitting. Growing use of the windows environment makes such software ideal for teaching and training. One of the most interesting products recently is Alexander the integrated remote sensing and GIS software from ITC in the Netherlands. It is the first piece of GIS software for the Acorn Archimedes/A3000/A5000 series of microcomputers. The advantages of Alexander are that it runs within a windows type environment, is intuitive, provides a familiar environment for the regular Acorn user, and is relatively low-cost. It also comes complete with a set of online tutorials and a facility to develop new tutorials. Together these options make the package ideal for education and training at a number of different levels. The CTICG (Computers in Teaching Initiative Centre for Geography) at Leicester University has also been responsible for porting GRASS (Geographic Resources Analysis Support System) onto the Acorn R260 UNIX workstation. In addition, IDRISI can run under the Acorn PC emulator, making the Acorn platform quite attractive for education.

Postgraduate

Although there are already many very well established postgraduate degree/diploma GIS courses, a number of new ones have been started in the past year e.g. an MPhil in GIS at Cambridge University. Clearly there is sufficient demand to support these new courses, at least in the foreseeable future. In addition, a growing number of postgraduate degree and diploma courses in geography, remote sensing and the environmental sciences are now also offering optional modules in GIS. For example, Aberdeen University provides a GIS module for postgraduate degree and diploma students in both the Rural and Regional Resource Planning (RRRP) course and the Environmental Remote Sensing (ERS) programme.

As more students become educationally aware of GIS and have the opportunity to take GIS courses at the undergraduate level as a part of their curriculum—as an option or a compulsory core course component in a number of subjects—then it is quite likely that many will seek further opportunities to take postgraduate courses and training in GIS. To be educationally effective such courses should seek to build upon existing knowledge of GIS or to develop introductory theory into practice. Given that many students may already have been exposed to GIS as a subject much lower down the education system, it may be necessary in the coming years to adjust the content of the postgraduate GIS course curriculum to meet the likely a priori knowledge of students. In the future it may then be possible for a higher education course to make many more assumptions about a student's knowledge at the point when they enter postgraduate education, thereby facilitating GIS education in much greater depth. In the long term, educational continuity and progression could therefore lead to many new opportunities for both the student and the tutor.

GIS and training

Training and short courses

Looking through the list of news items included in most newsletters, journals and electronic mail messages, it is hard to avoid noticing the growing number of GIS short courses and training sessions currently being offered by universities, vendors and GIS special interest groups. Typically these offer a number of different activities ranging from the introductory to the more advanced levels. While some aim simply to raise awareness many others provide "hands-on" experience with a commercial GIS software product, e.g. ARC/INFO, IDRISI, or SPANS, or even larger systems.

Training, however, does not equate with, nor does it replace, the need for a sound education in GIS—even if the intricate workings of a GIS system are increasingly being hidden behind a user-friendly graphic user interface (GUI). Nevertheless, training does have a very important role within the overall structure of the education system to provide appropriate guidance on using specialized hardware and software. In many ways it is an extension of education. Usually this means that training goes beyond introductory "hands-on" experience with a system and typically offers exposure to, and experience and familiarity with a specific GIS system.

GIS training courses often cater for people who are being offered new career opportunities in GIS or who wish to enhance their existing GIS knowledge, and therefore require in-depth working knowledge with a particular GIS system. Such people may not necessarily have a background in either geography or GIS. Instead they wish merely to use the technology as a tool, either out of necessity or as part of their GIS education. Some may already have access to GIS software but want to learn more about its operation, or even be at a stage when they want to begin to use the software for a particular application.

Graphic User Interfaces (GUI)

Increasingly, software developers are turning to the use of a windows-like environment as the user interface for most GIS systems, helping users to interact more effectively and efficiently with the GIS software. It also helps them to learn how to use a system more quickly and to make use of more GIS software through a common style of interface.

In the past, interfaces to GIS software have often required a good working knowledge of a command language through a Command Line Interface (CLI). Often they have been difficult for the inexperienced user to get to grips with. In addition to the requirements for knowledge about GIS, use of such systems often needed considerable additional knowledge of computers, operating systems, files, and database structures. To this end, the availability of Graphic User Interfaces (GUI) is helping the "usability" of GIS (a term used by Hearnshaw and Medyckyj-Scott, 1991) in a number of different ways. First, as with any other software, research evidence has revealed that the "learning curve" is greatly reduced when a user is confronted with a GUI. It is intuitive, easy to use, operates like other software for a single system, but moreover is user-friendly. It facilitates use of applications software without having to know what goes on behind the screen or inside the box. While this situation is not always perceived to be an advantage, on the other hand it does mean that a system user is able to spend far more time interacting with the data and the information than with the operating system.

Other advantages of a GUI apply not only to the potential long-term user of a system but also to the personnel faced with training a new user who may have only limited knowledge of computers and software. A GUI offers considerable potential for GIS system training through its ease of use. In the future this type of interface will provide an important link between the user and the applications software to aid in system familiarization and applications.

GIS tutors

GIST, the GIS Tutor from Birkbeck College (Raper and Green, 1989) has proved to be an extremely useful piece of educational and training support software both for the potential GIS user interested in learning the basics of GIS theory and also as a reference for the seasoned user. GIST is also particularly useful as a "resource-based" learning package which can be used in a number of different ways with a wide variety of age groups and at many levels of education. Although the software permits only limited user interaction, it successfully combines theoretical, animated and graphic components of GIS theory into a stimulating and interesting format. Implementation on the Apple Macintosh also helps to make it an attractive package. A new version of the software to be released very soon will offer an even more useful introductory guide to GIS.

GIS syllabus and curriculum

Coggins (1990) and Unwin (1990) have discussed the development of GIS syllabuses and curriculums for secondary and higher education respectively. In the meantime, the NCGIA Core Curriculum from the US has been widely adopted in the UK and Europe. Now there is also talk of a European GIS Core Curriculum being developed, with a distinct European flavour and including case studies. Whether it will be entirely new or a modification of the existing NCGIA Core Curriculum still remains to be seen.

This whole area is still an important one for consideration in the future. With the opportunity now to develop educational progression and continuity, from secondary education through to training, it is more than likely that development of a GIS syllabus and curriculum will again become a topic for discussion, especially in relation to the EC.

A decision on what might be included at each level in the education system would be an invaluable step forward and would also help to reduce the possibilities for repetition, overlap, and confusion at different levels within the education system. It would also provide a good opportunity to develop a broad overall structure and framework for GIS education. However, it should be a collaborative effort between teachers, local authorities, vendors and software developers to provide the most satisfactory solution.

Miller (1992) has recently described something along these lines: a proposal for a new four-year undergraduate curriculum designed to be the foundation for what he terms "the New Geographer". This effort is a collaboration between ESRI (Environmental Systems Research Institute) and the University of Redlands. The curriculum seeks to provide for a multi-disciplinary focus and systems approach to environmental problem solving. It includes elements of Geography, Mathematics, Physical and Environmental Science, Computer Science, Humanities and Social Science, GIS, Business and Economics, and Environmental Design Studies.

Conferences, workshops, journals, newsletters and special interest groups

During the past year there have been a number of GIS and education sessions at major GIS conferences in the UK, Europe and the US, e.g. AGI'91, EGIS'91 and '92. These conferences have covered GIS developments in both secondary and higher education as well as in training. Although the education and training sessions have typically attracted relatively small numbers of people—probably because education and training are far less glamorous than sophisticated system applications—there nevertheless appears to be growing interest in both areas. In the future this will be particularly important in relation to three areas: GIS in school education, GIS education and training in the Eastern European countries, and new applications of GIS, e.g.

multimedia. The importance of GIS education and training in a European context will become very important over the next few years (Masser, 1991). Already a special interest group, under the wings of the EGIS Conference, has been established to cater for the future needs to set up a dialogue, provide a forum for the discussion and the dissemination of information.

While these sessions have previously tended toward a lecture type format, future conferences will probably see a demand for more "hands-on" workshop education and training sessions. There is perhaps a lesson to be learned here as far as conference content and space allocation at future conference programmes are concerned. There are numerous possibilities open, one of which could be for distance "hands-on" GIS education and training workshops accessing GIS software located at remote sites.

GIS Journals and Magazines, such as the *IJGIS* (the *International Journal of Geographic Information Systems*), *GIS Europe,* and *Mapping Awareness,* continue to be major sources of educational news and information with their sections on GIS in education and training. Likewise newsletters from special interest groups (the British Cartographic Society GIS Special Interest Group), vendors and software developers (e.g. GRASS CLIPPINGS; ARC News from ESRI) frequently offer news on GIS education and training. All of these will continue to be useful sources and a valuable means for disseminating information in the future.

Educational software and materials

Outside of the mainstream GIS software which is marketed commercially for use in higher education and training there have also been a number of other recent developments. Interest has been expressed by both the larger and smaller GIS software developers and vendors to provide "cut-down" versions of software for use at both the school and the undergraduate level. This software could be used for demonstrations by tutors and/or teaching exercises or for student "hands-on" work. Others are interested in the possibilities offered by GIS emulation and multimedia to provide GIS experience without the need for expensive software, hardware and peripherals.

There are still many gaps in the marketplace for both software and introductory teaching materials suitable for GIS education and training. This situation is gradually changing, however, and there are now a number of resource packs, workbooks and software packages in various stages of development and completion for both secondary and higher education and training. These include AEGIS (Freeman, 1991) for schools, and the NCGIA schools initiative to develop a schoolteachers' manual and student workbook (Palladino, 1992), tutors and multimedia. It is likely that many more such materials will become available in the very near future to cater for the current and future needs of both GIS education and training.

Responsibility

The question of who should be responsible for education and training frequently arises. Is it higher education, the vendors, the software developers, or all of them? Who is best qualified, equipped and resourced to undertake training and/or education?

For the most part it is difficult for institutions in higher education to cater for training on large commercial GIS. Currently, finances are simply not available to justify or support the purchase of the large number of expensive systems that would be required. However, with agreements like CHEST (Combined Higher Education Software Team) and educational discounts from other software suppliers, higher education can offer both education and limited training on some of the smaller systems such as IDRISI, SPANS and ARC/INFO. More often than not, however, specialist training with a GIS is best provided by vendors or systems developers who have the most intimate working knowledge of the larger GIS, and access to the required number of systems.

A GIS education and training body

In order to provide educational progression and continuity, however, there is still a need to bring together teachers, academics, software developers, vendors and local authorities. At present, for example, different levels in the education system are still far too isolated from each other and from the real world. School education is to some extent cut off from the rest of the education system, e.g. they are not included in the GIS conference circuit or on the Janet Email system.

Collaboration between interested parties must be encouraged but it must also be properly co-ordinated by some sort of umbrella organization. A recognized body is now needed to co-ordinate all the developments which are now underway. Already some attempts towards this co-ordination are coming about through the AGI (Association of Geographic Information), NCET (the National Council for Educational Technology), the GA (Geographic Association) and, similarly, the EGIS effort in Europe. However, positive and concerted action is now required to enable progress to be made on this front.

Finally

GIS offers a focus for geography, a tool, and also a means of teaching not only about geography but also about other disciplines. The development of educational progression in GIS education has the potential to help improve the GIS background a person has when entering a new and higher level in the education system. Ultimately it will lead to a better knowledge of the components and fundamentals of GIS and overall awareness in relation to geography. The structure of such an education system will also offer new and interesting ways to help those returning to education to acquire new knowledge and to develop new skills which are directly applicable when they return to the workplace. A well-developed education and training system will provide the framework for the future. Clearly, however, not everyone is going to become a GIS specialist nor, for that matter, should they. However, if we want to make the best use of GIS in the future then we must educate people accordingly so that they have the basics both to use and understand GIS—what it is, what it does, and what it can be used for. If people do not have the educational prerequisites for GIS how can we realize its full potential in the future?

The current interest being shown in GIS education and training is beginning to support the view that providing a sound education is vital. Recent developments now offer some promise for the future, but they must be co-ordinated.

GIS education and training in the AGI'92 Yearbook

The primary objective of bringing this collection of papers together in this book is to highlight some of the most recent issues and developments in GIS education and training. The order in which the papers are presented seeks to develop some feel for the potential of developing educational progression and continuity from the secondary school, through undergraduate and postgraduate education, to training. While these papers only examine a very small cross-section of all the possibilities one could cover in the limited space available, they nevertheless provide a valuable and up-to-date insight into and perspective on GIS education and training.

References

Cassettari, S., 1991, GIS and the National Curriculum, in *Proceedings of the AGI'91 Conference,* Birmingham, England.

Coggins, P.C., 1990, Horses for Courses: Education in GIS, in *Proceedings of the Mapping Awareness'90 Conference,* London, England: Blenheim Online, pp. 19-1–19-5.

Department of Education and Science (DES), 1991, *Geography in the National Curriculum (England),* March, London: HMSO.

Freeman, D., 1991, Development of an Education GIS for the National Curriculum, in *Proceedings of the AGI'91 Conference,* Birmingham, England.

Gilmartin, P. and Cowen, D., 1991. Educational Essentials for Today's and Tomorrow's Jobs in Cartography and Geographic Information Systems, *Cartography and Geographic Information Systems,* **18**(4), pp. 262–7.

Gittings, B.M., and Mounsey, H.M., 1989. GIS and LIS Training in Britain, in *Proceedings of the First AGI Conference,* Birmingham, England, pp. 4.4.1–4.4.4.

Goodbrand, C., 1991, Educational and Geographical Information Systems, in, Heit, M., and Shortreid, A., (Eds), *GIS Applications in Natural Resources,* Colorado, USA: GIS World, pp. 61–63.

Goodchild, M.J., and Kemp, K.K., 1990, *NCGIA Core Curriculum for GIS,* Santa Barbara: NCGIA, National Center for Geographic Information and Analysis, University of California.

Green, D.R., 1991a, Education versus Training: A Call for Comment from the Commercial World, *Mapping Awareness,* **5**(7), pp. 14-16, 41.

Green, D.R., 1991b, GIS Software for Schools: What about it? *Mapping Awareness,* **5**(9), pp. 34-36.

Green, D.R., and McEwen, L.J., 1989, GIS as a Component of Information Technology Courses in Higher Education. Meeting the Requirements of Employers, in *Proceedings of the First AGI Conference,* Birmingham, England, pp. C.1.1–C.1.6.

Hearnshaw, H., and Medyckyj-Scott, D., 1991, How Usable is Your GIS? in *Proceedings of the AGI'91 Conference,* Birmingham, England.

Heywood, D.I., 1992, *The Institute for GIS in Education,* Personal Communication.

Heywood, D.I., and Petch, J., 1991, GIS: A Toybox Approach, in *Proceedings of the AGI'91 Conference,* Birmingham, England.

Masser, I., 1991, Promoting GIS Awareness: The European Dimension, *Mapping Awareness,* **5**(9), pp. 9–13.

Miller, W.R., 1992, Creating the New Geographer, Education News, *ARC News, Winter,* **14**(1), p. 18.

National Curriculum Council, (NCC), 1991, *Geography in the National Curriculum,* York, England.

Palladino, S., 1992, Personal Communication.

Raper, J., and Green, N.P.A., 1989, GIST—A New Approach to a Geographical Information System Tutor, in *Proceedings of the First AGI Conference,* Birmingham, England, pp. C.4.1–C.4.6.

Unwin, D.J., 1989a, Curriculum for Teaching Geographical Information Systems. Report to the AUTOCARTO Education Trust Fund, RICS.

Unwin, D.J., 1989b, Geographical Information Systems and School Geography: Some Essentially Random, but Cautionary, Thoughts, Communication to the Geographical Association GIS in Secondary Education Working Committee, Sheffield, England.

Unwin, D.J., 1990, A Syllabus for Teaching Geographical Information Systems, Research Report No. 2, Midlands ESRC Regional Research Laboratory, University of Leicester.

Unwin, D.J., and Dale, P., 1989, An Educationalist's View of GIS, in *Proceedings of the First AGI Conference,* Birmingham, England, pp. C.3.1–C.3.8.

42

GIS and secondary education

Alan Wood
Kingston College of Further Education

and

Seppe Cassettari
Kingston Polytechnic

Secondary education

Secondary education, simply put, consists of programmes of learning running under a range of institutional and funding formats for the age group 11 to 16 with opportunities for additional study up to age 19. The picture is complicated by the existence of Further Education or the tertiary sector which frequently overlaps and integrates with secondary provision. Age-progression relationships are also less than straightforward with many pupils continuing to follow first-level secondary courses such as GCSE beyond the normally recognized completion age of 16.

Overall it is probably more appropriate to evaluate the place of GIS in secondary education through examination of the two broad streams of study, academic and vocational. These currently conclude with GCE A level and BTEC Ordinary National Level or equivalent and provide routes of progression to higher education or employment and continuing education via training and part-time study.

Within both of these streams teaching methodology and practice is frequently innovative. However, subject content and skills requirements, together with attainment levels, more normally result from reactive processes in response to requirements for progression. Future directions lie with curriculum initiatives from the Department of Education, academic bodies, higher education, examining bodies and industry in general. A cursory examination of the influences acting on the 11 to 16 programmes reveals that all areas are in a state of flux providing at the same time a degree of uncertainty and an opportunity for review and constructive change. The following addresses these developments and directions as they affect GIS and related disciplines.

Programmes for the 11 to 16 age group

Current practice within this sector is represented primarily by GCSE programmes supplemented by a vocational strand though Technical Vocational Educational Initiative (TVEI), City and Guilds and Royal Society of Arts options. All these are entering a period of transition influenced by the Education Reform Act of 1988 leading to the establishment of a National Curriculum for all ages up to the age of 16 and the creation of the National Council for Vocational Qualifications (NCVQ) criterion-based grading and grouping of skills.

Each in its own way reflects an appreciation of the need to rationalize the wide-ranging nature of subject content, skills requirements and standards offered by present academic and vocational "lead" bodies. Common points are criterion-based attainment targets set against programmes of study reflecting the need for core studies and associated foundation subjects; a cross-disciplinary approach and the recognition of general transferable skills.

The National Curriculum

The National Curriculum comprises core studies in English, Mathematics and Science, together with additional foundation subjects including geography and technology; two areas with implications for GIS (Cassettari, 1991a). For each subject the National Curriculum sets out a detailed programme of study with appropriate assessment arrangements. Included in this programme of study is a series of attainment targets, which are defined as "the knowledge, skills and understanding which pupils of different abilities and maturities are expected to have by the end of each key stage" (DES, 1990). A programme of study is defined as "the matters, skills and processes which are required to be taught to pupils of different abilities and maturities during each key stage" (DES, 1990). There are four-plus consecutive key stages which cover the years of compulsory schooling from 5 to 16.

The development of the National Curriculum for school children up to the age of 16 across Britain has provided individual disciplines with an opportunity to review their aims, content and methods of delivery. The proposals for Geography were published in early 1991 (DES, 1991) followed by the non-statutory guidance notes (NSG) for geography, which were published in May 1991 (NCC, 1991). The purpose of the NSG is "to give advice which reduces the workload involved in implementing the Order. It explains the key features of the attainment targets and programmes of study and offers advice on how to design a geography curriculum based on the Order" (NCC, 1991). Part of the NSG is the clear recognition that the National Curriculum is aimed at achieving a cross-disciplinary approach to the foundation subjects and a flexible approach to the teaching methods.

Structure of geography in the National Curriculum

The three aims of geography in the National Curriculum as set out in the NSG are:

- to help develop geographical knowledge and understanding;
- to introduce pupils to geographical enquiry;
- to help pupils develop a sense of identity through learning about the UK and its relationships with other countries.

There are five geography attainment targets (AT), each of which has ten levels of attainment on a continuous scale, covering all four key stages. At each level there are criterion based statements of attainment (SoA) which are more precise than the broader ATs and are related to one of the ten levels of attainment. The five ATs are:

- AT1 *Geographical Skills:* This covers the use of maps and fieldwork techniques;

- AT2 *Knowledge and Understanding of Places:* This covers the distinctive features, similarities and differences between places in local, regional, national and international contexts;

- AT3 *Physical Geography:* This covers weather and climate, rivers, seas and oceans, landforms, and animals, plants and soils;

- AT4 *Human Geography:* This covers populations, settlements, communications and movement, and economic activities;

- AT5 *Environmental Geography:* This covers the use and misuse of natural resources, the quality and vulnerability of different environments and possibilities for protecting and managing environments.

The four key stages relate to the ten levels of attainment in the following way:

- Key Stage 1: levels 1 to 3,
- Key Stage 2: levels 2 to 5,
- Key Stage 3: levels 3 to 7,
- Key Stage 4: levels 4 to 10.

For each key stage there is a programme of study (PoS), defining what should be taught to help pupils reach the appropriate targets (ATs). Both planning for and delivering the National Curriculum requires thought about both attainment targets and the programme of study together.

GIS in the National Curriculum

Attainment Target 1, Geographical Skills, is divided into two components—the use of maps and fieldwork techniques. It is intended that these skills are developed not in isolation but are integrated into the material that relates to the other ATs.

The use of the term Geographical Information System is specifically mentioned in AT2 in Attainment Level 10, under the section on the use of maps. Attainment Level 10(a) requires pupils to be able to "evaluate the effectiveness of a composite thematic map as a Geographical Information System".

While not specifically mentioned in the other Attainment Levels, many of the skills may be directly applied to the use of GIS. For Example AL 8(b) requires pupils to "synthesize information from different map sources to produce a sketch map which identifies important geographical features and reveals spatial patterns and associations within an area".

The ALs within AT1 are set out to introduce pupils to the use of maps and to develop their map work skills to a point where they will be able to understand the broader concepts of using GIS for problem solving. It might be argued that a more specific reference to GIS as a decision making tool might have been desirable.

The importance of AT1 as an integrating set of skills that cross the other ATs should not be underestimated. It is particularly significant that the National Curriculum has positively embraced the use of maps within geography. The specific reference to GIS within the ATs is an important progression to the development of map-based skills.

The introduction of the National Curriculum for Geography started with year groups 1, 3 and 7 in autumn 1991 and will be fully implemented by 1995. For many schools this will require considerable planning and a review of curriculum policies.

Cross-disciplinary skills with Geography

The cross-disciplinary theme is very strong within the National Curriculum. This is important to GIS which by its nature is an integrating subject ranging across the traditional discipline boundaries. The National Curriculum for Geography requires pupils to study the relationships between the ATs and their links to other subjects, including history, mathematics, science and technology. The NSG identifies the strong links between geography and information technology. It notes that "IT encourages pupils to handle information more effectively, to pose and test hypotheses, to communicate and present the results of geographical enquiries and to ensure and collect data about the environment". These are all elements that fall within the general definition of GIS.

It is recognized that many aspects of IT teaching, such as planning, continuity and progression, and resources need to be considered and may well be part of a school's overall IT curriculum policy. The NSG does, however, recognize that "some aspects of IT are best taught in geography". The example quoted is the increasing interest in earth observation from space as recognized in AT.

The NSG does specifically state that GIS "can provide schools with a means of using IT to handle spatially referenced data". It does not, however, expand upon this statement and may leave many teachers wondering where they can go for further advice.

Programmes for the 16 to 19 age group

Programmes within this sector can reasonably be divided into those which are essentially vocational, such as those validated and moderated by the Business Technician Education Council (BTEC) and those which are essentially academic as offered by the various examination boards for the General Certificate in Education (GCE) at Advanced level (The Associated Examining Board, 1991; University of London Examination Board, 1991).

GCE Advanced level

Syllabuses in this area are subject specific and vary in content and approach across the examining boards, although core content, option patterns and subject and general skills are observable in each. Modes of delivery focus towards externally set examination papers with project and coursework contributions playing a variable but minor role in final assessment.

Within this framework the place of spatial studies surfaces naturally in geography and environmental science. Each of these subjects illustrate the traditional nature of the discipline as an interface between the natural and human environments, and social, historic, economic and planning activities.

Closer examination of sample syllabuses reveals that skills in data collection, spatial analysis, interpretation and presentation are stressed and set against information sources such as maps, air photographs, remotely sensed images, economic and demographic statistics and fieldwork. However, computer-based data sources, techniques, general methodology and GIS are noticeable by their absence. This absence is puzzling given the emphasis on IT in subjects such as Business Studies, Design and Technology and Art and Design, where databases and CAD systems are specifically mentioned. This is not to say that awareness of IT, GIS and statistical software does not exist within the teaching of geographical subjects but the lack of a "top down" emphasis is surprising given the current growth of GIS in undergraduate courses and the fact that A level provides the dominant feeder into higher education.

Schools Examination and Assessment Council (SEAC) proposals are moving towards a more co-ordinated and modular framework across examining boards for the academic year 1994-95. This will hopefully offer the opportunity to redress the balance through the inclusion of IT studies in geography.

Vocational courses—BTEC

Vocational courses differ in their structure and are more usually module-based programmes grouped under broad generic headings and leading to full-time Diplomas (16 modules) and part-time Certificates (10 modules). Their derivation relates to the need to associate programmes of study with broad areas of industry in order to provide entry to employment following full-time study or facilitate continuing education and training through part-time courses set against a national standard. BTEC validating boards are made up of members from industry and education, ensuring a close link between need, approach and content.

Individual programmes are frequently designed to meet local requirements and may be compiled from a combination of BTEC standard syllabuses and centre-devised modules. Cross modular studies, IT, transferable common skills and student-centred resource-based learning are essential in all courses. A wide range of assessment methods appropriate to the skills being evaluated are encouraged and are frequently criterion-based, reflecting a move towards proposals being implemented through the National Council for Vocational Qualifications (NCVQ).

Identifying the inclusion of GIS within the vocational sector leads to those subjects naturally associated with the handling of spatially referenced data within the generic areas of cartography, surveying, planning, architecture and land use. Most recently courses with a GIS core theme have been approved at both national and higher national level.

National Vocational Qualifications (NVQs)

In 1986 the government established the National Council for Vocational Qualifications (NCVQ) to reform the system for vocational qualifications in England, Wales and Northern Ireland by creating a single, coherent and comprehensive framework based upon occupational standards set by industry (NCVQ, 1990). NVQs are nationally recognized qualifications linked to each major section of employment and set out to indicate clearly competencies covering knowledge, skills and the ability to apply them at work. Competencies and competence levels are decided by "lead bodies" made up of representatives from employers, trade unions and professional bodies, supported by the Training Agency of the Department of Employment, together with advisors with expertise in training and education. NVQ requirements are described in outcomes and emphasize achievement rather than learning processes, with the intent of allowing increased flexibility in course design and delivery.

NVQs are classified according to five levels, from I relating to routine work activities and foundation competencies, to IV and V which relate to complex and technically specialized activities including management and progression to professional status. Levels I and II relate to the 11 to 19 programmes, thereby overlapping with the secondary and tertiary sectors.

The Survey and Mapping Alliance, constituted from organizations with interest in the broad field of surveying and cartography, including the AGI and RICS, is currently considering the establishment of NVQs for the survey and mapping industry, which includes many aspects of GIS. Progress towards "mapping" the industry and defining competencies has not been as swift as many would have hoped.

Education and training for GIS

Education and training for GIS is now an established feature of conference programmes, journals and magazines. The discussions have typically focused on curriculum issues including overall design (Unwin *et al.,* 1989); planned programmes, for example NCGIA; relationships between education, training and awareness (Clark, 1991); employer requirements (Rhind, 1990) and specific practical trails (Freeman, 1991).

There is little doubt that a healthy and vigorous debate on the issues surrounding education and training have an important part to play in the development of GIS. However, the majority of those currently involved in the discussion are already committed users and able to see its value as an integrative approach to handling spatially referenced data in the context of decision-making and management. What is required is a broadening of the debate to include the more traditional geographer.

Outward looking initiatives within the secondary sector are relatively small in number but perhaps significant in their impact. The NCC statement on GIS in geography is a major and potentially precipitative step forward. The AEGIS trials support the viability of GIS studies at this level given the correct choice of resources and examples (Freeman, 1991)

Successful modules covering digital mapping and GIS have been incorporated within BTEC Ordinary National courses in cartography at Kingston College of Further Education (KCFE) since 1989 and will form the core of new Diploma and Certificate courses in GIS from September 1992, completing the final link in an integrated framework to degree level developed in conjunction with Kingston Polytechnic (Cassettari, 1991b). KCFE have also been involved in providing GIS workshops for GCE A level programmes in geography and computer science. There is then a sense that the introduction of GIS in the vocational courses is helping to filter the subject through into the academic programmes.

Experience from Kingston, from the GIS Awareness Programme (GAP) reported on by Clark (1991) and the AEGIS project reviewed by Freeman (1991), provide pointers towards the problems and possible solutions in the raising of GIS awareness in the secondary education sector:

- General papers and articles or advertising targeted at the uninformed seem to produce little reward by way of increasing interest, but personal contact through visits, seminars and exhibitions are more effective. The GIS community therefore needs to consider ways of establishing a physical presence in secondary schools.

- There needs to be a common and straightforward language register in relation to GIS principles, functions, operations and commands to make it easier for teachers to assimilate the principles and the technology into their teaching.

- The GIS community needs to appreciate the current knowledge and resource platform available in secondary education in order to provide the necessary tools to bridge between existing practices and GIS technology and methodology.

- An appreciation is required of the needs of education for limited and achievable off-the-shelf exercises supported by simple explanatory documentation suitable for pupil consumption in the natural timetable slots used for teaching.

- There is a need for on-going support and co-operation between vendors, employers, trainers and educators to promote GIS education and training, perhaps through the establishment of outward looking local centres of interest in GIS.

The future for GIS in secondary education is potentially bright but there are a number of problems that threaten the wider adoption of the technology, in particular the amount and pace of change in the sector, the lack of overall awareness and the limited resource base from which most institutions start.

References

The Associated Examining Board, 1991, ''1993 syllabuses for GCE Advanced Level and Advanced Supplementary Level''.

Cassettari, S., 1991a, GIS and the National Curriculum, in *Proceedings of the AGI '91 Conference,* Birmingham.

Cassettari, S., 1991b, ''Geographical Information Systems at Kingston'', *Cartographic Journal,* **28**.

Clark, M.J., 1991, GIS Awareness The technical and educational challenge, in *Proceedings of the AGI '91 Conference,* Birmingham.

Department of Education and Science, (DES), 1990, ''Geography for Ages 5 to 16''—Proposals of the Secretary of State for Education and Science and the Secretary of State for Wales, London: HMSO.

Department of Education and Science, (DES), 1991, *Geography in the National Curriculum (England),* London: HMSO.

Freeman, D., 1991, Development of and education GIS for the National Curriculum, in *Proceedings of the AGI '91 Conference,* Birmingham.

National Council for Vocational Qualifications, (NCVQ), 1990, ''A Brief Guide'' Central Office of Information.

National Curriculum Council, (NCC), 1991, Geography non-statutory guidance.

Rhind, D., 1990, ''Research, education and training in GIS—Where the AGI fits in'', *AGI Yearbook 1990,* London: Taylor & Francis, pp. 313–7.

University of London Schools Examination Board for GCE, 1991, ''Regulations and syllabuses. June 1993 to January 1991'.
Unwin, D.J., 1989, ''Curriculum for Teaching Geographical Information Systems'', *Report to the Education Trust Fund of AUTOCARTO for the RICS*, London.

43

GIS and secondary education in the United States

Steve Palladino
The National Center for Geographic Information and Analysis (NCGIA)

Introduction

While GIS education is a major growth area in US universities and colleges, the presence of geographic information system activities in the nation's schools is rare. With the rapid growth in GIS utilization in both the private and public sectors in the past decade, it is disappointing that this revolution has had relatively little impact on secondary or primary education. This article explores the place of GIS in the secondary schools (grades 7-12, ages 12-18), the nature of the US educational system, ongoing GIS activities in the schools, and present efforts by the National Center for Geographic Information and Analysis (NCGIA).

The place of GIS in the school classroom

Three key educational themes often mentioned are the push for increased use of information technology, the emphasis on teaching students to think analytically, and the call for renewed instruction in geography. Though GIS alone cannot satisfy these themes, it can be a major component. The desire for computers to play a more important role in the schools is made ever easier to fulfil as micro-computers continue to become both less expensive and more powerful. With an abundant supply of micro-computer-based GISs and improved user interfaces, it is reasonable to assume that a computer-based exploration of GIS is within the grasp of many schools. A GIS can participate as both a storage medium and a powerful analytical tool for geographic data in the improvement of students' analytical abilities, and it can aid instruction in geography, as well as in science, technology, and other subjects.

As these educational themes find an ally in GIS, the issue of how best to make use of the potential of this technology must be confronted. Three levels of GIS implementation can be identified, each with its own needs for instructional materials and different expected results. The most basic level is that of simple exposure to the GIS as an information technology and its place in society. The goal of this exposure is to make students aware of GIS applications, the fundamental capabilities of the systems and the occupational opportunities they present. An intermediate level of implementation involves the use of GIS as an instructional tool to support existing learning objectives. In this mode the GIS is a method of data storage, retrieval, and simple analysis, rather than the center point of the classroom activity. At the final, most GIS intensive level, the systems are a major focus of study or are an integral part of an analysis oriented class project.

Each of these levels presents an unique pedagogic challenge and calls for varying combinations of instructional materials and for appropriate computer hardware and software configurations. At all levels there is a need for written materials designed with the expressed purpose of explaining GIS in general terms to secondary school students.

These materials should include an introduction to spatial data, use of GIS to store and manipulate this form of data, various applications areas using GIS, basic GIS functions such as buffer, overlay and area computation, and ways in which GIS can impact the students' futures. These general introductions to GIS could be presented in forms other than text alone. Media such as overhead projector transparencies, video cassette, video disk, hypertext, and simple software modules could be employed.

At the intermediate level of GIS implementation, there is a need for a simple GIS package or for a Pseudo-GIS designed for educational purposes. A Pseudo-GIS performs only simple geographic data display and query, but omits excessively complex data handling and analysis. It serves purely as an educational tool, much like the type of inelegant microscope used in a school biology course, but prepares students for work with a more complex GIS. Both types of packages should emphasize ease of use through graphic user interfaces, simplified menus and educational modules. The AEGIS package being developed for use in UK schools is an example of an attempt to provide an appropriate GIS for educational purposes. The most intensive level, GIS-based analysis, can employ fully-functional GIS packages, but still requires supporting instructional materials. Here too, student friendly packages will most likely meet with the greatest success.

GIS instruction at any of these three levels could take place in a variety of courses. Obvious choices are geography, earth science, biology and environmental studies. Other courses which could easily incorporate some aspect of GIS into their structure are history, economics, computer science, physical science and mathematics. Since practically all disciplines consider elements that are in some way spatially referenced, a fertile imagination will be able to construe some use of GIS in just about any course.

Obstacles do exist in the path of expanded GIS presence in the schools. A primary barrier is the scarcity of materials designed specifically to adapt GIS concepts to secondary school requirements. Increasing teacher awareness of GIS is fundamental. Without support materials, however, even motivated teachers often do not have the resources to initiate study with or about GIS. Also an important issue is that of hardware and software availability and cost. Many school budgets are ill-prepared to bear the financial burden of GIS in its standard pricey manifestation.

The situation in the United States

An overview of the current educational environment in the US is provided as a context for discussion of GIS potential in the schools and of existing activities in this country. Unlike England and Wales where the Education Reform Act 1988 mandated a National Curriculum, a comprehensive national curriculum does not exist in the US. In fact, very little curriculum control has been attempted at the national level due in part to the federal system of governance. Here the states retain control over education and have often relinquished curriculum decision making to the local school district level. Thus GIS will not find its way into the US schools through inclusion in a national curriculum like it has implicitly, with some creative thinking, at various levels in the geography curriculum, and explicitly, Attainment Target 1 for Attainment Level 10 for geography in the English and Welsh schools (Cassettari, 1991).

Without central control over curriculum it would seem that an impossible heterogeneity in courses would result. Some semblance of order, however, does exist. The states of California, Texas and Florida establish textbook adoption guidelines for all their schools. Since these large states provide a cohesive market, publishers follow their content requirements which results in *de facto* textbook standards for the nation (Gaile and Willmott, 1989). A result of this pattern is that there are few texts for subjects which are not strong in these adoption states. Since geography and other spatially oriented topics which might support the introduction of GIS are lacking in the course offerings of these states, there is a corresponding dearth of text material which would support GIS activities or to which discussion of GIS could be appended.

Geography, one possible home of GIS education (Kemp *et al.,* 1992), is a natural place in the schools in which to introduce GIS activities. Unfortunately this model, used in England and Wales, will not work well in the US where geography is one of the under-represented subjects. When it is taught, geography is often incorporated as part of a course labeled "social studies". In 1989, Hill noted that less than 10 per cent of secondary school students were enrolled in geography courses. Unlike the case in the UK where successful battles were fought in the 1980s to keep geography as a distinct and favoured discipline, geography in the US is struggling to be resurrected after a long interment (King, 1989). Favorable conditions exist in public opinion for a resurgence in geography, but this important subject still faces many battles for recognition in a curriculum environment dominated by history. Very few social studies teachers or social studies specialists in state education departments are trained in geography (Gardner, 1986). However, geography education has made some important strides in the last few years, beginning with the Geographic Education National Implementation Project (GENIP) in 1985. This broad-based project defined a framework of five fundamental themes: location, place, relationships within places, movement over the earth and regions. The establishment of themes was part of an attempt to help communicate geography to the diverse community of social studies educators and others whose support was vital to the strengthening of geography in the schools. Another grass roots effort that is having a significant impact is that of the National Geographic Society's network of Geographic alliances which are active in almost every state. These state Alliances seek to "mobilize teachers, school administrators, and education policy makers to push for improved geography curriculums in their states" (Grosevenor, 1989). Despite these constructive developments, geography education still has a long way to go in the US.

This weakness in geography coupled with the common deficiency in technology and vocational education makes the introduction of GIS in the schools more challenging. Without the natural personnel and curricular bases from which to introduce GIS, alternate strategies must be considered. Fortunately, though it is always geographic and technological in nature, GIS is adaptable to many existing course objectives. As GIS becomes a stimulating topic and tool in US classrooms, it is likely that the teachers and students who are exposed to it will desire increased involvement with the geography, science, computing, math and environmental models inherent in a GIS view of the world. GIS can be an important promoter of the integrated, stimulating education experience for which some in the US education community are calling.

Given this educational environment, it is not surprising that there appear to be very few cases of GIS use or even exposure in the US secondary schools. Schools that have begun to utilize GIS tend towards the extraordinary. In Ohio, the exclusive, private University School, is using SPANS in an effort to manage their near pristine, 200 acre school property. They hope to expand these GIS management efforts to the neighboring state parks. In Virginia, the Thomas Jefferson High School for Science and Technology, a special public school for precocious teens, is using IDRISI and pcARC/INFO in its geoscience department. These students, who all learn to program in Pascal and other languages, not only use GIS, but some are even creating their own GIS programs. Both of these cases, however, represent the fringe. They are well endowed in computer facilities, inspired teachers and students of high caliber. These schools exemplify the situations in which the most involved level of use, GIS based analysis, can take place. In Oregon, the State Department of Education has instituted a program called Workforce 2000. This program seeks to train community college (two-year colleges) and high school students in the use of modern technologies. Students are learning the use of Global Positioning Systems and a GIS, GeoSQL, for forestry management. At the university level, a North Carolina faculty member has been exploring the use of GIS to enhance teaching of geology in that state's high schools. Initially it may be most common, as in this case, for individuals in the university, government, or private industry sectors who are intimately involved with GIS to instigate activities in the schools. Nevertheless, concerted efforts by motivated teachers are required to make GIS use in schools a reality. This level of required effort possibly explains why, while there are probably other cases in which some limited GIS exposure is taking place, for the most part GIS has had very little impact on the US schools.

Prime culprits in this poor showing are the lack of both teacher and general public awareness of GIS and a lack of materials designed to expose students to GIS. Popular awareness is slowly increasing with the growing role of GIS in our daily lives. However, efforts need to be made to provide teachers and students with some basic GIS materials and software and a framework for their use. Creative projects and lesson plans which incorporate GIS in a variety of courses need to be developed and tested. Despite the lack of momentum, secondary education and the university communities in the US are warming up to the idea of GIS in the secondary schools. However, the strongest push has recently come from vendors and software developers. It appears that their marketing senses are picking up on the vast potential of GIS in pre-collegiate education. Interest has been expressed for supporting secondary educational activities with MapInfo, GRASS, ArcView, and the IDRISI-based Global Change Database Project. Although they are excited about the possibilities, communication with these companies indicates a desire on their part for a framework to guide development efforts.

The NCGIA secondary education project

The NCGIA, as a national research center, is attempting to fill this structural and materials vacuum. The NCGIA was established by the US National Science Foundation, as a consortium of three universities, to take a leadership role at the academic level in basic GIS research. The Center was also given an educational mandate. Due to the rapid growth and development of the GIS field, there was a lack of individuals prepared for careers in GIS research or applications. The Center responded to this need by developing, in conjunction with GIS researchers worldwide, a Core Curriculum in GIS (Goodchild and Kemp, 1990). This curriculum is a one year sequence of courses for the upper level undergraduate and the graduate student. This successful project has encouraged and aided GIS education programs at many universities in North America and elsewhere.

Having played a major role in university level GIS education, the NCGIA sees a role for the Center at the pre-collegiate level. The need for some form of framework in which GIS will find acceptance and utility in the schools is apparent. To this end, the Center has begun a secondary education project which will support GIS implementation in the secondary schools. The Center is acting as a communication node for interested parties in the universities, schools, and the GIS industry. In addition, the Center has embarked on a year-long instructional materials development project by collecting and developing materials that provide basic instruction in GIS and that support GIS use as a tool for exploring existing subject areas. In order to provide truly useful products, secondary school teachers will be intimately involved in the project mainly through participation in a week-long workshop where they will act as consultants for the evaluation and creation of materials most appropriate to their instructional needs. The workshop will include a short course in GIS to provide a foundation for these consultation efforts. Following this intensive week of involvement with GIS concepts, software, and educational materials, the teachers will produce GIS oriented lesson plans. The teachers will be selected from the local area in order to establish a continuing partnership between the teachers and university researchers. Beyond teacher enrichment, the outcome of this project will be the publication of two documents: a project summary and an instructional materials handbook.

The project summary will provide a detailed review of the workshop and the entire classroom materials development process. The summary will be a guide for similar co-operative efforts between teachers and the universities and institutions in their local area which have GIS expertise and equipment. The instructional materials handbook will be distributed broadly as an educational publication. It will include a student workbook with accompanying teachers' manual, teacher created lesson plans and related materials possibly including items such as a video introduction to GIS, tutorial HyperCard stacks, and educational software. The handbook will also provide a listing of available GIS materials, software packages, and other resources.

A related project being proposed by the NCGIA in conjunction with a national physics research center will bring secondary school teachers to the university campus for six weeks in the summer. In an effort to bridge the gap between what is taught in the schools and what takes place in university research, they will work alongside faculty members on various research projects. Those teachers involved with GIS research and application will have the opportunity to become intimately acquainted with GIS. They will be prepared to act as spokespersons for the integration of GIS into the schools and to help in addressing some of the impediments that require long-term effort. The project will support networking between these teachers and the university researchers for an initial period of three years with the goal of encouraging the teachers to expose students and other teachers to GIS and the sciences. Also a long term goal of this project is collaboration between these teachers and the faculty members in the development of some significant instructional resources.

Conclusion

GIS use in the US secondary schools is only in the beginning stages. Which levels and forms of GIS implementation will gain popular acceptance in the schools is yet to be seen. The NCGIA hopes to aid the creativity of educators and GIS developers by providing an initial base from which to build.

Acknowledgement

These efforts are supported by a grant from the National Science Foundation/NSF (SES-88-10917). This support by the NSF is gratefully acknowledged.

References

Cassettari, S., 1991, Introducing GIS into the National Curriculum for Geography, *Proceedings of the Association for Geographic Information Annual Conference,* Birmingham, March.

Gaile, G.L. and Willmott, C.J., (Eds), 1989, *Geography in America,* Columbus, Ohio: Merrill Publishers, pp. 1–9.

Gardner, D.P., 1986, "Geography in the School Curriculum", *Annals of the Association of American Geographers,* **76**, 1, 1–4.

Goodchild, M.F, and Kemp, K.K., 1990, *NCGIA Core Curriculum in GIS,* Santa Barbara: National Center for Geographic Information and Analysis, University of California.

Grosevenor, G.M., 1989, The Case for Geography Education, *Educational Leadership,* November, pp. 29–32.

Hill, A.D., 1989, Geography and education: North America, *Progress in Human Geography,* **13**, 4, 588–98.

Kemp, K.K., Goodchild, M.F. and Dodson, R.F., 1992, Teaching GIS in geography, *The Professional Geographer,* May, **44**(22), pp. 181–91.

King, R., 1989, Geography in the school curriculum: A battle won but not yet over, *Area,* **21**, 2, 127–36.

44

GIS and higher education

Dion Vicars
Oxford Polytechnic

GIS is the culmination of the development of several disciplines, which is probably why it has taken so much longer than the apparently similar disciplines of CAD and desk-top graphics to develop. The principal problem is that mapping is organized by topology rather than geometry. GIS is not only technology though. It is about applications and their use. These developments have enabled a huge range of disciplines to be made accessible to computer users.

Displays at mapping exhibitions are always impressive demonstrations of the flexibility of the software but they belie both the craft, which has made them so, and the depth of the application. The accessibility of the map-based output conceals the professional level of detail involved. In learning GIS the student has to grasp this point as well as the advanced GIS technology. Students on MSc courses and professionals undergoing training have the background of their first degree to cover the professional aspects. They are in a position to understand the end product and use it as a starting point to learn the relevent GIS techniques. Undergraduates, though, have to learn both at once. This dual learning is achieved by starting from the end product through the medium of demonstrations and graded case studies; they must be able to experiment with several end products to find out what it is about first. This has a fundamental effect on course design and the use of equipment and software. The course must progress all aspects simultaneously, so a basis of graphics workstation use and elements of the application areas are consolidated in the first year giving a basis for development in a similar cycle of more advanced material in the second and third years. This process of learning cycles is particularly applicable to combined subjects, such as Planning or Geography and GIS, where it is essential for the GIS material to be embedded in the course so that the technology is supporting and facilitating the concepts rather than being the end product in itself.

For undergraduate teaching, the progression of detail should start from accessible demonstrations of the GIS functions, through modelling exercises and examples. At this point, collaborative research pays off most handsomely because real situations demonstrate tangible advantages and uses of GIS, an essential element of a sound GIS teaching programme and the basis from which GIS construction must start. Ideally a ''kit'' of GIS parts would allow prototype applications to be built quickly but at the cost of some finesse, allowing the overall construction process to be appreciated and leading to a fully detailed implementation employing the full range of detailed construction tools to meet requirements which have been systematically specified.

The Graphic User Interface (GUI) is an established way to use graphics systems of all kinds because of the intuitive interaction of the mouse and the screen. The fundamentals of the mechanics are the same for the whole range of graphics workstations even though there are several variations in style between the contenders: Mac systems, Windows, Unix, Domain etc. Any introduction to Graphics and GIS has to begin with examples in this mode of operation, in GIS— although the GUI is good for handing graphics and the cartographic aspects of compiling a map or any derived graphic. The implications behind GIS construction means that the level of the interaction is a command level but with the mouse as an uncomfortable intermediary! The side effects are that crucial functionality can be hidden unless the program has safeguards and context dependence built into it. There still does seem to be a case for the good old fashioned typed Command Line Interface (CLI). At least all elements of the function have to be considered in context even if the typing to achieve a result is long and arduous!

The two mapping systems currently enjoying popularity in the higher education arena are of these two extremes. The SPANS system from the Canadian company TYDAC is entirely menu driven, while ARC/INFO is fundamentally controlled by command lines but can be configured to operate in a menu environment with a certain amount of skill in macro-language programming. SPANS is essentially raster, of image, based but now allows for additional vector data to be added which is the reverse of ARC/INFO being vector based with the additional use of raster images on workstations. The advantage of such a system is that it is amenable to the more detailed environment of Land Information Systems as well as general geographic modelling for which both systems offer a full set of tools. Both run well on PCs and have been ported to other platforms.

The main difference between them is that the Combined Higher Education Software Team, CHEST, have chosen ARC/INFO as the preferred higher education software together with Bartholomew's data for the UK for a five year period which began in September 1990. The terms of supply are tailored to the requirements of higher education and in are strong contrast with commercial applications. The range of use varies enormously; from "massed access" for beginners' practical classes over a single term to detailed use by dissertation or research students. The CHEST agreement has fundamentally changed the terms on which such software is supplied to educational establishments so that all supplies must match these conditions to sell into this sector. Other essentials are the availability of technical helplines and user groups to proliferate educational material and case studies and to share ideas and expertise. The availability of many copies of the software as they are needed means that the whole of an academic institution can be talking the same GIS language with corresponding possibilities for collaboration.

An innovation such as GIS takes considerable time to be integrated into any undergraduate course. The "knowledge" has to spread though the support staff and the academics to be integrated into the study areas. This is a long process; the courses must be altered and interfaced to the essentials of the discipline, all of which is only achieved through considerable academic planning. Any such courses which have not begun by September this year will not have had a cohort complete their course by the time the ARC/INFO licence has expired. As the terms of its continuance are not yet negotiated, institutions need to consider carefully now which GIS best meets their needs. There are several other GIS implementations available on both the PC and Mac platforms. Atlas GIS is an American product which is available on both. It has a well designed user interface and appears to be marketed as a "package" with little the need for helpline support.

Map outlines are at least as important as systems and one can assume that the licence on Bartholomew's data will be subject to the same licence conditions as the accompanying software but it too will become installed in projects and teaching material. Our American colleagues do not have this problem as is obvious from the wealth of non-copyright map data which accompanies any American GIS software. The British product INFORMAP though, is another "package" program and is supplied to educational institutions with some map outlines of their choice. Ordnance Survey material can be digitized for educational use although it requires some commitment to do so. The end product is data whose maintenance cost is predictable.

It is important to support GIS undergraduate teaching with industrial contact in as many forms as possible e.g. research projects, consultancy, and industrial placement. Teaching lacks integrity and enthusiasm without it. One particularly constructive method is that of the Teaching Company Scheme funded by the DTI and most of the research councils. Those with most obvious relevance to GIS are probably NERC, ESRC and SERC but this does not preclude others. The essentials of the scheme are that a small team of good graduates work on a project which is relevant to a company or group of companies. Academics are involved to manage the project and supply specialist technical expertise. For medium and small companies 70 per cent of the complete cost is funded, leaving them with a completed project for 30 per cent of its cost, ownership and royalties being negotiable. The DTI has recently updated the terms for these awards and is currently seeking to increase the number of companies in operation.

A degree in GIS needs to cover several areas of application, with a few of them in depth, as well as the areas of computing which are effectively applied in establishing and specifying user needs and implementing them as a database. The relatively new topic, Human Computer interaction, too is very relevant; menu interaction does not automatically solve the problem of making any program accessible, let alone a GIS. Cartography is concerned with the quality of map bases and visual communication, another crucial element in GIS output and consequently an essential in GIS education.

The study and application of Geographical Information Systems seems to have a lot in common with engineering in that theory and application enhance each other. Is it now time for the AGI to take the initiative in considering the basis of GIS engineering as a professional qualification in the same way as the engineering councils?

45

Teaching Earth and Social Science in higher education using computers: Bad practice, poor prospects?

Dr David J. Maguire
ESRI (UK)

Tell me, I forget
Show me, I remember
Involve me, I understand

The basic premise of this chapter, neatly encapsulated in the above proverb, is that practical GIS (Geographical Information System) teaching should be an important part of higher education syllabuses in all branches of earth science and many branches of social science. Unfortunately, however, with notable exceptions, current practice falls woefully short of expected and easily achievable standards. It is also suggested that earth and social science students should be exposed to the basic principles of computers and information technology (IT) simply because of their importance in contemporary science and society. These two views have been explored at length in the literature (Shepherd 1985; Unwin 1991) and are often referred to as teaching *with* computers and teaching *about* computers. This paper will concentrate on the former, since it is where the real possibilities and problems lie for GIS.

An understanding of mathematical, statistical and computer concepts and techniques is today unquestionably an important part of scientists' education and training (see Gold *et al.*, 1991 for detailed arguments relating to geography). Although a comparatively new field, GIS is rapidly becoming accepted as an addition to this list. Yet with few exceptions, these subjects are universally disliked and, where possible, avoided by teachers and students alike. This paper explores the reasons behind this distaste and offers some suggestions about how GIS can improve the situation.

Why are computers under-utilized in teaching?

All the evidence in the literature, the discussions at conferences and the author's experience, as former Co-Director of the UK Computer Board Computers in Teaching Initiative Centre for Geography, suggests that computers are little used for teaching science and social science in higher education establishments. There are five main factors which influence the use of computers in teaching science and social science:

technology; teachers; management; experts; and students. All probably contribute to this situation in one way or another.

Blame the technology

The history of teaching earth and social science using computers has been inextricably linked to technological developments in information technology (Maguire 1989). In the 1960s and 1970s the major difficulty for teachers who wished to use computers in their teaching (and research) was the lack of hardware (Figure 1). With the advent of minicomputers in the 1960s and microcomputers in the 1980s this lack became less of a problem. Teaching using computers was, nevertheless, still restricted because of the paucity of software. The advent of general purpose software for statistical analysis, spreadsheet calculation, cartography, geographical information system development, word processing and simulation, has now alleviated this difficulty. Today, as several surveys have reported (Unwin 1974; Maguire 1989) there is adequate access to both hardware and software in most departments for teaching earth and social science using computers. The barriers of expensive, difficult to use and unavailable technology should no longer apply.

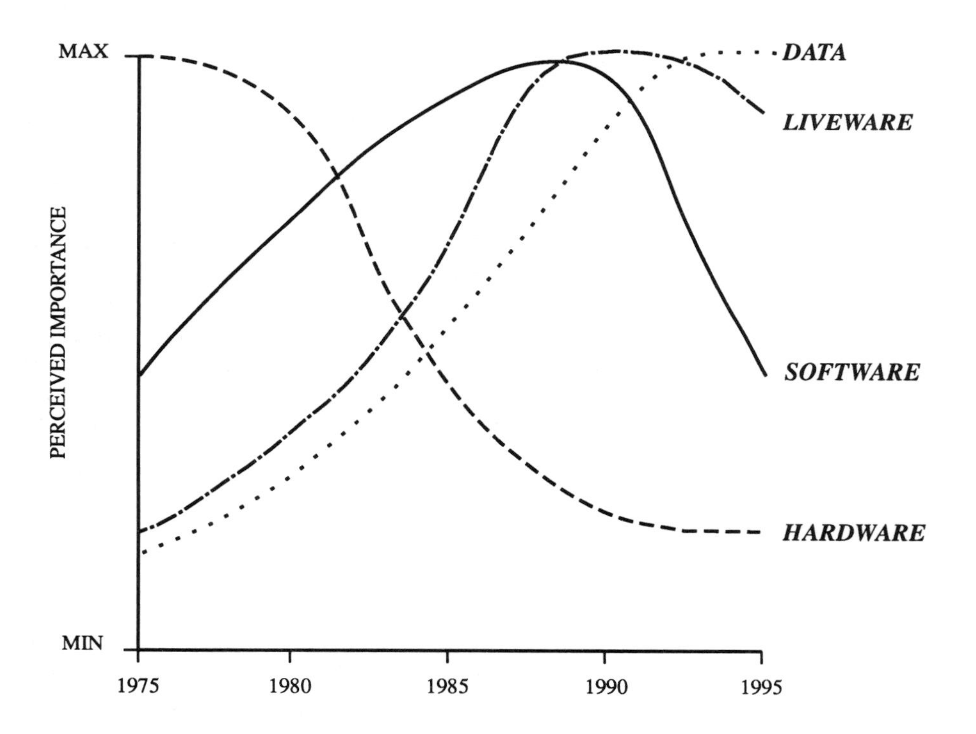

Figure 45.1. The changing role of various key elements in teaching geography with computers.

There is, of course, more to teaching than just providing a suitable computing environment. In order to teach earth and social science using computers it is essential to have appropriately trained personnel. While in the past this has been a significant problem, as courses on using computers ranging from one or two days to one-year Masters programs have proliferated, it is today secondary to the need for data sets and problems on which to work. This combination of data and problems, together with suitable written descriptions, exercises and, possibly, macros, has been termed "courseware" by Unwin (1991). The development of such materials can be very time-consuming and it is this more than any other material aspect of teaching which is limiting progress.

Blame the teachers

Today, the almost universal method of teaching using computers involves the dreaded practical class. For example, in the case of geography it is the norm for first and second year undergraduates to be taught under three streams: physical, human and practical. The names and content of curricula may vary from institution to institution, but the approach and aims are consistent. Such practical courses are universally disliked by both staff and students. The reasons for this are many and include the following: they often occupy more time than other lecture based courses (often three to six hours a week in blocks of two or three hours, instead of two to three hours per week in one hour blocks); the material is often examined by time consuming coursework assessment, whereas other courses use end of year written papers; the staff who teach such courses are often young, inexperienced, or temporary appointments press-ganged into teaching in the absence of other volunteers; finally, the courses frequently adopt a rigid statistical test driven, cookbook style with a low imagination and stimulation content (if its week six it must be correlation, if its week 10 it must be multiple regression!).

Because computers are separated from the systematic branches of subjects and are used in wholly different teaching, it is easy to see why their relevance is often questioned. It is only when computers are fully integrated into systematic and applied aspects of disciplines (for example, social geography, engineering geology, field survey) that they will become truly accepted by teachers and students. This means using computers to teach topics such as, ecosystem development, salt weathering, the spread of bubonic plague, and the optimum location of industrial outlets. Teaching just descriptive and inferential statistics and mapping is far from adequate. Furthermore, computers must be used by specialist systematic teachers in their areas and within the existing teaching framework.

Blame the management

It is easy to blame the teachers for these problems. In part they should take some of the blame for their lack of imagination and unwillingness to expend the effort to rectify it, but it is unfair to criticize them for everything. The current rewards for people in the teaching profession are extremely poor and the prospects appalling. Coupled with this, management is weak, performance monitoring is almost non-existent and there is a paucity of funds for staff development. Thus, there is a failure to recruit good new staff and existing high quality staff are prone to leave or become disinterested. For poor staff the situation is even worse, since there is little opportunity for staff development and retraining. The situation is further complicated because the rewards for teaching in comparison to research are negligible under the current regime (especially in the areas of GIS and modelling where there is a large amount of research money for good staff).

Blame the experts

The content of geographical curricula is primarily research driven. That is to say, higher education lecturers teach what they research and lecturers research what lecturers find stimulating and fundable. Most undergraduate textbooks seek to set out the basic principles and practice of part of the discipline. Unfortunately, and for some reason this is especially the case for computer-oriented aspects of earth and social science, much research tends to be highly theoretical. This of course compounds the problems experienced by those who question the relevance of computing and quantitative methods.

Blame the students

Some of the blame for the low level of knowledge gained by students must reside with the students themselves. In general the material associated with computers is more difficult to understand than other regional and systematic aspects of earth and social science, though this need not necessarily be so if computers are used properly. Unfortunately, and increasingly, undergraduates often lack the necessary scientific and mathematical education to prepare them for such material.

How can lecturers be persuaded to use computers to teach Earth and Social Science?

Recognition of the need to utilize computers in teaching is not new. However, previous projects have not met with much success. As long ago as the 1960s, in the UK the NDPCAL (National Development Programme in Computer Assisted Learning) and Computers in the Curriculum projects sought to promulgate interest in this issue. Currently, there are many projects both in the UK and elsewhere which are addressing this issue. For example, in the UK the Computers in Teaching Initiative (CTI) has established around 20 reasonably funded discipline-specific centres to stimulate the use of computers in their respective disciplines. The Geography centre has met with considerable success producing a regular newsletter and some publications. However, there does not appear to have been a massive increase in teaching using computers. The reasons for this are those outlined above and it will take more than a single initiative to resolve some of these problems.

There seem to be two key requirements for the success of computers in teaching earth and social science. First, the government and education authorities must establish an appropriate career and reward structure for academics. This will encourage high quality teachers and managers to join and remain within the profession. A key element of this reward structure must be the identification of teaching as a valuable activity.

The second key requirement might be provided more easily and immediately. There is still a need for the development of demonstration projects for which the costs and the benefits of teaching using computers are fully assessed. Though there are many problems associated with assessing the value of new teaching methods, there are at present remarkably few, if any, well-documented examples of the successful or unsuccessful use of teaching geography using computers. Certainly, there are none known to the author in which the costs and benefits have been fully determined. It is here that GIS can really come in to its own. There can be few earth and social science subjects which are as visually stimulating or inherently practical as GIS. A selection of well-chosen case studies could really demonstrate the enormous advantages of computers in the earth and social sciences. Such case studies might deal with subjects such as the managing utility pipe networks, environmental monitoring, town planning, retail store siting and health care resource allocation. Stimulating subjects taught by lecturers who also believe in the value of their subject are two of the essential ingredients for successful teaching.

Conclusions

In spite of the fact that, in the author's opinion, computers can make a significant improvement in the quality of teaching earth and social science, there is little tangible evidence that they are in widespread use. The reasons for this are many and include the lack of support for teaching by higher education management and the failure of so-called experts in the field to provide good demonstration projects and cost/benefit assessments. As student numbers increase and the value ascribed to teaching continues to decrease this trend seems set to continue. Introduction of GIS into syllabuses, taught by the right people, in the right way may help to improve the standards of computer and substantive earth and social science teaching.

References

Gold, J. *et al.*, 1991, *Teaching geography in higher education: A manual of good practice,* Oxford: Blackwell.

Maguire, D.J., 1989, *Computers in Geography,* London: Longman.

Shepherd, I.D.H., 1985, Teaching geography with the computer: Possibilities and problems, *Journal of Geography in Higher Education,* **9**, 3–23.

Unwin, D.J., 1974, Hardware provision for quantitative geography in the United Kingdom, *Area,* **4**, 200–4.

Unwin, D.J., 1991, Using computers to help students learn: Computer assisted learning in geography, *Area,* **23**, 15–34.

46

GIS teaching without a GIS?

Jonathan Raper
Birkbeck College

Introduction

Over the last five years Birkbeck College has pioneered a series of new initiatives in the teaching of GIS both for university students and professional updating learners. These developments include:

- May 1988, following the recommendations on GIS education and training in the Chorley Report, a series of quarterly short courses in GIS was inaugurated: over 40 have now been held involving over 500 people;

- December 1988, the author and Nick Green completed version 1.04 of the GISTutor package which was designed to help professionals on the short courses explore GIS concepts: over 400 have now been sold;

- October 1989, a new course unit called Principles of GIS (PGIS) was inaugurated for undergraduates and GIS professionals (taught in the evening) and was used to test Core Curriculum course materials from the US National Centre for Geographic Information and Analysis (NCGIA);

- October 1990, a Masters in Geographic and Geodetic Information Systems (GGIS) was launched by a University of London consortium led by University College and Birkbeck College.

Since it is now five years after this start in GIS teaching, it seems worthwhile to take stock of the experience gained from these projects and to offer some suggestions for the way forward.

The basis of GIS teaching at Birkbeck

Birkbeck GIS teaching has developed from two key principles: first, the belief that there are aspects of data modelling and spatial theory which underlie all GIS, and, second, the belief that commercially available GIS are specific implementations of that body of underlying theory. Clearly, GIS teaching should cover both of these areas *and* link them together. However, while the spatial theory has proved straightforward to teach in a generic way, inevitably the implementation issues have not. This leads to a central dilemma in GIS teaching: how to teach the computer implementation of GIS without over-emphasizing a particular approach.

In an ideal world the solution to this dilemma would be to teach the implementation details of a wide range of GIS, giving the learner experience of each. Since neither time nor money makes this possible there seem to be three distinct approaches to the problem:

- Use a ''simplified'' GIS which illustrates the implementation basics;
- Teach the implementation details of one or two of the market-leading commercial GIS;
- Simulate the workings of a GIS in a tutor and develop course materials on implementations.

At Birkbeck all three of these approaches to the teaching of implementation are used, but they are combined in different courses in different ways. To this author all these approaches are appropriate means of teaching the implementation mechanisms used in GIS and can be classed as education. Systematically teaching the details of implementation using a commercial GIS, in a way which goes beyond what is needed to expose its workings as a GIS can be classed as training. This is the distinction drawn at Birkbeck between education in implementation and training per se.

At the highest level, GGIS Masters students are taught the principles of spatial theory and the details of various implementations using all three of the approaches given above. Given the intercollegiate nature of the Masters course it has been possible to illustrate the implementation issues with a wide range of commercial GIS and other spatial analysis tools including Laser-Scan, ARC/INFO, and Smallworld GIS. Since many Masters in GIS are aiming for employment in the GIS sector it has seemed appropriate to provide this teaching, since implementation of database design, spatial structuring and spatial analysis can place substantial limits on theoretically attainable targets in spatial theory.

In the undergraduate Principles of GIS (PGIS) course students are not exposed to a full commercial GIS, but to a ''simplified'' GIS (MapGrafix) with an extremely easy to learn interface (although MapGrafix has developed substantially through the lifecycle of the course). This strategy aims to acquaint students with personal experience of the implementation details through a database design and construction task involving them in decision-making. Experience shows that this is a productive exercise and PGIS students emerge from their practical work able to identify clearly the aspects of a spatial data handling and theory which are difficult or time-consuming to implement.

In our short courses the only implementation details which are taught are those which can be demonstrated by an expert or simulated since time does not really allow ''hands-on'' experience. This may change as there are now some extremely user-friendly GIS emerging, which could perhaps be taught in a morning. However, the degree of customization and preparation required in advance may still leave this a substantially ''guided'' experience, which may even leave the learner feeling that using a GIS is easier than it actually is! Accordingly, on short courses the ''simulation'' option for teaching implementation details and spatial theory is the norm at present.

A variety of GIS "courseware" has been developed at Birkbeck to fulfil this "simulation" requirement, ranging from pure GIS based learning tools such as the GISTutor to applications of GIS in geographical problem solving such as the Scolt Multimedia Project. These two courseware products are briefly described in the rest of this chapter (references to full descriptions are given) along with details of how they are integrated into the various course curricula. Finally, details of the UGIX GIS user environment developed at Birkbeck are given to illustrate the work that has gone into designing a generic means to access GIS and spatial databases.

GISTutor

Background and description

The Geographical Information Systems Tutor (GISTutor) was developed by the author and Nick Green as a means of introducing the generic principles of GIS and some examples of their implementation (Raper and Green, 1989; Raper, 1991). The intended users were those with a basic knowledge of geography and cartography who wished to understand how a GIS worked and what it was capable of doing. The first version (1.04) concentrated on core GIS concepts and included the following eight sections in the tutor module:

- Capture (Data capture in raster and vector mode),
- Edit (Checking for errors and correcting them),
- Structure (Compressing and spatially structuring geometric and attribute data),
- Restructure (Scale change and vector-raster format conversion),
- Manipulate (Transformation and projection change),
- Search (Querying geometric and attribute data),
- Analyse (Spatial analysis algorithms and applications),
- Integrate (Logical overlay of co-registered spatial data).

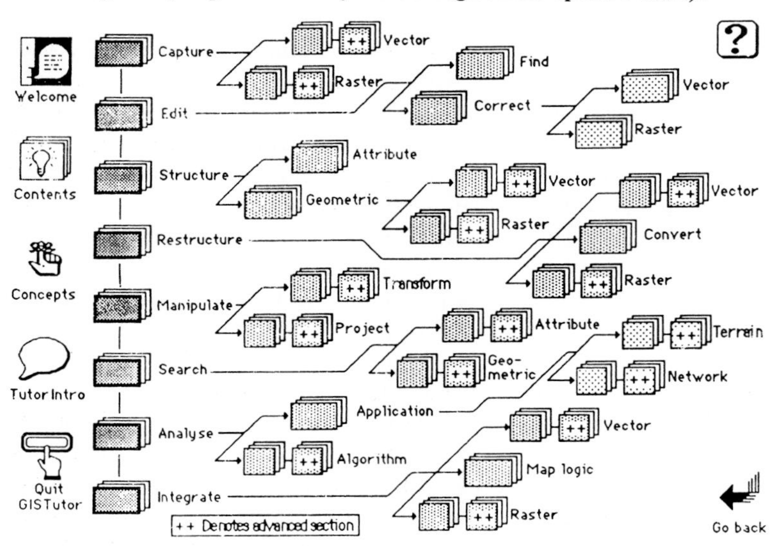

Figure 46.1. The Tutor module map index for GISTutor II (other modules are accessible from the 'Welcome' card).

The Tutor module has been greatly expanded in version II (see Figure 1), and has new modules covering GIS implementation, directories and applications. An 'introduction to spatial data concepts'' section has been added to give more background on GIS which experience of version I shows cannot be taken for granted for many learners. The concern in this introduction is to show that a GIS is a system for creating and manipulating a model of the real world. The decisions governing the creation of the model and the choice of measurement strategies for the source data are (arguably) just as important as the subsequent decisions about representation issues such as the choice between raster or vector data structures. Hence, the introduction describes basic data modelling processes, defines raster and vector data forms, illustrates Cartesian co-ordinate systems, details data sources and describes the hardware and software components of a GIS.

GISTutor was developed in Hypercard on the Apple Macintosh, and version II has been ported to the Windows environment for MS-DOS. This type of system is based on "Hypertext" concepts which permit the associative linking of ideas, and it allows the user to define their own path through the materials presented. This, however, implies that information to be presented must be clearly structured and easy to navigate. The full range of graphical effects and animation capabilities of Hypercard were employed to achieve the maximum visual effect. The GISTutor was also designed to be exclusively "mouse-driven" and requires no commands to be entered at the keyboard.

It was observed from questions about the use of early prototypes that many users who are new to the subject matter find it difficult to access the material without some form of apparent structure. Hence, in the construction of GISTutor it was decided to use a project-oriented metaphor for the organization of the eight topics (see above) starting with data capture and proceeding through structuring operations to various forms of query and analysis. This in effect, narrows down the user's choice of links between units of information from all links physically accessible, to a hierarchically organized system of sequentially linked topics. However, the user can also move directly to a "map" index at all times, enabling an advanced user to explore a higher level of links within the concepts presented.

GISTutor also maintains a record of the topics (sections) and sub-topics (sub-sections) whose last cards have been reached, flagging them to the user by highlighting the buttons on the section header card and the map. This record is maintained however the user chooses to browse through the material. In order to facilitate browsing or following a train of thought, cross reference buttons are also provided to link associated subjects, for example, line thinning in the sub-section on "raster editing", and vectorization in the "conversion" sub-section.

Curriculum use

The GIS data handling and spatial theory topics included in the Tutor module can be seen as a short GIS curriculum in itself, or can be readily integrated into a GIS course. At Birkbeck, typically, the GISTutor is used in a practical session after some associated teaching which corresponds to a particular section in the Tutor module. This material can also be projected onto a large screen using a computer projection system; this facility has proved extremely useful when teaching animated multi-step procedures (e.g. generalization) or dynamic adjustments (e.g. projection change). In this situation the whole group can see the animation and hear the teacher's explanation, and later each student can run through the section again in their own time.

In teaching the short courses the authors have found that use of GISTutor by a small group with student:teacher ratio of 10:1 is a productive learning environment, as the instructor can respond to individual queries as they arise. Since the GISTutor simulates certain GIS operations in raster and vector modes using the Hypertalk scripting language, implementation aspects of GIS can also be readily illustrated without the student actually using a GIS.

The students can also customize their own notes by printing out the screens using a report generation facility, annotating them later as appropriate. GISTutor also contains a report generation facility which, on completion of a session, allows the teacher to check on the progress of the student through the various topics. This report is also used to track the interests and aptitudes of the students and helps to indicate where to improve and develop the subject matter.

Scolt Multimedia Project

Background and description

One of the key problems with much multimedia courseware is that it offers the learner few opportunities actually to interact with the material displayed on the screen. As a result the learner is often reduced to a highly passive role, when the courseware could be much more challenging. This is a special problem for the development of GIS courseware, since the effective use of a real, functioning GIS often relies on interactive access to, and feedback from, the spatial data. This problem can be considered to be one of the key deficiencies in the design of existing multimedia courseware developed for geographical applications.

Given the experience of the GISTutor project in developing animated demonstrations of spatial theory, it seemed logical to apply this experience to the development of multimedia courseware for geographical teaching. Hence, the Scolt Multimedia Project was initiated to embed simple spatial analysis tools within multimedia courseware, allowing the learner to extract spatial information from the maps, imagery and documents available within the system (Raper, *et al.,* 1992).

This work is being conducted as part of the UK Enterprise in Higher Education Initiative (EHEI) which aims to encourage innovative teaching methodologies. This work in the Department of Geography at Birkbeck has focused on the need to relate GIS technology to actual problems which the student encounters in field studies. Hence, the courseware being developed concerns the changing environment of a stretch of sensitive coast in north Norfolk (including Scolt Head Island—hence the name) where sea level rise and coastal erosion threaten rich agricultural land and leisure developments. Students access the system and carry out assignments prior to attending a field trip in the area.

The "organizing metaphor" for the system is a matrix of time/space combinations which are chosen to reflect different periods of time (today, turn of the century—100 years ago, the last millennium—1000 years ago and the end of the Ice Age—10 000 years ago) and three different spatial scales covering different regions of north Norfolk. Information on environmental process, population and economy themes is available for almost all these combinations. This diagram acts as the primary information organizer; in each time/space "combination" the three themes provide a key to the available information.

The system is designed to accommodate a wide range of primary data including statistical information, maps, air photos, scanned archival material and photography. The user can navigate to the level of this data by choosing themes and selecting spatial location from the appropriate keymap. The spatial analysis tools can then be used to extract or measure a particular feature in this primary data; in Figure 2 the extent of dunes and associated vegetation have been output from user digitizing on an air photo.

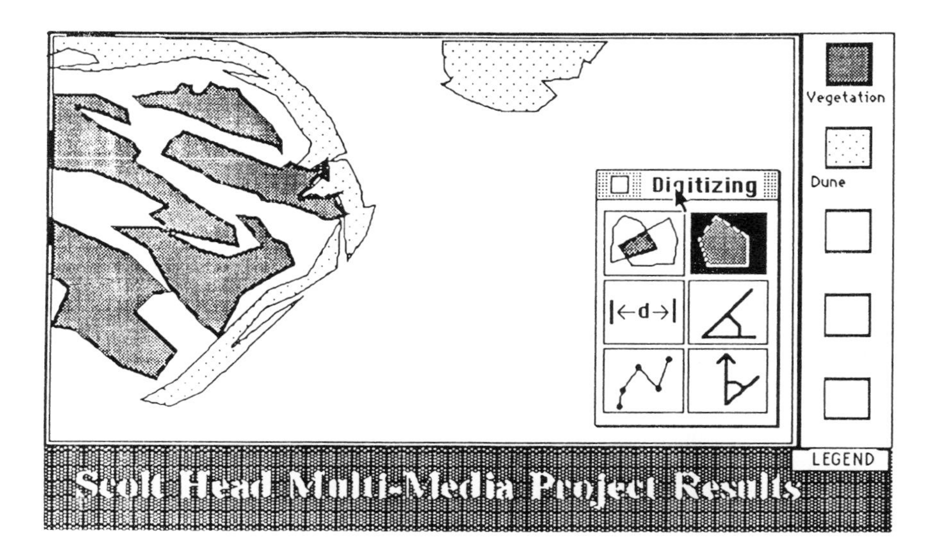

Figure 46.2. Results of dune/vegetation digitizing from aerial photographs of Scolt Head area.

Curriculum use

In the development of the Scolt Multimedia Project (SMP) courseware, the design process aimed to link theoretical classroom lectures in an introductory course unit called ''Britain's Geographical Development'' with practical fieldwork exercises in the north Norfolk area of east England. What was required was a system which could prepare the students for their visit by giving a general background to the area, *and* which could pose specific problems for each student to solve prior to the fieldwork exercises. Specifically it was desirable to encourage the student to compare the ground measurement of features with their characteristics when extracted from topographic maps or air photos.

The danger for this kind of system, however, is that if the student does not interact with the information, then they become passive observers, perhaps only absorbing information at a superficial level. By incorporating generic spatial analysis tools it is hoped that students will be encouraged to check certain facts for themselves. Building questions into assessments requiring use of these tools has also motivated closer inspection of the materials! The results of this elementary spatial analysis by the student can also be printed out and incorporated in assessed work. Throughout, the student is concerned only with the spatial operations and not their implementation in a GIS.

Universal Geographical Information Executive

A strategic objective developed in parallel to the GISTutor and SMP is a wider design approach to the improvement of the interface to GISs. In general GIS systems are difficult to use and a significant learning overhead to most systems lies in training associated with the operating system and the command structure. These considerations have provided the blueprint for the development of a GIS user environment named the Universal Geographic Information eXecutive (UGIX) being developed at Birkbeck College. UGIX is made up of three distinct modules:

1 User interface (including screen dialogues, task choice and command processor);
2 Help (the complete GISTutor information system);
3 Expert system (high level advisors, such as generalization tolerance selection).

UGIX is being implemented as a generic user environment which communicates with any underlying GIS through a series of application programming interfaces, and is described in Raper and Bundock (1991, 1992). It provides an illustration of how the user could eventually access a GIS through a generic iconic interface based on a spatial organizing metaphor rather than through a conventional command interface. This would dramatically lower the learning requirements for GIS, and would reduce the need for training in specific GIS.

Conclusion

This chapter has hopefully illustrated that the contention in the title is already attainable, i.e. GIS teaching can be carried out without a GIS. This is not only achievable but desirable; GIS already change extremely rapidly and effort spent learning one system may soon be wasted. It is the underlying data handling techniques, spatial theory and implementation constraints which endure, and it is these that are taught in GISTutor and SMP, and fostered by UGIX.

Acknowledgements

The authors gratefully acknowledge the permission of the Cambridge University Aerial Photography Library to use aerial photography of Scolt Head Island in the creation of the SMP courseware. The Enterprise in Higher Education Initiative have financially supported this and other projects in the Department of Geography at Birkbeck College. Our thanks also go to Apple Computer (UK) who donated a laboratory of Macintosh computers to the Department of Geography, Birkbeck College enabling much of this work to be developed and tested. Finally thanks in particular to Nick Green (co-author of GISTutor) and David Livingstone (co-author of SMP) for their vital contributions to the work reported here.

References

Raper, J.F., 1991, Using computer demonstrators and tutors in GIS teaching: Lessons from the development of the Geographical Information Systems Tutor, *Cartographica* **28**, 3, 75–87.

Raper J.F. and Bundock, M., 1991, UGIX: A GIS-independent user interface environment, *Proceedings of Autocarto 10*, Baltimore, MD, 24-29/3/91/, 275–295.

Raper J.F. and Bundock, M.S., 1992, UGIX: A Layer Based Model for a GIS User Interface, in Mark, D.M. and Frank, A.U. (Eds) *Cognitive and Linguistic Aspects of Geographic Space,* NATO ASID 63, Las Navas del Marqués, Spain, July 8-20, 1990, Dordrecht: Kluwer, pp. 449–75.

Raper, J.F., Connolly, T. and Livingstone, D., 1992, Embedding spatial analysis in multimedia courseware, *Proceedings of the European GIS Conference,* pp 1232–7.

Raper, J.F. and Green, N.P.A., 1989, Development of a hypertext based tutor for geographical information systems, *British Journal of Educational Technology,* **3**, 164–72.

47

Towards the twenty first century in GIS education

B.M. Gittings, R.G. Healey and N. Stuart
University of Edinburgh

Abstract

As we enter the second decade of GIS education it is worth examining what has been achieved and look forward to the developments that we should expect and press for in the future. The present educational routes for the GIS professional are reviewed and for those contemplating further education, the present market requirements for GIS skills are summarized. A number of important trends relating to the technical and managerial aspects of GIS are outlined and shown likely to cause additional demands for education in these areas.

The existence of an institutional infrastructure to support GIS professional standards and development is examined and the ability of existing professional bodies to cater for the professional development of the diversity of GIS practitioners is challenged, together with the legitimacy of any one set of standards. This chapter concludes with an explanatory account of how the University of Edinburgh has responded to these changing educational needs with the re-evaluation of its own courses and the launch of an advanced modular MSc course in GIS.

Introduction

The market for GIS Education and Training is now well established in the UK, being serviced by a range of vendors, consultants, the Regional Research Laboratories, Polytechnics and Universities (Gittings, 1989a; Gittings and Mounsey, 1989; Rhind, 1990). A variety of options are available ranging in duration from short awareness courses, through package-specific training courses to post-graduate diploma and degree courses (nine to 12 months) and full three year undergraduate degrees.

Short awareness courses are often provided at little or no charge by the large computer vendors (interested in stimulating sales of "corporate GIS solutions") or the consulting companies (interested in encouraging interest in their services), leaving little room for selling courses of this kind. Further, the potential capacity of existing undergraduate and post-graduate courses is probably around 200 places annually, suggesting that competition will become stiff and that only those institutions prepared to move with the times and with a significant critical mass of staff and equipment will continue to be successful. In effect, at all levels, the market for education is probably approaching saturation, despite the fact that there remains a shortage of skilled personnel in the short to medium term. This is not to say that we are predicting any loss of momentum within the industry; it is simply the case that the supply of graduates from the available courses is beginning to draw close to demand.

It is therefore worth looking into the future and assessing how the market will develop. What do we expect of the GIS professional? We would argue that GIS professionals (as against technicians) are likely to continue to be recruited from the graduate community, either having completed a degree in which they have had some exposure to GIS, or undertaken a broader degree followed by an intensive postgraduate course in GIS.

What do we expect of these graduates, both in terms of education and experience? Are the current courses adequate? How must these courses develop to continue to provide quality graduates and how do we ensure professional development in employment?

This chapter examines the need for GIS professionals who will ensure the maintenance of standards and provide expertise and managerial skills. It is these individuals who will lead GIS into the next millennium. To engender such professionalism, an integrated and supportive institutional framework is required; this must bring together the educational establishments, responsible for instilling a detailed theoretical and practical grounding in the field, and professional institutions who will guide the GIS practitioners through their careers.

Education vs. training

Much has been said in relation to GIS education vs. training, and the stances adopted tend to reflect the types of courses offered by the institutions concerned (Petch, 1989; replies by Gittings 1989b and Maguire and Unwin, 1989; Petch and Haines-Young, 1991).

In this chapter, we discuss the opportunities for the GIS professional, who requires a broad understanding of theoretical concepts rather than a wholly practical training in terms of teaching functional operations. We would argue that while there is a need for technician-level training, the primary requirement of the industry is for the GIS professional. In common with the other professions, these GIS professionals should have a university-level education. The question is whether a university education in GIS itself (rather than the more traditional disciplines of geography, computer science, engineering, etc.) is desirable or even achievable.

In the UK, there are two *de facto* educational routes for the GIS professional. First, there is the GIS-related degree and second, a more general degree with GIS-related work-experience and on-the-job training. A GIS or GIS-related (LIS, SIS, etc.) degree course could involve either an undergraduate degree course (as offered by Kingston Polytechnic) or a postgraduate Master's course (as offered by Edinburgh, Leicester, Nottingham, etc.) following an undergraduate qualification in a related degree.

Arguably, the route through an undergraduate degree in a related field, combined with a Master's degree in GIS provides a better education both in terms of the breadth of material covered and the degree of specialization achieved at the completion of the course. The disadvantage is the additional year required.

Our experience indicates clearly that the ability to gain work experience between the undergraduate and postgraduate components ensures a more mature and aware professional at the end of the day. Individuals returning later in life for a postgraduate qualification have identified the limitations of their knowledge and wish to fill these gaps in addition to consolidating existing skills.

The second route which has been available to the GIS professional is through work-experience with appropriate seminars and training courses along the way. Prior to 1985, this was the only possible route. There remains no alternative to this on-the-job approach for the many professionals who are well established on their career path, having achieved a degree-level qualification which is not GIS and are unable or unwilling to take the time out to consolidate their GIS education in terms of a further degree. However, the Salford University approach to GIS education, using correspondence-based courses, may be of assistance here.

In any educational programme, clearly there will be some training component such is necessary in terms of engendering sufficient competence with individual software packages so as to be able to undertake exercises illustrating the concepts being taught. Thus training in a range of practical techniques is an important part of a course, and is certainly of appeal to industry in the short term. Any training undertaken as part of a course increases the appeal of the graduate on the employment market because the employer saves the costs of that training. However, employers will readily admit that it is the broad grounding in the fundamental concepts and issues which make an employee a valuable asset in the longer term.

Current market requirements for GIS skills

Experience over a number of years in both favourable economic conditions and more recessionary times indicates a clear message in terms of the GIS job market. This message is that the greatest shortage is of graduates who possess not only the relevant theoretical understanding but in-depth technical expertise of a practically demonstrated kind in areas such as algorithm design, software engineering methods and database programming/inter-facing. This reflects perhaps two rather different shortcomings in computer science courses and advanced level GIS/digital mapping/remote sensing courses respectively. Computer scientists in general emerge from university courses with little or no exposure to problems of spatial data handling, while graduates in the mapping sciences have, until very recently, had little emphasis in their courses on the fundamental role of database management in analysing their class of problems. Prospective employers must then choose between computer scientists who lack any appreciation of the relevant application area (i.e. GIS), or applications specialists, who lack the computing background to develop the software tools! The eventual choice of candidates would depend on the particular requirements of specific posts and the time available for in-house training.

It can be argued that this demand for technical GIS/computing skills, rather than knowledge of applications and a detailed understanding of one or more of the major packages, reflects the immature nature of the GIS industry where many products are still actively being developed and jostling for position in the market place. Nevertheless, it seems likely that this situation will persist for some considerable time, while products are still being developed for niche markets and while major applications of the technology still require extensive modification/customization of existing software. Both user interfaces and customization tools are improving over time but at present this is occurring after a significant lag time with respect to the mainstream of the software industry as a whole.

Additional future requirements of the GIS graduate

Considering the current development of GIS technology and its limited penetration into major market sectors such as local government at the present time (Campbell and Masser, 1992), there are major implications for GIS education in future. These can be divided into technical and managerial aspects, although the two are closely related.

Technical skills relate first to skills with GIS themselves and second to the understanding of how GIS fits within the IT strategy of an organization. Based on the arguments above, we would expect a graduate of a GIS course to be technically proficient in the use of GIS technology and aware of its limitations. Future graduates will find their credibility depends not simply on technical competence but also on their critical awareness of the limitations of the data and the methods they apply and how this is transmitted through the products which they create.

Issues arise in terms of spatial analysis using a GIS. The possible financial and other penalties resulting from error propagation in overlay analysis, need to be appreciated as much by managers as by the specialists in the present day research community!

Increasingly, different and additional skills must be provided. These relate to the growing role of GIS within the overall information management strategies of large organizations. GIS graduates will be required to have an understanding of the role of structured design methods such as SSADM within large IT development projects in more conservative organizations, and object-oriented system design methods in the more adventurous. An appreciation of the impact of distributed database management, open systems, high speed networking and telecommunications will also be required if they are to contribute to the correct positioning of GIS within long-term IT strategies. Without this broad understanding of the overall content of information management, GIS professionals will either be sidelined or the technology will not add value to corporate data to anything approaching its full extent.

Managerial aspects of GIS databases are obviously closely related to these more technical issues. If the corporate information resource is to be held within a GIS, the full range of implications must be understood by all graduates in the field to avoid costly errors and loss of credibility. Prominent among these issues are questions of data ownership and access, where public agencies are involved, together with legal issues that arise in cadastral and a variety of other application areas.

Finally, there must be an awareness of the political implications, both internally and externally to organizations, of both centralized and distributed control of large spatially referenced datasets, which, by their very locational specificity, are likely to contain information of a confidential or sensitive kind.

The GIS graduate of the future will bear a heavy responsibility to maintain the highest professional standards of expertise and integrity. He/she must be an enthusiastic advocate of the genuine benefits of the technology, while demonstrating a keen critical awareness of its wider political or social implications in particular circumstances.

GIS education of the future must rise to the challenge of providing graduates of this calibre and breadth of understanding. Technical expertise is the bedrock of a forward looking advanced programme in GIS, but it must also impart an attitude that is open to new ideas, self-critical and able to evaluate the rapidly changing relationships between technology and society that characterize our information age.

Professionalism and GIS

If GIS is to succeed as a technology its exponents must adopt an appropriate level of professional standards, that is they must be able to demonstrate knowledge and competence in the subject area. Equally important, in terms of professionalism, is the ability to identify the skills of others and to pass clients to those most appropriate to deal with their business (for example as would operate within the legal profession). The guardians of the professions are the Professional Institutions. The two main professional institutions who have demonstrated a growing commitment to GIS are the Royal Institution of Chartered Surveyors (RICS) and the British Computer Society (BCS) as Kennie and Mather (1991) have noted.

Buchanan (1991) notes that the AGI is not constituted to fulfil the role of a professional body. Herein lies a problem. Despite intentions which were voiced during the foundation of the AGI, which suggested a very close alliance with RICS and the provision of professional services for AGI members through that organization, this has not come to pass. Thus the AGI has grown as an organization with corporate and institutional as well as personal membership as a GIS facilitator, rather than as a guarantor of professional standards. Without restriction on membership (through examination) standards cannot be enforced. Although there are dangers in the "closed shop" approach of some of professional institutions, there are also significant benefits for the members in terms of recognition and continued professional development.

So if the AGI is not, and is unlikely to become, a professional institution, where does this leave GIS practitioners? The BCS and RICS (and some others as described by Kennie and Mather, 1991) provide such professional services, but there is some considerable risk that many GIS professionals may feel their interests fall between the remits of these societies (that is computing/IT versus surveying).

The British Computer Society has a well established special interest group for GIS, but treat it as an application, rather than an area within their core remit. Importantly, they are not willing to regard a GIS degree as a valid subject in terms of exemption from their professional examinations.

Kennie and Mather describe how the RICS is undertaking a major recruiting drive into GIS, but despite a broadening net, there tends still to be an emphasis on traditional areas, much more appropriate to the sub-discipline of LIS. Logan (1991) argues that the Land Surveyors Division of RICS can provide the necessary infrastructure and can act as a professional body for the full breadth of GIS. Indeed it is currently evaluating the GIS Master's course at the University of Edinburgh with regard to allowing this course to give exemption from the professional examinations. However, Logan (1991) warns that land surveyors will have to broaden their horizons; but to what extent is this realistic? The professions are renowned for their conservatism, especially in relation to new technology.

Thus we have argued that the BCS is dealing with the core computer science element, and the RICS is dealing essentially with LIS. There will have to be some movement in the viewpoints of these institutions with respect to each other if a co-ordinated level of professionalism is to be achieved across the GIS field. This broad range of GIS professionals and the positions adopted by the institutions in relation to GIS results in a real educational challenge. The requirement is for a course or range of courses which provide sufficient flexibility.

The response from the University of Edinburgh

In 1985, the University of Edinburgh was the first institution in Europe (if not the world) to offer Master's and Diploma courses in GIS. This was a response to a growing awareness in the UK that there was an acute shortage of skilled personnel to support the emerging field of GIS.

When the Chorley Committee reported in 1987, they confirmed there was a critical need for education and training courses to provide a pool of skilled personnel quickly. The one-year MSc and nine-month Diploma programmes were very effectively producing employable graduates with both a sound knowledge of basic principles and practical expertise on commercial systems.

Against this background of major expansion in the subject area of GIS and a diffusion of awareness, the staff in Edinburgh considered their aspirations for the future of their GIS courses. It is likely that we will see the peak in demand for GIS education and training, at least in Europe, in the near future. For this reason, we were convinced that if we wished to maintain our reputation as a leading centre for GIS education and to ensure the continuing high quality of our students in future, it was essential to lay the foundations now for an advanced course in GIS education.

Petch and Haines-Young (1991) go some way towards recognizing the need for two different types of GIS course; namely courses for the GIS specialist, interested in developing the technology itself, and those for the applying of GIS within a particular field (the applications specialist). The Edinburgh team had identified the importance of this distinction, and any new course structure would have to comprise a basic foundation onto which these different courses could be built. Further, since future applicants for higher degrees in GIS may be expected to have greater competency with computers and more knowledge of basic principles, there is scope to develop an advanced course of GIS education.

A new course structure must promote flexibility. It has to permit students who already have a good grounding in the basics of GIS, for example those seconded from the GIS industry, to be exempted from certain basic components and allow these individuals to take further options in order to enhance and consolidate their expertise. The optional component has to be significant and varied to permit students from a host of different backgrounds to chart their direction through the course in such a way as to gain the maximum benefit. Further, capable students, with strong research interests should be able to replace a taught option by undertaking independent project work and submitting a project report.

A detailed critical evaluation of the existing curriculum (which had run for six years) was undertaken. Account was taken of other curricular initiatives, for example that of the NCGIA (Goodchild and Kemp, 1990) and the RICS AutoCarto Education Trust Curriculum (Unwin, 1989).

It was felt, however, that change should not occur only as a reaction to criticisms of the existing course, but should also be pro-active in representing our collective ambitions for a better course in the future. We wished to continue to promote excellence in GIS education and to provide graduates the sound combination of theoretical knowledge and practical expertise.

In addition to this internal evaluation of weaknesses in the existing course structure, feedback was gathered from both present and past students. Elements that students felt needed more attention were identified. This information was weighed against the advice on academic aspects of the course given by our external examiners, an evaluation of the skills of our recent graduates from industrial employers and the comments of the research councils and the professional bodies. The intention was to produce a course which would be appropriate for accreditation by, for example, RICS.

Initially it was important to consider suggestions purely on academic merit and not restricted by staff, resource or institutional limitations. Thus an idealized course would be developed from the more exciting ideas. A list of lecture courses was compiled that emphasized our view of the core areas underlying GIS, together with topics that demonstrated the diversity of the field and its areas of application.

The final structure involves three "base" modules, which would normally be compulsory and represent the areas of fundamental basic knowledge. Additional to these the students choose from a menu of option modules. Where students come to the course with a demonstrable competence in one of the base courses, exemption would be offered, with the student then able to choose a further option. This solution was felt to be a considerable improvement as it offered students significant flexibility to construct a package to suit their own intended career paths (Stuart, 1992).

The research component of the degree was also ripe for change. While the benefits of project work were clear, it was felt that the existing dissertation format was too academic and promoted an ethos which was quite different to anything students would be expected to adopt once employed. More appropriate would be a staged and self-managed project which would culminate in a technical report in industry or a research paper in academia. Thus a two part submission is now required of the student's which encompasses a concise academic paper and a longer piece of technical writing. This will promote greater selectivity, objectivity and conciseness. Another advantage, in terms of the student's career is the likelihood of publishing the academic paper component (something which has proved very difficult in the past, involving considerable effort reformatting and editing of the dissertation). This will have wider benefits; it will bring a large amount of research (up to 20 man-years per annum) into the public domain for the first time.

Thus the Edinburgh response has involved a complete revision of existing courses, increasing modularity, making curriculum changes and adapting to our perceptions of the requirements for graduates in the coming years.

Conclusions

As the GIS sector matures, there is the need for people who are experts in the technology itself and others who are experts in its application to particular areas. These are distinct requirements and the educational resources are different for each. The next ten years will see a rationalization of existing GIS education and training opportunities. Training is likely to be serviced by specialist independent training organizations (as exist in the wider IT sector at present) and by vendor courses.

The GIS education sector, which we regard as critical for the development of GIS professionals, will also undergo rationalization. This will be due not only to increased competition, but also to increased consumer expectations. Those institutions with a well developed resource infrastructure and a critical mass of staff, which can offer a range of GIS programmes, are likely to continue and remain successful.

As educationalists, we play an important part in ensuring professional standards throughout the industry and the higher education sector must be much more responsive than they have been in the past to the changes in the industry. Setting the educational balance is critical; courses should be neither too academic nor too commercially orientated. These programmes set educational standards for individuals and the profession as a whole and should not be influenced by the goals of individual companies.

A route to continued professional development through the institutions is not clear. The AGI is not a professional institution and therefore cannot provide the services and regulatory functions that might be expected of such a body. However, the existing professional institutions (BCS, RICS, etc.) which associate themselves with GIS do not cover the entire spectrum. Despite moves by these institutions to encompass GIS, it is likely that a number of GIS professionals will find themselves without a home. This can only have a detrimental effect on the industry in the longer term, and potentially divert the focus of GIS as a field in a less than natural direction. Thus there either has to be some movement in the positions of the institutions, in terms of the breadth of their acceptance of the GIS professional, or a new institution has to be developed (perhaps linked to the AGI) which takes the initiative. Whether there is willingness to take action in the short term is not clear.

References

Buchanan, H., 1991, A Young Professional's View of the GIS Industry, *Proceedings of AGI '91,* Third National Conference and Exhibition of the Association for Geographic Information, 3.19.1–3.19.4.

Campbell, H. and Masser, I., 1992, The impact of GIS on Local Government in Great Britain, in, Rideout, T.W., (Ed), *Geographical Information Systems and Urban and Rural Planning,* Edinburgh: The Planning and Environment Study Group of the Institute of British Geographers, pp. 14–24.

Gittings, B., 1989a, Education and Training—The Missing Link? *AGI Yearbook 1989,* London: Taylor & Francis, pp. 323–4.

Gittings, B.M., 1989b, Letters—AGI News, *Mapping Awareness,* **3**(4), 3–4.

Gittings, B.M. and Mounsey, H., 1989, GIS and LIS Training in Britain: The present situation, *Proceedings of AGI '89,* First National Conference and Exhibition of the Association for Geographic Information, 4.4.1–4.4.4.

Goodchild, M.J. and Kemp, K.K., 1990, *NCGIA Core Curriculum for GIS,* Santa Barbara: NCGIA, National Center for Geographic Information and Analysis, University of California.

Kennie, T. and Mather, P., 1991, Do GIS Specialists need the Professional Institutions, *Proceedings of AGI '91,* Third National Conference and Exhibition of the Association for Geographic Information, 3.20.1–3.20.8.

Logan, I.T, 1991, Professionalism in GIS: The Role of the Royal Institution of Chartered Surveyors, *Proceedings of AGI '91,* Third National Conference and Exhibition of the Association for Geographic Information, 3.18.1–3.18.5.

Maguire, D.J. and Unwin, D., 1989, Letters—AGI News, *Mapping Awareness,* **3**(4), 4.

Petch, J., 1989, Letters—AGI News, *Mapping Awareness,* **3**(3), 8.

Petch, J.R. and Haines-Young, R., 1991, Training for GIS, *AGI Yearbook 1991,* London: Taylor & Francis, pp. 189–91.

Rhind, D., 1990, Research, education and training in GIS—Where the AGI fits in, *AGI Yearbook 1990,* London: Taylor & Francis, pp. 313–7.

Stuart, N., 1992, Experiences from redesigning an established course of GIS Education and Training, *Proceedings of EGIS '92,* Third European Conference on GIS (EGIS ' 92), Munich, Germany, **1**, 23–26 March, pp. 48–57.

Unwin, D.J., 1989, Curriculum for Teaching Geographical Information Systems. Report to the Education Trust Fund of AUTOCARTO for the RICS, London.

Part IX

Organizational and Management Issues in GIS

48

Organizational and managerial issues in using GIS

Heather Campbell
University of Sheffield

Introduction

Technical progress in the last few years has removed many of the barriers which inhibited the development of geographic information systems (GIS). It is generally agreed that the potential of this technology to store, manipulate and display spatial data is considerable. However, the introduction of GIS technology involves the complex process of managing change within environments which are typified by uncertainty, entrenched institutional procedures and individual staff members with conflicting personal motivations. Given these circumstances personal, organizational and institutional factors are likely to have a profound influence on the extent to which the opportunities offered by GIS will be realized in practice (Audit Commission, 1990; Campbell, 1991; Department of the Environment 1987; Openshaw *et al.,* 1990; Willis and Nutter 1990).

This part combines an overview of organizational issues and their relationship to GIS with five examples of strategies adopted by a variety of organizations to handle this aspect of system development. The overview provides a framework for the subsequent case studies by drawing on the literature concerning the implementation of information technology in general and applying this to GIS.

The structure of the overview is as follows. The first two chapters review a number of important assumptions about the nature of technology and organizational issues. This provides the basis for exploring the organizational implications associated with implementing GIS in the third chapter. As a result of this evaluation the next chapter seeks to identify the conditions required for the effective implementation of GIS. The final chapter draws these issues together.

The nature of technology

Assumptions concerning the nature of technology are fundamental to understanding the issues involved in the adoption and implementation of computer-based systems. The Public Policy Research Organization of the University of California at Irvine has carried out a series of studies of computer use in local government in the United States (see, for example Danziger *et al.*, 1982; Danziger and Kraemer, 1986;

International City Management Association, 1989). Central to the work of these researchers is the concept of computer technology as a package which includes not only hardware and software but also people, personal skills, operational practices and corporate expectations. The term computer package is valuable as it emphasizes the contribution of resources other than simply equipment to the successful operation of automated systems. However, the computer package is not envisaged as an independent entity but rather to be embedded within the human and institutional context within which it is located. This raises a number of issues concerning the nature of the environment and more particularly the type of organizational issues which are likely to influence the experiences of users.

Organizational issues

A useful framework for the analysis of organizational issues is provided by the findings of a study which investigated the use of geographic information on British local authority planning departments (Campbell, 1990a). This in turn draws on the Irvine Group work in the United States as well as research undertaken in the private sector (see, for example, Hirschheim, 1985; Mumford and Pettigrew, 1975; Pettigrew, 1988). These investigations recognize the critical role performed by social and political processes in the general development of the computer package within an organization as well as the adoption of a specific technology such as GIS. The framework suggests there are three groups of organizational factors. These are the organizational context, the personalities of the individuals involved and the degree of organisational and environmental stability. Evidence from this work indicates that these factors are likely to have a decisive influence on both the initial decision to adopt a technology such as GIS and the subsequent process of implementation and utilization. The subsequent discussion will explore each of these sets of issues in turn.

The organizational context

The first set of factors emphasizes the individual nature of each organization into which technology is introduced and the impact these characteristics have on the initial adoption and subsequent development of computer based systems. These contextual factors are sub-divided into two levels, the internal organizational context and the external environment.

The internal organizational context refers to the characteristics of the organization in which the computer-based system is located. These include features such as the organizational structure, administrative arrangements and procedures for decision-making. Many organizations are sub-divided into a great many sections, as a result the adoption and subsequent implementation of technology is not simply embedded within one context but must also take account of the individual characteristics of each unit. The diverse range of environments to be found within a single organization is therefore likely to inhibit the process of gaining commitment for the introduction of a technology such as GIS. This in turn has implications for the subsequent implementation and utilization of the system. Existing administrative arrangements may also hamper the introduction of new technology, while the potential of a system such as GIS to enhance accessibility to information is likely to be viewed with suspicion as it opens up the decision-making process of individual units to greater scrutiny. Consequently sub-sections may refuse to allocate resources to such a project or at a later stage prove unwilling to agree standards or share data.

Features of the external environment also influence the experiences of organizations. Computer manufacturers and suppliers obviously have a significant role to play, particularly in terms of their research, development and marketing strategies as well as their post-sales support and training programmes. However, a number of other agencies can also influence the pace of GIS adoption and the effectiveness of implementation. These include other organizations in the same field, central government, professional opinion and perhaps even broader trends in society. These elements provide the personal and more general communication channels through which knowledge of new developments and opinions as to their value are transferred. Such conditions may also influence the availability of skilled staff.

The Chorley Report drew attention to central government's role in removing a number of barriers inhibiting the take-up of GIS (Department of the Environment, 1987). It suggested central government has a role to play as a major supplier of attribute and digital data and also through the encouragement of training and research. The wider issue of national policies towards information provision also raises more general questions of civil liberties, such as matters associated with privacy and confidentiality. At the most general level economic conditions are likely to affect the availability of funds both for the initial purchase of systems and their subsequent development. Uncertainty over annual budgets tends to discourage long term investment, although there are cases where the introduction of technology is perceived to be a means of reducing costs. Many organizations are limited by the manufacturer of the existing computer facilities and the associated structure of the data base to the range of products they consider it practical to implement.

The internal and external environments provide the background against which the introduction and implementation of technology is embedded. The detailed nature of the key characteristics varies between organizations but the underlying concept remains useful. The emphasis placed on the influence of contextual factors should not, however, be assumed to suggest rigid determinism as these elements interact with the individuals present within a given organization.

People

Individuals within organizations tend to differ considerably in their motivations and values. Furthermore it is unlikely that their personal goals will necessarily coincide with those of the organization in which they work. In many instances the benefits resulting from the introduction of new technology are not shared evenly with the challenge to the existing pattern of interests perceived by some as a threat and others as an opportunity. It is likely bargaining over the control and distribution of benefits associated with the introduction of a computer-based system will take place at a corporate level, within separate sections and, where appropriate, between agencies.

Key individuals often play an important role in both the initial acquisition of new technology and the subsequent process of achieving effective utilization. The initial purchase of, for instance a GIS, is frequently associated with the activities of a champion, while their role in relation to implementation is often critical. This individual must possess the necessary ability, willingness and intimate knowledge of the workings of the organization to fight successfully the inevitable political battles. Frequently this individual is a member of middle management. Consequently if they are to be effective it is also important they command the support and respect of senior staff.

Generally interest in computerization is not confined to the individuals most directly involved. Staff throughout the organization will be concerned about the personal rather than institutional implications of new technology. In certain circumstances they may be antagonistic, particularly where their past experiences were not entirely favourable. Consequently there is no automatic link between the presence of information systems and their use. These issues are especially important in relation to the implementation and utilization of technology but are also likely to have a significant impact on adoption in situations where senior decision makers share these negative attitudes.

Individuals within the same organization possess very different skills, views of the activities for which technology is useful as well as varying in their willingness and inclination to use computers. The interpersonal relationships between users and technical specialists have a significant impact on the development of information systems. In addition, a lack of mutual understanding between senior management and computer experts can affect the development of technology. Failure on the part of system designers to appreciate fully the needs of users can lead to wasted and redundant resources.

The activities of a key individual and the general interrelationships between personalities within an organization affect the adoption and utilization of technology. The introduction and subsequent outcome of computerization is therefore influenced by the interaction between a complex set of human and contextual factors. However, conditions are not static, as the third element of the framework emphasizes.

Instability

The degree of instability present within a given organization as well as the level of change in the external environment have an important influence on the adoption and subsequent implementation of technology. The impact of instability on the development of computer-based systems appears to be complex. The decision to introduce technology or modify an existing system is usually prompted by some change in the organizational context. Promoters of GIS may exploit changing conditions to highlight the value of innovations in information technology. However, when the level of instability crosses a certain threshold it becomes less likely that senior management will compound this situation by the introduction of new methods which will themselves engender a degree of uncertainty.

The development and subsequent maintenance of systems is also likely to be affected by modifications to, for instance an organization's internal structure or the resignation of key personnel. The latter in particular can have a very profound impact as new working relationships have to be developed, while considerable experience and knowledge may be lost. All organizations face a measure of instability but in certain instances the degree of volatility is such as to either inhibit resources being made available for the introduction of technology or to disrupt the implementation and utilization of an existing system. These comments suggest change is not just limited to the nature of the available technology but is an inherent part of the context in which the activities of organizations are embedded.

Implications for GIS

The framework of issues outlined above demonstrates the importance of viewing technology in broader terms than simply equipment as well as the significant impact organizational factors are likely to have or the adoption, implementation and utilization of information technology in general. There seem to be few doubts that these issues will have an important influence on the development of systems such as GIS. The multi-user nature of many GIS imply that detailed consideration of the appropriate managerial strategies to adopt is even more important in the case of this technology than some other computer-based systems. The costs of data collection and conversion together with the relevant hardware, software and personnel have led many organizations to favour a corporate approach to implementation on simply pragmatic grounds. Furthermore it is suggested that isolated applications inhibit the process of integrating spatial data from a variety of sources which is regarded as one of the main facilities offered by a GIS. As a result the multi-user nature of GIS is likely to add further complications to the management of new technology whether in the context of several sections within the same organization or a multi-agency system.

The concept of an increased level of information sharing is frequently associated with the idea of developing a multi-user system (Masser and Campbell, 1992). It is particularly important in this context that consideration is given to the role of geographic information in the decision-making processes of the organizations involved as the value of a GIS is dependent upon the usefulness of the information generated. It therefore follows that in the case of a multi-organizational system it will be important to have a clear understanding of user needs in all the contexts involved in the project. Furthermore GIS has the capacity to assist with a wide range of activities including operational, strategic and managerial tasks.

The organizational characteristics of GIS therefore imply that the implementation of a GIS is likely to be accompanied by an attempt to extend corporate activities and therefore have significant implications for the development of administrative practices within organizations. This issue is particularly significant as alterations to established procedures are liable to be perceived as a challenge to the existing ownership and control of information and more importantly as a threat to the *status quo*. However, even in instances where a multi-user approach is not adopted it will be important to take account of organizational issues with respect to the development of the system and have a thorough understanding of the role of geographic information in the activities of the organization.

Conditions for the effective implementation of GIS

Given the nature of technology and the likely impact of organizational factors on the development of GIS this section seeks to introduce a more prescriptive element into the discussion. A number of studies based both in private and public sector organizations and examining a range of computer applications have sought to provide guidance to users. (Audit Commission, 1990; Beath, 1991; Bjorn-Anderson *et al.*, 1986; Crosswell, 1991; Danziger and Kraemer, 1986; Hedberg, 1980; Mumford, 1983; Robey, 1987). The findings of these studies are largely summarized by Masser and Campbell's three necessary and generally sufficient conditions for the effective utilization of GIS (Campbell, 1990b; Masser and Campbell, 1992). These three conditions are outlined below:

- The existence of an overall information management strategy based on the needs of users and the resources available within the organization such as staff skills;

- The personal commitment and participation of individuals at all levels of the organization; and
- A high degree of organizational stability with respect to personnel, administrative structures and environmental considerations.

The likely impact of organizational issues on the implementation of GIS is marked when one considers applying the above conditions to a multi-user environment.

The development of an information management strategy is crucial if organizations are to avoid a mismatch between the information required by users and the data generated, while unrealistic assumptions about the resources and skills available are also likely to cause operational difficulties. Effective utilization, however, depends on the existence of an overall information management strategy in each of the organizational units involved which, in turn, assumes general agreement or priorities and the resources to be devoted to the project. The second condition emphasizes the importance of involving and gaining the commitment of staff at all levels and in all the participating organizations. Users have a critical role as they have the fullest appreciation of their context and needs and as a result it is important they are placed at the centre of the process. Much of the literature stresses the importance of champions. In the context of a multi-user system it is clear this individual must have the respect and trust of all the sections involved. The third condition suggests the need to have a high degree of organizational stability to cope with the management of change which is associated with the introduction of new technology. This stability may be difficult to achieve, particularly as external factors beyond the control of those involved with the project may hamper progress. In the case of a multi-user system the instability of even one organization is likely to cause sufficient uncertainty to threaten the success of the whole operation.

Conclusion

The overview provides a general framework for the consideration of the nature of organizational issues and the conditions which are necessary for the effective adoption, implementation and utilization of GIS. Given this framework the subsequent case studies examine a number of managerial approaches which attempt to satisfy these conditions. The examples are drawn both from public and private sector organizations and include a range of GIS applications. Keith Chamberlain examines the development of an information management strategy in the context of the ever-changing conditions of local government in Hertfordshire. He focuses particularly on identifying user needs with respect to geographic information. Derek Taylor concentrates on the strategies that are being developed in Lancashire County Council to ensure the commitment and participation of users in a multi-agency project. Graham Wild's discussion focuses on the third condition by reviewing Gwynedd County Council's attempt to utilize GIS to exploit the change and uncertainty currently present within local government. Peter Elkins' chapter complements these discussions by examining the approach that Southern Water PLC is developing to ensure that momentum is maintained after the initial phase of system implementation. In particular he explores the changing context for system development, the ongoing process of identifying information priorities and the process of maintaining staff commitment. The final contribution by Helen Mounsey and Nick Pearce draws these issues together by outlining the framework for GIS implementation by a firm of management consultants.

There is general agreement that the potential of GIS is considerable. However, the overview and the subsequent case studies demonstrate that to avoid resources being wasted users and researchers must give consideration to the organizational implications of introducing GIS. Few systematic studies have so far been undertaken, although work is currently underway at the University of Sheffield to explore a number of these issues in relation to the experiences of local government (see Campbell, 1992; Campbell and Masser, 1991).

References

Audit Commission for Local Authorities in England and Wales, 1990. *Management Papers: Preparing an Information Technology Strategy: Making IT Happen*, London: HMSO.

Beath, C.M., 1991, Supporting the information technology champion, *MIS Quarterly* **15**(3), pp. 355–72.

Bjorn-Anderson, N., Eason, K. and Robey, D., 1986, *Managing Computer Impact: An International Study of Management and Organizations*, Norwood, New Jersey: Ablex.

Campbell, H., 1990a, The organisational implications of geographic information systems for British local government, in *Proceedings of the European Conference on Geographic Information Systems*, Amsterdam, 10-13 April, Utrecht: EGIS Foundation, pp. 145–57.

Campbell, H., 1990b, The use of geographical information in local authority planning departments, Thesis submitted for PhD in Town and Regional Planning, University of Sheffield.

Campbell, H., 1991, Organisational issues in managing geographic information, in Masser, I. and Blakemore, M. (Eds) *Handling Geographic Information*, London: Longman, pp. 259–82.

Campbell, H., 1992, GIS implementation in British local government, paper presented at NATO Advanced Workshop on Modeling the Use and Diffusion of Geographic Information Technologies, Sounion, Greece, 8-11 April.

Campbell, H. and Masser, I., 1991, The impact of GIS on local government in Great Britain, *Proceedings of Association for Geographic Information Conference* Rickmansworth: Westrade Fairs, Birmingham, 20-22 November, pp. 2.5.1–2.5.6.

Crosswell, P.L., 1991, Obstacles to GIS implementation and guidelines to increase the opportunities for success, *URISA Journal,* **3**, pp. 43–57.

Danziger, J.N., Dutton, W.H., Kling, R. and Kraemer, K.L., 1982, *Computers and Politics: High Technology in American Local Governments*, New York: Columbia University Press.

Danziger, J.N. and Kraemer, K.L., 1986, *People and Computers: The Impacts of Computing on End Users in Organisations*, New York: Columbia University Press.

Department of the Environment, 1987, *Handling Geographic Information*, Report of the Committee of Enquiry Chaired by Lord Chorley, London: HMSO.

Hedberg, B., 1980, Using Computerized information systems to design better organizations and jobs, in Bjorn-Anderson, N. (Ed.) *The Human Side of Information Processing*, Amsterdam: North-Holland, pp. 19-37.

Hirtschheim, R.A., 1985, *Office Automation: A Social and Organizational Perspective*, Chichester: John Wiley.

International City Management Association, 1989, *Computer Use in Municipalities: The State of the Practice, 1988*, Washington, DC: International City Management Association.

Masser, I. and Campbell, H., 1992, Information sharing: The impact of GIS on British local government, paper presented at 19 Workshop of the National Centre for Geographic Information and Analysis, San Diego, 27-29 February.

Mumford, E., 1983, Successful systems design, in Otway, H.J. and Peltu, M. (Eds) *New Office Technology: Human and Organisational Aspects*, London: Francis Pinter, pp. 68-85.

Mumford, E. and Pettigrew, A., 1975, *Implementing Strategic Decisions*, London: Longman.

Openshaw, S., Cross, A., Charlton, M., Brunsdon, C. and Lillie, J., 1990, Lessons learnt from a post mortem of a failed GIS, in *Proceedings of Association for Geographic Information* Conference, Brighton, 22-24 October, Rickmansworth: Westrade Fairs, pp. 2.3.1–5.

Pettigrew, A.M., 1988, Longitudinal field research on change: Theory and Practice, paper presented at the National Science Foundation Conference on Longitudinal Research Methods, Austin, Texas, 14-16 September.

Robey, D., 1987, Implementation and organizational impacts of information systems, *Interfaces,* **17**(3), 72-84.

Willis, J. and Nutter, R.D., 1990, Survey of skills needs for GIS, in Foster, M.S. and Shand, P.S. (Eds), *AGI Yearbook 1990*, London: Taylor & Francis, pp. 245-303.

49

Customer culture, information systems and local government

Keith Chamberlain
Hertfordshire County Council

Local government is going through one of the most dramatic periods of change in its history. I am not referring to structural change, although that is certainly challenging. The ''wave'' of cultural change that is presently affecting many local authorities is even more fundamental.

Information systems for cultural change

As my contribution to the AGI yearbook I wish to examine the fundamental role of local government, and the role of information in the delivery of local government services. Also I propose to focus on the future rather than the present, and to set out the way in which I see information systems, and in particular Geographic Information Systems, being used in the future.

During the 1970s and 1980s information systems were being developed by computer (and other) professionals for ''users'', who were normally other professionals, often at middle management level. In my view, the greatest change in computer marketing lies in the substitution of the word ''customer'' for the word ''user''. During recent years the emphasis of local government has moved away from the direct provision of services towards an ''enabling'' role, along with the onset of business units (or cost centres) reporting not through bureaucratic structures but directly to management boards, and delivering for the needs of customers. Furthermore customer interests are being advanced through the greater emphasis on performance, linked to such initiatives as the adoption of the Citizens' Charter and the widening of the scope for Compulsory Competitive Tendering (CCT) into ''white collar'' services.

I believe that this new culture provides major opportunities, as well as challenges, to those of us involved in information systems in local government. The devolution of service delivery responsibility to ''business units'', along with greater management accountability concerned very much with ''performance'', lies at the heart of the change.

The nature of the challenge

The challenge lies in providing the managers of these business units with the information they need to deliver their own particular services. Information is becoming increasingly recognized as a key resource, and I suggest, as important a resource as finance and accommodation.

A typical business unit manager, in my view, requires four different types of information[1]:

1 Contextual Information

This is information about the "business" climate, the organizational climate as well as forecast changes and their impact. Typical examples would include demographic change or the impact of political decisions.

2 Opportunity Information

This relates to new opportunities arising from changing customer needs and aspirations, or possibly potential new customers. Typical information would include the results of customer surveys or other market research.

3 Management Information

Here we include the backbone information needed to gauge performance, and in particular financial information on budgets, spending, and commitments, on personnel and staffing (particularly training records and needs) and client records.

4 Task Related Information

This will relate specifically to the nature of the business itself. A business involved in delivering a Forward Planning service in a County Planning Department would need a wide range of demographic, land use and employment data for example.

Much of this information can be "brought to life" by the use of GIS. Yet in order to achieve such benefits, the various databases need to have a clearly defined geographic dimension, such as grid reference or postcode, to enable them to be interrelated, modelled, and displayed on the map. To mention just a few simple examples:-

Contextual: modelling the existing and future "customers" of client based services such as Education and Social Services by mapping Census of Population statistics.

Opportunity: distributing, geographically, the service aspirations of library users by linking addresses to mapping via the Post Office's Postcode Address System and General Postcode Directory.

Management: displaying and modelling accommodation costs on a property by property basis using digitized property data on an Ordnance Survey digital map base.

Task Related: mapping and modelling changing land use patterns and linking these to demographic patterns and maps of planning constraints for Structure Plan monitoring and review.

However, the key question is how to ensure that the breadth of this necessary information is assembled, organized, and itself delivered in the most cost effective way. The solution lies within the organization itself and will vary according to local circumstances.

If I may use my own authority for illustration, Hertfordshire has, during the past two years, been moving away from a ''service provision'' culture towards one led by the meeting of customer needs. More particularly within the Planning and Environment Department a number of business units have been established to deliver, or enable, service delivery. One of these is specifically charged to deliver a range of information services including intelligence provision, IT, surveys, as well as GIS and more traditional mapping. This ''Information Unit'' is now well established and operates both as a contractor to other, ''client'', units and as a client in its own right for developing and promoting the strategic development of its services on behalf of both the County and District Councils.

Many of the new local authority ''businesses'' will aspire to ownership of their own datasets (particularly management, and ''task related''), and where the datasets are unique this may well be the right and proper solution. Yet each local system should be governed by data standards, e.g. for spatial referencing, to enable data exchange and enable the very real benefits of GIS and other ''corporate'' initiatives.

The importance of management and organization

During recent years there have been quite remarkable developments in information technology, and in particular the dramatic pace of change in the availability and use of micro-computers. The recent work of County Planning Officers' Society[2] is clear evidence of these developments. Those of us responsible for information systems and services have almost *too* much choice.

In my view the issue has, for many years, been one of organization and management linked very much to the role of information in the decision-making process and in service delivery, rather than technology. Any GIS is a major investment in terms of data, computer technology and people. At a recent workshop organized by County Planning Officers' Society, Peter Woodsford stressed the importance of the corporate nature of the information resource and the need to ensure that proper information management procedures are in place. As he vividly quoted ''As a hard-pressed, but successful GIS Project Manager remarked *en passant*—all you have to do is to organize your data and organize your people''.

In Hertfordshire County Council we accept and acknowledge the importance of a sound organizational base. As part of devolved management the newly formed Planning and Environment Department is seen as the lead department for GIS. It has overall responsibility for liaison with Ordnance Survey on behalf of the County Council, and plays the leading role in terms of mapping and map production and for policy analysis and general GIS services. Yet other service departments are enabled and encouraged to exercise their own initiatives for ''asset management'' based GIS within a very general corporate framework. This framework is being set out within a corporate GIS strategy which is presently being put in place and which stresses the importance of:

- identifying the information needs of customers;
- an organizational network of GIS users;

- a means of sharing all geographic databases by common geographic referencing; and,
- an awareness of the links between the range of technologies and procedures.

Conclusion

The new thinking within many local authorities with devolved decision-making responsibilities provides an excellent opportunity to "grasp the nettle" and to demonstrate the "added value" provided by efficient and effective information.

Notes

1 This list is developed from ideas set out originally by McFarlan, F.W., 1984, "IT Changes the Way You Compete", *Harvard Business Review*, and used by John Ward at the Cranfield School of Management, 1989.
2 County Planning Officers' Society, 1990, "Computers in Planning".

50

Lancashire: A green audit—organizing and developing a GIS to undertake an environment audit

Derek Taylor
Lancashire County Council

Introduction

Local authority environmental auditing is a very new phenomenon. The first completed audits appeared as recently as 1989. Since then, and despite the fact that they are not required to do so by law, many authorities have taken up the challenge. They see audits as a valuable way of assessing the quality of their local environment, identifying their own impact upon it and informing policy and practice review. Further stimulus will be given by the new British Standard (BS7750) on Environmental Management Systems, which requires regular audits to be undertaken as a part of its implementation. The DOE, local authority associations and the Local Government Management Board, are also investigating the feasibility of applying the forthcoming EC Eco-audit Regulation to local government in the United Kingdom. At present confined to site-specific, private sector industrial processes, the Regulation may not be easily adaptable for local authority purposes. However, the fact that testing the concept has ministerial backing is significant. It is the first tangible recognition that there is a need for a consistent, nationally-based framework for public sector auditing. Whatever emerges is bound to produce an upsurge in practice.

Audits can be of three kinds:

- an analysis of the environmental conditions existing in the area administered by the local authority—referred to usually as a State of the Environment Report;
- a review of the environmental impact of the Council's own policies and activities—known as an Internal Audit;
- a review of the management system and procedures employed by the local authority to organize and deliver its environmental commitments and responsibilities.

Lancashire County Council is tackling all three, but this chapter deals principally with the first component. The preparation of a State of the Environment Report requires the collection, storage and analysis of a considerable amount of data from a wide variety of sources. As a result it was decided in Lancashire to utilize geographic information systems technology. In so doing the main issues have involved organizing and gaining the co-operation of the different agencies holding environmental data and organizing the data into a common format. This chapter mainly concentrates on the first of these issues. However, before examining these issues the types of data sets needed to conduct a Green Audit will be outlined.

Green audits and GIS

Lancashire's State of the Environment report, referred to more usually as the "Green Audit", has two, linked aims:

- First, to provide Lancastrians and all interested parties, with the first-ever complete picture of the state of health of their own surroundings;

- Second, we hope to use this increasing awareness and knowledge to help generate the steps needed to improve and sustain our local and the wider, environment.

Potentially, the scope of the Green Audit, and related data needs, is enormous. Assessing the state of a complex system like the environment requires that it be broken down into more specific and manageable components. Having done this, specific "indicators" are identified for each component which can be measured or assessed, to build up an overall picture of environmental quality. For example, air quality cannot be measured as a single phenomenon. Levels of individual pollutants which influence quality, however, can be measured—e.g. SO_2, NO_x, low level ozone, particulate matter and radon. Each of these particles is an indicator of air quality. Legal standards help further in assessing overall quality levels. The composite picture revealed by each component provides a snapshot of the overall state of quality of the local environment.

Eleven components were analysed:

- structure (geology, soils, climate, vegetation, population);
- air quality (gaseous and particulate pollutants);
- water quality (surface, coastal, underground, drinking);
- waste (arisings, disposals and impacts);
- noise (sources of complaints);
- energy (generation, use, impacts, renewable sources, conservation measures);
- land and agriculture (land cover, land quality and use);
- wildlife (extent and quality of habitats, protection and threats);
- landscape and townscape (quality, protection and threats);
- open space (amount and condition);
- transport (extent of network and use, impacts).

It is clear from this list that in undertaking a Green Audit we would be required to collate data from a great many sources. However, while we had some idea about the range of data we wanted to collect, we did not know whether or not it all existed or in what format much of it was held. At a very basic level one of the primary objectives of the Audit was to identify what environmental information actually existed. As a result of the need for a highly flexible system it was decided to purchase a GIS. The advantages of this technology were perceived as follows:

- it would allow us to store/access a wide range of information of many types whose only obvious common feature was that they all relate to places on, under or above the Earth's surface—the earth, in this case, being Lancashire;

- it would enable us to relate information on the basis of its location. In particular, to provide information on what exists at particular locations in terms of a whole range of quantitative and qualitative attributes. As a corollary, we needed to be able to identify locations in the county where one, or a combination of, environmental conditions apply;

- data would come from many agencies in an even greater diversity of forms. A GIS would accept data in different formats and enable us to interface with other computer systems;

- we needed facilities for handling time-series data for monitoring change over time;

- flexibility of output was a consideration because we wanted to use the database in as wide a range of environmental and problem-solving situations as possible;

- the output could be produced in an easily-understood format. We were looking for new ways of presenting complex information that people could readily assimilate, understand and use.

The following comments focus on our efforts to ensure the effective implementation of a GIS in the context of being reliant on a multiplicity of agencies to supply the source data.

Organizing the people

The Green Audit exercise would not have been successful without the commitment and co-operation of people both inside and outside the County Council. As a result an organizational structure was established which attempts to link all these elements (see Figure 1).

Within the County Council there is commitment at the highest level; from the Leader of the County Council and from the Chief Executive/Clerk who have both given the initiative the highest priority and taken a great interest in the progress of the work. This attitude has been of vital importance in ensuring that adequate resources were made available and that there have been no major impediments to the smooth running of the project.

Management of the Green Audit initiative within the County Council is carried out by an Officers' Management Group (OMG) formed of chief officers from the key council departments such as surveyors, analysts, education and, of course, planning. It is chaired by the Chief Executive/Clerk and oversees all aspects of policy and progress on the Green Audit. It acts as the link between officers and council committees and between the County Council and the outside world.

As far as elected members are concerned, the first point of referral is to the Planning Committee. The main committee for decision-making, however, is the Council's Policy and Resources Committee. As its name suggests, this is where the major decisions on policy and member co-ordination are taken. The Leader of the Council chairs the Committee.

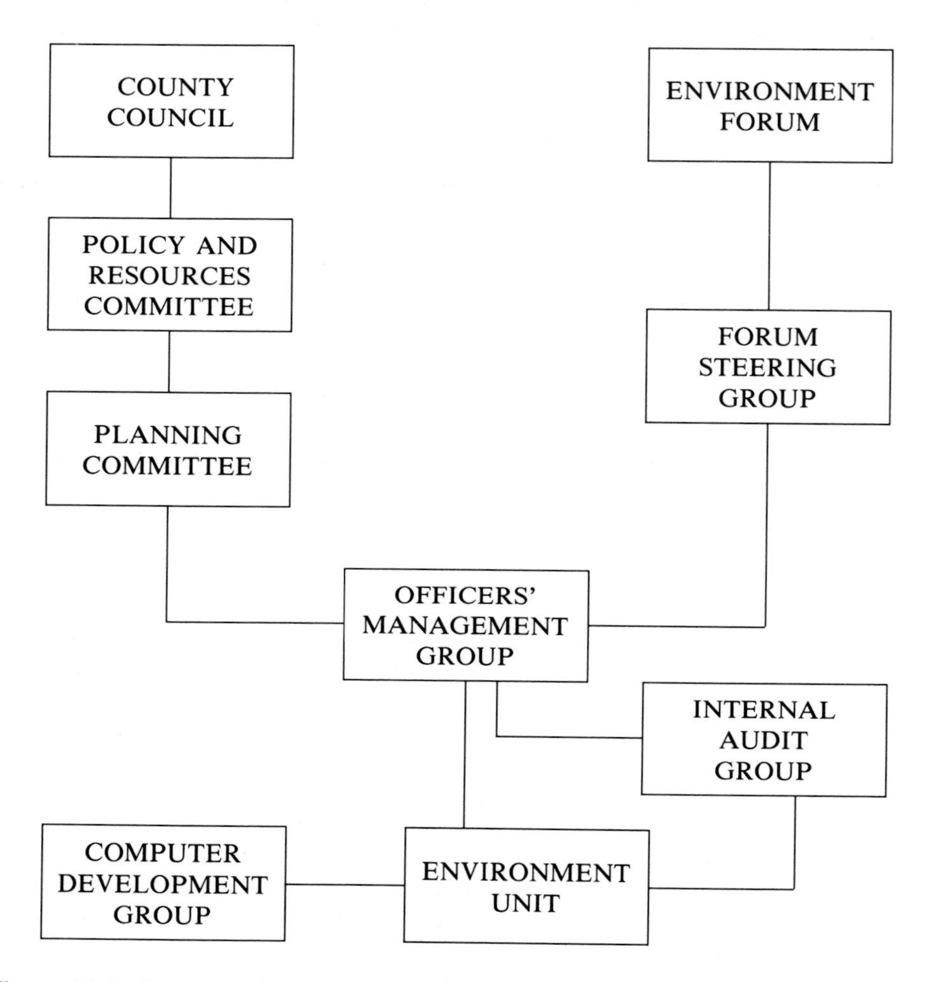

Figure 50.1. Structure for Lancashire County Council's Green Audit

Outside the County Council there are many organizations which have statutory responsibilities for environmental protection and which hold much of the data on environmental conditions. It was obviously important to involve them at the earliest stage in the exercise. The Lancashire Environment Forum was therefore set up in December 1989 and comprises organizations drawn from central and local government, industry, academic establishments and the voluntary sector. It now has a membership of over 70 groups and is, again, chaired by the Council Leader.

The tasks of the Forum have been first to guide and co-ordinate the production of the Green Audit, second to translate this audit into an Environmental Action

Programme and then finally to implement this programme through the work of individual agencies. The Forum is also responsible for up-dating the State of the Environment Report for May 1994.

Throughout its existence the Forum has been an inspiration to all those involved. Its tasks have been carried out in a true spirit of co-operation with the unifying goal of improving environmental quality in the county. A small steering group of key Forum organizations, chaired by the Chief Executive/Clerk, deals with day-to-day Forum business.

The Forum's first task was to provide the data necessary to compile the Green Audit. All the information required was publicly available though in hundreds of different locations and formats. An initial questionnaire established the existence of data and details about how and where it was stored and provided a contact name and number which would later be followed up.

Detailed work on the initiative is carried out by the Planning Department's Environment Unit. The Unit is five-strong and its head reports directly to the OMG and, through that group, to the Forum. Its posts were created specifically to undertake Green Audit work. The task of developing the GIS fell to the Computer Development Group which is also based in the Planning Department. There is very close liaison between these groups with one technician and a GIS workstation now located in the Environmental Unit. Expertise and confidence in handling a GIS have taken time to develop with hands-on experience over months, rather than weeks, being essential.

Organizing the data

All the material needed for the Green Audit has been derived from publicly-available sources. These included statutory registers, published reports, files, record-systems, computer databases and maps. Given the range and the enormous diversity of formats in which the information was held, this proved to be a considerable task.

The situation has demanded a gradual, evolving approach which can be summarized into four phases, although there has been a degree of overlap. A preliminary stage identified the parameters of the database—e.g. basic map scales, geographical boundaries, structure of the data library. Equally important was the process of deciding in detail which Green Audit information was to be collected, where it was held and its format and availability. Forum members were surveyed by questionnaire to solicit these facts.

The next stage ran from March 1990 for about twelve months and encompassed the collection, organization, analysis and presentation of the data. With a few notable exceptions, this was done manually. The GIS was used to assemble and organize data in a few cases, which greatly assisted the process of familiarizing staff with the equipment.

The third and major developmental stage, from about April 1991 to the present, has involved the input of all major GIS datasets (nearly 100 of them). It has also witnessed a considerable advance in operational expertise. Enhanced ability, coupled with an ever-widening database, has made it possible to undertake increasingly sophisticated analysis and output. Data input is a major task of any GIS exercise. GIS experts and vendors often gloss over this, rarely mentioning how long it takes, or how boring and problematic it can be. Three main factors have had to be addressed in Lancashire's case—data formats, timescale and storage capacity.

Getting the data into the right format has taken the most time. Fortunately, the organizational arrangement of having the equipment as well as environmental and computing staff in the same place has proved a great help. For many datasets, the data-formatting process has taken almost as long as the digitizing activity which followed. As far as the latter was concerned, the time required depended, obviously, on the nature, scale and complexity of the dataset. Thus, inputting the soil associations, agricultural land classification, and other multi-polygon datasets covering the whole of a county the size of Lancashire required about two man-weeks in each case. More simple datasets like planning policy designations, ancient woodlands, land liable to flooding and mineral working sites took a couple of days each. Really complex datasets were not tackled in-house, for instance, solid geology and OS 1:50 000 contour lines were purchased pre-digitized, while some other formatted datasets have been donated by Forum partners. Storage problems arose quite early on, and were relieved temporarily by the purchase of add-on facilities. A permanent link into the authority's main frame has brought even greater capacity and relief, but the issue will no doubt return.

The fourth, and final stage, concerns what happens next. After two years, we are at the point where we should be able to reap the benefits of the system in terms of analytical capabilities providing the basis for the ongoing maintenance of data sets and education through providing the general public with access to the data.

Conclusion

Both GIS operation and environmental auditing are relatively new work areas in local government circles. They require the development of specific skills and strengths. Lancashire County Council has consciously attempted to marry the two together, and, after two years of intensive ground work—and, no doubt, many mistakes—the benefits are beginning to emerge. A clear purpose and the full commitment of the County Council's corporate management system have been vital, but two other factors have underpinned the operation from the outset. Both are as important as each other, and, in this instance, they need to be very closely interwoven. These are the essential requirements of defining clearly the roles of the people involved, (and the working relationships between them) and of ensuring that the data processing aspect is properly planned and executed: organizing the people and the data. Using GIS as a key element in this organized approach is proving to be a considerable bonus.

Note:

Derek Taylor BA, MA, MRTPI, is Head of the Environment Unit, Lancashire County Council. The views expressed in this article are his own.

Copies of *Lancashire: A Green Audit* are available in full and summary form (price, including postage, £35 and £5 respectively) from: The County Planning Officer, Lancashire County Council, East Cliff County Offices, PRESTON PR1 3EX.

51

Corporate GIS and organizational development: A local authority perspective

Graham Wild
Gwynedd Council Council

Introduction

In most accounts of GIS implementation, and there is a rapidly growing portfolio of these in UK local authorities, the system is introduced in order to perform current tasks more efficiently. In most cases an attempt has been made to identify and quantify the benefits, and, if these outweigh the costs over a reasonable time-span, resources are released. GIS is perceived as a sound investment in the way it increases efficiency and reduces cost in a number of identified areas of activity. This was the route taken in Gwynedd, and such is the potential power of GIS that it can generally be justified in this way

In addition to tangible, measurable benefits, however, much has been made in a more generalized way about the "added benefits" of corporate GIS (Mahoney, 1989) over and above those realized simply by doing present tasks better. The way in which it recognizes the value of the information resource and imposes strategies to its management represents a separate level of benefit, more difficult to measure but nonetheless real. Economies of scale, cost sharing, better quality of information, and speed and flexibility are clear examples of these benefits. On the whole, top management is less impressed with them, which are often seen as a somewhat unreliable bonus to the tangibles.

It can be argued, that there is yet a further layer of benefit which puts GIS into a totally new role: that of facilitating change in its host organization. The suggestion is that GIS can be viewed as an "intervention" in organizational development (French and Bell, 1990). It can be seen as a means of facilitating the process of change in the culture, processes or structure of an organization. This would suggest a hierarchy of roles accorded to GIS which would parallel the pyramidal model of organizational structure (see Figure 1).

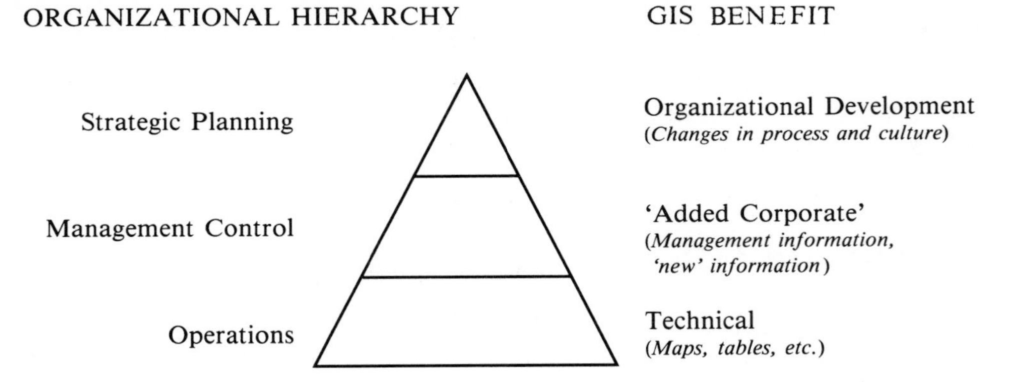

Figure 51.1. Pyramidal model of organizational structure

This chapter, attempts to set out a few observations about the third, wider, role of GIS in organizational development. It is grounded in personal experience of promoting and implementing a corporate GIS project in Gwynedd, and subsequently in strategic management in our Chief Executive's team. Instead of starting with the perfectly reasonable question "What are the most favourable organizational factors for success in GIS?" (Campbell, 1990) it asks the question "How in a changing environment can GIS help me change the organization?" Given the degree of reorientation which is taking place in local authorities at present, this would seem to offer a useful alternative perspective to the GIS issue, and may open up a new area of discussion.

GIS in Gwynedd

The story of the Gwynedd GIS project seems a fairly typical one. A group of map-using specialists from a number of departments got together to assess the impact of the Ordnance Survey's digital mapping programme. By a process of mutual briefing and persuasion, a project was developed on the wider basis of GIS. This was promoted in the organization through a blend of soft-sell awareness activities and the development of the cost-benefit case. Having secured resources, the focus shifted to the need to make the investment pay through sound project management.

Perhaps a key factor in the Gwynedd GIS project was the determination from the start to view itself as determinedly corporate, with a keen eye to the added benefits of such an approach. What was less clear at the outset was whether the corporate benefits were valued by the organization as a whole. Having established a GIS unit with ARC/INFO running on Sun 2 machines, the urgent need is to develop a portfolio of applications, construct the requisite data sets, and deliver a reliable and high quality service. Our immediate objective is to operate effectively within the framework of an emerging business plan. However, in developing in this way, with its focus on specific and the technical, there is a danger that the search for the wider corporate and organizational benefits will become increasingly marginalized. In short, a resource with the potential to change culture, may be used merely as a source of cheap and clever maps.

GIS as an organizational development intervention

Organizational Development is frequently misunderstood. It has acquired the "soft" and simple meaning of "change in an organization". However, in the context of management OD has a very specific definition:

> Organisation[al] development is a top-management-supported, long-range effort to improve an organisation's problem-solving and renewal processes, particularly through a more effective and collaborative diagnosis and management of organisational culture (Wendell *et al.,* 1990).

OD therefore is a conscious, planned, and structured intervention to steer an organization or part of one away from its "natural" development path. Whether or not it is perceived as OD, most organizations from time to time go through a similar process of conscious change.

The pressures for change in local government have never been greater. In particular, government reforms are biting deeply into the way local government operates, changing its orientation and accountablity, and setting it new tasks. The needs and expectations of the community are continually changing and public attitudes to environmental issues are strengthening. All of these issues suggest a need for a radical review on the part of local authorities of the way they operate and the technical and managerial resources they now need. What potential role, therefore, has GIS in this readjustment? In the words of the management theorists, can GIS be viewed as an "intervention" to bring about a change in the way the organization functions? Margulies and Raia (1978) offer a framework for assessing the key attributes of effective OD intervention. Corporate GIS appears to answer most of the criteria:

- It is planned.
  Corporate GIS is by implication, if not definition, a planned activity. While single-purpose GIS can be set up overnight on an *ad hoc* basis, a convincingly corporate GIS can only be achieved through careful preparation and negotiation, and implemented in a planned way as a medium to long term project.

- It is problem-orientated.
  Effective GIS is problem-orientated, the powerful tool being deployed to address a specific set of information needs within an active management process. Applications are developed to deal with specific problems related to information quality, integration, accessibility and presentation.

- It reflects a systems approach.
  GIS operates in a way which facilitates understanding and helps construct a rational interacting framework by which problems can be structured.

- It is an integral part of the management process.
  GIS can and should be totally integrated into the process of management. It is in essence a management support system and should be developed as a component of the "normal" management process. Indeed, the end-user need not be aware that GIS is being used.

- It is not a "fix-it" strategy.
  Because of the timescale involved in developing fully corporate GIS and its high start-up costs, it has few attractions as a quick-fix, but is essentially a medium to long term strategic solution.

- It focuses on improvement.
  The justification of GIS investment has to be improvement measured against the explicit or implicit objectives of introducing the system. Whatever method employed, usually cost-benefit, the measurement of "improvement" is critical.

- It is action-orientated.
  Corporate GIS is about action and change at all levels. The change to more systematic work patterns, the emergent skills, the redefinition of jobs, the development of more important strategic roles for information and its presentation, and the very visible hardware systems are profound changes in organizational life. GIS has a strong impact on all those who come into contact with it.

- It is based on sound theory and practice.
  While there is scope for ineffective, even damaging implementation, GIS offers solutions which are firmly grounded in sound theory and practice. The emphasis on the accuracy, timeliness, and relevance of information, logical relationships, effective statistics, responsibility, openness and sharing in the use of information, and responsiveness in systems design is evidence of this firm base.

The clear conclusion is that a corporate GIS can provide a powerful development tool. Perhaps it can even change cultures. It brings with it new values, new opportunities and new perspectives; it can provide innovative decision support; it can change attitudes, question working practices, create new areas of skill and knowledge, promote imaginative thinking, and transform the way in which information is communicated. However, as discussed earlier, GIS is rarely implemented for such high-minded reasons. It is usually introduced for more humble and mundane purposes. Nevertheless, the wider organizational impacts of corporate GIS should be taken into account, particularly at a time when strategic change is on the agenda of local authorities.

GIS and change in local government

While many local authorities are taking the initiative in seeking more effective ways of delivering services and continually review the process of local democracy, the main challenge results from imposed change. This includes compulsory competitive tendering and its extension into central services, quality management and customer care, community care, the "charter" movement, and the restructuring of local government itself. Simultaneously there is increasing concern about strategic planning, the effective role of members, the development of business cultures, human resource management issues and public relations. Added together these set a formidable agenda for organizational development for the next decade in UK local authorities. What role can corporate GIS have in facilitating the process?

The answer lies in the way in which corporate GIS can transform the management of the often neglected information resources of a local authority. The Local Government Training Board (1991) emphasizes the role of information in tackling the key issues facing local government, and stresses the importance of a strategic approach to information management. Corporate GIS offers a strong response in that it embodies an approach which answers the need for co-ordination, sharing, quality standards, compatibility, flexibility, user-friendliness and security; all key elements of a corporate information strategy. In short, by adopting the GIS concept, an authority is also embracing a strategic approach to information.

In addition to this general effect, GIS also offers a great deal to those involved in specific areas of change. This can be demonstrated by looking briefly at three of these, namely, strategic planning, the development of the member role, and customer orientation.

Strategic planning

Over the last decade there has been a steady progression from the traditional model of local authorities as loose federations of service departments to tighter, co-ordinated, strategically managed, integrated organizations with common goals and a co-ordinated effort to achieve them. This reorganization has created demands for new types of information. The strategic planner needs broad pictures, simply drawn which explains how successful the efforts of the authority are in achieving its key medium and long-term goals. While conventional lists and tables will continue to have value, GIS offers new ways of presenting complex information drawn from a number of sources which answers entirely different questions. More importantly the map-picture is a powerful means of presenting such information.

The allocation of financial resources provides a useful example. A great deal of financial information is geographic, i.e. it is possible to relate much of it to the place it is spent. While we usually know how much we spend and what we spend it on, do we really know *where* we spend it? GIS opens the doors to a new range of information which explains the deployment of resources geographically and can relate it, for example, to measures of need. Is our resource allocation equitable? Do our bus services serve the people who are most dependent on them? Do rural areas "do better" than the towns? These types of questions are critical to effective strategic planning, and sets political debate in a less anecdotal and parochial setting.

The development of the member role

The elected member has three broad roles:

- Participating in the running of the authority (the Strategic Planning role),
- Managing the delivery of services (the Board Member role),
- Representing the needs and interests of his or her community (the Constituency role).

GIS has much to offer all three of these roles. As discussed above, it has the potential to create new types of strategic information. Its role in supporting the management of services is the central argument in the decision of most local authorities to adopt it. However, it also has considerable potential to support the representative role of elected members.

The development of applications tailored to the needs of members which present intelligence about their constituency and its setting in a simple and attractive way would do much to counter inherent *ad hoc* decision-making, based more on prejudgement than reality, which is often displayed in the council chamber. While it may be naive to assume too strong a causal relationship between understanding and "good" decision-making, a commonly understood and accepted factual base is a better starting point than the usual jumble of unstructured half-truth and anecdote. The concept of a "constituency atlas" can become a cost-effective reality through GIS. The member service could be extended in a modified form, to provide a public information service. Here the power and flexibility of GIS could be exploited to provide a highly attractive product. Indeed this concept is entirely consistent with the values of openness and sharing which many authorities are currently adopting.

Customer orientation

The movement towards the "business culture" in local authorities driven by competitive tendering, trading accounts, service level agreements, and the production of service and business plans is now a component of cultural change throughout UK local government, and is likely to remain so. In terms of information management the emphasis has shifted from operational needs to marketing needs. While the former is still important, identifying and understanding the needs of the customer is becoming critical. Whether within or without a competitive tendering environment, the specification of services and the monitoring of their performance is the key to effective management. The ability of GIS, particularly by means of census data and address gazetteers, to generate customer information at any level from strategic overview to detailed operational output gives it a unique role. It offers a key tool in the reorientation of local authority service management. Its capability for analysis, modelling, and communicating will be increasingly valued as the shift towards the customer is sustained.

Conclusion

This brief exploration of the potential of GIS to operate at the top level of organizational management suggests that its potential as an instrument in organizational development is substantial. In a period of radical change, realizing this potential is of special urgency. By implication it emphasizes the benefits of a corporate approach, further weakening the case for single use systems. These systems would not only miss out on the economies of scale, and the added benefits of information sharing, but also on the potential to address major strategic issues in corporate management. An organization adopting and effectively implementing corporate GIS will change in the way it perceives the value of information, manages that information, and is able to communicate it to the people who need it. To that extent at least it can be said to be truly culture-changing.

References

Campbell, H., 1990, The organizational implications of geographic information systems for British local government, *Proceedings of the First European GIS Conference*, Amsterdam, 10-13, April, Utrecht: EGIS Foundation, pp. 145-57.

French, W.L. and Bell, C.H., 1990, *Organizational Development*, Eaglewood Cliffs, NJ: Prentis-Hall.

Local Government Training Board, 1991, Information Management in Local Government: A Discussion Paper.

Mahoney, R., 1989, Should Local Authorities Use a Corporate or Departmental GIS?, *Mapping Awareness,* **3**, 2, May/June.

Margulies, N. and Raia, A.P., 1978, *Conceptual Foundations of Organizational Development*, New York: McGraw-Hill.

52

Stability demands acceleration

Peter Elkins
IT Southern

Introduction

Maintaining the acceleration

Projects and businesses thrive when they are accelerating steadily; GIS implementations are no exceptions. Many organizations have moved from the evaluation stage, through a sometimes painful cusp, to commitment and out into the assembly phase. It is now vital that management and GIS "champions" gather those forces necessary to maintain implementation acceleration, as this is essential for stability. To drift, as any good navigator will confirm, is to lose your way and possibly founder.

To encourage progress, a context is presented to the often confusing and contradictory ideas and experiences currently besetting GIS implementations. Many of these are not new to IT Southern's GIS team, so in addition, appropriate warning "flags" are raised to help the unwary.

Fragmentation to critical mass

Cohesive forces

A "critical mass" is a significant condition for getting things moving; it is required to change one situation to another, more progressively. On the other hand, fragmentation dissipates energy and limits progress.

For those responsible for implementing GIS projects, 1992 been heartening, in that the cohesive forces of good management have developed that "critical mass" in many organizations. The ideas, activities and directions of many GIS-focused businesses (those containing true end-users), have come together very effectively.

External forces, such as the Computerized Street Works Register (CSWR) and the concept of a National Street Works Gazetteer have focused the mind wonderfully. These coupled to guiding messages from the AGI and Local Government Management Board have helped to provide a sense of real-world perspective to the GIS deliberations. In Southern Water, back in the mid 1980s, early "ripples" of these future developments were sensed through well structured teams, brought together to implement GIS as a part of a wider information technology (IT) strategy.

Resting on our laurels?

To many, GIS has always been seen as a quiet revolution. Good signs are now here; meaningful tranches of data conversion have been achieved and now support real business decision-making. Assets, areas, properties, attributes, have been accumulating apace. Progress in the utilities has been especially encouraging.

This stabilizing picture is good news, but it's still too early to celebrate. In the main, progress has been made in GIS supporting detailed, disparate, and low-level activities. Although, at IT Southern, this has required meticulous planning, yet more energy will be needed to continue building the GIS data resource for full corporate use; this will be the case for many organizations. Several implementations are at the threshold of delivering true GIS facilities and many GIS vendors are responding positively to this challenge. Encouragement should now be given to these cohesive forces as they bind the fragments of the many GIS plans into that vital "critical mass".

Pressures

Current business pressures are, in the main, similar to those felt in the early 1980s, reincarnated perhaps, dressed in a new vocabulary, but essentially the same; they are, however, more intense in terms of the required managerial visibility, effectiveness and efficiency. Pressures can be directed to good effect; they can produce that needed GIS acceleration. Some interactions of pressures can be solidly persuasive toward GIS and some seriously destructive. It is important to understand which "mixture" one is dealing with.

Conflict, contradictions, and confidence

Current GIS pressures often appear to be in conflict and sometimes contradictory. These need to be unravelled to make sense of the situation and to allow each source issue to be addressed with an appropriate effort.

In water and environmental focused businesses, the demands of regulatory authorities are clearly significant drivers. GIS can respond by allowing high-level or aggregate views of business information, which are completely unified with the geographic base from which it is derived. This coherence results in confidence in the information supplied to regulatory bodies. However, has there been enough progress made towards achieving this unification of base data and business information?

A key question continues to be, "Is your information audit-proof?" This is an important question for the utilities, particularly for those who have responsibility for asset management planning, and especially accountants dealing with financial assets. We should, of course, be able treat such a threat to our GIS bailiwick with unruffled confidence. To take such a stance, however, we must have laid down base data in a competent manner and have GIS applications to enable the data-to-information process, and its reverse, well proven. To this challenge, is there still unruffled confidence or do I detect just a little nervousness? Maybe a little fear is healthy.

In conflict with the rigours of audit proofing, we have the apparent contradictory demands for flexibility at the GIS application, systems design, and implementation levels. A technical and business solution is required which can cope with all of these rigours, and do so against the inevitable business turbulence, restructuring, and cultural change, which GIS itself can induce.

The pressure to maintain flexibility of a strategic design is clearly stoked from within. It is the flexibility, underlying such implementations as the one within Southern Water, that has allowed both stability and the adjustments needed through water industry privatization. Healthy liaison between Information Technology departments and business end-users has been critical to success. Co-ordinating the understanding as to what is achievable at each stage and communicating this to GIS end-users has been a key activity. Identifying liaison staff early in the process has been particularly helpful.

A rigorous approach does therefore, seem consistent with the requirements for flexibility. Are the implementation plans in hand to achieve this?; these are best assembled at outset.

A pressure to be certain—dogma?

Opinions on GIS projects still tend to polarize. For example, do we redraw the map records first, before we start the data capture, or do we convert as-seen? There is clearly a pressure to be certain before making the next move. Such polarizations of the argument have been sensed by many as a confrontation to their personal confidence. This has often impaired progress in the decision-making. Most, I believe, would favour getting existing geographically-based records converted to as soon as possible and onto a GIS; at Southern, we agreed this approach after seeking a balanced opinion from a Regional Working Group. Again, a well-identified team to determine the most rational course was valuable.

Other organizations will have their own views about data conversion, depending on the basic state of the paper geographic records and the knowledge about the business/assets base, which these represent. In the final analysis, the decision reduces to a professional assessment, set against the business risks. The message is clear; let good sense prevail and avoid the stress of self-imposed dogma which can freeze the creativity needed to achieve the best solutions.

Recession

At times of financial stress, business priorities will be re-evaluated, which is only good business sense. However, sound principles do not automatically change as a result of economic recession. The application of geographic principles within business and IT environments is still demonstrably sound, and so further examination is best left outside the areas of doubt generated by recession.

Many organizations moving toward GIS implementations are relatively stable at times of recession and this fact should be regarded as an ally when promoting plans. It is a managerial responsibility to identify those aspects of a GIS strategy which will be affected by the current economic conditions and those which will not. Identification does, of course, require that we know the difference.

We must be careful not to stabilize GIS projects such that they do not have a forward path. Effective and rigorous communication of priorities between project managers and GIS developers will prove worthwhile.

From data to coherent information

What are the drivers?

The base-data, used to relate business activities geographically, will always need compilation to provide useable information. Combinations of raw data are now forthcoming from GISs and these are significant indexes of implementation success; utilities are making especially good progress here. What are the current drivers, however, which demand information-from-data, and hence GIS acceleration?

Basically, new information produced from raw data equals new opportunities for GIS end-users. As a GIS develops, this information exists, just awaiting business needs and objectives to precipitate it. To maintain progress, data analysts need to be pro-active in presenting the information opportunities which can assist decision-making. Having well established links between analysts and GIS end-users is an effective arrangement.

In achieving information from base data, there comes the conflict of GIS infrastructure versus the deliverables of specific applications and what importance to assign to each. Just as we need to provide IT infrastructure, e.g. communications links to allow departmental GIS facilities to progress, we must be sure that the right data infrastructure continues to be built, facilitating high-level information production from base data. GIS infrastructure is a critical investment as it encourages benefits, vertically, throughout an organization.

Of key interest continues to be the value added by those individuals receiving GIS generated information. For example, are decisions taken directly from this information, or do individuals just perform further manual collation, without any resulting decision? Clearly, GIS, especially when corporate, can both streamline organizational structures and increase the range and intensity of business-situation analyses.

Striking a balance between organizational streamlining and the range of supported activities will be difficult and stressful. However, we must speed up evaluations of this balance.

In addition, the traditional concept of data and information ownership will have to change. We must "unbundle" ownership into responsibility for meaning, accuracy, currency, security and availability. At Southern Water, there has been a major step in moving the responsibility for maintaining base asset data into divisional centres and away from the sudden and conflicting pressures of the water operational environment.

But who is available to see?

There is pressure on organizations, across private, public and government sectors, to run "lean and mean". This can mean fewer available staff to digest information. The right level of data distillation will be critical to the success of GISs, particularly as integration with other systems take place, i.e. as GISs migrate toward true corporate functionality.

Data "deluge" must therefore be avoided as practice shows that under these conditions, no one will be able to unravel its potential message. Critical distillation is needed—critical to the business activity, mindful of who will be making the decision, and sensitive to the level in the organization where the decision will take place.

Understanding the underlying data structures, the aggregation processes, and the mechanisms for correlating data sets will be key to success. It will be important to make sure this understanding exists prior to building a GIS; no doubt a firm lead will be taken by the AGI to assist this.

Keeping the past up-to-date

Data distillation and aggregation applies not only to "what is" but also, to "what was". We must, therefore, attend to the past. This is no reactionary stance; it's sensible, and pragmatic, as anyone engaged in monitoring planning applications will, no doubt, agree. Historic data and its own aggregations must not be forgotten. Again, this business requirement demands a good grasp of GIS data management.

Cultural change

Time

Platitudes on cultural modifications, resulting from GIS, are legion and are beginning to sound a little threadbare. Like platitudes in general, however, most continue to be true. It is often the limitation of the culture which defeats a GIS project. So what needs to be put in train now, to ensure an organization is pro-active in controlling its own destiny? How do we arrange and deploy the human resource? It would be pretentious to attempt to set down tenets, but, from experience, certain actions and ideas do deserve attention; most revolve around "time". GIS implementors must plan for the fact that a culture takes time to grasp implications, time to understand, time to gain comfort with a new vocabulary, and time to change. Time may be the limiting factor in achieving the desired project acceleration.

Complexity and flexibility

There must be recognition that the computing environment for integrated GIS solutions can be equivalent in complexity to very large non-GIS applications, such as corporate billing systems. The computing platforms and environments together with management procedures should be evaluated and treated commensurately.

Flexible GIS strategic designs need to be maintained to allow for rapid adjustments resulting from changes in business objectives. A notable case within the water industry has demanded the complete reorganization of the waste water function, not in years, but months! A culture with high calibre managers to control such change is vital; in these situations a networked GIS, which can be easily reconfigured, will be invaluable.

Do not forget the value of the paper copy. The tactile experience can lend a considerable comfort in the "almost impossible" working conditions some GIS end-users can experience when dealing with a burst main on a wet winter's night. The paper may have to be water-proof!

Keeping it to yourself

History shows that we are all potential hoarders of informal graphical data sets. Such data sets disclose key facts about the businesses and organizations, which need to be supported and of course prospered. Having encouraged staff to reveal those fabled "little black notebooks", which have abounded within the utilities and local government (and no doubt elsewhere), those driving implementations must now ensure that the information is actually transferred to the GIS!

However, let's be realistic; personal knowledge about assets, land, property and their relationships will always be present and always be needed. The degree to which this situation is encouraged and the percentages of personal data retained will prove an interesting next stage to cultural change. But beware, keeping it to yourself may hamper your acceleration.

Conclusions

Without doubt, pressures are on GIS implementations; pressures both constructive and destructive and pressure to deliver promises. As for the "critical mass", good signs are present. So, conflicts and contradictions must be settled to ensure that systems can deliver the information required in sensible time frames. Above all, the forces, which can accelerate projects, must be directed to maintain stability.

53

GIS selection and implementation— Not simply a question of technology

Helen Mounsey and Nick Pearce
Coopers & Lybrand

GIS literature abounds with advice on system selection, so why write another paper on the same topic? In practice, much that is written to date focuses on the technical issues which surround system selection and implementation, including product functionality and fit to requirements, and IT architectures on which to base GIS. But issues which surround the management and organization of GIS selection and implementation can also impact on the success (or otherwise) of the project. This chapter examines issues which, based on our experience, form the keys to successful implementation.

Project lifecycles

A quick examination of the current client list of any major GIS vendor would reveal the number and diversity of organizations currently implementing GIS. The scale of their projects will vary in scale, from full corporate implementations to single departmental pilots, in cost from tens of thousands of pounds to multi-millions, and in timescales from a few months to many years. Superficially they will have little in common beyond the use of the same vendor's software. In practice, their projects will share a common lifecycle of stages, which are broadly:

- initial introduction of the concept of GIS into the department or organization;
- a study to establish the potential for GIS in support of the organization's business;
- system selection to choose the most appropriate system; and
- a cycle of implementation and review to evaluate the success of the project and guide future developments.

Each stage requires more than an understanding of technology to ensure successful completion. What are the critical issues of organization and management?

First, find a project sponsor

Interest in GIS within an organization rarely originates from the top, yet senior management commitment is critical to the long term success of the project. Usually the initial enthusiasm is found at lower levels, and is often sparked by an introduction to the exciting technology of GIS. For the long term success of the project however, this enthusiasm must be translated into a sound understanding of its impact and benefit on the business. Projects which have been in the technology lead are those which have come under the closest scrutiny when the high costs do not appear to be balanced by any tangible benefit.

Senior management commitment is needed for three reasons. First, GIS is a long term investment, often modelled over a 10-year time period. Most businesses will evolve throughout this time, placing changing demands on the technology. Senior management are (hopefully) those with the long-term business vision, far-sighted and uniquely placed to understand how their business will evolve with time. The project team must be clear that their task is to translate this vision and incorporate it into a long-term GIS strategy. Everyone's common aim must be a GIS which evolves to meet changing business requirements, rather than a white elephant.

GIS is primarily an integrating technology, based generally on large corporate databases. Many benefits are dependent on a full corporate implementation of the system. Although for financial or political reasons, the proposed project may be planned initially as a departmental pilot, corporate vision is a necessary prelude to project rollout, and in some cases can avoid duplication of effort. Senior management can play an important selling role across departments.

Senior management are also the cheque signers, and will sanction an investment which may ultimately total many millions of pounds. Without the commitment of senior management, funding may be difficult, projects liable to termination and the long term commitment of resources necessary for the success of the project in jeopardy.

A well-organised management seminar is often the first step to selling GIS within an organization and identifying a sponsor. Seminars should be highly focused and draw on the experience of external users, suppliers and consultants if appropriate. A wide range of key senior staff should be invited to attend, including those with little apparent interest in or use for the technology at this stage. Topics for a management seminar may include:

- What is a GIS?—a short review of the nature of GIS and their ability to handle spatial data. Technical detail is not required at this stage;
- What can GIS offer my organization?—an overview of the potential applications of GIS within the organization, supported if possible by practical examples from similar users;
- How will it impact on the organization?—a realistic statement of the costs, benefits, timescales, risks and organizational impact of the technology; and
- Where next?—a clear proposal for the way forward for the introduction of GIS.

It is worth spending some time on this stage. GIS are not easy to implement, especially in an environment where corporate priorities can easily change. A good project sponsor can identify and remain focused on realistic and achievable long-term objectives. Commitment of other departments and groups within the organization to these objectives is typically one of the keys to successful GIS implementation.

Next, a feasibility study

GIS have often been excluded from corporate information systems strategies, and left to develop independent of any other IT initiatives. They are frequently championed by users who have received little or no support from corporate IT departments. However, there is a growing realization that GIS is a particularly complex form of information system, and subject to the same development methodologies. Organizations who include a requirement for the use of structured IT development methodologies in their information systems strategy will find that these can be used for GIS implementation, usually with few modifications.

A feasibility study often forms the first stage of such methodologies, and is used to confirm in greater detail the potential for GIS within an organization. Feasibility studies are designed to confirm:

- the existing and future applications of GIS throughout the organization through a user needs study, and the development of a range of business models which reflect the activities and databases within that organization;
- the case for investment in GIS, including a detailed breakdown of the costs of hardware, software, data and staff resources, the benefits of GIS and the risks involved in system implementation;
- an outline of the alternative IT architectures needed to support the required applications; and finally
- a timetable for implementation, including interdependencies with any other ongoing projects.

Much has already been written on the methodology for feasibility studies, including the use of business modelling techniques to determine user requirements, and of investment appraisal to identify full project costs and benefits. Such technical issues are not the subject of this chapter. Suffice it to note that a good feasibility study will take a hard look at the potential for GIS within an organization. It will require time and staff resources to complete, and it is tempting, especially for a small pilot implementation, to ignore the requirement for one. In practice, our experience confirms that this will serve only to undermine the long-term success of the project. The findings of the study underpin the remainder of the project in establishing implementation timetables and priorities. Any weaknesses in the documentation of user requirements or the investment appraisal may reveal themselves in late or non-delivery of system functionality and in project overspending later in the implementation timetable.

Some thought should be given to the composition of the project team. Feasibility studies should be conducted by staff who understand the organization's business, but who are also aware of the potential of GIS. They are demanding of time and resources, and are often best undertaken by a project team composed of internal staff and external consultants. Internal staff can rarely be released from their existing posts in order to devote sufficient resources to an all in-house study. External consultants may be willing to undertake the whole study, but their findings are unlikely to gain commitment and buy-in from all parts of the organization. The best study teams comprise a mixture of skills from both sources: a project manager, formally trained systems analysts for business modelling and users to confirm the completeness and accuracy of the results. External consultants can be included for their specialist advice and skills in GIS, and internal staff to confirm the potential for GIS within the organization. The two parties can work together to enable a transfer of skills and knowledge from consultant to staff, to form a solid foundation for continuing GIS development within the organization.

System selection: The next hurdle

An agreed and prioritized statement of user requirements should result from the feasibility study, together with an understanding of the costs and benefits of GIS implementation within the organization. Following approval to proceed by senior management, the system selection and procurement stage can begin.

Competitive system selection and procurement may take some time to complete. Many potential users of GIS are surprised to discover a requirement for at least four months (four weeks to complete, approve and distribute the documentation which forms the Invitation to Tender, six weeks for suppliers' queries and response, and six weeks for evaluation and benchmark testing). Methods of selection and procurement vary widely between organizations. Some are bound by strict rules, while others exert complete freedom of choice. However, some general guidelines designed to ensure the success of this stage can be identified.

Ideally, system procurement is based on documentation included in the Invitation to Tender. The quality of this important document is often overlooked. Suppliers will be asked to tender the costs of delivering a solution which is a combination of hardware, of standard software and of customization. Even in a non-competitive tender, costs can only be accurately assessed through a detailed understanding of requirements. A clear and mutually agreed picture of the product to be delivered at the procurement stage will often save much misunderstanding later in the implementation.

Returned tenders need to be assessed in a logical and structured manner. In a competitive tender, this will support discrimination between suppliers. In any tender, a technical evaluation will provide an understanding of the product to be delivered, its strengths, and its weaknesses.

While GIS are increasingly focused on narrow market segments, within a segment technical differences between GIS are narrowing all the time. Thus commercial selection criteria assume increasing importance in discriminating between products. Supplier credibility, viability, commitment to the user and to the industry sector in general are all important. With so many GIS vendors seeking a share of a still restricted market, the financial success and stability of potential vendors assumes an increasingly important place in system selection. Costs cannot be ignored, but must be balanced against the other factors. Lowest cost tenders may not be the best, indicating, for example, an imperfect understanding of user requirements. Technical, commercial and financial criteria are often weighted equally in the final evaluation exercise. A seemingly "expensive" tender which includes realistic allowances for support, development and training is likely to prove a better investment than cheaper options.

Benchmark tests are often seen as an unnecessary evil. However, they offer an opportunity to explore the proposed GIS configuration in a controlled and structured manner, and to understand in more detail its strengths and weaknesses. Moreover, they can be used to establish performance figures for later inclusion in the contract for system supply and development. They are recommended in some form even for the smallest GIS pilot implementation, and for major developments where functionality and performance are critical can play a key role in system discrimination.

A period for contract negotiation to cover system supply and implementation forms the last stage of the GIS selection process. The contract forms the document on which the initial system success will rest, including timescales and implementation priorities, resources and terms of payment. Given the importance of this document to the long term success of the project, three key requirements emerge:

- that sufficient time be allowed in the procurement timetable for contract negotiation (usually a minimum of four weeks);

- that for larger systems, contract negotiation is highly complex, and benefits from the input of a specialist IT lawyer; and

- that the contract is fair and workable to both parties. Negotiations need to result in a "win/win" situation, as both parties need to work with each other for the continuing success of the project and their mutual benefit.

And finally, to implementation

Development and implementation of GIS is undoubtedly the hardest stage in the introduction of this technology. GIS are not turnkey systems, and in reality there will be many months of hard work between system delivery and production running.

One key lesson emerges from all GIS implementations in our experience: a good project manager is critical to the long term success of the project. The skills required to deliver a fully functioning GIS are no different from those required for any other form of information system development. Good project management skills are more important than a detailed understanding of GIS. The right person can allocate available resources in a planned manner to ensure a controlled workload. Expectations for the new technology will also be running high. A project manager will need to strike a fine balance between managing expectations and instilling a practical view of possibilities on the one hand, and maintaining enthusiasm when things develop more slowly than anticipated (which they will) on the other. "Little victories" are important—small but frequent successes are an important factor in retaining team enthusiasm.

Two other groups of people are critical to the long term success of the project. First, the project team who support the manager. As at the feasibility study stage, good project teams consist of a mixture of IT staff (systems analysts and programmers) to develop applications, and user representatives to test, document and ensure their successful introduction.

End users form the other group who are critical to the outcome of the project, but who are often ignored. GIS is ''end-user'' technology. It will be used throughout the organization by non-specialists, and in some cases may change dramatically the nature of their work. Better, then, to get them involved in the project at an early stage.

The safest way to anticipate how the culture and working patterns may change within an organization as a result of the introduction of GIS is to undertake an organizational impact review as part of the implementation phase. This will help to establish the changes in individual roles and responsibilities. Sometimes this will need a corporate perspective to establish how the organization should adapt as a result of the introduction of GIS. Working practices may have to change in order to reap the benefits of shared data and technology. Border skirmishes and the NIH (Not Invented Here) syndrome have no part to play in the corporate implementation of GIS, and are best avoided by recognition and planning at an early stage.

Finally, regular progress reviews are central to the success of a GIS implementation. Variously disguised as post implementation assessments, appraisals, evaluations or project audits, they play a key role in the continued success and development of a GIS project. Carried out at regular intervals, they can be used to:

- measure the success of the project against the original objectives—are the benefits that were predicted being delivered?

- consider the possibilities for expansion in the light of changing business requirements and technological opportunities.

Although well established in other areas of information systems development, audits do not seem to have been widely applied in GIS projects. Documented examples are currently very hard to find—whether because they are not widely undertaken, or because the systems under review have not emerged in a favourable light is not clear. Nevertheless, the need for them is clear. Commonly believed to be necessary only in the early life of a project, in practice they should be carried out at regular intervals throughout the whole project life. Rather than being held in response to crisis, they should be regularly planned to ensure problems are identified and addressed at the first opportunity and before they become too serious. The results will help to focus future investment and support the continuing development of GIS to meet the requirements of the organization.

Conclusion

''Success'', when used as a measure for GIS implementations, is not gauged in black or white. Most projects will achieve something, and few will be totally without value. However, the current economic climate controls investment more stringently and demands full returns. The technological aspects of GIS implementation are increasingly well understood, the organization and management issues perhaps better so. But both contribute to the success of the project. In this chapter we have sought to identify some of the more major aspects of organization which, if addressed, can enhance the long term success of the project.

Part X

Mainline Issues in GIS

54

Technology review

Chris Corbin
IT Southern

Introduction

Geographical Information Systems depend upon a vast range of information processing technology for their success. The range of technology involved extends from telecommunications at one end of the spectrum, through computer hardware and peripherals to software at the other end. The software not only enables the physical components to be brought together in a coherent way to achieve a useful function, but also performs a vital enabling function with regard to turning data into information. Tracking this range of technology is a daunting task especially as the information technology industry is still young and fast moving.

The collection of technology papers brought together in this section of the AGI yearbook concentrate on only a few of the technical issues confronting the geographical information user of today. These chapters review both existing technology and technology that is under development. In considering these different technology domains common threads can be established. By considering these common threads one can gain an insight as to where the technology may be leading us, when and whether it will be of benefit.

Standards

One way of measuring the maturity of a technology is to consider the number and type of international or national standards that are in place appertaining to that technology. Geographical Information Systems are still in their infancy and it is therefore not surprising that currently there are few standards that relate specifically to the handling of geographical data.

Standards predominantly evolve out of human experience and knowledge and, as such, take a considerable time to be defined, especially when compared to the rate of change experienced within information technology. Because the standard making process involves people that represent organizations, there is often a tendency to protect the vested interests of those organizations during the process of formulating standards.

The past year has seen significant progress being made with regard to the National Transfer Format (NTF) with the publication of the British Standard BS 7567. The presence of BS 7567 is indeed a glowing tribute to all those involved in bringing this achievement about.

It has taken seven years for NTF to reach the status of a national standard from the initial inception of the working party in February 1985 to the publication of BS 7567 by the British Standards Institution in June 1992. Sowton later in this section describes the history and philosophy adopted in the development and publication of BS 7567.

Another notable event that has occurred during the past 12 months is the publication of the DIGEST, the transfer format used extensively within the military communities. The DIGEST supports raster and matrix transfers as well as three vector data models and, although comprehensive, is not as "transparent" as BS 7567 which supports the transfer of a wider range of application models.

It has taken eight years for the DIGEST to reach its current stage of development and availability from the initial inception of the DGIWG in June 1983 to the publication of Version 1 of the DIGEST in June 1991. Ley describes the background to the DIGEST, makes comparisons between the DIGEST and the various increments that NTF passed through, as NTF moved towards BS 7567 status, as well as noting the differences in philosophy between the civilian and the military sectors.

Both Ley and Sowton clearly demonstrate the processes through which standards develop and the effort that is required. It is particularly pleasing to note the convergence between the civilian and military communities with regard to these two transfer standards which has resulted in BS 7567 accommodating both the existing users of NTF and the DIGEST. BS 7567 can be categorized as an evolutionary standard as both the immediate past and the future have been accommodated. This approach as Sowton notes, has been recommended by CERCO (French acronym for European Committee of the Heads of Official Mapping Organizations) to the CEN (Comité Européen de Normalisation) technical committee TC 287. It is hoped that in the interests of all geographical information users that the current initiative to create a pan European Transfer Format (ETF) results in a standard that is evolutionary and that the process is completed within the next 12 months.

A notable difference of approach between the AGI and the DGIWG (Digital Geographic Information Working Group) is that of defining feature codes and attributes as part of the transfer standard. The approach adopted by at least one of the AGI standards committees has been that these definitions are separate standards and are not part of a transfer standard, even though they may be referenced during a transfer, as described by Sowton. Ley on the other hand describes the DGIWG approach where a Feature and Attribute Coding Catalogue (FACC) has been defined and included within the DIGEST. The FACC is rigorously controlled by a DGIWG committee that meets approximately every six months as described by Ley.

DIGEST Version 1.1, which is due to be published in the summer of 1992, is mainly a change of presentation. Version 1.1 will be structured as BS 7567 into a number of volumes, one of which will be the FACC definitions. Here again this can be considered as a positive step and perhaps, preparing the way for the DIGEST to be published by one of the national or international standard organizations in the future.

On reflection both parties have acknowledged the requirement for such definitions; the difference is how to go about the standard making process and even here the difference is minor. Perhaps in the military sector the need is more pressing and as suggested by Ley is easier to achieve, than in the civilian sector, where the range of applications is extremely broad. In the civilian community the organizations that represent the various disciplines or sectors would need to be involved in such a standard making process. As such this is not a task that could be accomplished by the AGI alone. The approach adopted by the AGI was therefore correct in concentrating on publishing BS 7567 in its current guise.

Transfer standards such as BS 7567, although essential, support the bulk transfer of data. For many geographical information users this is an overkill as there is a growing requirement for interactive data transfers. As Haworth describes the quantities of data now being accumulated, e.g. EOS (Earth Observing System) tend to make bulk transfers impractical. Instead of taking all the data to the user, the user is taken to the data. To enable interactive transfers to occur and for these transfers to be made internationally if required, there is urgent need for the feature codes and their associated attributes to be defined. Such definitions underpin rule-based technology as described by Quinn *et al.*, e.g. the definition of a road. Creating these definitions will not be without difficulty as described by Land,[1] however a number of national initiatives which are now well under way should assist, e.g. the Land and Property Gazetteer, the Street Gazetteer, etc.

The growing use of monochrome, grey scale and full colour raster data is also leading to increasing pressure for standards in this area to be defined. BS 7567 permits the transfer of raster data as externally defined files within vector levels 3, 4 and 5. A number of raster transfer standards currently exist and one such working standard, ASRP (ARC Standard Raster Product), has recently been published. Gennery, later in this section, provides both the background and a technical description of ASRP. ASRP is part of the DIGEST family of standards.

Standard issues arise in nearly all of the accompanying chapters and a common theme that arises is the time they take first to be published, and second the time they take to appear within products. Dickinson and Reid, in reviewing colour scanner technology demonstrates the consequences of the lack of standards by detailing the number of file format conversion routines that have to be provided. *De facto* standards such as the Tagged Image File Format (TIFF) have been evolving over the past six years with input from a range of different application areas.

Haworth, in reviewing database engines, highlights the importance of international standard interfaces with regard to protecting the high investment that has and continues to be made in data. One such standard considered by Haworth is SQL which has been evolving over the past six years. It will be a further four years before the requirements of the geographical information user will begin to be accommodated by SQL. Here again the total elapsed time is likely to be of the order of 10 years when SQL-(96?) will have been ratified and published. The AGI has played an important role through its standard committees by intercepting the SQL standard revision process early enough, to ensure that the geographical information users requirements will be met.

Arnstein in reviewing windowing technology within the graphical user interface also describes the various standards that are evolving. Here again the *de facto* standard, X-Windows has been evolving over the past six years.

Lake, in reviewing parallel computing technology, introduces the concept of standardization at product level, where standard industry components are being exploited in parallel computer systems. The industry standard components themselves conform to the published international standards. This approach has many advantages, one of which is that only the higher order standards related to parallelism need to be defined, a similar situation to geographic information systems.

Standard making, can be a way of developing a market and one such example is described by Ley where he describes the publication of the Digital Chart of the World (DCW) which is structured to the Vector Relational Format (VRF). DCW, which will become available in May 1992, is targeted at the PC user community. Lang[2] also reports that one of the goals of the DCW project is to create a generic standard, an interesting example where the military communities are moving into the civilian sector not only to influence standards but also to develop the market as a data provider. One of the advantages of the DCW project may be the resulting growth in awareness of the potential of GIS, which in itself would be of benefit to the whole of the GIS community.

Raster technology

Raster technology has been available for some time but has taken a long time to be exploited within GIS even though exploited elsewhere in the information processing community. Devereux and Mayo suggest the reasons for this delay are a combination of the:

- vested interest of both the system and data suppliers in vector technology, e.g. the availability of the Ordnance Survey large scale digital mapping vector products, i.e. a data supplier;
- impact of pilot projects;
- GIS community underestimating the cost of the data conversion activity;
- view that problems existed with raster data but not vector data;
- purchasers of GISs over-specifying their immediate operational requirements.

This line of reasoning is interesting which, if true, has probably held back the exploitation of geographical information systems. It is supported indirectly by the original thinking incorporated within the NTF specification where raster data was assigned an independent level (level 0). BS 7567 (NTF) has rectified this in that raster data has now been integrated into levels 3, 4 and 5.

It is also interesting to speculate as to what has been the predominant force that has changed the situation with regard to raster data—satellite imagery and remote sensing? This may be a technologists view! Evidence that raster data has indeed now come of age is the proliferation of articles appearing within the GIS media on the subject, e.g. The Integrated Poster of Europe.[3]

The major problem that still confronts the majority of geographical information users is that of data conversion and capture. Raster technology has enabled, and will continue to enable improved methods to be used during the data conversion process. Developments such as automatic vectorization, as described by Devereux and Mayo, combined with knowledge-based systems techniques described by Quinn *et al.* have great potential for bringing about significant savings when converting or capturing data.

Colour scanner technology which enables a particular data set, e.g. roads, rivers, vegetation, etc., to be stripped off colour maps, drawings and photographs, has now reached a stage of development where the technology is both reliable and cost effective to exploit. Dickinson and Reid provides a review of colour scanning technology which gives an insight into how the technology achieves this data stripping function by considering one particular colour scanner product. Once the data set has been separated, intelligent vectorization can then be used if required as described by Devereux and Mayo.

Parallel processing

The range of disciplines that are exploiting geographical information systems continually grows as the people involved become aware of the capability of geographical information processing systems and the technology reaches a point where it is cost effective to do so. Many of these application areas involve:

- modelling real world systems, e.g. networks such as those managed by the utility companies or those involved in the transportation industries,
- analysing data, e.g. environmental monitoring using satellite data,
- retrieving data from across a range of databases,
- managing data.

The pressures upon geographical information users are continually growing and demand that the decision support activities be processed in real time.

The technology currently available does not enable this requirement to be met and may not do so before the end of the decade. Lake reviews the current position of parallel computing and provides an insight into the technology itself, the forces driving future developments and the potential areas where parallel processing could be exploited. He makes the point that a number of parallel system architectures exist which handle particular aspects of parallelism. The parallel system architecture that would be the easiest to exploit is that of data parallelism by coupling a database management system to the parallel computer system. An example of this is the Oracle Parallel Server, where Oracle's RDBMS (Relational Database Management System) exploits the Meiko Computing Surface and has been on general release since October 1991.[4]

Database engines

As information technology advances, the range of real world problems that can be solved and human activities that can be supported grows ever wider. Some of the likely consequences of this development are that:

- the quantity of data held electronically is growing almost exponentially;
- the range of data types held electronically is broadening;
- much of this data will be captured automatically without human involvement.

As a result, the information technology that is currently available is operating at the limits of its capability both technically and economically, i.e. the technology involved in managing and turning this data into information. Due to this situation the data is often discarded or lies idle. This then fuels further technological advancement, almost a self-sustaining cycle.

Examples of where information technology is fuelling this process are:

- telemetry systems monitoring real world systems, e.g. a utilities electric, gas, water or telecommunications network, air pollution in towns and cities,
- remote sensing/satellite imaging, e.g. EOS, the Earth Observing System,
- point of sale systems, e.g. online supermarket checkout systems, digital telephone systems, etc.

The geographical information user may, while performing a spatial enquiry, access data that resides within many databases which may be distributed throughout an organization or across organizations. These databases will:

- vary in size, age, quality, completeness and currency;
- be created and maintained by a range of applications which may or may not be integrated and or related to the geographical user;
- probably have been designed such that the geographical information users' needs will not have been taken into account;
- owned and maintained by a variety of organisations;

and as such the information technology must enable such spatial enquiries to be met without recourse to continual rework of the data, for example convert the data to the latest technology or the need to restructure the data before it can be exploited. A fundamental requirement of the geographical information user is that the technology must be evolutionary and support data of varying ages which maybe held across a broad range of database types.

These requirements and constraints have been considered by Haworth, who reviews database engines and the benefits these will bring to the geographical information user. These include not only those already being achieved using the current database engines but also those that may accrue in the future. It is apparent that the next generation of database engines are not likely to appear before 1996 at the earliest. This time-frame should not only intercept the latest SQL-95/96 (Standard Query Language) standards, which will embrace the spatial extensions, but also the likely needs of the geographical information user to interchange data dynamically.

It is interesting to note the background of one of the database engines described by Haworth, in that the concept of database engines has been under development for over 20 years. The original concept of CAFS dates back to December 1969 but it did not appear as a product until 1979 when ICL announced the CAFS 800 system. CAFS 800 was a dedicated database engine of which 14 were sold within the first two years. It was not until 1982 when ICL announced CAFS-ISP that the product became integrated into the ICL mainframe family operating under VME when the exploitation of the product then took off within the ICL VME customer base. In 1991 ICL quietly announced the availability of SCAFS, the Relational Database Accelerator for Unix which runs under SQL and is XPG3 conformant. After 20 years the product has potentially reached a position where it could be exploited by a wider user community via the use of RDBMS products such as INGRES and Operating Systems such as UNIX. The INGRES Search Accelerator is a form of data parallelism, as described by Lake. Such developments are of key importance to the geographical information user whose transactions, as described by Haworth, tend to involve considerable database searches. The database engine technology is here today and can be exploited, while the massively parallel database engines as described by both Haworth and Lake are developed and brought to the market.

Knowledge based systems

The exploitation of GISs could be greatly extended if the processes involved in:

- creating and maintaining data,
- data retrieval,
- data analysis,
- decision support,

could be computer assisted by utilizing human expertise and knowledge that is held within the system. This would provide benefits which would include:

- ease of use, hence opening up GIS to a wider body of users,
- data integrity and data quality would be maintained to an acceptable set of criteria which underpins the decision support process,
- improved use of the data,
- improved decision support,
- the freeing of human resource to tackle other activities,

resulting in a faster return on the investment being achieved.

Quinn *et al.* reviews the current state of the art with regard to knowledge based systems and considers how these can be exploited to achieve the above-mentioned benefits. Another example of where the exploitation of a rule based system combined with a parallel processing system is being researched is that of placing labels related to a feature on a map. This has applicability not only for the cartographer but also for the general geographical information user, where following an enquiry, the system dynamically generates vector data from alphanumeric data and places the labels correctly for maximum clarity and accuracy, for example underground pipe work or ducting. Work performed by Jim Mower, an assistant Professor of Geography and Planning at the State University of New York, Albany, quotes significant performance improvements to be gained by the use of parallel processors.[5] The time taken to place a single label on the map took 30 minutes using a Sun Microsystems work station compared to three minutes using a Thinking Machines parallel processor.

This comparatively simple example would indicate that parallel processors certainly have a role within GISs in the future and that the performance of the technology has to be significantly improved for use in applications demanding dynamic placement of spatially related information. Such techniques can be exploited in applications other than GIS applications, wherever there is a need to place one item in relationship to another, for example microprocessor design, layout of printed circuit boards, etc. This wider applicability is beneficial in that it supports the ongoing development of the technology.

The graphical user interface

One way of measuring the success of any particular technology is to consider how widespread the use of that technology has become. An example of this is windowing technology. Over the past 12 to 24 months the benefits of windowing technology have begun to be appreciated by the personal computer user community. Technology that was previously only in the domain of the UNIX workstation community has now spread to the MSDOS community. Such technology behaves as a catalyst in that it is masking the differences that exist between the other technologies such as operating systems.

For information technology to deliver the maximum benefit to the user it needs to support a wide range of functions that any particular user or group of users perform during their daily cycle of activities. The trend is for this support to be provided through a desk top workstation through which the user has access to a range of tools, one or more of which maybe related to geographic information processing. As time has progressed the range of tools has grown. These tools have been developed and produced by a wide range of organizations, each with their own characteristics. The user is then presented not only with the problem of remembering all the various user interfaces to each of the applications but also of stitching them together to meet a particular requirement. Arnstein describes how windowing technology is assisting to overcome this problem.

Arnstein not only describes the graphical user interface technology and its development but goes on to consider the benefits to the geographical information user and the developer of geographical information processing tools. A major benefit to the application developer is that window handling can be separated from the application which results in improved portability, less code to develop and maintain, etc.

The benefits to the geographical user will be in the medium to long term. One such benefit would be to remove the current differences the geographical information user experiences, for example when requesting the system to display the attributes of a particular graphical object. The system may respond by opening up another window and connecting the user directly to an RDBMS with its own interface, which may have been designed predominantly for character based systems.

Arnstein suggests that, like the SQL standard ISO 9075, the graphical user interface standards that are evolving do not take account of the geographical information user community. As a result these requirements will have to be accommodated in later iterations of the graphical user interface standards, and as in SQL, each supplier will be developing their own extensions.

Summary

A number of threads have been detected while considering the current status of information technology and its applicability to the geographical information user. These threads include:

- standards, e.g. BS 7567,
- use of mass-produced components within products, e.g. massively parallel computer systems,
- evolutionary, e.g. BS 7567, Database Engines, Parallel Computers,
- ability to exploit data accumulated in the past as well as opening up new opportunities for the data that will be accumulated in the future, e.g. Database engines,
- scalability, e.g. massively parallel computer systems,
- ease of use, e.g. knowledge based systems, GUI,
- automation, e.g. knowledge based systems, raster to vector conversion.

The review shows that irrespective of the information technology domain one considers standards are:

- fundamental to the successful widespread exploitation of the respective technology;

- dynamic, i.e. evolutionary;
- taking on average between six to 10 years to be ratified.

The International and National standards in themselves tend to reflect the past, the established bedrock, rather than the future. The future is represented by the ratified standard, if there is one, plus proprietary extensions; standards can be used as a product differentiator. There is no doubt that currently the standard making process is too slow to bring the benefits claimed.

The AGI has so far been instrumental in bringing standards forward which are in the interest of all geographical information users irrespective of whether they are members of the AGI or not and has taken a positive role in European standard making. All geographical information users should, in their own interest and that of their employer, support the standard making process by:

- actively supporting the representative bodies that are involved in the standard making process by becoming members, e.g. the AGI;
- make GIS purchases against a minimum set of standards;
- exploit existing standards where ever possible;
- ensuring their requirements are known and understood by those involved in the standard making process;
- ensuring that the systems acquired are "modularized" so that standards can be "plugged in" relatively easily.

As a result of responding in this way, the time period to produce standards and for these ratified standards to appear in products and services will be reduced.

An important point to note from the accompanying technology chapters is that the technology itself is equally applicable to both the large and small geographical information user; a system can be configured to match the initial requirement and then be expanded incrementally as and when the need grows, or can be accessed across a communications network.

As has been described geographical information systems for their success are not only dependent upon a wide range of technology, but also other application systems, as much of the data that a geographical information user is interested in resides in other systems, i.e. databanks. If the other application systems do not also use the appropriate technology, then irrespective of whether the GIS technology is available, the progress of geographical information processing will be impeded—data is as important as the technology.

Conclusion

The information technology required to meet both the requirements of today and the future, as they are understood today, of the geographical information user, is either available or under development. It is suggested that the technology is keeping up with the aspirations of the geographical information users but perhaps it is the human and organizational aspects which are not, as a result of which the GIS technology available today is either not being used or is being used but not to its full potential. These areas are explored further within other sections of this yearbook.

Notes

1 Land, N., 1991, "The Classification of Spatial Data—A proposal for a National Standard", *AGI 91 Conference Proceedings*, pp. 3.16.1–3.16.7.
2 Lang, L., 1992, "Digital Chart of the World Becoming a Reality", *GIS Europe,* **1**, 1, February, pp. 58–60.
3 Van der Laan, F., Mak, A. and Nijpels, K., 1992, "Poster of Europe Illustrates Raster/Vector Integration in GIS", *GIS Europe,* **1**, 2, March, pp. 28–32.
4 Jakobek, S., 1990, "Scalable Computing in the 1990s", Meiko—publication number X0081-00N100.04, September; 1991, "The Meiko Computing Surface: An Open Systems Platform for Oracle", Meiko publication number X0081.00N101.02 October.
5 Alexander, M., 1991, "Mapmaker takes High-Tech Road", *Computerworld*, 8 April.

55

British Standard BS 7567—Electronic transfer of geographic information (NTF)

M. Sowton
Chairman NTF Steering Committee

The standard

After six years of development the National Transfer Format (NTF) has been published as a British Standard (BSI, 1992). Those involved in the production of this British Standard should take great pride in having progressed in just one year from a draft for public comment, written between March and July 1991, to a published standard. The final drafting incorporating the comments received, some of which were substantial, took under three months.

The resulting standard incorporates all the practical comments received since Version 1.1 and will be available in three parts:

Part 1 Specification for NTF Structures
Part 2 Specification for implementing plain NTF
Part 3 Specification for Implementing NTF using BS 6690

BS 7567 will be managed by the IST/36 Geographic Information Technical Committee of the British Standards Institution. This committee incorporates the AGI Standards Committee. It is through the link between the AGI Standards Committee and the NTF Steering Committee that the latter committee will continue to manage NTF. This means that BS 7567 is technically managed through the existing structure of the steering committee and working groups but ultimate control is exercised through the IST/36 Technical Committee of BSI.

Development

A detailed account of the development and purpose of BS 7567 will be found in the author's paper in the proceedings of the Mapping Awareness Conference in February 1992 (Sowton, 1992a). It should be remembered that the philosophy behind NTF was to provide a flexible and neutral format for the transfer of geographic information with varying levels of complexity. BS 7567 is medium, system and application independent. It was specifically designed so that a transfer did not have to be more complex than was warranted for the data being transferred but provision was made for any known data model to be accommodated.

A brief tabulation will give a résumé of the stages through which NTF has progressed. It is worth remembering that throughout its development the format has been used and while the needs of users indicated changes which were necessary, these changes had to be implemented without disruption to other users—a neat balancing act but a process through which NTF has matured into the new British Standard.

February 1985	Working Party to produce National Standards for the Transfer of Digital Map Data established.
August 1986	First draft of NTF published at Autocarto London.
November 1986	Public meeting held at the Royal Institute of Chartered Surveyors —need for transfer levels accepted.
January 1987	Version 1.0 published.
January 1989	Version 1.1 published.
July 1989	Responsibility for NTF transferred to AGI from Ordnance Survey (OS).
June 1990	Consultancy to implement NTF using BS 6690.
June 1991	Version 1.2—DIGEST Compatible—completed but not published.
July 1991	BSI draft for public comment issued.
December 1991	Drafting contractor appointed.
January 1992	Pre-publication draft BS 7567 issued.
March 1992	Camera-ready documents for BS7567 passed to BSI.

It is not possible to give a realistic estimate of the effort which has gone into the creation of BS 7567 from its inception as NTF. However, there are a number of major contributors without whose effort and support the enterprise would have foundered; these include:

1 Ordnance Survey,
2 Military Survey,
3 Laser Scan,
4 Alper,
5 Geobase Consultants, and
6 AGI.

In addition there are numerous collaborators who have tested the results, taken part in working groups and directed the work through the steering committee. To each and everyone credit is due. It is hoped that BS 7567 and its subsequent use is an acceptable tribute.

Users of NTF

At the last count there were nearly 300 registered users of NTF which does not include users of systems which incorporate NTF interfaces but who are not registered other than through their system supplier. How many of the registered users are actively transferring data using NTF is not known but certainly a growing number of systems requiring OS digital map data are specifying NTF. In March 1993 the Ordnance Survey Transfer Format (OSTF) format is to be withdrawn by OS and new customers are being encouraged to receive data in NTF. The initial pressure to create NTF came from the utilities who found OSTF unsuitable for transfers of their own data among themselves and thus an unknown number of active users of NTF exist in the utilities. As a result an increasing number of suppliers are offering an interface into NTF mainly at levels 2 and 3. A user requirement for the fully topological model in level 4 does not yet exist in the UK outside the Military Survey.

Outside the UK there have been a number of requests for information about NTF and other standards have been based upon it. Until recently the Italian Cadastre were using NTF level 2 for the transfer of this data but pressure to use DIGEST indicates that this format is likely to be adopted if the DIGEST FACC is modified to incorporate a section for their features (see Ley's chapter, this volume).

BS 7567 and other standards

In early 1991 the AGI considered that there would be advantages if the two main standards likely to be used in the UK were made compatible. Thus the new Version 1.2 of NTF was modified quite extensively to ensure this compatibility. In fact an additional level was added so that when necessary a full topological model (which implies that all areas are defined and without overlaps) needs to be transferred it can be done within BS 7567.

More recently a report (Lomax *et al.*, 1992) was commissioned by AGI with funding from the Local Government Management Board which was undertaken by Smith Associates to determine the degree of compatibility between NTF, DIGEST and EDIGeO (French Transfer Standard). The main aim of this report was to identify the degree of compatibility between the three standards with the intention of using this information to support the work of a recently established CEN Technical Committee. The report identified criteria which ought to be incorporated into a new format for Europe and evaluated the three standards against them. The comparison was not done to establish any one standard, a modification or a combination as pre-eminent but rather to identify their underlying philosophies and the advantages and disadvantages each contains. It is hoped that having established how far existing standards have moved towards compatibility that this will act as a catalyst to produce an acceptable European standard into which national standards can interface.

CEN

In February 1992 the Technical Board of CEN (Comité Européen de Normalisation) formed a Technical Committee—TC 287 to undertake the definition of standards for geographic information. Part of the work of TC 287 will be to determine the constituents of a transfer standard for Europe and consider ways in which this can be introduced without disrupting enterprises in geographic information already transferring data using existing standards.

Since no decisions have been made about how this work will be done it is not possible to say whether proposals already made by CERCO (French acronym for European Committee of the Heads of Official Mapping Organizations) to use a central format into which national formats would interface will be accepted as a basis from which to develop a European Transfer Format (ETF). A description of this concept will be found in the author's paper "From National to Normalised Transfer Format" (Sowton, 1992b).

Conformance testing

At present there is no internationally agreed standard for conformance testing of independent transfer standards. However, work on this important consideration is being carried out by a Commission of the International Cartographic Association under the Chairmanship of Professor Harold Moellering. Progress to date is that a draft list of criteria for the evaluation of standards has been produced. It is intended that a final list of criteria should be published early in 1993. How the result of this work will be implemented has not been decided but at this stage there is no intention to produce a table of standards showing how they conform. The list is intended to indicate the best practice to which a standard should aspire.

Raster and other external data

In the early versions of NTF raster data was separated from the vector transfers into level 0. With the advent of systems which combine both raster and vector applications it is considered that this separation is no longer appropriate.

Since raster and other non-vector data is likely to be supplied in highly specialized and source specific formats which are industry standards the translation of the actual data into an NTF format is unnecessary. Consequently in BS 7567 provision has been made for external files to be defined using NTF header and descriptor records within levels 3, 4 and 5. These external files may contain data such as raster, grid or even vector data in, for instance, DXF format. Thus level 0 is no longer included. It has not yet been decided which standards will be supported within external files but it is proposed that a list of commonly used industry standards will be issued.

NTF/plain versus NTF/BS 6690

BS 7567 offers the options of encoding data in the original NTF format (NTF/plain) specified in versions up to and including 1.2 and using BS 6690. BS 6690 is the British Standard version of ISO 8211, the International Standard, and is identical in all respects. The reason for offering two alternatives in BS 7567 is that an immediate change from the well supported NTF/plain to NTF/BS 6690 would cause too much disruption within the industry.

The introduction of BS 6690 encoding will markedly improve the efficiency of use and the effectiveness of the standard but a change of this magnitude can only be introduced gradually. Eventually users will recognize the benefits of NTF/BS 6690 and make the change when the perceived benefits outweigh the cost and upheaval of introducing a new interface for NTF/BS 6690. NTF/plain would be phased out and eventually withdrawn over a period but only as a result of market preference (Sowton, 1992a).

Feature coding

In other standards a high degree of importance is placed on the provision of feature and attribute coding. IST/36 policy is that feature and attribute coding should be standardized separately from the transfer standard as dictated by market needs. The philosophy behind BS 7567 remains that it is not part of a transfer standard to become involved in the application of data and how it relates to the real world. The function of a transfer standard is to make provision for all the data in a transfer to be received without loss. If the sender and receiver have agreed coding systems these can be referred to in the header and there is no requirement to transfer the coding manual with the data. Thus agreed coding systems such as OS 88, NJUG 11, FACC (from DIGEST) and DX 90 (from IHO) can all be specified and transferred through BS 7567.

Conclusion

The publication of BS 7567 will provide a mature and stable standard for the transfer of geographic information through which the market for geographic information within the UK will be enhanced. It is anticipated that the experience in creating BS 7567 will assist the CEN TC 287 to create a workable solution for the transfer of data within Europe and AGI will continue to support this activity.

References

BSI, 1992, British Standard BS 7567 Electronic Transfer of Geographic Information (NTF).

Lomax, A.M., Kellagher, H.A.B. and Clifton, T.R., 1992, "Comparison of Standards for Geographic Data Transfer" Smith Associates TR-92/108/1.0; Available from AGI.

Sowton, M., 1992a, "The National Transfer Format: The path towards a British Standard", *Mapping Awareness 92 Conference Proceedings*, (Ed.) Blenheim-Online pp 321–31.

Sowton, M., 1992b, "From National to Normalised Transfer Format", *Proceedings DRIVE Workshop—European Digital Road Map*, (Ed.) Daimler-Benz Stuttgart.

56

The Digital Geographic Information Exchange Standard—DIGEST

Richard Ley
Military Survey

Introduction

The widespread demand for geographic data in computer-readable form within the defence communities of NATO led to the formation, in 1983, of a focus of expertise on standardization matters: the Digital Geographic Information Working Group (DGIWG). Originally four countries (France, West Germany, UK and USA) were involved but since then the DGIWG family has grown to eleven nations with Belgium, Canada, Denmark, Italy, the Netherlands, Norway and Spain joining the founder members. It must be stressed that the DGIWG is not a NATO organization but comprises nations which are members of NATO. Furthermore, the DGIWG is charged with encouraging the use of its standards in both the civil and defence worlds.

One of the major aims of the DGIWG is to ensure that different national geographic information systems can exchange data effectively and efficiently with no loss of information. The term exchange covers data transfer within and between national production agencies as well as between an agency and user and between two users. Moreover, this exchange mechanism must not be system specific but must allow member nations to select their own system configurations. In other words, the mechanism must ensure full interoperability within and between different national systems. The concept of interoperability, with respect to digital geographic information, demands not only well-defined data models but also consistency in terms of data accuracy and structure, feature coding, geo-referencing, data, spatial resolutions and precision.

The conversion of these concepts into a framework document was the largest single task facing the DGIWG. Indeed, DGIWG's policy and technical staff have expended over 30 man-years of effort on digital geographic information exchange mechanisms. The major technical input has come from the panel of experts which is an informal working level group of GIS data specialists. Both collectively and individually they have been active in this specialist field for about 10 years. Thus, they represent one of the most experienced single bodies working on geographic data transfer standards in the world today. Furthermore, they are a truly international group of experts. The publication of the DIGEST (the DIgital Geographic information Exchange STandard) is the culmination of this work (DFMA, 1991).

Technical description of the DIGEST

The DIGEST is a family of internationally agreed standards and is designed to establish a uniform method for the exchange of digital geographic information based upon a common logical organization of any geographic dataset, independent of data type; it is a layered specification. Consequently, the DIGEST not only provides clear definitions of data models, data organization and data structure but also includes sections on data quality, feature coding, recording standards and the necessary supporting information. Moreover, the DIGEST is reliant upon the International Standardization Organization (ISO) for many of its constituent standards; it only defines new standards where there are no appropriate ISO ones.

The DIGEST specification applies to raster, matrix and vector data. The raster model supports multi-colour graphics or colour separates in either red, green and blue components or colour-coding in unsigned binary integer format. When these image bands are combined, they provide a multi-colour raster image. Matrix data are arrays of non-radiometric information (e.g. soil types or elevation) pertaining to points at regularly defined intervals (e.g. grid or graticule). Thus, the matrix data structure is identical to that of the raster model. Both data types do not depict features *per se* and so are devoid of any "spatial intelligence".

Vector data is supported by three different data models, which correspond to different levels of "spatial intelligence". These are, in increasing levels of "spatial intelligence", spaghetti vector, chain-node vector and full topological vector. Spaghetti data supports neither shared boundaries nor nodes at points of intersection. In its strictest interpretation it can be used only to represent line and point features. However, area information may be captured by the use of "closed lines" which circumscribe an area. The chain-node data structure is characterized by the provision of a logical node whenever two lines intersect and so allows both shared boundaries between features and the logical segmentation of features by nodes. Both the chain-node and spaghetti data structures utilize both spatial and feature (optional) pointers to depict the relations between spatial and feature records.

The most complex is the full topological vector which supports point, line, area and complex features. It provides a complete and logically consistent topological representation of the two-dimensional space covered by a given dataset. Two implementations of this data type are presented in the DIGEST. The feature-orientated implementation contains well-defined topological relations between topological entities (faces, edges and nodes) and features. It demands a many-to-many mapping between entities and features. Consequently, it requires high levels of processing power and RAM. This implementation supports full producer-to-producer exchange. The second implementation of this model is provided by the Vector Relational Format (VRF). It also utilizes faces, edges and nodes to represent vector data but linkages are supported by a number of relational tables. Processing and RAM requirements are much lower and this implementation can be supported by PC-based operations. This implementation facilitates producer-to-user and user-to-user data transfer. The Digital Chart of the World dataset (DCW) which is to become available in May 1992, is structured in VRF as it is aimed at PC-users.

Cartographic text annotation (related to specific or general features) is an integral element of datasets. However, it is an aspect of symbolization that is independent of the particular data model used for a given dataset. Consequently, the structure used to support cartographic text in the DIGEST is defined so that the logical consistency and completeness of the vector data models are maintained irrespective of the presence or absence of cartographic text in the dataset.

The DIGEST makes ample provision for the transfer of data quality descriptors. Quality statements are mandated at various levels (e.g. dataset, geometry, feature and attribute) for such topics as source, accuracy (positional and attribute), logical consistency, completeness and currency. This allows recipients not only to evaluate the quality of the data but also to integrate data into receiving databases without ambiguity.

One of the underlying aims of the DIGEST is to reduce the difficulty of exchanging digital geographic information between different users and different computer systems. Consequently, there are sections in the DIGEST covering such aspects as encapsulation, character representation and media standards. Two encapsulations of the DIGEST are included. The ISO 8211 (ISO, 1985) version is more appropriate for tape transfer whereas the ISO 8824 (ISO, 1987) encoding is preferred for telecommunications transmittal. Character representation for names and raster data must follow ISO 2022 (ISO, 1986b) whereas all other character information must use the character set and method of encoding defined in ISO 646 (ISO, 1983; i.e. ASCII). Data volumes are defined by ISO 1001 (ISO, 1986a) for magnetic tapes and ISO 9660 (ISO, 1988) for CD-ROMs. As can be seen, all of these rely heavily upon current ISO standards and are biased towards existing user requirements. As such they are not comprehensive but can be expanded as the need arises.

Coding specification

The DIGEST introduces the concept of a prescriptive coding scheme for vector data models through the Feature and Attribute Coding Catalogue (FACC). This provides a standard scheme for documenting those features and their attributes which are commonly found in digital geographic information. In addition, a data dictionary may be used where particular applications require the definition of a few specialized features and attributes not yet contained within the FACC. The FACC is under strict change control to avoid ambiguity and misinterpretation. It is a dynamic document and is expanded to meet new user needs as they arise. Its major benefit is that it allows the efficient exchange of digital geographic information without the permanent need for carrying a comprehensive, dataset-specific, data dictionary with each transfer. This renders many-to-many data transfers far more elegant and efficient as it requires only one look-up-table.

This concept of a standardized coding scheme is not unique to the DGIWG. The French EdiGeo standard includes such a mechanism, as does the International Hydrographic Office's exchange standard (SP57). Furthermore, the Italian Cadastre is interested in defining a scheme for their use. Conversely, the UK's National Transfer Format (NTF) follows a different line in that it has evolved around current practice; it is a responsive, flexible exchange standard. Typical data transfers utilizing NTF are one-to-one exchanges of data accompanied by their specific data dictionary. Many-to-many transfers can only be supported by building and maintaining several sets of look-up-tables, an expensive requirement. Following the prescriptive rather than responsive route to coding specifications does limit the user's flexibility; he or she must be far more formal in his procedures and coding. However, it does permit far greater exchange potential with guaranteed data integrity, minimal data transformations, reduced data volumes and overall decreased management costs.

Compatibility between DIGEST and NTF

The different approaches of the DIGEST and NTF towards data dictionaries is reflected in their respective philosophy towards transfer formats. The DIGEST explicitly defines a separate model for each data type whereas NTF is intended to be a general transfer format capable of supporting a wide variety of user-defined data models. Users are free to choose which level they employ for a given data model and there is no formal correlation between data models and levels. Nevertheless, irrespective of these underlying philosophies, a technical evaluation between NTF and the DIGEST is possible; this evaluation was undertaken in 1991 (Kallagher, 1991). This resulted in a number of recommended changes to NTF. Most of these are included in the new Version 1.2 of NTF (Rowley and Sowton, 1991). This evaluation has been updated and the following section compares DIGEST 1.0 with NTF Version 1.2 for the three data types, namely raster, unstructured vector and structured vector.

For raster and matrix data, the simplicity of the data structures renders the two standards essentially compatible. Furthermore, in considering unstructured vector data, the DIGEST spaghetti model is generally compatible with NTF Level 2. However, there are some instances where important information would be lost in translating from DIGEST into Level 2 and it may be more appropriate to translate into NTF Level 3 "complex spaghetti subset". The limit of one attribute per feature imposed by NTF Level 1 would result in significant loss of information when translating from DIGEST spaghetti into NTF Level 1.

In terms of structured data, the DIGEST topological and chain-node data models are explicitly defined and presented as separate options whereas the NTF Level 3 data model is in effect a "superset", from which users may choose a "subset" for a particular application. Thus, by suitable definition of an NTF Level 3 "chain node subset", the NTF Level 3 data model may be compatible with the DIGEST chain-node data model. However, it is not possible to translate a general NTF Level 3 dataset into a DIGEST chain-node without significant loss of data. Level 3 provides a "partial topological subset" but this does not support the full topological data model which is mandatory for DIGEST topological datasets. This is catered for by Level 4. However, although this level allows a DIGEST full topological dataset to be translated into NTF, the reverse is not necessarily possible as NTF Level 4 does not demand a complete and consistent topological representation of data.

All standards are dynamic and NTF Version 2.0, as outlined by Sowton, introduces some minor modifications to this evaluation of data models (Sowton, 1992). The main one is the dropping of a specific NTF Level 0 (the raster data model) in preference for an undecided number of industry standards. However, it is assumed that the DIGEST raster will be one of these and so this change should not compromise compatibility between the two standards.

In conclusion, there now exists a well-defined family of levels and "subsets" within the draft NTF Versions 1.2 and 2.0 which provides compatible profiles for the DIGEST raster, spaghetti, chain-node and full topological data models. Problems arise when attempting to transfer general NTF datasets into the DIGEST equivalents. Ideally Level 3 needs to be expanded to include a "full topological subset". These problems are intensified due to the different approaches of the two standards to coding specifications. However, if a user defines his data dictionary as FACC and selects the DIGEST profiles within NTF, then it should be possible to exchange data in one standard into the other without loss of information.

Towards an international data transfer standard

The defence communities of NATO have been using digital geographic information since the mid-1970s. In the 1980s, the use of this type of information spread into the civilian sector. The 1990s is witnessing not only an accelerating increase in both data availability and the range of applications in the two communities but also the widespread exchange of data between them. For example, DCW which was produced by the defence community will be available to all. Despite this convergence between the two communities, there are still fundamental differences in both communities which affect their respective approaches to standardization.

First, in defence there is only one set of data producers—the national defence mapping agencies. In the civilian sphere, producers range from the national mapping agencies, through the large utilities to small contractors supplying limited data to single users. Second, defence's requirements for digital geographic information has always been international. Civilian datasets tend not to transcend national boundaries and there are still very few applications which require multi-national data, although this must certainly grow. Third, within NATO it was realized from the outset that its requirements could only be satisfied by international co-operation and standardization. Within the civilian sector, national standards have preceded any attempts at international standardization. This has introduced vested interests into the international negotiations. Fourth and perhaps most significantly, the military can impose mandatory acceptance of standards far more readily than is possible within the civilian and commercial communities. Consequently, it is much more difficult to establish internationally agreed standards within the civilian sector.

The 1990s will witness a decade of maturing national and international standards for the transfer of digital geographic information with ISO recognition before the end of the millennium. The DIGEST is the one standard which has multi-national support. Moreover, it has been put into the public domain and is freely available to all (contact Military Survey; DMA, 1991). Its applicability to civilian demands has been realized by several countries. The French EdiGeo standard and the emerging Belgian standard are based upon DIGEST. The Italians believe that it can also fulfil their needs. The new draft NTF Version 1.2 and 2.0 permit a DIGEST profile and, with an accompanying FACC-defined data dictionary, will allow full transfer of data without loss of integrity. Thus, within Europe, there is a growing convergence rather than divergence of national standards. To strengthen this move, the DGIWG has proposed that the control of DIGEST should pass from defence into the civilian sector. DGIWG has approached the Comité Européen des Responsables de la Cartographie Officielle (CERCO), with the intention that CERCO set up a DIGEST organization with representation from the defence and civilian mapping agencies of Europe and North America. If this option is followed, it is this body which will determine the future evolution of the DIGEST and its associated FACC plus any expansion of the participating membership. If international pedigree and standing are the basis for ISO status, then the DIGEST must be a major contender for the international transfer standard for digital geographic information.

Acknowledgements

The author would like to thank the DGIWG, which has produced the DIGEST, and Smith Associates who, under a MOD Contract, conducted the technical evaluation of the DIGEST and NTF. However, the author accepts the responsibility for any inadequacies contained within this chapter and the views expressed here are his and do not necessarily reflect those of either the DGIWG or the UK MOD.

References

Defense Mapping Agency (DFMA), USA (on behalf of the DGIWG) 1991, DIGEST, Edition 1.0, June. Distributed within the UK by Director General Military Survey, Elmwood Avenue, Feltham, Middlesex, TW13 7AH.

ISO 646, 1983, Information Processing—ISO 7 bit coded character set for information interchange, Second Edition.

ISO 8211, 1985, Information Processing—Specification for a data descriptive file for information interchange.

ISO 1001, 1986a, Information Processing—Files structure and labelling of magnetic tapes for information interchange Edition 2.

ISO 2022, 1986b, Information Processing—ISO 7 and 8 bit coded character sets—Code extension techniques. Edition 3.

ISO 8824, 1987, Information Processing systems—Open Systems Interconnection—Specifications and Abstract Syntax Notation One (ASN.1).

ISO 9660, 1988, Information Processing—Volume and File Structure of CD-ROM for information interchange. Edition 1.

Kallagher, H.A.B., 1991, *A Report on the Compatibility of NTF and DIGEST.* TR-91/123/1.0, London: HMSO.

Rowley, J. and Sowton, M., 1991, National Transfer Format—Progress Report. *Mapping Awareness,* **5**, 5, pp. 25-27.

Sowton, M., 1992., The National Transfer Format—The path towards a British Standard. Conference Proceedings. *Mapping Awareness*, pp. 321-32

57

A working raster standard—ASRP (ARC Standard Raster Product)

David Gennery
Military Survey

Introduction

Over recent years a number of defence systems have come to rely on Digital Geographic Information (DGI). Many advanced aircraft, weapon, intelligence and information systems now need data of this type to enable them to carry out their roles; and many more systems with similar data requirements are under development. Similar demands are recognized in many NATO nations, not just the UK, and the benefits that can be obtained from co-ordinated productions and data exchange can be used to accelerate data provision. Raster data was, at the outset, recognized as the only data type that would allow the demand for map display capabilities to be met; vector data, although more suited to user requirements is much more labour intensive with considerably longer production times. ARC Standard Raster Product (ASRP) is the result of many years practical development and negotiation by members of the DGIWG (Digital Geographic Information Working Group) to identify a raster product that is not only DIGEST compatible but economic to produce and distribute, sufficiently flexible to meet the wide range of international uses and compatible with existing products already being issued by DGIWG nations. ASRP is a virtually seamless product that meets these criteria while also retaining a high level of quality that enables user post-processing activities to be applied without undue loss of perceptual image quality in either feature retention or text legibility

ASRP (DGMS, 1991) is the result of a collaborative effort by the DGIWG nations to define an exchange standard for raster data for both producers and users of the data. It has been designed to meet a wide range of user requirements from the most stringent fast-jet cockpit displays to the much simpler desk top planning aids. Its design has also taken account of different existing national production procedures and provides a compatible solution that utilizes the best qualities of each.

Background

ASRP development has been a four-year effort by the UK, first to provide a national product and second to convert that product to an internationally agreed standard. ASRP is the culmination of these developments through products known as SRP1 and SRP2. Standard Raster Product 1 (SRP1) was a simple colour coded image of any base map to the same projection and datum as that graphic. No seamlessness, legend images or product standards were to be applied. This product was recognized as being problematical to users as it necessitated considerable user post-processing to tile and restructure the data. From this initial product SRP2 was developed. The development of Standard Raster Product 2 (SRP2) began before SRP1 was formally published with the intention of providing a more qualitative product. SRP2 was designed such that it could be produced from a range of existing hard copy sources as well as digital raster images; maps/charts, reproduction material, The Federal Republic of Germany's Standard Raster Graphic (SRG) product at 200 dots/cm resolution and also the United States' (DMA) product Arc Digitized Raster Graphic (ADRG) at 100 dots/cm resolution. ADRG provided the concept of seamlessness (i.e. The Arc System), the provision of legend images, but posed the problem of converting RGB coded data to Colour Coded form. It was this problem that led to the development of Transition Codes and the Extended Colour Coding technique. Presentation of SRP2 to DGIWG nations eventually led to its adoption as the basis for ASRP.

The specification

The DGIWG International Specification for ASRP is now placed under the custodianship of the Director General Military Survey MOD (UK) on behalf of the DGIWG. The Specification Version 1.0 is concerned primarily with the image data and descriptive files. Development and negotiations are still continuing on aspects pertaining to media, packaging, product maintenance, product referencing and importantly, data capture specifications.

It is generally understood by all users of raster data that it is an unintelligent data type whose prime purpose is to replicate hard copy maps on display screens. It was once stated that raster data is "the quick and simple solution to digital display data". Years of development have proven this not to be the case and even the more sophisticated specifications such as ASRP must be supported by carefully tested data-capture specifications. These specifications are necessary not only to enable harmonization of colours between printing inks on maps of varying series but to state scanning and processing parameters required to recognize and record the individual colours, lines, text, printing screens etc. used in map production. Without qualitative data capture specifications perceptual data quality can easily be destroyed.

An area of continued specification development is that pertinent to the handling of printing screens on source graphics. It is desirable for users to have such screens replaced by solid colour to reduce their post-processing costs and any subsequent data scintillation that is apparent on some display configurations. Whether this process can be viably incorporated into the production process without excessive overheads is still to be fully assessed.

General description of ASRP

ASRP is provided in a transfer format and users may need to post-process the data to meet specific system requirements. It is provided in a reference system that enables a global seamless dataset to be achieved without necessary transformations, sizing or cutting and patching that would be required were the data retained in map sheets to their original datum and projection.

The actual area of mapping provided by an ASRP dataset is known as the "map image" and incorporates all outsets. The map images, one per dataset, will be exclusive of marginalia and notes which will be provided as separate "legend images" within the data volume. The map image will be produced from one or more base graphics and will be identified by a unique reference. Map insets will be repositioned to their correct geographical location, to the same scale as the original source series and will be held as either separate map images or incorporated within others.

ASRP data will be divided into datasets such that one or more datasets fit onto a single unit of the distribution media. The content of a dataset is defined by geographic co-ordinates which typically coincide with chart and map neatlines. A dataset will include data from one or more scanned base graphics.

Incorporated into each individual dataset are several data files that contain:

a. Header information: Information about the contents of the dataset;

b. Source information: Information about the individual base graphics within the dataset (including accuracies, original datum and datum shift parameters, and projection information);

c. Quality information: Information on the accuracy/quality of the map image;

d. Image information: The pixel data of the dataset.

The ARC system

ASRP data is designed to carry any map/chart series that is required for electronic display but to a common reference system, i.e. WGS84 (World Geodetic System 1984). The equal ARC-Second Raster Chart/Map (ARC) system is the projection and co-ordinate system used. This data, being virtually seamless, ensures that raster graphic data from adjacent images abut exactly to provide unbroken coverage under this common projection system.

The ARC system divides the Earth's mathematical computing surface, known as the ellipsoid, into 18 latitudinal bands called zones, with zones 1-9 covering the Northern Hemisphere and zones 10-18 covering the Southern Hemisphere. One zone in each hemisphere covers each of the polar areas. Each non-polar zone covers part of the ellipsoid between two latitude limits and completely encircles the earth.

ARC projection distortion

Transformation into ARC zones involves a modest distortion due to the projections used. This distortion is apparent when large areas of the map images are viewed or printed in a rectangular format, but should be indiscernible when displayed on a monitor screen. This distortion is visual only and does not affect pixel co-ordinate accuracy.

In each non-polar zone distortion is seen as stretch (at the pole ward limit) or shrink (at the equatorial limit) in the east-west direction. There is no distortion along a latitude near the centre of each zone. The stretch and shrink does not exceed 18 per cent, exclusive of the overlap extent. Stretch in the overlap region is typically less than 25 per cent, but for small graphics and high latitudes, it may be significant. Distortion in the polar zones is less than 10 per cent.

Data

Vertical: The vertical datum for ASRP is the same as the vertical datum of the base graphics.

Horizontal: The horizontal datum for ASRP data is the World Geodetic System 1984.

A set of transformation parameters may be provided should the user desire to transform the data back to the original datum system. The constants for the WGS 84 ellipsoid are:

a = 6378137.0 metres
b = 6356752.3142 metres
e2 = 0.00669437999013

The Extended Colour Coding Technique

The Extended Colour Coding (ECC) technique aims to provide a maximum of "useful colour" information in a minimum volume of data. Normal colour coding techniques provide a single code for the combined red, green and blue intensities which make up a colour. ASRP adopts this principle where there will be one colour code for each colour on the original graphic but also introduces the use of transition codes, colours geometrically positioned between selected primary colours in a colour co-ordinate space. These transition codes (e.g. tones of grey between black and white, tones of pink between red and white, etc.) provide an anti-aliasing effect and may be incorporated to aid the definition of text, fine lines, and areas of high contrast.

The inclusion of transition codes also improves the subjective image quality and the flexibility of onward computations. The allocation and generation of transition colours will vary from one base graphic series to another so as to attain the best quality result with the minimum of colour changes. Transition codes are additionally beneficial in that they aid to remove some of the inherent problems that arise from map scanning:

Anti-aliasing

Use of transition codes at the edges between primary colours reduce the high contrast and the serrated effect caused by pixellation of a smooth line.

Dot patterns

Printing screens and dot hachures are more easily retained as recognizable patterns when supported by transition colours.

Image rescaling

This requires intelligent resampling. ECC provides a higher perceptual image quality for the resolution, therefore providing a higher level of information upon which to base down sampling techniques.

The use of standard colour coding techniques, including ECC, not only allows for the removal of arbitrary variations within the colours on the original base graphics but enables the cross reference to standard printing colours, the CIE (x,y,Y) reference values of which will be provided in the dataset header information together with nominal red, green and blue intensities.

Image quality

The pixel resolution used by ASRP is 100 dots/cm where, at a scale of 1/50 000 a distance of 0.5 km is represented by a nominal 100 pixels (or one cm). A combination of resolution and colour provides an ability to generate acceptable and desired quality in both visualisation and computational effectiveness. It is estimated that ECC provides a perceptual quality of 100 dots/cm resolution that is equal to colour coded data at 150 dots/cm.

Visualization refers to the perceptual image. Measuring the perceptual quality currently relies upon the techniques of "If it looks good it must be good" and quality will also have a bearing on the computational effectiveness which must also be measured if the product is to meet a range of requirements that may involve additional processing activities. As long as the identification of primary codes and the selection of transition colours is undertaken in a controlled production area then the likelihood of having poor quality data is minimized. The adherence to a well defined capture specification is essential in maintaining such quality. Data will also be supplied within ASRP detailing the relationships and rederivation between selected colours.

Data volumes

Given a resolution and a product scale the calculation of data volumes for a geographic area is relatively straightforward. However ASRP is to be provided in Run-Length Encoded form. Early trials have shown that, on average, little reduction in data volumes is achieved on the majority of map series (1/50 000–1/1 000 000) that contain dot screens. This is partly due to the use of ECC but primarily due to the feature density shown on maps in NW Europe. Map series that have no screens have enabled a 2:1 compression to be achieved.

Media and packaging

Final developments which are currently in hand are linked to media, packaging, labelling and data maintenance procedures. Interim procedures are soon to be announced and initial distribution is almost certain to be on magnetic tape with map images relating to the source graphic sheetlines from which it was scanned or produced.

In the longer term it is anticipated that optical disc will be the preferred and possibly only media. With this change of media a redefinition of image sizes, maintenance procedures, supply procedures and referencing systems will be required.

Summary

ECC has been developed such that it can be produced from existing products based on the standard Colour Coding system at higher resolutions or from RGB data at similar or higher resolutions. It is also possible to convert ECC data to Colour Codes of RGB forms.

ECC data can be criticized for sacrificing some data flexibility and consistency when compared to RGB data. However this is not considered detrimental to final use for users who wish simply to display the data at the product resolution. For those who wish to post-process the data with a large reduction in resolution or numbers of colours then it must be emphasized that the appropriate application of specifically designed post-ASRP processing techniques must be employed. Where this advice is acted upon there is reason to believe that ASRP will form a valuable standardized mapping baseline across a range of raster mapping systems.

References

''The Digital Geographic Information Working Group International Specification for ARC Standard Raster Product'', Director General Military Survey (DGMS), 1991, Version 1.0, September. Issued and maintained by DGMS, Elmwood Avenue, Feltham, Middlesex, TW13 7AH on behalf of the DGIWG.

58

Colour raster scanning technology: A review

Peter Dickinson, *Kirstol Limited*
and Geoff Reid, *Dainippon Screen Engineering Europe*

Introduction

In addition to the already substantial and growing number of Geographical Information System (GIS) implementations, there is an increasing equivalent demand for data capture. This demand arises from the requirement to convert into digital format the many millions of existing paper-based topographical maps, charts and diagrams that currently exist around the world. The data conversion and capture costs probably represent the largest proportion of the total budget for the implementation of a fully automated GIS (Adnitt, 1991).

Until recently data capture has been largely carried out by manual digitizing and while continuing to play an important role for certain aspects of data capture, new technology by way of high resolution large format colour raster scanning is set to establish a major role in data capture for GIS.

Colour scanning technology, we would suggest, has now reached a stage of development such that it can be profitably applied to the data/capture conversion process so as to have a significant impact in reducing costs in the overall conversion process. It is the intention of this chapter to look at colour data capture technology and examine in particular the advances being made and how this can be of benefit the users of geographical information.

Background

In GIS or map making applications, it is often required to capture particular features from a variety of sources. The features may be roads, waterways, pipes and services etc., which are usually required to be separated into layers so that individual types of feature can be updated, overlaid on other backgrounds or composed with other data to form a new map. The source material may be maps, charts, aerial photographs, satellite images, etc., and as these may well be in colour they will probably show more detail than if in monochrome.

Most colour maps have their features printed using only a few colours so that they are best scanned in Colour Coded mode; i.e. eight bits/pixel are used to represent up to 255 different colours. Colour Code discussed later has a considerable advantage over RGB in that the file size is a third less than the same document scanned in RGB 24 bits/pixel.

Colour aerial photographs may contain much more detail and range of colours and so to preserve information are best scanned in 24 bit RGB mode, i.e. eight bits Red, eight bits Green, eight bits Blue; whereby 16.7 million colours may be represented. Subsequent processes can work on the best possible data and extract the required features and reduce the number of colours later.

Colour data is written as a pixel and is read through three colour filters—Red, Green and Blue (RGB). The RGB colour model is called additive because overlaying red, green and blue light produces white light. Colour images consist of three planes of pixels that represent the traditional RGB values. Storing eight bits of information on each red, green and blue value results in a 24 bit colour image.

The price of hard disc storage, memory and processing power has been falling steadily over the past decade making the manipulation of colour raster data a cost effective proposition. The reducing cost of processing power means that raster data may now also be converted to vector format fairly easily.

The modern colour scanner has overcome the disadvantages of earlier types of scanner requiring three passes/scans of the original for RGB measurement, i.e. one scan of red, green and blue—hence the difficulty in registration when overlaying the three colours. Development of the colour scanner is such that both RGB and colour coded reproduction can be obtained by a single pass/scan of the original. Additionally, selectable resolution and variable scanning pitch in very small increments suit desired image quality or available data storage, all contributing to a more productive and accurate colour scanning facility.

Advantages of colour data capture

Different types of feature within the map may already be classified by colour layers in the original map or source material e.g. contours—brown, water—blue, roadways—red, forestry—green, etc. Some modern scanners, as well as scanning in monochrome and RGB, can also produce colour coded data. That is to say each of the colours used in a printed map is assigned a colour code number i.e. a typical map may be printed with 10 colours so 1 = white, 2 = black, 3 = red, ...10 = light blue, for instance.

In the past, the only practical method of extracting the information required was to manually digitize the source material to separate out the layers or features required. Alternatively, monochrome scans could be used with operator guided and assisted methods to extract the required information layers. There were few colour scanners on the market and the large amounts of data they produced could not easily be handled by the then available technology.

Scanners produce raster data—in fact nearly all input or output devices (scanners and plotters) operate with raster data. Pen plotters are the prime exception that operate with vector data. Scanners of the type under discussion produce a raster image which is ideally suited to applications such as GIS. The image is initially held in raster (or bit map) form which can be used as a raster background or as a map base. A raster image can be cleaned up if necessary using simple editing techniques such as thresholding.

Many of the programs developed to manipulate and process data within geographical information systems use vector data, so it is a requirement that facilities be available within a scanner configuration or the associated GIS to convert the raster data to vector data format.

The main advantages of raster format are:

- A raster image is ideally suited to applications in GIS where programs are available and used to overlay essential utility information in the form of vectors.

- It is easy to output from a raster input device, i.e. a video digitizer or scanner; on to a raster output device such as a visual display monitor or graphics printer.

- The display of raster data is usually faster than the display of vectored data because the cathode ray tube monitor screen is a raster device and thus a vector-to-raster conversion does not have to be performed.

Vector format on the other hand, involves the use of directed line segments instead of pixels to make up an image. A vector image is made up of shapes that are made up of line segments. By their nature, vector data are already highly compressed.

- Many computer applications in geographical information (GIS), desktop publishing and other graphics systems are now able to handle colour raster data.

Thus it can be seen that if the source material is scanned and data is output in colour coded format, the source material has already effectively been separated into useful layers, and editing is carried out on raster data i.e. colour coding, and hence output, is only what is required. This can make subsequent vectorization much faster and can be handled by semiautomatic digitizing (Adnitt, 1991) with a much reduced level of operator intervention, thereby increasing productivity.

Thus the benefits of colour scanning include:

- Automatic Data Capture of colour feature layers,
- Close facsimile of printed product,
- Elimination of operator digitizing errors compared with manual digitizing,
- Higher productivity compared with manual digitising.

Features of a good quality colour scanner would include:

- Point recognition i.e. precision at pick-up/training prior to scanning;
- Single pass scanning, substantially increasing map detail conversion productivity;
- Versatility and ability to scan in other modes such as greyscale and binary;
- Operator selectable resolution of up to a true high resolution of 1000 dpi. A variable scanning pitch that is user selectable in, say five micrometer increments (approx 0.0002″) over a range from 25 micrometers to 9 900 micrometers to suit desired image quality or available data storage space.
- Auto-calibrated scanning speed to suit computer/workstation which is processing and receiving the data.
- In the event of only certain areas on the map being required to be scanned, a cropping (trimming) function can designate a quadrilateral area in 100 micrometer increments, without scanning an entire map. Alternatively, if it is expedient to scan in a large area of the map, software resident in the workstation can be used to crop out the required area from the raster image file.

The time taken to scan a map will be a very small fraction of the time taken to manually digitize a complicated map. Scanning times vary according to map size and scan resolution. Typically a map of 1200 mm × 1200 mm scanned at a resolution of 380 dpi (150 lines/cm) would take 30 minutes to scan in colour coded mode. This would produce an uncompressed raster file size of approximately 340 MBytes. The file size could be significantly reduced if the data file was output in run length encoded format, e.g. the Dainippon Screen Scanner (ISC-1200PD) produces a raster data file of approximately 80 Mbytes. Likewise the workstation may be used to convert the file to other compressed or uncompressed file formats and this would typically take a further five to 10 minutes.

Typical raster file sizes for scanning different map sizes scanned at different resolutions with no compression and not run length encoded using the ISC-1200PD colour scanner are:

1200 mm × 1200 mm map scanned at 1000 dpi or 400 lines/cm

RGB 24 bits/pixel	= 6.9	GBytes
SCD eight bits/pixel	= 2.3	GBytes
Greyscale eight bits/pixel	= 2.3	GBytes
Binary (black and white)	= 0.288	GBytes

1200 mm × 1200 mm map scanned at 380 dpi or 150 lines/cm

RGB 24 bit/pixel	= 1022	MBytes
SCD eight bits/pixel	= 340	MBytes
Greyscale eight bits/pixel	= 340	MBytes
Binary (black and white)	= 42.6	MBytes

600 mm × 600 mm map scanned at 380 dpi or 150 lines/cm

RGB 24 bits/pixel	= 255.5	MBytes
SCD eight bits/pixel	= 85.2	MBytes
Greyscale eight bits/pixel	= 85.2	MBytes
Binary (black and white)	= 10.6	MBytes

Typical scanner configuration

The typical scanner configuration comprises of a scanner attached to a workstation that may or may not be connected to a local or wide area network. Figure 1 shows a typical ISC-1200PD configuration.

The functions performed by the attached workstation include the following:-

- control of the scanner;
- acceptance of the image data from the scanner;
- raster editing where required on the image file;
- data conversion to other file formats; e.g. DXF, IGES.
- copying data from disk to transportable media such as tape or optical disk;
- despatch of data across a network to other systems.

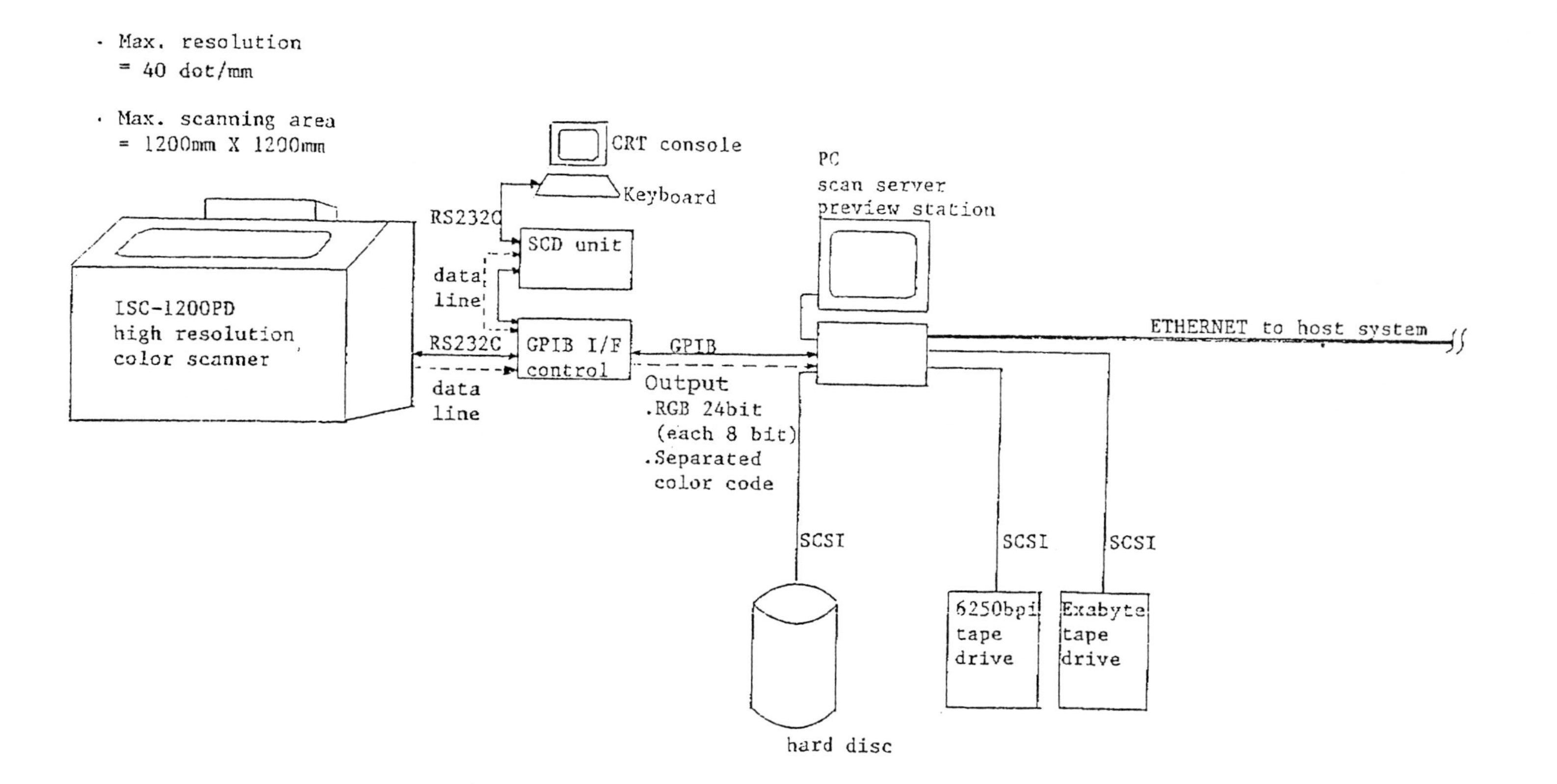

Figure 58.1. Typical configuration of colour scanner

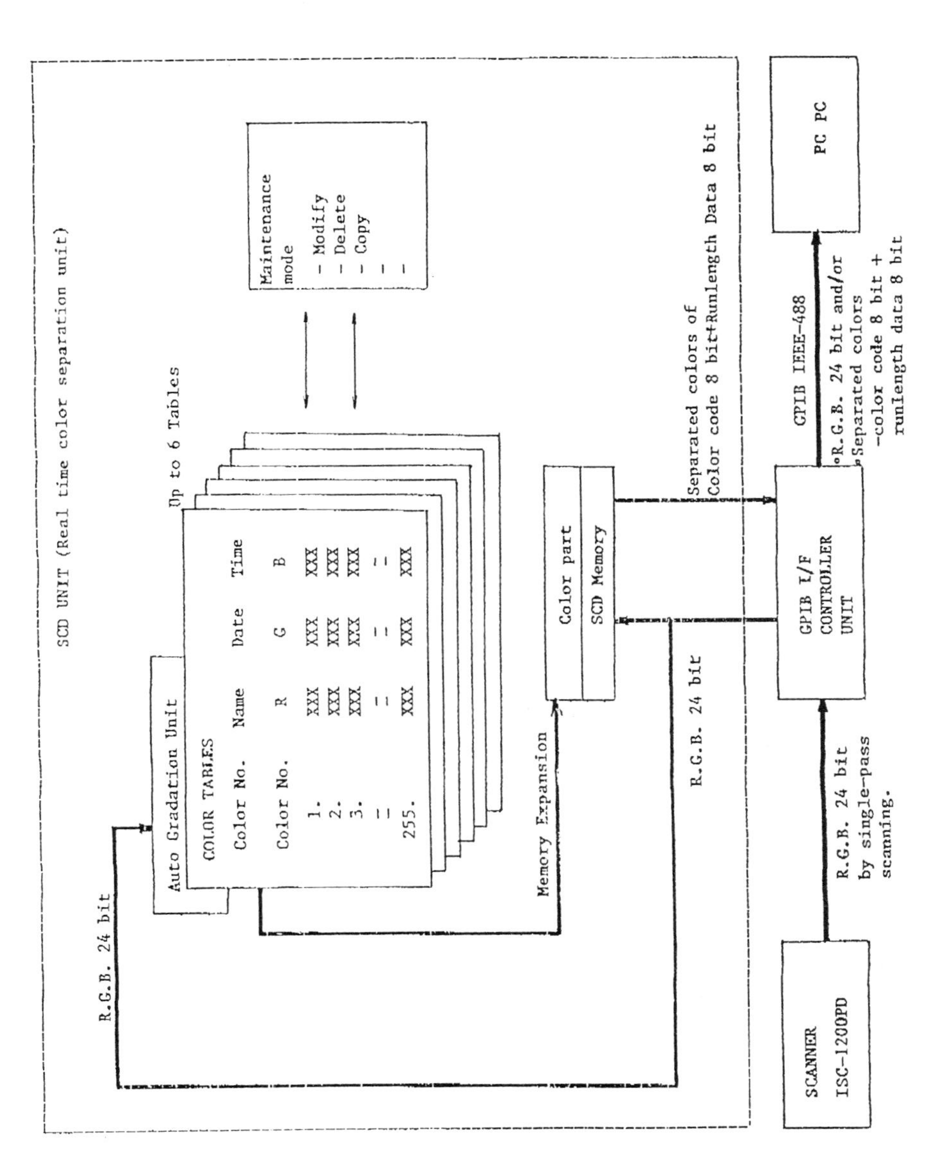

Figure 58.2. SCD unit (Real time color separation unit)

If powerful enough, the same workstation could be used to manipulate and compile new map data or to convert from raster to vector data. However, it is more usual to use a separate workstation or a network of work stations to do this task.

File formats

Scanner configurations typically support a range of raster formats which may include: RLE or Run Length Encoded format, which may be converted to various standard run length encoded formats such as RLC, RLE IN (Intergraph) etc. and TIFF or Tagged Image File Format. As described by the Aldus manual (1990), TIFF

> was originally developed and published by Aldus together with Microsoft, Hewlett-Packard, Microtek, Datacopy, DEST and New Image Technology in 1986. TIFF Version 5 published in August 1988 supports bilevel (black and white), greyscale and colour images. The depth (number of bits per pixel) for greyscale and colour images is not limited by the TIFF specification, although the recommended depths are 4 or 8 bits for grey images, and 8 bits per primary colour (red, green, blue) for colour images.
>
> Standard data compression techniques are defined within the TIFF specification. Two bilevel compression techniques are supported: CCITT Group 3 and a simple run-length encoding (Macintosh PackBits). For grey and colour images a form of LZW (Lempel-Ziv and Welch) encoding is used. Although Aldus has acted as the primary author of the TIFF specification, TIFF is not owned by Aldus and there are no licensing fees for using it.

CCITT group 3 (CCITT, 1984a) and 4 (CCITT, 1984b) standards were developed by the Consultative Committee on International Telephony and Telegraphy (CCITT) for facsimile transmission.

LSLI—Laserscan Standard Image Interchange Format

The PCX/PCC Graphics File Format was one of the earliest attempts in the PC world to enable storage and standardization of graphic images. A standard file format was necessary, both to allow the movement of images between applications and to provide file compression to save disk storage space. The PCX/PCC graphics file format is an example of a method used in industry which became a standard by default. As it has been around for such a long time, the PCX/PCC graphics file format is probably supported by more graphics application programs than all other (PC compatible) graphics file formats combined.

The workstation controlling the scanner can easily convert the image file to different file formats after the scan has been examined on the screen and once any raster editing has been performed. Some formats only require a simple change to the file header, while others require the whole image file to be processed or converted into a tiled format.

It is perhaps worth mentioning that TIFF and its various derivatives are becoming the most utilized industry standard formats for data image file interchange and storage. Aldus (1990) claim that the major goals for TIFF are that it be extensible, portable and revisable.

Colour scanning development

There are a number of colour scanners on the market that can either reproduce a colour image or selectively separate individual colours. In order to consider the developments and the technology involved within colour scanners the Dainippon Screen ISC-1200PD Scanner will be used as an example.

Colour scanners are used in the reprographic business where pre-press image processing systems handle raster images in YMCK data for printing with process colour inks. When required to print in colour or when using pigments or colorants, a primary colour is defined as a colour that subtracts a primary colour of light and reflects or transmits the other two. RGB is thus converted into Yellow, Magenta and Cyan. When three pigment primary colours are mixed, all light is absorbed giving black—hence YMCK (Yellow, Magenta, Cyan and Black).

The facilities offered by the type of scanner under discussion result from the requirement for separation of specified colours as used in the wall coverings, textiles, flooring and related businesses where inks can be used in an infinite number of combinations. In the separation of specified colours, the RGB spectral density of the original is not converted to YMCK dot percentage data as with pre-press scanners; each scanned pixel is converted to a code that corresponds to a specified colour table number (combination of particular RGB spectral densities).

Colour data capture—general

The ISC-1200PD colour scanner contains an integrated Self Colour Discrimination unit (SCD). The SCD unit is designed to import RGB 24 bit data and to compress the data into colour coded eight bit runlength data.

Before commencing the scanning activity it is first necessary to program the SCD unit with the user-defined colour tables to recognize certain colours in a map and separate them so that only those required colours are extracted from the map. An additional function of the SCD unit is the "Colour 0" facility, which may be enabled or disabled during scanning. (If "Colour 0" function is disabled, all colour data scanned in from the map original will be forcibly assigned to the closest matching user-defined colour).

A colour table is composed of user-defined colours and RGB values corresponding to each defined colour. Up to 255 colours and respective RGB values can be defined and programmed in a colour table as shown in Figure 2.

There are basically two ways in which to program the colour tables into the SCD unit. One way is to transmit RGB values of respective colours in a map to the SCD unit by using the colour pick-up facilities of the colour scanner, i.e. placing a map original on the drum (up to 1200 × 1200 mm size) and colour values picked up by "training" the optics to pick up the specific colours used in the map and generating a colour table. The alternative is to directly key in pre-determined RGB values into the SCD unit. Having once established the pick-up data and programmed the colour tables into the SCD unit, all of the scanned data in a combination of RGB 24 bit data sent from the scanner is separated in real time into the required number of colours according to the colour tables.

Colour separation by the SCD

After a set up of colour tables is implemented, selection of one colour table is made ready to scan and expand the table in the memory of the SCD Unit. During the scanning of a map (single pass scanning) RGB 24 bit data is transmitted in a rapid speed from the scanner, processed through the expanded memory and converted into any of assignable colour codes to what its RGB value is closest in the colour table. Separated colour data is compressed in real time and output to a PC based platform (Pre-view station) by GPIB Interface. These processes all taking place in the hardware to activate a high speed separation.

Colour code reassignment by "Colour 0" function

Scanning accuracy in terms of colour separation is critical at a border of one colour with another colour. This might well result in unmatched colours migrating to one of the designated map colours. "Colour 0" is a function to intentionally define an area of uncertainty around each colour of the user defined palette, so that any scanned pixel that is not immediately recognized as one of the map colours, by the SCD Unit, is assigned to the "Colour 0" bin, i.e. a blank colour gap is left between the colours in the map's palette. The amount of colour gap can be set by designating "Colour 0" as a percentage, and by using optional Colour Code Reassignment Software and "Colour 0" function, very accurate colour separation is possible.

"Colour 0" is registered in the SCD Unit when the SCD memory is expanded. Practical use of "Colour 0" is to clean an image with Colour Code Reassignment software. By enabling "Colour 0" and doing a scan with "Colour 0" expanded, the "Colour 0" codes can be detected and manipulated by software operating on the image file.

For example some of the fringing pixels surrounding black characters and text in a map may be read in as grey (RGB values). If "Colour 0" is enabled these pixels will be assigned to "Colour 0". They can then be assigned to a white background, cleaning up the image afterwards by Colour Code Reassignment Software. With "Colour 0" disabled all scanned RGB data are forced to be assigned to one of the user-defined colour codes set in the colour table. There is no space or gap of undefined colours in the expanded colour palette in the SCD memory. If "Colour 0" is enabled, there is a space of undefined colours set in the colour table and the gap between colours in the palette is expressed as a percentage which must be input before expanding the memory.

Note that if "Colour 0" is enabled, it works for all colours in the map. Fringing pixels round other features (e.g. red roads) may also be set to "Colour 0". So image cleaning or Colour Code Reassignment software must be limited to user-defined areas or take account of the colour of adjoining pixels or features.

Standards

The subject of format standardization is uppermost as a user requirement. A working raster standard is available and implemented within the Defence Industry (see Gennery, this volume). The recently ratified BS 7567 (see Sowton, this volume) standard also permits the exchange of raster files. Currently standards related to raster data are evolving, including digital data exchange specifications which enable colour image data generated by Colour Electronic PrePress systems (CEPS), Computer Graphics or CAD Systems to be freely exchanged.

Conclusion

Colour raster scanning together with its attendant hardware and software technology is now at a stage where it has become a viable and cost effective solution alternative for certain aspects of AM/FM/GIS projects.

References

Adnitt. N., 1991, ''Data Capture—The Technology, The Issues'', *AGI 91 Conference Proceedings*, Paper 2.2

Aldus Manual, 1990, ''Welcome to the Aldus TIFF Developer's Tool Kit'', September, pp. 1–3.

Consultative Committee on International Telephony and Telegraphy (CCITT), 1984a, Recommendation T4, ''Standardisation of Group 3 Facsimile Apparatus for Document Transmission''.

Consultative Committee on International Telephony and Telegraphy (CCITT), 1984b, Recommendation T6, ''Facsimile Coding Schemes and Coding Control Functions for Group 4 Facsimile Apparatus''.

59

GIS implementation: A new look at the raster alternative

B.J. Devereux and T.R. Mayo
University of Cambridge GIS Laboratory

Introduction

For the last 12 months many of the leading GIS vendors with established vector systems have been striving to reach the point where they could provide current users with the capacity to handle raster data. Most new systems appearing on the market have an increasingly strong raster capability to the extent that the distinction between raster and vector systems is becoming increasingly blurred. More and more organizations concerned with the marketing of raster data are becoming concerned with the development of standards which will ensure market accessibility to their products and, as Strand (1992) suggests, interest in so-called raster data viewers is growing extremely rapidly.

This chapter will examine the reasons for an early predominance of vector-based technology from a UK standpoint and the user community's rather delayed interest in raster products. It will then consider some of the developments which have led to the recent and very rapid uptake of raster technology. Finally, it will speculate on future developments in GIS which will increase the importance of raster data and its processing still further.

Technological background to the raster/vector divide

In the mid-1980s organizations contemplating the installation of GIS were faced with a choice between vector-based systems requiring manually digitized cartographic data or raster-based systems requiring automatically scanned vector survey documents. In a large number of circumstances it is likely that the decision to automate a manual, paper-based, administrative process would have raised a substantial array of issues and the choice between raster and vector would have been given only marginal consideration.

The conventional wisdom was to implement a pilot project using a vector system. The pilot approach would have been viewed as a mechanism for ensuring that long term investment in GIS technology could meet organizational requirements and that in the extremely unlikely event of problems being encountered, large-scale investment would not have been wasted. Convincing arguments for the vector approach would have been difficult to come by. The preponderance of systems with vector capability would be one and the existence of the Ordnance Survey's vector-based, digital mapping programme would endorse the view.

Raster technology, on the other hand, would have had a less convincing pedigree. Numerically fewer systems would present themselves as acceptable candidates. The absence of a widely publicized raster scanning programme for basic Ordnance Survey data would reinforce a negative view and questions might be raised about the accuracy, quality and functionality of raster data. A decision to ignore the raster alternative could be justified still further by looking at the large number of pilot systems using vector data structures already in progress.

Looked at today, the conventional wisdom surrounding yesterday's planning of GIS programmes may be questioned. The "safe option" which pilot studies seemed to offer has not been without its problems. Involvement in a pilot study immediately results in commitment to a particular brand of GIS technology which commits investment in hardware, staff, data and perhaps most important of all, data structures. In retrospect this commitment is often sufficiently large to preclude realistic evaluation of alternative manufacturers and methods even if performance of the pilot was not completely satisfactory. The limited spatial scope of many pilot schemes is also a problem because it can significantly mask the true costs of data acquisition and hence programme implementation. The current status of digital data availability is such that data can often be acquired at low cost for a small study area but may be either unavailable or massively expensive for a larger region.

Probably the major difficulty pilot schemes present to decision makers is the fact that they provide little or no insight into the way technology is changing. Rhind (1991) discusses the constraints on development paths imposed by early investment in technology in the context of Ordnance Survey digital mapping. The raster or vector dilemma provides a classic example of the same problem faced by many data users throughout the UK. Early investment in expensive vector methods and data structures created a strong resistance to acceptance of raster techniques and this, more than anything else, has delayed take up of raster technology.

These doubts about the effectiveness of raster data were often expressed as doubts about the extent to which raster-scanned maps could meet user requirements. In the early days of raster processing it is likely that many of the doubts expressed were justified but a closer look at most users' requirements for cartographic data suggests that vector-based systems also had significant problems in this respect.

User requirements and cartographic data formats

The primary function of most commercial GIS is to display organization-specific infrastructure such as plot boundaries, pipes, cables, water mains, or other facilities ("foreground data") against a map background and to use the foreground data to interrogate an associated database. In many cases the background is simply required to provide a visual context for the foreground data. In some applications however, there may be a need to measure the location of foreground objects in relation to the background survey. Distance from the road centre line to a mains cable would be an example. A relatively smaller number of specialist applications involve the need to simultaneously manipulate several background layers.

Given these requirements it is clear that from a user point of view the greatest demand for structured digital data lies with foreground information which may need to be captured in many layers. The main requirement of background data is first that it exists and second that it is recorded in such a way that it may be registered with the foreground data and provide accurate position measurements. For many applications structured background data is not a first order priority. Without doubt the most commonly used type of background data is Ordnance Survey 1:1 250 and 1:12 500 sheets.

By an accident of history the mode of supply of background information from the Ordnance Survey is in vector format. By the time Adams (1982) was proposing a raster alternative to vector Ordnance Survey data the commitment to the latter was made. Paradoxically, the natural output of the original vector process is highly structured, very accurate and very expensive information. At the time the original commitment to vector was made, raster processing was probably a desirable but not an acceptable alternative because raster technology was not sufficiently well developed.

The ramifications of this observation are important. Despite the efforts of the National Joint Utilities Group to reduce the cost and increase the availability of background data by recording less feature codes, the GIS community was committed to reliance on vector data. The result is a database of digital maps which provides many users with a level of structure and functionality which they may never need and which is extremely expensive. The costs of the programme have been a burden to the Ordnance Survey and even if they can be written off as a once only investment as Rhind (1991) suggests, many user organizations would find it extremely difficult to justify expenditure on the data at a price which reflects the true value of the information.

It is also noteworthy that of the 220 000 survey sheets to be digitized there is still a significant number of 1:2 500 maps not available. The current projected finishing date is still in some doubt because the programme is driven by demand, but 1995 now seems to be the most common date quoted.

From the preceding it is clear that early development of GIS could have been expedited much more effectively if access to appropriate raster technology had been available for rapid capture of background data at the start of the 1970s. Unfortunately this was not the case.

The need for raster processing capability

The major expenditure of effort devoted to creating a national digital map archive, will, despite all of the problems experienced in the past, provide the UK with an unrivalled base for manipulation of geographical data. The major task which still faces a large part of the GIS community today is capture of foreground data. By definition this is the responsibility of individual organizations. It will now be argued that the shortcomings in raster technology which limited its exploitation in the 1970s have been overcome and successful future development of GIS depends heavily on taking advantage of the many benefits raster processing can offer.

The requirement is to capture information drawn on large scale survey sheets kept in a paper archive. Foreground information is usually relatively sparse in relation to the background. It is frequently hand-drawn or stencilled in pencil or coloured pen as lines, symbols and text. The documents are often damaged, torn or faded. Almost always the data generated must be structured and feature coded.

The vector approach to capturing these data implies digitizing from the source documents, manual editing to remove errors then re-registering the vectors with

Ordnance Survey backgrounds. Experience with the background task at Ordnance Survey has demonstrated without any doubt whatsoever that this is an expensive, time consuming and difficult process. The "raster alternative" of Adams (1992) is now both desirable and acceptable.

The traditional objections to raster have been related to the cost of raster scanners, the poor resolution and large volumes of data they generate and the lack of "intelligence" in raster data when compared to vector. The technological advances which remove these objections will now be reviewed.

Hardware technology and raster scanning

The most obvious development which has led to improved raster processing capability has been the introduction of high performance workstations offering what was once regarded as mainframe capability at very low prices. Modern disk storage systems typically enable gigabytes of data to be stored in units which will sit on or underneath a desk. Processors with 16 megabytes of memory are the norm and perhaps most important of all, the raster display capabilities offered by these devices have improved dramatically. Early displays based on vector processors have given way to raster refresh systems capable of displaying eight-bit data in colour on pixel grids with dimensions approaching 1024×1024 pixels.

The availability of workstation technology and its high-bandwidth bus structures and parallel interfaces has enabled improvements in CCD array technology to be exploited to the full and the resolution of raster scanning devices has increased dramatically. Drummond and Bosma (1989) review a wide range of low cost scanners, many of which are capable of scanning large format documents at very high density. Woodsford (1991) notes that it is now commonplace to scan documents at densities between 500 and 1000 dots per inch (DPI). Even the lowly fax machine operates with a standard density of 200 DPI. Once scanning densities get much above 500 DPI the visual appearance of raster scanned maps is such that they can barely be distinguished from their vector counterparts.

The development impetus gained by scanning technology shows no sign of diminishing in the near future. Two dimensional CCD arrays are now being used in a variety of scanning applications and are likely to provide lower cost, higher performance systems in the future. The improved data handling capability of personal workstations has also enabled the introduction of colour scanners which scan images using RGB filters and generate 24 bit data. Fitzgerald (1991) provides an interesting description of advanced, colour raster scanning/plotting technology currently in use at the Mapping and Charting Establishment.

Even though most of these devices are capable of near instant recording of cartographic documents and typically generate megabytes of data in seconds, it is still feasible for system designers to contemplate storing raw image data in eight-bit form without thresholding. This enables a range of image analysis techniques to be used in GIS which can provide major improvements in system performance.

Image analysis and raster data

The need for high data throughput limited the range of image analysis techniques used in early raster GIS. Many systems used extremely simple thresholding procedures to separate map lines from background and stored the resultant representations as bitmaps in a compressed form. In illuminated flat-bed scanners the use of single thresholds for entire maps often caused poor quality results due to variations in the illumination of documents during scanning. This also slowed the scanning process because users would invariably have to spend time experimenting with different thresholds to ensure acceptable quality across the entire document.

Major progress has been made in the development of adaptive thresholding processes which involve the computer updating the threshold in different parts of the scan depending on local variation in contrast between linework and background. These techniques can be particularly effective in removing the problems of line raggedness and fading which were significant in the use of raster in early GIS.

The capacity to carry out more operations on eight-bit data before storage has also introduced the possibility for a degree of colour separation even when using a monochrome scanner. Many documents used in GIS have foreground data drawn in colour over the background map. Certain colours produce a quite distinctive spectral response when scanned in monochrome and these can be sliced out at the thresholding stage to give a separate foreground and background map. Experiments carried out by the authors have demonstrated that land parcels picked out in red against Ordnance Survey backgrounds can be handled in this way. Limited cleaning of the resulting raster foreground data is required and this can be done using fully automatic erode and dilate procedures.

The need for efficient transmission of raster data by fax machines has improved raster processing capability in GIS by stimulating improved methods of data compaction. Run length encoding schemes developed under the Consultative Committee on International Telephony and Telegraphy (CCITT) are now well established for dealing with thresholded scans of map data. They involve recording sequences of pixels with the same value in the image as a count followed by the value. Because the linework in most maps is sparse in relation to the background, individual runs tend to be long and large compression ratios can be achieved with no loss of information. Even greater compression can be achieved by using two dimensional implementations of the principle where changes between successive scan lines in the image are also incorporated into the coding scheme.

Prospects for direct and efficient encoding of eight-bit greyscale data to a degree where it can be used routinely without a major cost overhead have increased with developments in Huffman coding procedures (see Gonzalez and Wintz 1987). Here the principle is to examine the image histogram and identify the frequency with which each value in the data occurs. Data values which occur most frequently are encoded with a low numeric value which can be stored compactly. Look up tables enable the image to be reconstructed.

Probably the most important area of image analysis software development which is increasing the value of raster in GIS is raster to vector conversion within integrated data structures. Developments in this area offer the key to creation of truly integrated raster/vector systems and progress will be reviewed in the next section.

Raster processing techniques for automatic vectorization

Early attempts at raster to vector conversion adopted an "all or nothing" approach to the problem. Raster scans were fed in at one end and spaghetti vectors without feature codes came out the other. Experience with such systems very rapidly revealed their shortcomings. The ability to create algorithms for recognizing and extracting vectors from survey maps demanded a level of sophistication not yet attained in studies of computer vision and artificial intelligence. The result was generation of vector datasets which often cost more to verify, edit and feature code than manual digitization would have done in the first place. It is extremely unlikely that such systems will ever provide an effective tool for vectorizing foreground data.

Much more promise has been seen in semi-automatic systems where data is vectorized from a raster scan displayed on a computer screen. The Fastrak system (Fulford, 1981) constituted an early example of the principle and was a precursor to the modern Vtrak marketed by Laserscan.

Research at the University of Cambridge Department of Geography has attempted to extend the concept of selective raster vectorization still further. The research programme has two prime objectives:

1 To create a formal model of the digitizing process. Use of the model which is described in Devereux and Mayo (1990) enables identification of those raster to vector conversion tasks the computer can achieve and those tasks which require operator intervention.

2 To identify a range of procedures which can be used by an operator to vectorize the different parts of a map. The concept of a toolkit for dealing with different types of cartographic features makes this approach unique and extremely efficient.

A simple example of one technique from the toolkit relates to vectorization of straight linework. The operator points approximately to each end of a line segment displayed on the screen. The system takes over and interactively identifies and locates the line segment in question. Because the line is located by the computer, the operator can work extremely rapidly without worrying unduly about the precise position of each cursor hit. Nevertheless the vectors which are generated are fitted accurately and reproducibly to the linework in question.

Field trials have demonstrated that the line work on 1:2 500 maps of urban areas can be captured with considerably greater positional consistency than that achieved with a traditional vector table. Time savings of around 30 per cent have been achieved over traditional digitizing. The process can deal effectively with solid, dashed and chained lines which are composed of straight segments. For sinuous lines a line-following tool is also available with similar characteristics.

For printed character and symbol data other tools are at an advanced stage of completion and work in a similar fashion. The operator points at the symbol or at either end of the character string and the system processes the raster data to recognize the word or object in question. Because all of the tools are used interactively and the resulting vectors are drawn over the raster, any mistakes by the computer system are recognized immediately by the operator. This results in very high quality vector output and reduces editing costs dramatically.

Building on these basic tools effort is now being devoted to design "supertools". A good example of the utility of such tools is provided by digitization of polygons. Using the line identifying tool the operator can vectorize a rectangle or near rectangle such as a plot boundary by pointing in the vicinity of its corners. A "polygon identifier" is an example of a supertool which requires one cursor hit inside the polygon and leaves the computer to identify the vectors which define its boundary.

A supertool for this application is now well established and can deal with polygons which are stippled and contain information such as plot numbers.

Apart from the obvious benefits these and similar techniques can offer for reducing the costs of cartographic data capture, their availability can also help to ease implementation problems in GIS. Use of raster scanning in the first stages of project implementation is of major benefit because it starts to show a demonstrable return on investment very early on. Completion of a seamless map data base showing both foreground and background data can be completed very quickly and in many circumstances can provide a significant part of the functionality required from the system. As staff confidence grows in the new technology, selective vectorization of foreground data can be pursued in a piecemeal fashion with high demand areas being dealt with first. Ultimately, once the foreground data has been dealt with, the background can be separated from the foreground and migration to structured Ordnance Survey background data can take place if this is appropriate.

Raster standards

From the previous discussion it is clear that many of the problems which used to be associated with raster data handling in GIS have been resolved and the use of raster data is going to grow rapidly. Clearly it is important that effective standards can be introduced for data handling, interchange and storage.

Both internal handling and exchange of raster data is deriving benefit from developments in open systems design and protocol. Vendor-specific tools for writing raster handling software such as SUNVIEW and DECWINDOWS are rapidly converging on a standard version of X11 accessed via a common menu interface such as Open Look or Motif. These developments will undoubtedly stimulate exchange of both data and software within the GIS community although further progress is still needed before users can expect straightforward portability of software and data between rival manufacturers' platforms.

Already numerous organizations are committed to particular raster storage formats. The Tagged Image File Format (TIFF) introduced by the Aldus corporation seems to embody most of the data compression techniques described above. Like many proposed data formats it is potentially so complex that its widespread use in GIS could lead to significant problems. In particular the costs to vendors of providing clients with software to deal with the full implementation could well limit its usefulness, despite its popularity with many scanner manufacturers in the publishing industry.

Probably the best prospect for effective transfer standards lie within the AGI working group currently investigating possible specifications for a raster National Transfer Format (NTF). Early indications suggest that they will propose a simple but flexible format for raster data which enables compact storage and straightforward decoding of image data with separate files to define geometry and transformations. One can but hope that further exploitation of raster technology will not be limited by failure within the GIS industry to agree on standards of this type.

Future developments in raster technology

Most of the emphasis in the discussion so far has been placed on raster representations of digital maps. This is because the authors believe that this is the major issue facing most current users of GIS. Looking to the future it is clear that new types of data are becoming increasingly useful and many of these are inherently raster in nature.

Ordnance Survey digital terrain data is now available for most parts of the country at a scale of 1:50 000. The data is made available in grid form and requires raster display capabilities for its exploitation. As demand to process height information within GIS grows it will become increasingly clear that new software capabilities will be needed for full exploitation of its potential. Fortunately, procedures for vectorization (conversion to contours) of this data type are well understood and widespread availability of hybrid raster/vector processing techniques should not be long in making their appearance.

For some considerable time, remotely-sensed data has been playing a key role in specialist GIS within the science and research community (Devereux, 1988). Widespread uptake by the main users of GIS has been limited by low spatial resolution of satellite imagery and problems with the rectification of very high resolution data collected in aircraft surveys. A new generation of high resolution, spaceborne sensors will soon be providing data which is eminently suitable for use as backgrounds in many GIS. Progress with geometric correction techniques (Devereux *et al.*, 1990) for aircraft data now means that more organizations are looking to airborne imagery to fulfil their needs. Recent work carried out by the National Rivers Authority has generated data of this type for the south coast of the UK and provides a case in point.

Perhaps one of the most exciting possibilities held by both satellite systems such as SPOT and aircraft systems such as those described above is the possibility of using automatic, stereo-matching of overlapping images to generate accurate and detailed raster models of both terrain height and the height of features on the terrain (Norvelle, 1992).

Conclusions

Early technology has resulted in a commitment to vector techniques in a large number of GIS systems. The user community has a strong financial and organizational commitment to these techniques. Rapid advances in raster technology are now providing improved functionality and lower cost methods which can provide major impetus to the uptake and growth of GIS. Such developments are being brought about by cheaper and more powerful hardware, advances in raster scanning technology, development of raster to vector conversion techniques and improving standards for handling raster data. New technology will increase our reliance on raster data especially as use of remotely sensed information becomes more widespread. Provision of integrated systems with combined raster and vector capability is a major responsibility facing GIS vendors if existing and new users alike are to take advantage of the opportunities offered by raster processing.

Acknowledgement

The authors would like to acknowledge the support of Alper Systems Ltd. for their support of research into raster processing techniques at the University of Cambridge Department of Geography.

References

Adams, T., 1982, A raster alternative for Ordnance Survey digital data, in Rhind, D. and Adams, T. (Eds), *Computers in Cartography*, British Cartographic Society, Special Publication, **2**, 117-26.

Devereux, B., 1988, Remote Sensing: Expensive toy or cost effective tool?, *Proceedings of the 1st Mapping Awareness Conference*, Oxford, 16.1-16.6.

Devereux, B., Fuller, R., Carter, L. and Parsell, R., 1990, Geometric correction of airborne scanner imagery by matching Delaunay triangles, *International Journal of Remote Sensing,* **11**, 12, 2237-51.

Devereux, B., Fuller, R. and Roy, D., 1989, The geometric correction of airborne thematic mapper imagery, *Proceedings of the NERC 1989 Airborne Remote Sensing Campaign*, 19-34.

Devereux, B. and Mayo, T., 1990, In the foreground: Intelligent techniques for cartographic data capture, *Proceedings of AGI 90*, 6.2.1-6.2.9.

Drummond, J. and Bosma, M., 1989, A review of low cost scanners, *International Journal of Geographical Information Systems,* **3**, 83-95.

Fitzgerald, B., 1991, Mapping for the military—A time of change, *Cartographic Journal,* **28**, 13-15.

Fulford, M., 1981, The Fastrak automatic digitising system for line drawings, *Pattern Recognition,* **14**, 65-74.

Gonzalez, R. and Wintz, P., 1987, *Digital Image Processing*, Addison Wesley.

Norvelle, F., 1992, Stereo correlation: Window Shaping and DEM corrections, *Photogrammetric Engineering and Remote Sensing,* **58**, 1, 111-15.

Rhind, D., 1991, The role of the Ordnance Survey of Great Britain, *Cartographic Journal,* **28**, 2, 188-99.

Strand, E., 1992, Data viewers giving a new look to geographic information, *GIS Europe,* **1**, 1, 22-24.

Woodsford, P., 1991, Cartographic and GIS developments at Laser-Scan, 1987-91, *Cartographic Journal,* **28**, 43-46.

60

Knowledge-based systems for Geographical Information Systems

J. M. P. Quinn
ICL, Manchester

A. I. Abdelmoty
Heriot-Watt University Edinburgh

and M. H. Williams
Heriot-Watt University, Edinburgh

Introduction

Before 1980, interest in artificial intelligence was generally confined to a small number of university departments and on the whole was regarded as a rather speculative research activity. However, around 1980, the Japanese identified that the technology was developing to support AI applications, and that the market for such developments was enormous. Indeed, they predicted that knowledge-based systems would be the main application area for computer systems in the 1990s. This was the main motivation behind their Fifth Generation Computer Programme.

This programme triggered enormous interest in the development and application of AI techniques and systems around the world. Major research programmes focused on developing tools, techniques and applications for AI. Commercial interest rose dramatically as the potential of these new systems began to be appreciated.

The most important type of application is that known as a knowledge-based system. This is a system in which the knowledge relating to some particular decision process or task is captured and stored explicitly in some form. When the system performs the task concerned, it interprets the knowledge and arrives at appropriate conclusions. The most common form of knowledge based systems is the expert system. The knowledge stored in a knowledge-based system is usually acquired from experts in a particular field, and encoded in some knowledge representation formalism. The most common form of representation formalism is a set of facts and rules.

Knowledge-based systems have been applied to problems in a wide range of disciplines—from medical diagnosis to diagnosis of faults in equipment, from planning computer configurations to aiding civil engineering design, from prospecting to forecasting oil drilling operations, etc. It is not surprising therefore that the technology should be applied to problems in the area of Geographic Information Systems.

A Geographic Information System is a system for modelling and manipulating geographical data representing the real world. As such it may be regarded as the application of computer technology to the making and use of maps. The way in which a GIS is used can range from answering the simple query of ''where is...?'' which involves finding a particular location, to answering a complicated query of 'what if...?'' which might involve complex modelling and analysis of time and space. Since a knowledge-based system is a system which uses human knowledge and reasoning in handling various problems, the application of knowledge-based systems to GIS involve the use of the knowledge and reasoning of the map maker and the map user in handling geographical data.

Knowledge-based systems

A knowledge base is a collection of knowledge representing a particular area of interest. There are a variety of different ways of representing knowledge, including:

1 As a set of facts and rules, varying from simple logic formulas to complex production rules;

2 As a complex data structure, including semantic nets, frames, objects and decision trees.

The rule-based approach is the more common one in general use. In this approach one has a combination of facts and rules. For example,

IF: the patient has signs and symptoms s1 AND...sk, and certain other conditions t1 AND...tm hold

THEN: conclude that the patient has disease di.

This representation assumes problem domains that are categorical; that is, answers to all questions are either true or false. However, many expert domains are not categorical. Typical expert behaviour is full of guesses (although highly articulated) that are usually true, but there can be exceptions. This uncertainty can be modelled by assigning some qualification to both facts and rules. Such qualifications can be expressed by descriptors (for example, true, highly likely, unlikely, impossible), as a probability (a number between 0 and 1) or as a degree of belief expressed as a real number in some interval—for example, between −1 and 1. Thus the implication of the rule above might be expressed as,

THEN: conclude that the patient has disease di with certainty x.

For some applications, a knowledge base will consist of many facts and few rules —as for example, in a knowledge base representing a train timetable. For other applications, it may consist of more rules than facts—as for example, in a knowledge base for diagnosing medical diseases.

When a query is applied to a knowledge base, the system will check to see whether it can be satisfied from the set of facts stored in the knowledge base. If it cannot, the system will inspect the rules to find a matching rule which can be applied to produce a solution which satisfies the query. Executing such a rule may involve further searches through the set of facts or the application of further rules. This process is known as inference.

Application area of KBS in GIS

Four application areas of KBS in GIS can be recognized, namely, map design, geographical database management, geographic decision support and feature extraction (Robinson and Frank, 1986).

Map design

Two areas of map design are particularly active for the application of knowledge-based systems, namely, map generalization and name placement. Map generalization is the transformation that a map undergoes to be represented at a different scale. For example, for a map at 1:25 000 scale to be represented at 1:200 000 scale, it has to undergo several changes, such as widening of rivers, railways and main roads, elimination of streets and smaller roads, disappearance of buildings or their grouping using a certain symbol (Muller, 1990). MAPEX is an example of a rule-based system for automatic generalization of cartographic products (Nickerson and Freeman, 1986).

AUTONAP (Ahn, 1984) is an example of a name placement expert system. This system emulates an expert cartographer in the task of placing feature names on a geographic map.

Geographical database management:

The main property characterizing geographical databases is the fact that geographical data are spatially referenced, that is, an important element in their definition is their location in space. The complexity of this additional property comes from the diversity of relationships that it incurs between different data entities.

The almost infinite number of relationships that exist and their complexity, (exact definitions of spatial relationships are tricky to formulate, e.g. near and far) has impeded initial storage of such relationships in a database. This fact meant that only some of those relations which are obvious at the time of building the database are stored and the rest are left to be deduced at query execution time. Consequently, an efficient query processor for handling geographical data is essential.

There have been several attempts to apply KBS techniques to this area. One example is LOBSTER (Egenhofer and Frank, 1990) which aims to provide an intelligent user interface to spatial databases.

The use of rule-based systems has also been explored for handling spatial operations (Wu and Franklin, 1990), for example, the spatial overlay operation. Overlay is an operation whereby a specified set of conditions occur together, e.g. desirable areas for cottages might be defined as those areas that have a forest vegetation cover, have well drained soils and have a south-facing exposure but with non-agricultural land use criteria.

Geographic decision support:

Work in the area of geographical decision support systems is aimed mainly at building systems that model time and space to aid in future planning and analysis of different problems related to GIS. For example, evaluating site suitability for land-use activities such as urban, agricultural or industrial, was a problem for which systems such as GEODEX (Chandra and Goran, 1986) and URBYS (Tanic, 1986) have been developed. Expert knowledge from land-use planners is used to build the knowledge base of such systems.

Automatic feature extraction

The process of map digitization results in the representation of the map as a set of points and lines[1], known as the digital map. While all the map information is still present in this digital form, in the sense that the map can be reproduced again from the digital file, the spatial information conveyed by the paper map is now implicit in the geometric representation of lines and points. Thus in order to answer queries using this information, a process of feature extraction from the digital map file is needed, followed by the building of a geographical database.

Considering the immense amounts of data to be digitized[2], and the iterative process of updating the maps and redigitizing, the automation of the process of feature extraction and consequently of building the geographical database is highly desirable. The deductive nature of this feature extraction process has lent it naturally towards a rule-based approach. For example, Schenk used a rule-based system for the extraction of linear map features such as contour lines (Shenk and Zierstein, 1990). Hadipriono *et al.*, (1990) have developed a knowledge-based expert system for the extraction of drainage pattern types. Palmer represented the map as a grid (triangular tessellations) to represent objects with their elevation and used the rule-based language Prolog to conduct symbolic analysis in order to demonstrate how valleys, streams and ridges could be detected (Palmer, 1984).

De Simone used pattern recognition to interpret map features depending mainly on geometrical properties such as the width between two lines, and the degree of straightness or curvature of a line (De Simone, 1985). By comparing the values for these evaluated properties against a pre-defined range, different map objects were recognized.

FES is Forestry Expert System (Goldberg *et al.*, 1984) used expressly to analyse multi-temporal Landsat data for classification of land cover and land cover changes of interest to foresters. Using a multi-temporal Landsat image database, production rules are applied in two phases. First, production rules are used that involve change detection inference coupled with a reliability measure. The second phase generates decision rules regarding the current state of the image.

The rest of this chapter is devoted to a more detailed discussion of the application of knowledge-based systems to the area of feature extraction from digital maps.

KBS in automatic feature extraction from digital maps

A map is an abstract model of reality, designed to facilitate the extraction and analysis of spatial objects using their metrical and topological properties and relationships. Having maps available on computers promises far wider uses of these maps. However, if the digital map is stored and processed as a one dimensional list of features the two dimensional contextual qualities of the map are lost and little information can be interpreted. A common solution to this problem is to introduce feature codes to enable some selection of the features for processing. However such manual attachment of codes is both time consuming and costly. Furthermore the map user is limited to the classification supplied by the map user, i.e. to the repertoire of feature codes.

Hence the process of automatic feature extraction enables the retention of the two dimensional qualities of the map in digital form without the use of feature codes or human intervention. This is the first process of building a geographical database, which carries the representation of the real world entities and their relationships into a form suitable for the user to query and analyse.

For a human the process of reading and understanding a map is associated with forming a mental image through visual cognition of explicit spatial facts, such as location, direction, distance, etc. This process is usually a rule-based process. For example, to identify a road on the map, the fact that there are two roughly parallel lines with a clear area in between is first visually recognized, and since a road has two parallel road sides, and a road is usually a clear area, the fact that a road exists is deduced. This may be formalized as a rule as follows:

IF: two lines are roughly parallel AND
the area between them is clear AND
the minimum distance between them > some limit

THEN: conclude that the lines can be interpreted as a road

and built into a knowledge-based system for feature recognition. This strategy may be taken a step further and used to extract other facts about neighbouring objects such as the rest of the road network or the houses that lie on either side of it.

Spatial properties of geographical entities and the relationships between them that can be exploited in the process can be classified as either geometrical, such as the parallelism of road sides or logical, such as the existence of a house on a road side. The geometrical properties and relations were mainly used when attempting feature extraction from maps, as they can be expressed as mathematical formulae which can be easily formulated. The aim of this chapter is to show that the use of human intuition and expertise in the process of map reading can be used to formulate logical facts based on the relations between the map objects rather than considering the properties of objects separately.

Automatic feature extraction from digital maps as a rule-based system can be approached using the following three processes.

1 The recognition of logical rules used in identifying the object to be extracted,
2 The identification of facts that result from the satisfaction of those rules,
3 The establishment of an efficient search strategy for the execution of those rules.

However, although for a human the recognition process seems a straightforward one, it is not the case when this process is automated. This is mainly due to the following points:

1 A human treats the map pictorially as a whole, not as individual geometrical lines and points, and hence many relations that can not be formally stated are used;

2 Many properties and relations that might require an immense computation time (such as parallelism of two lines and the clarity of an area) are much easier to be inferred visually than to be proved mathematically;

3 The human mind tends to recognize incomplete information through sophisticated recognition capabilities combined with intuition and experience;

4 Maps usually carry some errors which can be eliminated quickly by the trained human eye but which would require a separate lengthy computation process on the computer.

Examples of difficulties encountered in the recognition process are:

1 How to distinguish between features which might look very similar on the map: for example, between a road and a canal or between a small house and a large double/triple garage (big enough to be similar in size to a house).

2 How to classify different map features: when is a road a street and when is it a highway? What distinguishes a terraced house from a detached one?

3 How to recognize features from a multi-layered map: for example, recognition of the underground network or a utility network superimposed on an urban map.

An example using feature codes

MapGin is an experimental system (Quinn, 1989) which models a feature coded Ordnance Survey digital map as a set of objects: roads, junctions and houses according to a set of simple built-in rules. These rules use the text, included as a feature code in the map, in the naming of roads abstracted as road centre lines, and numbering the individual houses. These houses may be detached, semi-detached or terraced houses. In the case of the semi-detached or terraced houses the outline is split up into individual houses by various means.

MapGin has been used to provide a graphical user interface to PLANES, ICL's Geographic Information Management System with MapGin providing the set of house objects and PLANES providing Local Authority type data for each of the objects. This allows the user to ask questions such as "show all the properties in the area with a planning application". PLANES determines which properties have planning permission and MapGin checks if they are in the selected area and if so highlights them.

One of the problems with the MapGin approach is that the rules are coded into the program, which is written in an imperative language for performance reasons. This implies that if the rules need to be changed to improve the analysis then the program has to be rebuilt. The Deductive Object-Oriented Database project (DOOD) was started at Heriot Watt University to combine the best features of object-oriented techniques and deductive databases so that the rules could be held external to the program and the system would be powerful enough to process the huge amount of data quickly. The project has been running for about one year and the results look promising (Abdelmoty *et al.*, 1992).

Conclusions

The contribution of knowledge-based systems to the processing of geographic information has been growing over the last few years as the computing power available has increased. Practical knowledge-based systems will emerge over the next few years that are capable of handling the enormous volume of geographic data that is available. It will however, still be necessary to be selective in choosing the area in which they operate, either in taking the Ordnance Survey data and generating partial addresses as with MapGin or tackling the problem of handling the multilevel distribution networks of the public utilities.

Acknowledgements

The authors wish to express their thanks to ICL for its support of MapGin and the work associated with it. In addition they would like to thank the Science and Engineering Research Council for their support for the research on Deductive Object-Orientated Databases, and to ICL and Ordnance Survey who are industrial partners in this project.

Notes

1 This is called the vector form of a digital map as opposed to the raster form.
2 There exists over 250 000 map sheets of 1:1 250 and 1:2500 scale in Britain alone.

References

Abdelmoty, A.I., Williams, H. and Quinn, J.M.P., 1992, "A Rule-based Approach to Computerized Map Reading", To be published, (1992).

Ahn. J.K., 1984, *Automatic Map Name Placement System*, Troy, NY.

Chandra, N. and Goran, W., 1986, "Steps Towards a Knowledge-based Geographic Data Analysis System", Optiz, B. (Ed.) in *Geographic Information Systems in Government*.

De Simone, M., 1985, "Data Structures and Feature Recognition: From the Graphic Map to a Digital Database", PhD Thesis.

Egenhofer, M.J. and Frank, A.U., 1990, "LOBSTER: Combining AI and Database Techniques for GIS", *Photogrammetric Engineering and Remote Sensing,* **56**, June (6), pp. 919-26.

Goldberg, M., Alvo, M. and Karam, G., 1984, "The Analysis of Landsat Imagery Using an Expert System: Forestry Applications", Proceedings AutoCarto.

Hadipriono, F.C., Lyon, J.C. and Li, T., 1990, "The Development of a Knowledge-Based Expert System for Analysis of Drainage Patterns", *Photogrammetric Engineering and Remote Sensing,* **56**, June, (6), pp. 905-9.

Muller, J.C., 1990, "Rule Based Generalization: Potentials and Impediments", Proceedings of the 4th International Symposium on Spatial Data Handling, **1**, pp. 317-34, IGU.

Nickerson, B.G. and Freeman, H., 1986, "Development of a Rule-Based System for Automatic Map Generalization", Proceedings of the 2nd International Symposium on Spatial Data Handling, Seattle, WA.

Palmer, B., 1984, "Symbolic Feature Analysis and Expert Systems", Proceedings of the International Symposium on Spatial Data Handling, pp. 465-78.

Quinn, J.M.P., 1989, ".....towards a Geographic Information System", *ICL Technical Journal*, **6**, May, 3.

Robinson, V.B. and Frank, A. 1986, "Expert Systems Applied to Problems in Geographic Information Systems: Introduction, Review and Prospects", *Computers Environment and Urban Systems*,**11**, (4). pp.161-73.

Schenk, T. and Zierstein, O., 1990, "Experiments with a Rule-Based System for Interpreting Linear Map Features", *Photogrammetric Engineering and Remote Sensing*, **56**, June, (6), pp. 911-17.

Tanic, E., 1986, "Urban Planning and Artificial Intelligence: The URBYS System", *Computer, Environment and Urban Systems*, **10**, pp. 135-46.

Wu, P.Y. and Franklin, R., 1990, "A Logic Programming Approach to Cartographic Map Overlay", *Computer Intelligence*, **6**, (2), pp. 61-70, Canada.

61

Parallel computing for Geographical Information Systems

Tom Lake
GLOSSA

Introduction

In this chapter, I present an overview of developments in parallel processing and their potential effects on technical computing in general with a minor look at the implications for Geographical Information Systems (GIS). I believe that, given an exposition of the present situation in parallel computing, many readers will be better able than I to envisage the specific impact on GIS. Furthermore, I would suggest that GISs are essentially parallel in their nature and that the development and wide availability of massively parallel systems will influence the boundaries and the nature of the GIS discipline.

Guy Haworth's chapter in this volume considers parallel database support. We hope that our pair of chapters will convince you that the necessary developments are in place to underpin parallel GISs. The adoption of this technology will then depend on the far-sightedness of the GIS suppliers and users.

What could be more appropriate for the study of GIS, the science of distributed systems with myriad components, than a computing machine itself based on these principles? For the past few years it has been clear that parallel computing would increasingly offer the most cost-effective route to high performance. Commercial suppliers are now strongly indicating their support for this view. It will be important for those now starting projects to plan for the adoption of parallel computing and accordingly to shape their investments so as to preserve them into the era of parallel computing.

The development of parallel computing

In the last year it has become clear that parallel computing will be the conventional approach for engineering and technical computing. Major suppliers have shown how they will be approaching the goal of a Teraflop machine—a machine capable of 1012 floating point operations (additions, multiplications, comparisons etc.) per second by 1995. In the US, Cray Research have announced plans for a massively parallel machine by 1993. Intel, already established with their iPSC/2 hypercube supercomputers, have announced their Paragon architecture and plans for reaching the Teraflop region. Thinking Machines have announced the CM-5 architecture; Alliant and IBM have also made announcements, along with many other suppliers.

In Europe, we have a lead in the development of the appropriate architectures with the Inmos Transputer and the development of indigenous suppliers for large parallel systems based on this component, such as Meiko, Parsys, Telmat, Parsytec and many others. The German company Parsytec has announced an architecture based on the T9000 Transputer (due to be delivered this year) which will reach the Teraflop region of performance, while Meiko, Parsys and Telmat have recently announced a common architecture for their machines.

The commercial developments cited are taking place because parallel architectures offer the most cost-effective means of reaching high computing performance. The building of powerful systems by using massive replication of mass-produced microprocessor components is now established as the best practice. It is effective not only in respect of computing power, but also as applied to disk storage, as massively parallel arrays of Winchester disks mass-produced for PC application provide highly effective and resilient storage for large-scale computer systems.

Many of the parallel systems named here are built with conventional microprocessors as their basic computational unit. Others again have bespoke processors. In addition they need to have effective interprocessor communication. The realization of very high rate communication mechanisms which can be built into large systems has made it possible to reach the very large system performance now being registered.

Much publicity naturally attaches to the prospects for the largest machines, suited for use as computing servers to communities of scientists and engineers. Careers will be made by buying and selling them, but the impact of parallel computing will be felt as much by the workstation user, who will also see much increased performance. Changes in the conception and capabilities of workstations will reach very many people. The T9000 architecture being delivered this year by Inmos comprises not only the T9000 processor, which is predicted to deliver 20 MFlops (20 million floating point operations per second), but also an interconnection processor capable of connecting up to 32 such processors and switching data between them at their full communication rate. These same communication processors will be ganged and stacked up to interconnect the largest systems. This will make for a very economical workstation architecture of very considerable and extensible power. There are many alternative approaches in what will be a very lively and fast-moving market over the next few years.

Developing microprocessor power

Readers will not be unaware that the development of parallel computing takes place at a time when new single microprocessors are increasing in performance very rapidly—perhaps doubling in less than two years. This means that the possibilities for large-scale computing are expanding very rapidly indeed. Many previously unmanageable computations will be performed with this combination of technologies and their availability will change practice in many fields.

Looking at the current set of high-powered microprocessors, the Intel i860 with its notional but hard-to-achieve 80 MFlops peak rating has been available for some time. Texas Instruments have their 50 MFlop rated C40 with full features for communication. Fujitsu has recently announced a high-performance floating point chip with peak performance above 100 MFlops. DEC is producing a new 64-bit microprocessor known as Alpha with a rating of around 200 Mips while the Inmos T9000 is rated at 200 Mips and 20 MFlops.

These processors achieve their speed by retaining on chip, alongside the processor, copies of some of the stored data. Fast though they are, the store chips could not keep up with these processors otherwise. As any office worker knows, using private notes for changing public information can lead to confusion as different note books diverge. This is the basic problem that parallel computing vendors have to solve—how to keep the stored information of a computation up to date for every processor that needs it without sacrificing the speed of processing. Generally the more advanced the processor, the more it uses local, private copies of data to achieve its speed and the more care is needed if it is to be used in parallel.

The special care needed for parallel use can be provided by the manufacturer in the form of special adaptations to enable the processors to co-operate successfully or by the software which translates programs for the particular machine in question. Manufacturers such as Sequent and Alliant specialize in taking standard microprocessors with little facility for co-operation and adding hardware and software to make them co-operate successfully. In other cases, most notably with the Inmos Transputer, the processor is designed from the start for parallel use and comes with all the hardware and software necessary for parallel operation.

Manufacturers of the data parallel systems described in the next section such as AMT and the US company MasPar use bespoke processors, often packing many processors on each chip.

An important feature of recent developments is the property of scalability. Parallel architectures are now designed so that no bottle-necks, undue costs or resource limitations need arise as the system is expanded. The larger the system the more power it should be able to deliver, given a suitably demanding problem. It is this feature which makes it possible to offer the very largest systems at a cost proportional to their performance.

Parallel computation

What sort of applications are driving the development of parallel computers? Some machines are designed for large scientific and engineering computations. The calculations involved in designing aircraft, the prediction of weather and climate, the computation of the predictions of basic physical theory for the properties of the elementary particles, the analysis of oceanic flow, the analysis of geology from sonic exploration, the prediction of the behaviour of vehicle structures on impact, the prediction and analysis of the properties of drugs and other chemicals from their molecular structure, all provide suitable challenges. It is clear that large computers are of economic importance and are changing the norms of design.

The realistic visualization of illuminated scenes, the statistical categorization of large volumes of data, and the presentation of artificially generated moving 3-D pictures, all require very substantial computation at the workstation level. The economic transmission of moving pictures (picture phone), the analysis of satellite images, the analysis of perceived radar signals from hostile aircraft are applications requiring very large programmable computing power to be made available in an automatic or black box mode. These applications differ in system requirements as well as raw computational demand and present varied challenges of reliability, compactness, degree of packaging, connectibility. Of particular relevance to GIS are the visualization applications (even interactive walk-through of scenes and structures), the environmental applications, the manipulation of images in vector and raster forms, and the statistical analysis of massive data sets.

It is clear that the picture of a system as a myriad of communicating activities is a very natural one. Physical laws are often formulated in terms of local interactions producing long-range effects by the propagation of local effects. Large-scale computations—big problems—are by their nature going to involve myriad variables or features. If it were not so they would hardly be so demanding. If we can describe our problem in bulk terms and preserve that bulk aspect of its description in the program that we write to solve the problem then we have made the best preparation for making it possible to use a parallel computer to solve our problem effectively.

Architectures of parallel machines and problems

Unfortunately, there is as yet no generally accepted means of programming a parallel computer which will be effective for all the different architectures of parallel systems available. There are some architectures which focus on so-called data parallelism—the identical treatment of many different cases. These machines are extremely cost-effective and are likely to be used at least as components of the highest powered machines.

Other machines focus on process parallelism. A large program is written as many little programs communicating by means either of common variables or exchange of messages. Each process may treat a case or portion of the data or act to direct or co-ordinate the computation. Machines which bridge this divide are unlikely to be in common use until the end of the decade. In the meantime we must find an approach which preserves as much portability as possible.

One way is to utilize the parallelism without programming it ourselves. If we have a database system which makes use of the parallel computer and a transaction processing monitor which manages many transactions then the programmer need only write the code for single transactions and the system will do the rest. Even if the application is not so easily factored, a parallel database system could be made good use of, especially if the database offers good facilities for writing programs to manipulate bulk data. The language used to express operations on relational databases, SQL, unfortunately offers only a record-at-a-time interface for the programmer, although its operations allow bulk manipulation of the database. An opportunity to overcome this problem may arise with the definition of the SQL bindings of the new standard FORTRAN 90, which itself has both record types and data parallel array operations. It may well be that languages providing better facilities for bulk data manipulation, especially the extended relational algebra, might be useful in opening up relational databases to full parallel exploitation.

In a similar way graphic and visualization systems, and also standard mathematical libraries and packages of many kinds will offer a route to exploitation of parallel systems without requiring direct programming of the parallelism by the application programmer.

If we do decide to program for the parallel machine we need to think carefully about how to maximize the portability of the program. The style that we adopt to write a portable parallel program largely depends on the nature of the computation. If the program involves identical manipulation of large amounts of data then it can be written in a form which can be executed effectively on all parallel architectures. If the computation involves many different cases which must be treated very differently or in sequence, then it will not be possible to write it for a processor specializing in the data parallel style and a process oriented style should be adopted.

For processes which communicate among themselves, rarely it is possible to consider writing the parallel program as many independent sequential programs which communicate using the facilities of a parallel operating system. This is usually a development of UNIX such as Mach or Chorus. The parallel program could then be run wherever the parallel operating system was offered.

For programs which need to communicate more frequently and intimately a means of writing a unitary parallel program is required. Further distinctions may be relevant—it may be possible to predict the number and loading of the different processes while writing the program, in which case the program can be relatively easily ported or it may take an unpredictable course.

In a recent survey summarizing extensive research at Caltech, Fox divides computations according to their patterns of communication and synchronization (Fox, 1988). Ninety per cent of the applications surveyed fell into the classes which parallelized well. The most portable applications were the Synchronous or data-parallel computations—Class I. He estimated that 50 per cent of large scale programs examined could be cast in this form—that is the most portable form. Programs in other classes could not so easily be ported across parallel systems. These included the Loosely Synchronous computations which have a regular case of spatial structure but where the cases demand distinct treatment as well as the Asynchronous problems. Software approaches are now being developed to allow the more portable applications to be programmed portably (Fox *et al.*, 1992).

Outlook

We claim that the promise of parallel processing for GIS is great, admit that currently some difficulties remain in the exploitation of parallel computers for GIS, yet emphasize that the potential rewards for perserverance are very great. These rewards are both commercial and academic, since parallel computing has the capability to transform the practice and the conceptions of GIS.

Quite recently the Edinburgh University Parallel Computing Centre, which hosts both data parallel and process parallel computers, has announced the UK's first centre for Parallel GIS. We can mark 1992 as the starting point for the development of this field and look forward to the progress reports that will surely appear in future editions of this yearbook.

References

Fox, G.C., 1988, What have we learned from using real parallel machines to solve real problems? in *The Third Conference on Hypercube Concurrent Computers and Applications*, Vol. 2, ACM Press.

Fox *et al.*, 1992, *FORTRAN D Language Specification*, Kennedy, Dept of Computer Science, Rice University, TR90-141.

62

Database engines for Geographical Information Systems

Guy Haworth
ICL

Flashback

Melbourne, 9 April 1986, Simon Clarke reporting...

> The State Insurance Office (SIO) and police believe there may be a large-scale third-party insurance racket in Melbourne. Their investigations follow the discovery of more than 240 claims for whiplash injuries from addresses in an outer suburb area with a radius of one kilometre. Some 22 claims come from one small street and involve more than 40 per cent of the households there.
>
> The SIO had taken a 30 per cent increase in payments for some years and recorded a $671M loss for '84/5 on compulsory third-party insurance before it decided to take a closer look at the data. Now it believes fraudulent claims account for a "very significant proportion" of the losses. The 240 claims above could cost $5–12M on their own.

The SIO's core data-processing systems are not GIS systems in the accepted sense but the geographical element of their information was key to their discovery of a well-organized fraud. In fact, the SIO were only able to take a sideways look at their data and ask the right questions because they had just attached a "CAFS" file-searching engine to their central system.

How much money would they have saved if they had been able to identify the highly non-random geographical distribution of claimants years before? How much more would they have saved in staff time and payments if their customers had known they could analyse the data in any way they liked?

GIS on a grand scale

In January 1983, NASA launched IRAS, the "Infra-red Astronomical Satellite" into polar orbit 560 miles above the Earth. Within weeks, it became clear that the IRAS mission was the most comprehensive look at the universe ever undertaken. In only 10 months, the satellite increased the number of known extra-terrestrial objects by 40 per cent and identified 250,000 distinct radiation sources in the 600 megabytes of data sent back to base.

Back on the ground, the sheer volume of this raw data caused very real problems for the experts of the US, Holland and Britain. Until this mass of information had been analysed and interpreted from many angles, the secrets of the universe which IRAS had observed would remain unknown to humankind.

In the USA, the scientists went to work with the full might of their Cray supercomputers, taking 20 minutes of dedicated computer time with every question. In London's Queen Mary College, Dr. David Walker (Walker, 1985) shared the university's ICL 2988. However, he was able to scan the data in less than two minutes with a CAFS search engine and not surprisingly, he raced ahead of the American team developing his ideas iteratively on-line at will.

Dr. Walker not only discovered 10 000 galaxies but found that they were clustered in a particular direction. Now we know that we are actually situated in the outer suburbs of the known universe—filing interesting insurance claims.

Both of these stories indicate that a database engine, a system or subsystem designed to support database activity can, like CAFS, give orders of magnitude better access to the data. It can turn data into information into knowledge, and point to discoveries that could not otherwise be made.

The requirement for database engines

Before we look at specific technologies for and examples of database engines, we look at the requirements which are emerging today for database support. Organizations using GIS today include international organizations, national and local governments, the utilities and enterprises in the transport, retail and marketing sectors. All these organizations depend increasingly on their information resource to:

- reduce operational costs,
- improve effectiveness from stock-holding to customer service,
- support business development in new markets,
- create lasting competitive advantage in a rapidly changing context

They develop ever more comprehensive and detailed information models of their respective worlds of interest. They have also recognized the escalating demand for information access, driven by the following factors:

- growing volumes of information from increasingly many sources such as scanners, remote sensors, GPS, multiple databases, etc.
- growing interest in access of unformatted information such as text, maps, photographs, structured plans, multimedia, etc.
- the increasing trend to work on-line with quality colour desktop presentation, good communications, power, etc.
- the increasing number, literacy and ambition of information workers—TP, management support, statistical analysis, inference, etc.

This growing demand for information management is being met by a variety of technologies to acquire, store, access, analyse and present information and, by international standards, to reduce the visible variety of these technologies and help us work together.

Many technologies contribute to information management: "IRDS" data dictionaries, transaction management systems, distributed processing software, wide-area communications and so on. Here we focus on the database engine itself, the business of storing, accessing and analysing information.

Information storage

The volume of raw data stored on computers, judged by the sales of disks, has been growing at 25-35 per cent per annum. In the GIS field, special factors are driving this growth. We now find graphics and image alongside data and text; a page of 4000 characters needs 40 Kbytes stored as graphics and 400 Kbytes as an image. Also, GIS data collection has recently been revolutionized in many ways including digitizers, scanners and remote sensing satellites.

NASA will be going for the database record circa 2000 with EOS, their Earth Observing System: 2 terabytes a day for 15 years = 11 000 terabytes = 11 petabytes, with 10 000 scientists seeking access to the data. If you are collecting geographical data, it's likely that you want to know where you are. The recent advent of the USA's Global Positioning System, GPS, offers new orders of efficiency and accuracy to surveyors and navigators worldwide, enabling more effective data collection.

Wherever this GIS data comes from, it has been collected at enormous expense and is literally priceless. Clearly it needs to be stored with security levels which would be the envy of a bank; no organization today can expect to survive the loss of its databases.

Further, it needs to be retrieved in a suitably short time for on-line users, regardless of the data volumes or transaction rates involved or the completeness of the data input to specify the query. As we move away from the classic COBOL-like data record to the less structured and more complex objects of GIS, we find the user asking higher-value and more wide-ranging questions, questions which incidentally make indexing techniques less relevant and information access more difficult.

All the above points away from the casual collection of all too portable PC discs and toward the "database engine"—a large scalable facility managed professionally on behalf of the organization and supporting large, growing and essentially indivisible collections of data. It is becoming clear that the main and enduring role of the mainframe or corporate server is as a database engine.

Information access

On-line systems potentially give information workers immediate access to the data they need. Today, the wider availability of suitable packages and "object oriented" applications development tools makes it far more likely that we will be working on-line.

On-line transactions can be characterized in terms of frequency and complexity, as judged by the load they place on the computer system; every computer system has its frontier of performance, see figure 1. Classic TP has high rate and low complexity; GIS transaction are much more likely to have medium and high complexity.

Information professionals in the GIS field will have an iterative and investigative style of on-line working like Dr Walker at QMC. They will be familiar with the short and predictable response times of classic TP systems and may wonder why their systems cannot currently achieve the same with their more complex queries. Where

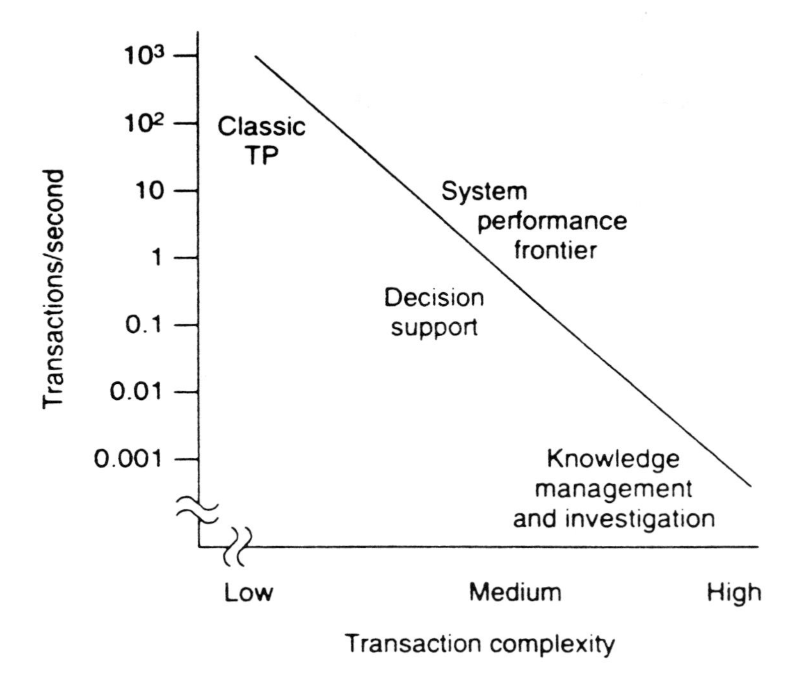

Figure 62.1. The range of transaction types.

"saga" transactions, like high-level query and statistical analysis, are the main workload, there will be a demand for response times closer to the classic two-second target.

Historically, as in the ICL, CAFS and NCR Teradata machines, the concept of parallelism has been employed to improve information access. Parallelism is relatively easy to use in the database field and we shall later look at the CEC Espirit II project EP2025, the European Declarative System, as a leading example of this approach.

Investment protection

Here, we examine how to protect your computer systems investment before and after exploiting database machines. We need to know how:

- best to prepare to use future database technology,
- to get best value from an investment in that technology.

The high cost and value of databases makes it vital that they are managed as long-life assets in an organization. Today, IT strategists are increasingly looking to build flexibility and survivability into their IT Systems by basing them on a framework of international standard interfaces. The "1992 effect" in Europe is just another step in the process of globalization which makes standards more important.

Standards allow existing systems to be insulated from and yet exploit new technologies as they become available. Here, we look at the standards which create opportunities but impose constraints on the introduction of the next database engines.

The ISO SQL standard has emerged as the latest interface between application logic and database management. Most new conventional DP systems are now based on relational databases supporting SQL. SQL, SQL-86, was formally defined for the first time in 1986 and has been evolved through SQL-89 and SQL-92 to meet new requirements.

Even so, SQL-92 will remain firmly attached to its DP-data roots. We will have to wait for "SQL3" (SQL-x where x > 1995) before SQL formally manages complex GIS objects and knowledge with the higher level of intelligence which is also required (see Quinn, Abdelmoty and Williams, this volume). Steven Dowers has been representing AGI on ISO/IEC (ITC1 SC21 WG3) SQL Rapporteur Group and two AGI papers (Dowers, 1991; Gradwell, 1990) are a good indication of the GIS requirements influencing SQL3 extensions.

One of the big unknowns in the database future is whether SQL and industry-supplied DBMS technology can be evolved sufficiently to meet the needs of those, including the GIS community, dealing with non-classic complex data. The slow pace of ISO development and the wide variety of unsatisfied requirements makes it unlikely that SQL will be the only interface to GIS data.

However SQL will need to be supported in the GIS context. The established Relational DBMS vendors are extending their products beyond ISO SQL to meet the most immediate demands and GIS systems will need to cross-refer to information behind the SQL interface.

The new wave of object-oriented OODBMS technologies, primarily developed for CAD use, show an alternative approach which will certainly be adopted by the GIS community. The risk of using OODBMS technology will in the long term be reduced by standardization, most likely based on the work of OMG, the Object Management Group.

The reality therefore is that in future, a GIS will need to get its data from more than one type of DBMS and via more than one interface. It is also likely, given that different agencies collect data on different aspects of the same geography, that the databases will be owned by more than one information supplier and be physically distributed.

With this in mind, X/Open and ISO have devised the OSI-TP architecture to enable a transaction manager to interwork coherently with several database "resource" managers. These might be distributed over a number of computers, sites and organizations. The fact that the data you need is locked in a variety of technologies need not be a barrier to accessing that data.

It is possible to use Distributed TP software to advantage to implement TP systems involving one actual DBMS—and one hypothetical database manager. Those systems can then be enhanced more easily to exploit new database platforms as they become available.

What are the implications of these technologies and standards for the next database engines? To give flexibility, they are likely to be a mixture of software and hardware. It is important that the various items of software can be effectively ported to the engine's hardware. The engine should certainly support the SQL interface and be able to support both the established RDBMS packages and emerging database technologies. The database engine will be a "server", providing data to a variety of "client" applications, and will therefore support the necessary interfaces for distributed application processing.

Database engines today

Suppliers are providing a wider variety of computers today; there is clear recognition that general-purpose computers are going to lose market share to computers designed for a specific purpose. Signals from product trends, development projects and current research indicate that the database engine will be the target product for more than one company.

ICL's CAFS engine, already mentioned, and the NCR Teradata machine are the most conspicuous examples of commercial products today. The CEC Esprit II project EP2025 EDS, the European Declarative System, is in many ways a logical development of the thinking behind CAFS and has features in common with the Teradata, Meiko and Ncube machines. I shall therefore review CAFS and EDS as two generations of related database engine.

The CAFS engine

The purpose of CAFS (Haworth, 1985) is to filter and analyse data coming off a disk at disk speed. The requirement for such a search engine can be simply stated. Suppose you know that the answer to your question lies somewhere in a Gigabyte file of data. You could search it all but you would prefer if possible to avoid this since many disk-accesses slow down processing. You create one or more indexes to the data, chosen to help the queries you have in mind as much as possible. These indexes focus the following file search when relevant but they are costly to create, store and maintain and clearly have their limitations. For example:

- a full word-level inversion of a text file may be three to four times the size of the original text;
- would you really create an index of:

  — houses with at least seven of a given set of 10 features,
  — plant over five years old,
  — men whose eyes are not green,
  — addresses containing the letters "EAD" somewhere?

So we know that to access data, which we will do often, we always have to retrieve some part of the file and may have to search all of it, just to find a fragment of information.

The CAFS search engine was therefore implemented as an integral part of ICL's Series 39 Corporate Server architecture. It recognizes the data in the bit-stream, evaluates fields, selects records of information meeting a given criteria and returns only the interesting parts of those records.

The CAFS engine produces response-time improvements of 10–1000, the equivalent of 10–30 years development of conventional computer technology. Significantly in the context of GIS, the "turbo factor" is greater the more unstructured, and less indexable the information.

The CAFS concept has recently been re-implemented on ICL's Series 39 and DRS6000 UNIX machines. Searching is now done at up to 8 MB/sec and new applications include "CAFS" support for the INGRES RDBMS.

CAFS in practice

We have two CAFS stories already and there is only space for a few headlines and literature references on others (ICL, 1985; Walker, 1985; Wiles, 1985).

Southern Water (Corbin, 1985), leaders in giving their end-users large-scale access to their information, won the Office Automation award in 1985 for the "best information storage and retrieval system"; 80 per cent of their CAFS traffic was new work and included queries on rainfall, water quality and plant.

The Inland Revenue's National Tracing System (Wiles, 1985) routes mail to the correct tax office even when only a few characters of handwriting are decipherable. At peak times, 20 transactions/sec access a duplexed 12 GByte file, some 50 million name/address pairs. Two-thirds of the transactions require an index-focused CAFS scan which delivers a response in two to three seconds. NTS uses 26 CAFS engines in parallel to achieve this.

CAFS has been invaluable for the investigative systems needed by the intelligence, defence and police organizations. "Location" is one of the three most important items of information in such systems. During the UK privatization of the gas, water and electricity utilities, forensic accountants checked out share applications against the "one per person" rule, foiling a £6M raid on one occasion where a ring of over 40 people shared each other's addresses.

Social security fraud has similar name/address characteristics; back in Australia, one pilot, based on only seven GB of data, overcomes the "state boundary" problem. When the system matched aliases and addresses, it sent identical letters to each alias asking them to attend the same meeting to discuss their situation; false identities disappeared without further prompting—no delays, no legal fees.

From CAFS to the next generation

CAFS has been widely used since its inception; the technology, features and benefits behind its success are worth noting and looking for when considering a second generation database engine:

- functions "shipped" to the best place in the computer architecture;
- use of inexpensive components and a new technology mix;
- use of parallelism (see Lake's chapter in this volume);
- quantum improvement in performance over specified workload;
- exploitation via interfaces already valued by the market;
- fit with customer's existing investments—hardware, software, data;
- improved performance for existing systems;
- new opportunities to create high value information from raw data;
- product scalable to meet growth in demand for performance.

In "relational terms", CAFS performs RESTRICT and PROJECT, two of the basic three operation on data records. Now the industry is moving to support the SQL interface fully with purpose-built database engines.

In the US, Britton-Lee introduced such a product some years ago and were followed by Teradata, now part of NCR, with more success. Both engines did little for standard transaction processing but chose to support decision-support and data analysis.

The EDS project

In Europe, Bull, ICL, Siemens and their jointly owned European Computer Research Centre, ECRC, identified a common interest in 1988 in building a database machine and put a proposal to the CEC for Esprit II funding. The initial objectives (Haworth *et al.*, 1990) of the EDS project EP2025 will be met this year and further work will continue into 1993.

The messages from CAFS' success provided a good template of requirements for EDS which are to:

- improve performance more than 10-fold from TP to investigative query;
- support valued interface standards, e.g. XPG/POSIX base, C + +, SQL;
- support established products and operate in client/server mode;
- maintain a near-linear relationship between size and power;
- adopt advances in hardware and software component technologies early;
- use parallelism to exploit the parallelism latent in data queries.

EDS exploits the fast-moving microprocessors and DRAM storage technologies on a large scale. Processors are increasing in nominal MIPS by 50 per cent per annum and DRAM chip storage is increasing in density by four every three to four years. At the end of 1992, the 80 MIPS processor will be in use and the 16 MBit chip will be adopted by the industry.

EDS aims for similar performance trends at the system level and the equivalent of several thousand TPCB transactions/second. In order to achieve this, EDS must be based on a distributed "share-nothing" architecture instead of the symmetric "share everything" architecture of current multiprocessor UNIX machines. In this arrangement of processors and stores, each processor "is near to" the part of the data in "its" store, the data layout is known to the DBMS, and the processors are asked to do the work related to their data.

To see how this works, imagine you ask a crowd of people if anyone has a birthday during the week. You have distributed your query to a "crowd machine"; each person has become a "processor" picking up part of the query, consulting their part of the "birthday database".

A key element determining the performance of a database machine like EDS is the communication system enabling the processors to work together. The Meiko Computing Surface uses a chess-board topology while the Teradata has its Y-net. The hardware and message-passing techniques of EDS have been developed only after extensive simulation by Siemens.

EDS will exploit parallelism in two ways. First, separate transactions can be executed simultaneously. All processing, even off-line batch processing, should therefore be thought of in transaction processing terms and divided up into a number of separate activities which can be run together. Second, a single transaction can be executed by a set of simultaneous processes. Parallelism within a transaction can be created by application logic or transparently by the optimizer within the RDBMS.

Just as ten CAFS engines can each look at 10 per cent of a file simultaneously, a hundred EDS processing elements could each look at 1 per cent of the data. In fact, any SQL query can be executed by a set of co-ordinated parallel activities as this query on plant-history and figure 2 illustrate:

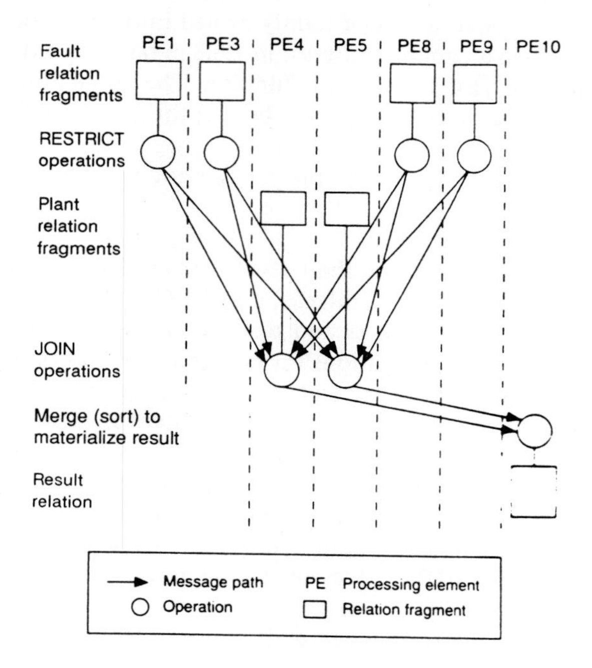

Figure 62.2. Parallel execution of a query.

Plant records (plant-id, location-id, plant-type ...) on processing elements 4 and 5

Fault records (plant-id, cause, ...) on elements 1, 3, 8 and 9

SELECT plant-id, location-id, cause FROM plant, fault
WHERE cause = electrical and plant-type = pump

By supporting the standardized POSIX and XPG operating system interfaces, EDS aims to be a platform for "UNIX" software and run RDBMS packages with much better performance.

In addition, the EDS research has developed database facilities for:

- support of user-defined data types and methods;
- support of complex objects and large objects;
- deductive database capabilities;
- general integrity constraints; and
- triggers (actions executed when a given event occurs).

With the above facilities, EDS will be able to manage complex information objects whose data content will not be simply retrieved but sometimes calculated or inferred from rules, e.g.:

- the age of a building;
- the fire-risk of a building, given a fire nearby;
- the safety envelope of an aircraft in flight;
- the predicted traffic load on a road, given a closure nearby.

The complementary concepts of object-orientation and deduction come together in the EDS database technology, enabling demanding GIS applications which require both to be supported naturally.

The EDS project demonstrates the potential of database engines based on a highly parallel architecture. You should be able to buy the database performance you are prepared to pay for; buy twice the hardware to double the throughput or halve the response times for complex queries.

The main partners in the EDS project are confident that the EDS approach to database has industrial potential. No products have yet been announced, but it is significant that all EDS-related projects proposed for CEC Esprit III funding, eight times oversubscribed, have been supported because of that potential.

Future scenarios

GIS applications require large-scale data management involving:

data acceptance:	integrity checking and calibration;
storage management:	distributing data to balance query load;
indexing:	characterizing the data by features useful for queries;
inference:	deriving higher-order information from raw data;
dissemination:	passing new data to interested parties.

Quinn *et al.* (in this volume) picks out map design, geographic database management and decision support, and automatic feature extraction as four key GIS applications. All of these are data and processing-intensive and most of the processing is within the database management part of the application. They will clearly benefit from the increased power of the database engines coming to market today. When these applications are required by mobile, real time or military situations with demanding availability and integrity requirements, the database engines required double and redoubles in required performance.

Perhaps this is a glimpse of the future

The Street Works Register

The government's Road and Street Works Act 1991 has resulted in the Street Works Register, a source of accurate definitions of roads, equipment above and below the surface, and repair history. The days of spatial data, hedged around with disclaimers on accuracy, are over.

The immediate beneficiaries have been local authorities and the utilities, particularly water companies who have to dig deepest to maintain their equipment. Their insurance claims for incidental damage to other unexpected equipment are radically down.

Significant savings on works costs amounting to many millions of pounds per year are being made and at the same time the country's street infrastructure is improved in terms of road quality and reduced delays to traffic and people. The database engines behind this advance provide a wide variety of information to all the agencies involved in street works and interested in definitive geographical data.

Heathrow, Friday 20.30

The computers of air-traffic control manage detailed data on each flight in the zone. Aircraft navigation systems radio in position data which is checked against past history and ground-radar data. The DBMS computes a safety zone for each plane. Flight paths on current instructions are simulated to check for near-miss situations.

The incoming shuttle from Manchester is diverted to an outlying station of terminal one. The DBMS infers that mobile steps and three buses are required. The nearest unassigned units arrive at the station just in time to meet the plane.

In the terminal, the displays show accurate forecasts of arrival and departure times, reflecting the fact that the DBMS behind them is fully informed of all delays and able to work out their impact.

Another "Gulf War"?

The last Gulf War was a sharp demonstration of the value of information technology and we can reasonably guess that the allied forces will have built on their success. Next time .....? Ground units communicate their GPS positions continuously, defining troop readiness and safety zones.

Overhead, remote sensors photograph the desert and send their results back to HQ. The DBMS picks out known features, registers the pictures, removes the images of friendly troops and, allowing for the time of day, checks the remainder for change against previous photographs. It infers that enemy troops are moving down a key road and that previously passable tracks of hard sand have been wiped out.

To avoid the dangers of information overload, the DBMS reports developments to the senior officers against the prioritized definition of their requirements.

Conclusions

Our ability to identify, acquire, store, enquire on and analyse data is increasing as never before, especially in the GIS field. Technologies are becoming available to manage a wider variety of data and to make intelligent inferences on that data.

The mainstream arrival of large-scale database engines is not far away. The experience of using the first such products tells us that they will radically change data management in the GIS field.

Abbreviations

C++	Object-oriented extension of the C language
CAD	Computer-Aided Design
CEC	Commission of the European Community
CAFS	Content-Addressable File Search (engine)
DBMS	DataBase Management System
DRAM	Dynamic Random Access (solid-state) Memory
DTP	Distributed TP
GB	Gigabyte, 10^9 bytes
GPS	(US.DoD) Global Positioning System
IRAS	Infra-Red Astronomical Satellite
IRDS	Information Resource Directory System (dictionary)
ISO	International Standards Organization
KB	KiloByte
MIP	One million instructions/second (of undefined type)
OODBMS	Object-Orientated DBMS
POSIX	Portable Operating System Interface for Computer Environments
Petabyte	10^{15} bytes
RDBMS	Relational DBMS
SQL	Structured Query Language
Terabyte	10^{12} bytes
TP	Transaction Processing
TPC	Transaction Processing Council, defines TP benchmarks
TPCB	The TPC's "B" benchmark for transaction servers
XA	The DTP interface between transaction and resource managers
X/Open	Worldwide body defining the Common Application Environment
XPG	The X/Open Portability Guide

Acknowledgements

My thanks to Chris Corbin, Steve Dowers, David Gradwell and Tom Lake for recent conversations and to my ICL colleagues for their contributions to CAFS and EDS.

References

Corbin, C.E.H., 1985, "Creating an end-user CAFS service", *ICL Technical Journal*, **4**, 4, (the "CAFS" issue) pp. 441-545.

Dowers, S., 1991, "SQL—The Way Forward", *AGI, Birmingham November 91*, pp. 3.13.1-5.

Gradwell, D.J.L., 1990, "Can SQL Handle Geographic Data?" A Presentation on the work of the AGI Standards Committee's SQL Working Party, AGI 1990 pp. D.3.1-12.

ICL Computer User Association, 1985 CAFS Special Interest Group, "CAFS in Action", November.

Haworth, G.McC., 1985, "The CAFS system today and tomorrow", *ICL Technical Journal*, **4**, 4, pp. 483-8.

Haworth, G.McC., Leunig, S., Hammer, C. and Reeve, M., 1990, ''The European Declarative System, Database and Languages'', *IEEE Micro*, Dec, pp. 20-23 and 85-88.

Walker, D., 1985, ''Secrets of the sky: The IRAS data at Queen Mary College'', *ICL Technical Journal*, **4**, 4, pp. 483-8.

Wiles, P.R., 1985, ''Using secondary indexes for large CAFS databases'', *ICL Technical Journal*, **4**, 4, pp. 419–40.

63

Windows in the open

David Arnstein
IT Southern

Introduction

Geographical Information Systems (GIS) are complex and the user interface is very important for making them effective. A good user interface builds confidence in the system and helps the user to learn how to use it more quickly. It also encourages the user to exploit the system more effectively. Graphical User Interfaces (GUI) have shown themselves to be particularly effective; even more so if a common (standard) one is chosen.

With the advent of window-based systems, users are now quickly seeing the productivity benefits of being able to run multiple applications on one desktop system. The productivity comes from the user being able to run a number of software applications (tools) concurrently, choosing the most appropriate for each task. The greatest productivity benefit is obtained when these applications or tools have a similar user interface and can inter-work, that is they can share data and otherwise communicate with each other.

GUI for accessing general software in a windows environment are becoming widely available and certain components have become international standards. The current functionality of these GUI is very applicable to GIS, although facilities specific to GIS are still missing. Fortunately most GUI are extendible. This chapter considers the role of the emerging standard graphical user interfaces and the consequences of using them within Geographical Information Systems.

GIS presentation

Commonly, database information is overlaid in diagrammatic form on a so called "map background". This background usually takes the form of a conventional map, perhaps reduced in detail to improve clarity on a normal (high resolution) visual display unit or monitor. Most often in the UK map backgrounds are either raster images obtained by scanning Ordnance Survey (OS) maps or drawn from vector data derived from OS digital vector maps. The type of background has an impact on:

- Size of filestore required,
- Speed of drawing,
- Method of panning,

- Zooming—relationship of widely separated objects,
- Cartographic standards,
- Visual representation of non-cartographic objects—lack of standards.

The user also needs ways of controlling the software, and areas of the screen to view associated information—text and images. The industry standard GUIs; Motif, OpenLook and Microsoft Windows 3, discussed in more detail later, are usually used by software developers to provide these functions.

Are the functions provided in these GUIs sufficient/appropriate for GIS? The GIS user has to be able to control the display of potentially vast amounts of information—from the extent of the geographic area covered and the information associated with the geographic "objects". The user may have to add or modify information. The following factors are some of those that affect the usability of a GIS through its user interface.

- Controlling the amount of information displayed,
- Selecting the geographic area to display,
- Panning and zooming the geographic picture,
- Defining action required—select object, pop-up associated information etc,
- Defining areas for analysis or search,
- Defining other selection criteria,
- Selecting how information is to be displayed,
- Showing areas/items that are locked against changes,
- Adjusting colours for aesthetics or user visual reasons,
- Importing and exporting of data (text, diagrams, images etc.) to other tools.

GUI history

A number of years ago graphical user interfaces, windowing systems, and graphics programming interfaces were not standardized and not available on many systems which were often character-based. Systems such as SUN Microsystems workstations that did have graphics capability and a GUI implemented their own proprietary facilities. Subsequently some graphics standards such as GKS (Graphics Kernel System, ISO 7942 1985) and CGM (Computer Graphics Metafile, ISO 8682 1987) were defined. There were still no standard user interfaces nor networking graphics standards. The success of the proprietary GUI and the widespread introduction of low cost networked graphical workstations generated a need for a common networked graphics based system.

At the Massachusetts Institute of Technology (MIT) in the early 1980s more and more people were starting to use graphics workstations from a variety of manufacturers and a network was being set up to unite them. The target was not only to provide data sharing but also to provide graphics access across the network to every student and lecturer. They wished to allow users access to workstations and other computers across the network and for them to run remote applications with local graphics display and GUI. Concurrent applications would be handled through a window system. Users would be able to run applications both on their own workstation and others. Thus they could run applications only available on a hardware platform different to their own. It would also allow users to run applications on a computer more suited to the processing, such as one with a high performance processor or fast access to a specific database. In addition it opened up the development of lower cost graphics terminals that could be used with any computer on the network.

Because there were no standard ways of achieving this networked environment, project Athena was established to tackle the development of a unifying mechanism. The project had several components:

Graphics networking

To support networking a special protocol was developed, now called the X protocol. This encompasses the whole of the user interface including the screen, keyboard and usually the mouse.

Programming interface

To provide a usable and consistent interface to the X protocol a programming interface was developed called the X library or Xlib. This provided programmers with low level drawing and input/output event handling facilities via a C language interface.

Toolkit support

An important requirement of the project was the provision of high level GUI that would use the underlying Xlib and X protocols. However, the project did not want to impose a specific GUI—it was too early to know what might become a standard and might stifle promising ideas. So a GUI support layer was built (the XT Intrinsics) which "understands" the concept of graphical objects such as buttons and scrollbars and makes building a GUI toolkit easier.

GUI

Although no specific GUI is defined, an example GUI toolkit was developed called the Athena widgets. This has been used for experimental programming and in the academic environment but is not used commercially. A number of other toolkits have appeared which implement various "look and feels" based on the work originally carried out at Xerox Parc.

Display mechanism

Programs using X Windows display their graphics by sending graphics instructions using the X protocol over the network (or within the local workstation) to a software module known as the X server which handles the screen (see later). The X server also handles the keyboard and mouse and informs the application about input. MIT generated sample X servers for a number of computer types. However, generally computer manufacturers and software suppliers have ported the sample code to particular computers increasing performance.

Window management

In order to use effectively the windowing mechanism a window manager is required. Again MIT supplied some sample window managers though there are commercial offerings which are more generally used.

All these facilities were made available free to the computer industry. Commercial enterprises took the software, ported it to their own hardware, added additional facilities and provided support.

The first generally available version of X Windows was X10 release 4 released in 1986. However this version was not widely taken up. There was a major revision of the X protocols for the X11 release. X11 release 1 became available in September 1987.

To control the releases and provide commercial backing for the continued development of X Windows, the X Consortium was set up in January 1988. Control of X Windows passed to the Consortium at X11 release 2 in March 1988. X Windows (X11 release 3) became widely available in 1989. The current release, though not yet generally available through commercial channels (as of February 1992), is X11 release 5.

A number of commercial companies make up the X Consortium which includes nearly all the major computer manufacturers. It commissions new facilities and controls releases.

It is important to understand that MIT did not set out to define a specific GUI—only the mechanisms required to support the software that implements one. So far it has been left to the industry to develop GUI specifications and GUI tools and the international standards bodies have not been able to choose one to adopt or to specify a new one. However, the underlying X Window mechanisms have been adopted as an international standard.

Partly because of its availability at low cost and its technical capabilities the X Window system was rapidly taken up by both the academic and commercial worlds and has become the dominant windowing system on UNIX (though it is not UNIX dependent) with many development tools and applications appearing in the marketplace.

The adoption of the X Windows based GUI on low cost Unix workstations represented a threat to the dominance of the DOS based PC as the desktop computer. In response Microsoft developed Windows 3. This has a similar look and feel to Presentation Manager on OS/2 and MOTIF on X Windows, although it is more limited than either. Windows 3 is now widespread on PCs.

Advantages of standard GUI

For the user:

- Application and presentation controls (where provided) are consistent between applications and as a result are easier to learn and for the casual user easier to remember.

For developers:

- Bigger market for development tools so more functionality and choice.

Advantages of X Windows based GUI

For the user:

- Co-operative applications,
- Cut and paste between applications,
- Display possible on low cost X terminals,
- Can run applications on the most appropriate computer in the network,
- Can integrate existing PCs into GIS network.

For developers:

- X Window-based products offer networking capabilities,
- Similar or same source code will compile and run on a variety of hardware platforms assisting portability,
- Assists integration with other applications/tools,
- Rich development environment.

Integration of PCs into networked GIS

Under DOS on PCs, Microsoft Windows 3 provides an excellent windowing environment and adds a degree of multi-tasking capability. It brings to the DOS environment similar benefits as X Windows brings to (mainly) Unix systems. It has some limitations such as, it is not as versatile, and nor is it a networked window system. DOS, itself, is not an ideal operating system on which to build networking facilities.

For that reason if a new GIS is being installed, a UNIX based system may be preferred. If the organization already has a significant number of PCs, possibly already networked, however, how can these be integrated? In this case a GIS built on X Windows has a great advantage. By attaching the PCs to the network using Ethernet and TCP/IP (if not already connected) and installing X server software on the PCs, the PCs can become workstations (access points) to the GIS and any other X Window-based software that becomes available.

Various X server packages are available, some of which run directly in DOS, effectively turning the PC into an X terminal, and others which run under Microsoft Windows 3. In the later case various options are available for managing the X Windows. Either a remote Window Manager can run, displaying its own windows mixed with native Microsoft Windows 3 windows or displaying its own windows within one Microsoft Windows 3 window. In this case the normal functionality of the X Windows manager is retained for the X Windows windows. Alternatively the X Windows windows can be managed by the Microsoft Windows 3 manager. This option results in a simpler environment for users who concurrently use DOS applications alongside X Windows ones, with the drawback that some of the usual X Windows' functionality will be lost. In the case of some PC X server software how the windows are to be managed can be user-selected at start up or can even be changed during a session.

It should be appreciated that, like Microsoft Windows 3, X Windows needs a great deal more resources than simple DOS. In general at least 4Mb of main memory and a fast processor is required to give reasonable performance. In addition any windowing system needs a high resolution monitor to be effective (in any case GIS normally needs very high resolution screens). In many cases X Window applications require to be able to display more than 16 colours which implies a relatively high resolution graphics card that can handle 256 colours.

The software that provides the TCP/IP support usually can also provide NFS (Networked Filing System) support. This allows PC users to access UNIX and DOS files stored on UNIX file servers as though they are held locally on the PC. This has the added advantages of file sharing such as reduced filestore, reduced maintenance and the possibility of central file backup.

For workstation users that require access to DOS applications, PC emulators are available at reasonable cost that provide PCs in a window. The workstations and PC emulators can share data with any networked PCs using NFS if made available.

For new users not requiring DOS, consideration should be given to using dedicated X terminals or RISC-based workstations which are likely to provide significantly greater performance for a similar cost. If access to DOS is required, consideration should be given to using RISC based workstations together with a PC emulator or PC add on board.

Limitations of GUI for GIS

Neither X Windows nor Microsoft Windows 3 provide tools that specifically support GIS functionality, e.g. there is no co-ordinate display widget or polygon bounding widget. However both environments are extendible and GIS tailored programming toolkits could be built.

Integration with general computing

GIS can be complete independent applications. They exist to store and deliver information. However, there are usually many other applications that need the same information or could benefit from the same or similar user interface. Such applications include Command and Control, Network Modelling, Census Data Analysis, CAD and even Customer Accounts. To support this need GIS is evolving into one component of larger systems that deliver a wide range of services.

UNIX workstations offer an integrated but open application environment that allows applications to be developed that sit on top of functional services such as expert systems, spatial databases, relational databases, user interfaces and real world models and supports the developing distributed and object orientated computing environments. This fits in well with the requirements of the new ways of implementing GIS and the new roles GIS are beginning to fulfil. It fits less well with the DOS/Microsoft Windows 3 environments. Whether or not Microsoft Windows/NT will prove a more suitable environment for PCs is yet to be seen.

Although there may well be a market for small low cost GIS, perhaps based on a stand-alone PC or UNIX workstation, in general GIS are large and form part of even larger systems. These large systems imply heavy investment in both software and data. To protect these investments they need to be built with Open Systems in mind. Again UNIX based systems and X Windows best fulfil this requirement.

The X Windows technology

X Windows is a package of software components and standards that provide a networked window user interface system. The standards are the critical components and have been adopted by the ANSI standards body. These standards define the message protocol that is used to send window managing and drawing instructions between the software components. These components may all exist on one computer or may be separated across a network. The way the messages are transported are not defined but TCP/IP transport protocols are usually used.

The software components used in the X Windows System fall into three categories.

X Window Server

One component drives the screen, the mouse, the keyboard and any other input/output device the user uses to access the X Windows-based applications. It responds to incoming messages requesting screen drawing and window creation or deletion, it manages colour and it send out messages to other programs about mouse movements and keyboard key changes etc. Since it provides this as a service it is called the X Window Server or X Server. An X server always runs on the computer or processor directly connected to the user's screen and keyboard.

Window Manager

Another component is called the Window Manager. Although the X server provides low level window control, in practice, to be useful, an X Windows environment needs a high level window manager. The window manager enforces rules such as whether or not windows may overlap and assists in the iconification of applications (this is where a running application window is temporarily "closed" or in Microsoft Windows 3 terms "minimized" and the application is represented by a small icon on the screen). It usually also provides a menu facility for accessing applications and certain X Window functions.

X Clients

The final category consists of the X Window user applications. X Window applications are called X clients as they are clients of the X server services—they access windows on the screen, the mouse and keyboard etc. via the X server.

X Window applications, or X clients, are written to use the X Window protocols. This is normally done by calling routines in the X library (Xlib) to handle input and output. Using the X library to provide a high level user interface such as OpenLook or Motif is difficult and normally applications are written using special toolkits. These make it easier to program items such as "on screen" buttons, messages, text and icons. Unfortunately there are no toolkits that currently support geographical user interface components.

The X Window Server and Window Manager provide the controlled environment for the applications and enable communication with the user. The Window Manager, though a very specialized one, is also technically an X client, just as are the user applications.

X Windows toolkits

Writing programs to use the Xlib interface directly is difficult. Provided as part of the X Windows distribution is the XT Intrinsics. This is a level of software that sits on Xlib and provides for additional functionality for developing a GUI toolkit. One such GUI toolkit, supplied in the X Windows distribution, is the ''Andrews'' toolkit. This is really only an example toolkit.

The main commercially useful toolkits are MOTIF, OLIT (OpenLook InTrinsics) and XVIEW. MOTIF provides the MOTIF look and feel whereas OLIT and XVIEW provide the OpenLook look and feel. XVIEW is SUN Microsystems toolkit that interfaces directly to Xlib whereas MOTIF (from OSF) and OLIT (from AT&T) interface to the X Intrinsics. There is also the OI (Object Interface) toolkit from Solbourne that provides both MOTIF and OpenLook look and feel.

The XVIEW, OLIT and OI toolkits are provided as part of AT&T's System VR4 UNIX. XVIEW and OLIT is provided with SUN Microsystems'' operating system. XVIEW is available on a number of other platforms.

The MOTIF toolkit is sold by OSF. Since OSF is supported by a number of manufacturers such as IBM, DEC and Hewlett Packard, the MOTIF toolkit and window manager is available on a wide range of hardware.

To provide a consistent user interface, a window manager with the same look and feel is required. These are generally supplied as appropriate, e.g. SUN Microsystems supplies OLWM (OpenLook Window Manager) and OSF supplies a MOTIF window manager.

Which toolkit to use can be a difficult choice. No look and feel has yet been standardized by any standards body. OpenLook is freely licenced at no cost and software produced using XVIEW or OLIT is also free of licence costs. The use of the MOTIF toolkit currently incurs a licence fee and under some circumstances applications built using the MOTIF toolkit incur licence fees. MOTIF, XVIEW and OLIT have C programming interfaces, OI has a C + + interface.

In addition to these toolkits there are design tools that use the GUI to allow the program designer interactively to build the user interface of a new program. This dramatically cuts down the time to design and code this part of the application. Some of these tools generate code for use with a particular toolkit, others can work with a variety of toolkits. The scope, ease of use and cost of these tools vary considerably.

The uncertain availability of the tools and toolkit across hardware platforms (computer architectures) can mean that choosing one particular toolkit will reduce the range of computers to which an application can be ported.

It should be realized that there is a great deal of similarity between MOTIF and OpenLook and, since programs developed with any X Windows toolkit use the same underlying X Windows protocol, all X Windows programs should co-exist peaceably. The only impact is that the protocols for cutting and pasting between windows and setting window colours have not yet been standardized so not all these functions will work between applications derived from different toolkits or running with an incompatible window manager.

X Windows marketplace

When X Windows was first introduced it had to compete with other emerging technologies. In particular there was SUN Microsystems' NeWS (Network Extendible

Window System) and the Next computer window system. Both of these have protocols based on Postscript while still implementing a client-server architecture. For full workstation based systems they offer greater performance and lower network loading than the X Window system. This makes it possible to use programs that run remotely across a relatively slow network link such as a dial-up modem. The improvements are achieved by downloading portions of the application as executable Postscript routines that assist the server, for instance, in drawing objects specific to the application.

In the event, these alternatives to X Windows were seen by the marketplace to be too closely associated with specific manufacturers and X Windows became the system most widely adopted, to the extent that it has now been enshrined in an ANSII standard.

SUN continues to believe in the superior technology of NeWS. To continue its support and integrate it into the (SUN) X Windows environment they provide an extended server that understands both the X protocol and the NeWS one. This allows both X Windows-based applications and NeWS-based ones to be used concurrently by a user. However, in the absence of equivalent servers and program development libraries on other systems, the running and use of NeWS-based applications is limited to SUN equipment.

There are also other advantages to Postscript, especially in the area of scaling the drawing of objects such as characters—the main purpose behind the original development of Postscript. Using Postscript it becomes much easier to provide fully scalable fonts. The advantages of Postscript have been recognized by the "X Window community" and there are moves to incorporate what is called Display Postscript into the X Windows protocol.

X Windows limitations

The X Windows System is very successful despite its limitations and some technical deficiencies, as proved by its widespread adoption. If an "open systems" windowing system is required there is currently no other choice. However, there are some limitations that should be recognized:

Speed of response

The underlying technology is raster image based. Images are transferred from the client application to the X server for display in raster form which makes for a big load on the connecting network. Even small changes require a significant amount of data to be transferred. This implies applications running remotely from the X server impose a significant load on the network. Even more importantly if the network is slow, because of the speed of data transfer, the response time of the system is much too poor.

Fonts

Because of the bias towards raster technology, fonts are used as bit images. This means that the storage required for a reasonable set of fonts is large. Current commercial implementations of X Windows require copies of the fonts to be available at each X server, multiplying the storage requirements. The latest release of X Windows (X11 release 5) provides for a font server that can deliver fonts over the network—this both reduces the filestore requirements and the difficulties of administration.

The basic programming interface

Xlib provides for very little support for drawing diagrams or pictures. The ability to draw simple lines of various widths in various colours/pixel patterns is of course provided. However, there is no provision for drawing even simple shapes such as circles. The various common GUI toolkits based on Xlib generally provide sophisticated facilities to create and manage general "objects" such as buttons or scroll bars but do not provide a high level drawing interface. Toolkits have recently become available that provide this high level drawing capability but they have not been standardized and so programs that use them may not be portable.

X Windows future

As described earlier, the X Windows System has been widely adopted because it satisfies the need for a portable, networking window system. It was developed in an academic environment where there is an expectation of constant improvement. Although now the X Window system is controlled and released by the X Consortium that pressure for ongoing improvement is still present. There is luckily a commitment to maintain backwards compatibility where possible. The types of improvements under development include the support of a font server, support for the high level 3D drawing interface called PHIGS (the extension to X Windows being called PEX—PHIGS Extension to X) and the ability to send video images (VEX?).

Part XI

Research and GIS

64

Editors' preface

N. Stuart and G. Clarke

This year's R&D section contains four chapters which illustrate the diversity of initiatives which have recently promoted either pure or applied research into GIS and show that GIS developments are continuing to diffuse into new areas.

The first chapter reports developments from the emerging field of route-finding and vehicle navigation. Popularized by the success of PC packages such as AutoRoute, the race has been on to develop and market in-car navigation. Two distinct types of navigation system can be identified: autonomous systems which hold all their route data locally and infrastructure-based systems which receive signals from an external system of beacons. The decreasing costs of optical disk storage has meant that autonomous systems can hold their own map information compactly in a limited space. A basic prerequisite for all such products, however, is a computerized road map, or more precisely, digital road network data correctly structured to permit navigation and containing relevant attribute information.

The chapter by Doug McCallum reports on the first phase of an EC DRIVE initiative which brought together a consortium of digital data producers, manufacturers of navigational technology and GIS consultants to create and test a prototype digital geographic database for car navigation. Digital road centre-lines and street names were obtained from Ordnance Survey OSCAR data and this information was combined with related traffic and services data held by the Automobile Association. An integrated road network database was compiled for several test areas in England and the data was supplied in a standard exchange format to two firms wishing to validate their in-car navigation systems. The results of this collaborative project were not simply that the navigation products performed successfully, but more importantly that the project required draft specifications to be formulated for acquiring, structuring and exchanging digital road network data, which seem likely to be adopted as the European standard. McCallum's paper provides a clear account of the benefits and difficulties in co-ordinating a major collaborative project in GIS and his comments highlight the critical role of exchange standards in the overall success of the project.

The second chapter reflects the rapid uptake of GIS in the retailing sector, with the author finding considerable interest in customized solutions which have developed as spin-offs from original research in urban and regional modelling. Martin Clarke explores the links between proprietary GIS and what may be better termed "spatial decision support systems" as he describes collaborative work between his company and Toyota GB.

A main argument of the chapter is that many of the standard analytical functions within proprietary GIS packages are not the most appropriate for issues relating to strategic planning in sophisticated service-based environments. In particular, he stresses the need to incorporate predictive modelling capabilities within GIS, and the importance of customizing systems to exact client needs. This is perhaps a pointer to future developments in GIS. A number of proprietary software houses have begun to realize the importance and potential of more ''advanced'' spatial analytical functions to modern day business planning (be they statistical or mathematical). This field will surely mature and provide one of the most significant advances in GIS technology in the 1990s.

The third chapter represents a topical contribution from workers who were at an ESRC Regional Research Laboratory. This initiative provided a strong base of well-founded laboratories across the UK, each with particular expertise and providing an important blend of theoretical and applied research. In the paper, Gary Higgs, Dave Martin and Paul Longley look at the use of GIS for examining the changing tax burden in Inner Cardiff following the replacement of the property-based rating system with the new ''community charge''. This is an interesting area of urban policy making which has not been previously addressed in the mainstream GIS literature.

The authors construct a street-level database from the 1991 Electoral Register which is then held in a spreadsheet package. The street network data has been digitized and stored in a GIS. The chapter discusses the need to link data from spreadsheets with a proprietary GIS and illustrates procedures for this transfer. We hope this discussion will be of interest to other readers who may have similar desires to merge data sets created outside GIS using location as a linking ''key''. The resultant system is shown to highlight the geographical patterns of ''gainers'' and ''losers'' under scenarios reflecting different systems of local taxation.

The final chapter shows some results of ''blue skies'' research into cartographic generalization. This topic was one of the recent Special Topics designated for urgent research attention by ESRC/NERC. While it is intuitively obvious that fewer features can be displayed on a 1:50 000 scale map than a 1:1 250 map, the process of generalizing map features is one that traditionally involves considerable knowledge and skill on the part of the cartographer. Attempts to automatically generalize smaller scale maps from large scale originals, which theoretically would allow mapping on demand at any chosen scale, have proved problematic. In this paper, Zhilin Li and Stan Openshaw argue that the Douglas-Poiker algorithm, commonly used in GIS for thinning down the number of points constituting a line, is really an algorithm for geometric data reduction and hence is not really appropriate for cartographic data generalization. They propose an alternative method based on the ''natural principle'' that less detail can be seen at smaller scales. Their preliminary tests of the new algorithm on map data seem encouraging and deserve further consideration.

While the four chapters here provide an interesting mixture of topical contributions, resulting from different research initiatives in the UK, inevitably this part can represent only a partial glimpse of GIS research activity. The AGI will continue to support R&D in the UK and publications such as this Yearbook and the Annual Conference Proceedings particularly aim to allow researchers to disseminate and exchange findings. We are aware of many other important research projects currently underway or recently completed and would end our preface by urging potential authors to submit drafts to us at their earliest convenience for possible inclusion in the next edition.

65

Creating a digital road map for Europe: Experience from the PANDORA project

D.G. McCallum
PANDORA Project Manager

Abstract

The Automobile Association, Ordnance Survey, Philips BV and Robert Bosch GmbH have collaborated in a DRIVE project to create and test a prototype navigation database. The project, PANDORA—Prototyping A Navigation Database Of Road-network Attributes—was managed by consultants MVA Systematica and was completed in early 1991.

The data requirements of future vehicle navigation systems such as CARIN from Philips, TRAVELPILOT from Bosch and the AUTOGUIDE scheme for London were examined. Digital street networks were extracted mainly from Ordnance Survey's large-scale digital mapping and the necessary road and traffic attributes were collected by the AA. These data were integrated into a specially designed prototype database for parts of London and Birmingham and the major interconnecting roads. Data were abstracted from the database and supplied to Bosch and Philips using the draft GDF standard developed in the DEMETER project. This dataset was then tested in field-trials using prototype vehicle navigation systems. The dataset was also provided to the DRIVE Project "Task Force European Digital Road Map" as Benchmark Test task number 12.

This chapter describes the project, dealing with its objectives and relationship to other European initiatives, the work undertaken, the standards utilized and developed, its results and conclusions, and the lessons learned with respect to provision of data for larger areas of Europe.

Background to European developments

In-vehicle navigation products are now emerging in Europe, the US and Japan. A basic prerequisite of all such products is a computerized road map or, to be more precise, a digital road network structured into links and nodes, together with other traffic and service-related attributes such as street names, one-way streets, banned turns, weight and height restrictions, plus a host of other items. However, creating and maintaining a digital road network for large geographic areas is an expensive and resource-consuming activity. In the past, in Europe, there have been many isolated and limited attempts at creating digital road networks for a variety of purposes such as prejourney routeing and for logistics planning (Neffendorf and McCallum, 1987). To date, however, none of these road networks has been sufficiently detailed or extensive enough to be able to provide route guidance over a large geographical area. Recognizing that a cohesive approach towards the development of a European digital road network was required, Bosch and Philips collaborated to produce the first draft of a standard termed the Geographic Data Files (GDF; Divisional Standardisation Department, 1988) within a project called DEMETER (Digital Electronic Mapping of European TERritory; Claussen *et al.*, 1988).

During 1988, considerable progress towards the establishment of a pan-European road network test dataset for vehicle navigation was made by CERCO Working Group VII, as part of the EEC's EUREKA PROMETHEUS initiatives. A benchmark test was proposed comprising two travel corridors running (1) approximately north-easterly from Rennes via Paris and Brussels to Rotterdam; and (2) thence to Arnhem, thereafter moving in a broadly southerly direction via Bonn, Basel, St Gottard and Chiasso to Milano. This test was subdivided into 11 ''tasks'', each dealing with a different area and aspect of data capture and with attributes reflecting the varied administrative, geometric and local travel conditions found in Europe. The UK was designated as task number 12, and the test area was defined as the London-Birmingham travel corridor. Bosch were developing a navigation product called EVA, but subsequently obtained the rights to exploit the US ETAK system in Europe, launching this system commercially in 1989, under the name TRAVELPILOT (Supchowerskyj *et al.*, 1988), first in Germany. TRAVELPILOT is being released in other EC member states and it is probable that future versions will include a route guidance capability. Philips, on the other hand, have been developing a route guidance navigation and information product called CARIN.

In parallel with these developments by the providers of in-vehicle navigation systems, the UK Automobile Association (AA), already renowned for its provision of traffic and travel information services, was becoming increasingly interested in the supply of data for emergent navigation products. This interest included the AUTOGUIDE system (Belcher and Catling, 1987), the UK version of the German ALI-SCOUT (Von Tomkewitsch, 1987) and (Hoffman *et al.*, 1988) for which a London AUTOGUIDE demonstrator was being developed. ALI-SCOUT has now been renamed EUROSCOUT.

The British national mapping organization, Ordnance Survey (OS), had also identified that provision of digital mapping data for navigation products could generate significant sources of future revenue. The OS was generally more advanced than most of the counterpart national mapping agencies (termed ''IGNs'') in Europe as it was already creating a very large national digital database of its various large and small-scale maps. OS had also been instrumental in the development of the UK's National Transfer Format (NTF) for the transfer of digital map data (Haywood *et al.*, 1986). Hence, OS was interested in prototyping the processes which were required to abstract digital map centre-line and other data from its national database. Because it had already researched and overcome problems associated with data capture, the OS was not especially interested in conducting further research on digital map data capture, which was being put forward as a possible DRIVE project by CERCO Working Group VII, of the European IGNs.

To summarize, therefore, by the late 1980s there were two key data provider organizations, the AA and OS, interested in prototyping data supply processes and two key system providers, Bosch and Philips, requiring digital road network data for their emergent navigation products. MVA Systematica, a computer services consultancy, which had previously undertaken separate assignments for AA and OS, and was known to both Bosch and Philips mainly because of its DRIVE planning work, acted as the catalyst in helping to form the PANDORA Consortium.

In 1988, the EEC called for detailed proposals for research projects to be submitted for financial support through its DRIVE initiative (Commission of the European Communities, 1988). DRIVE, or "Dedicated Road Infrastructure for Vehicle safety in Europe" to give its full title, is a partnership research programme between the Community and European industry, each of which will contribute a total of ECU 60 million (US $1 = ECU 1 approximately) over the three-year period which began on 1 January 1989. A second three-year phase of DRIVE, termed DRIVE II, began in early 1992. The objectives of the DRIVE programme are: (1) to improve road safety; (2) to improve transport efficiency; and (3) to reduce the negative environmental impact of road transport.

Following a meeting in September 1988 at which an outline of the proposed PANDORA project was discussed, it was agreed that a formal tender document should be submitted for part funding under DRIVE. The PANDORA Consortium comprised four partners, namely, from the UK, Automobile Association Developments Ltd (Prime) and Ordnance Survey, from Germany, Robert Bosch GmbH, from Holland, Nederlandse Philips Bedrijven BV, plus MVA Systematica of the UK as the sole sub-contractor and project manager/data integrator. Following a meeting in late September 1988 with the project manager of another proposed DRIVE project, termed "Task Force European Digital Road Map" (TFEDRM), resource and financial provision was made to allow for a small element of PANDORA participation in common elements of the work. After competitive bidding for DRIVE funds, the EEC selected the PANDORA project and issued a contract in late December 1988, for formal commencement of work in January 1989.

Following an extensive selection procedure, some 70 DRIVE projects were started in 1989 in the various EC member states. Two of these DRIVE projects were of special importance to the creation of an EC-wide digital road network for vehicle navigation. TFEDRM, led by Daimler-Benz and included Bosch, Philips, Renault, Intergraph and TeleAtlas, was a three-year project which had three main objectives, namely: (1) to define the technical and organisational requirements of a European Road Database; (2) to conduct the continental European Benchmark Test of the Digital Road Map comprising tasks 1–11 (excluding the UK); (3) to develop and define common European Digital Road Map standards. The other project, termed PANDORA, is the focus of this chapter and has, *inter alia*, provided the UK element (task 12) of the benchmark test. Through a series of meetings, close liaison both between these two projects and with other relevant DRIVE projects was effected.

Specific PANDORA objectives

The PANDORA project was directed at three DRIVE tasks: T328 (Digitized Road Maps), T329 (Study of a Road Database Management Structure), and part of T304 (Integrated Autonomous/Infrastructure-Supported Route Guidance Systems). To meet the DRIVE requirements, the PANDORA project had six primary objectives:

(1) To develop a comprehensive and reusable data model for a digital road network database for vehicle navigation;

(2) To provide the UK element (Task 12) of the Benchmark Test put forward by CERCO Working Group VII;

(3) To develop, prototype and demonstrate the methodology, systems and software;

(4) To demonstrate the correctness of the integrated data in field-trials;

(5) To apply, develop and publish standards for digital road networks;

(6) To determine legal measures for protecting data providers.

The objectives all contribute to the development of standards in digital road mapping which match user requirements, which are cost-effective to work to and which have the support of the major industrial parties. A fundamental component of the project was the data model, in which the basis for the standards and the integration work was stated in clear and unambiguous terms. In addition, the project was designed to verify the integrity of this data model and data through a series of navigation system field-trials.

Technical approach

Overview

Figure 1 illustrates the main technical processes and provides an indication of the main roles of the five organizations participating in the project. These five organizations are referred to throughout this paper as the PANDORA "Participants". The two data providers, the AA and OS, were responsible for obtaining and supplying the basic data to the data integrator, MVA Systematica. OS abstracted the required large-scale centre-lines and street names data from its national digital map database into an OS format known as OSCAR, finally supplying the data in National Transfer Format on magnetic tape to MVA Systematica. The AA obtained the required traffic and service-related attribute data either from its existing databases and records, or mounted special field surveys. Data were then referenced, edited and supplied to MVA Systematica on floppy disk. Both the AA and OS co-operated closely in all of the data supply activities involving the detailed numbering and referencing of network data and in the identification and resolution of any apparent omissions, anomalies or errors. The AA identified some private roads, not included by OS. The AA and OS were additionally responsible for the preparation of a deliverable defining proposed legal measures for the protection of data providers. MVA Systematica carried out the joint roles of data integrator and project manager. Data from AA and OS were first verified and, if errors were detected, were returned for correction and re-supply. Data which successfully passed the verification checks were added to a specially-constructed INGRES database running under UNIX and then integrated to form a cohesive AA/OS dataset.

Restructuring of the combined AA and OS data into the database specification called GDF-SDA (Geographic Data Files Specification Data Acquisition) then took place, prior to final supply to Bosch, Philips and the TFEDRM project, in GDF-EF (Exchange Format). Bosch and Philips used their own in-house systems for reading and verifying the GDF-EF data, reporting any apparent errors or inconsistencies to

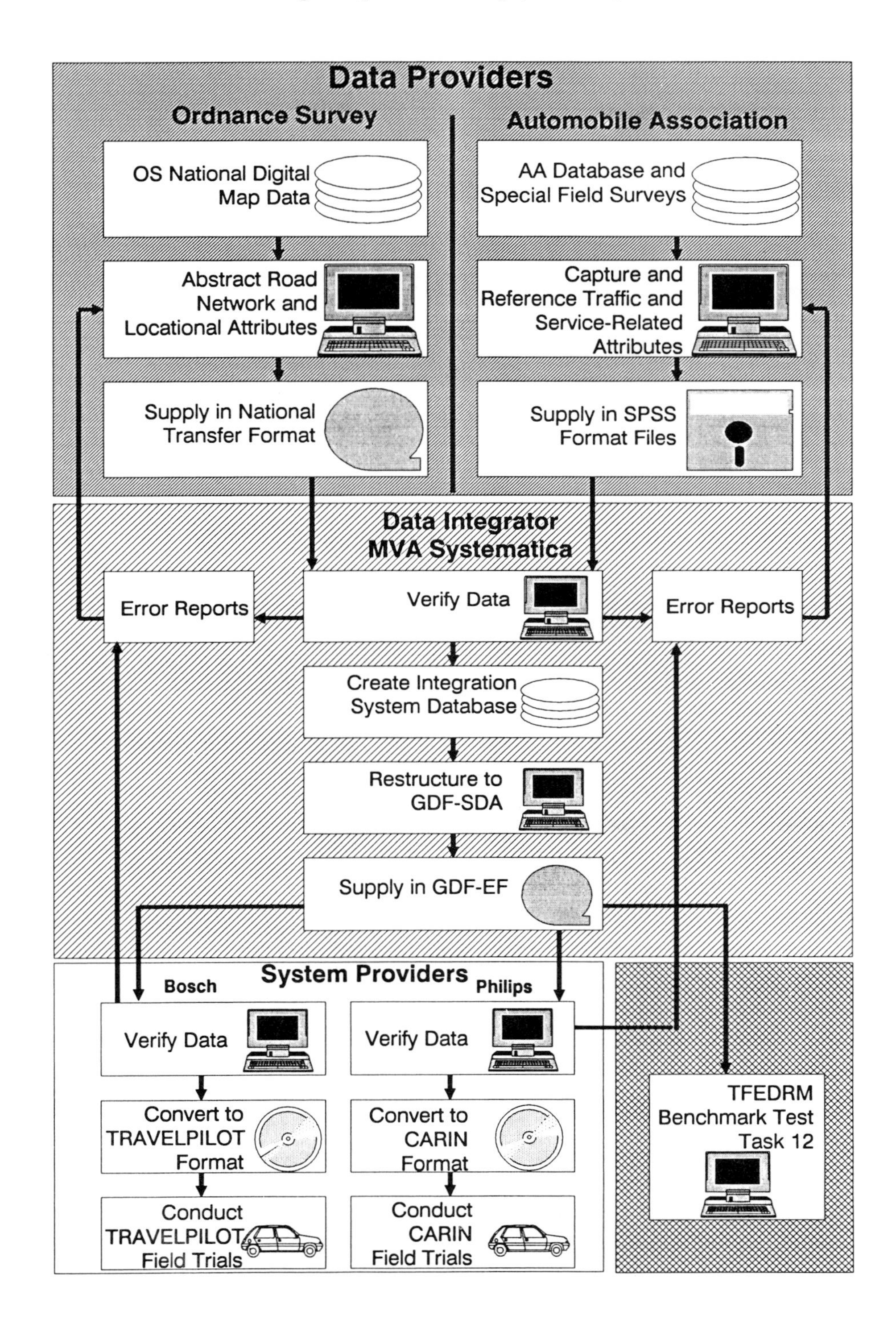

Figure 65.1. Flowchart of the Main Technical Processes in PANDORA showing Roles of Participants.

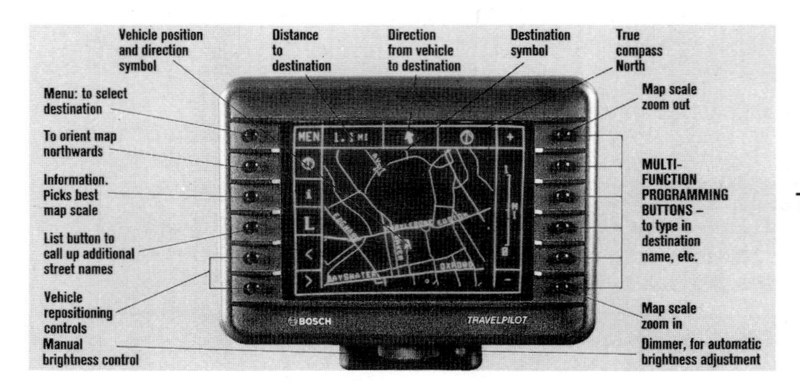

TRAVEL PILOT
(Bosch)

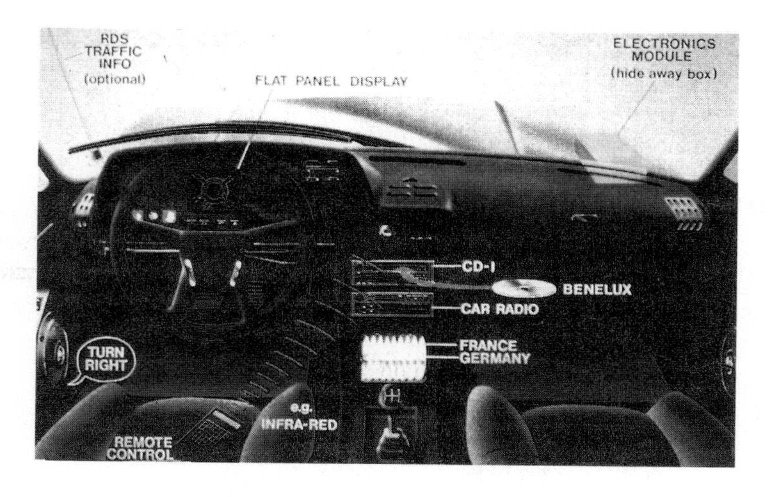

CARIN
(Philips)

AUTOGUIDE/
ALI SCOUT

Figure 62.2. In-vehicle Nativation Systems in Europe: TRAVELPILOT, CARIN And AUTOGUIDE/ALI SCOUT.

the data integrator. When all data had been successfully verified, they were then converted to the appropriate compact disc (CD) format for use in TRAVELPILOT and CARIN. Finally Bosch and Philips then mounted navigation field-trials of PANDORA data using specially-equipped vehicles. In addition, both Bosch and Philips conducted separate desk-based route selection tests using their route-finding algorithms. Figure 2 (a and b) illustrates the various elements of the TRAVELPILOT and CARIN systems, each of which uses CD for mass storage of the digital road network data, plus an electronic compass, together with dead reckoning and map matching (not shown in the figure). CARIN works with CD-I (Interactive), whereas TRAVELPILOT uses CD-ROM (Read Only Memory). The data requirements for the AUTOGUIDE/ALI-SCOUT system, shown in Figure 2(c), were also studied in PANDORA, although field-trials were not undertaken. In addition, Bosch and Philips, joint authors of the GDF, were responsible for the preparation of a deliverable concerning change requests for revision of the first draft of the GDF standard.

PANDORA trial area: OS data

Figure 3 is an OS plot showing the extent of the PANDORA trial area which lies between Birmingham and London, a distance of 170 km, ranging from heavily built-up urban and suburban areas through to rural countryside. The trial area is 125 km by 120 km, making 15 000 km^2 in total. There are six 5-km × 5-km square trial sub-areas (TSAs) derived from the OS's large-scale mapping (1:10 000, 1:2 500 and 1:1 250) and a further eight small-scale areas (SSAs) derived mainly from OS's 1:50 000 scale mapping. The AA provided full traffic-related attribute data for the three main interurban routes shown in Figure 3 and for every road and street within TSAs 1 through 4. The density of the local street network is indicated by the black lines plotted in Figure 3. Figure 4 is an enlarged OS colour plot showing the street centre-lines for TSAs 3 and 4 in the vicinity of Heathrow Airport. The area covered comprises two adjacent 5-km × 5-km "tiles", thus making an area of 50 km^2. Figure 4 gives the four road classifications adopted in PANDORA, namely Motorway, A-class, B-class and minor unclassified roads. Because they are plotted directly from the OS data, prior to supply and integration, none of the AA traffic-related attribute data is able to be shown at this stage in the supply process.

AA data

Figure 5 provides three photographs which illustrate how the AA undertook various elements of field survey data collection. The photograph depicting an elevated road directly on top of another illustrates the necessity for recording grade separation. The OS's OSCAR format files are two-dimensional, and do not therefore record the grade separation which is especially relevant at bridges, roundabouts and flyovers. In the prototyping exercise conducted in PANDORA it was not considered to be cost-effective to use video recording and transcription techniques. Consequently a combination of on-foot recording using paper forms and in-car recording by means of the very accurate distance measuring device shown in the photograph was used. These field survey records were combined with other service-related records abstracted from existing AA database and other records. The national grid references of individual attribute items such as AA classified hotels, restaurants, car hire companies, garages, traffic signs and many other tourist and information facility inventory items were then digitized. All AA attribute records were coded to accord with the OS's

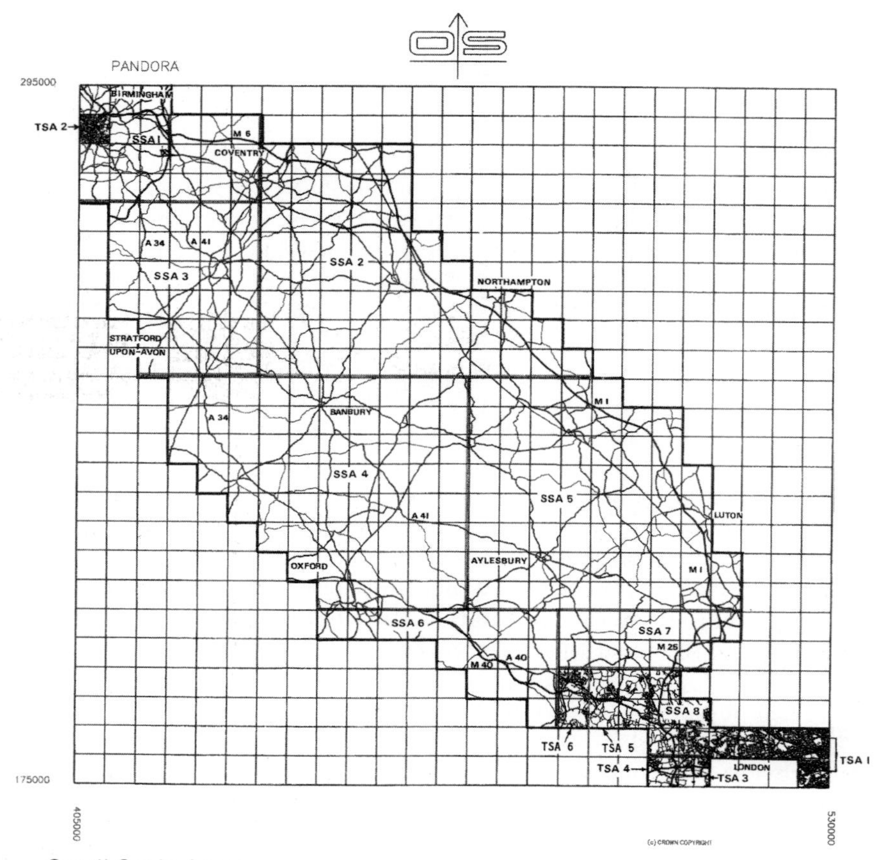

—— Small Scale Areas.

TSA 1 Trial Sub Area 1: Note TSAs 1-4 were fully attributed.

530000 National Grid Reference: These figures represent the number of metres north and east from the origin of the National Grid on the Adjusted Transverse Mercator Projection.

Figure 65.3. Plot showing the extent of the PANDORA Trial Area, associated Trial Sub-Areas (TSAs) and Small Scale Areas (SSAs).

(a) Grade Separation

(b) Recording on Foot

(c) Recording from a Car

Figure 65.5. Collecting AA traffic related attribute data.

OSCAR link (line) and node numbering system within each 5-km × 5-km tile. Following validation, all AA traffic and service-related data were then supplied to MVA Systematica, for further validation and integration in the INGRES database.

Integrated data: Pre-supply

Figure 6 provides a plot of the integrated AA OS data, prior to supply by MVA Systematica in GDF-EF. The plot was produced using the "Genamap Geographic Information System". Integrated road network data for TSA number 3, East Heathrow, may be compared with the original pre-integration OS centre-line data in Figure 4. The AA data illustrated in Figure 6 are only a subset of the entire dataset provided. Unfortunately, it was not possible to plot all of the AA data for a TSA in one diagram as there is so much of it that the plots would become too cluttered for clarity. Figures 7, 8 and 9 are colour photographs of Genamap screens illustrating how the integrated AA and OS PANDORA data may be viewed and utilized. In Figure 7, a "window" placed on central Birmingham enables the location of one-way streets, banned turns and compulsory movements to be viewed at a larger scale. In Figure 8, detail in the vicinity of "Spaghetti Junction" is revealed. Figure 9 shows how AA data on car parks can be used in the vicinity of a route.

Integrated data: Post-supply

Figure 10 is a further plot of integrated data, produced using Philips' internal equipment, following the successful reading and processing of the GDF-EF records supplied by MVA Systematica on magnetic tape. The plot shows a section of the Heathrow TSA number 3 post-supply data. As a result of post-processing requirements for the CARIN prototyping, the OS network geometry has been slightly simplified. The plot highlights one-way links explicitly identified by the AA, which are not evident in the pre-supply data plots given in Figures 4 and 6.

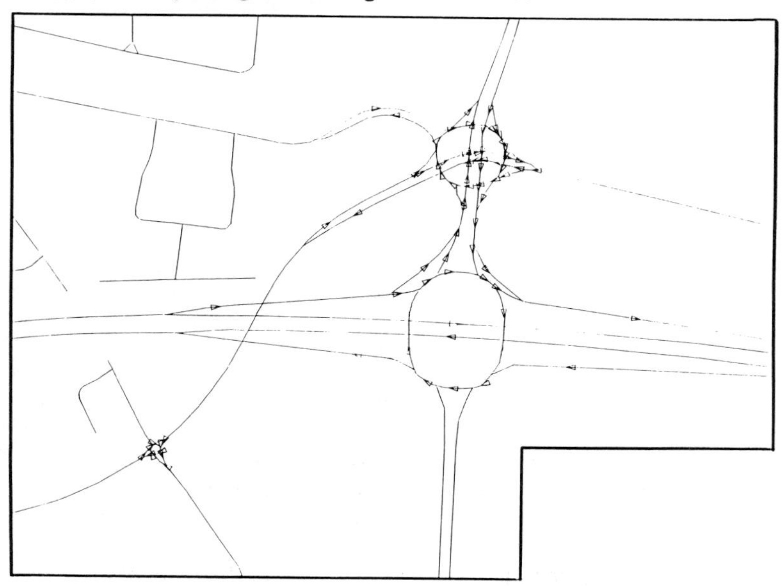

Figure 65.10. Plot of integrated data post-supply for part of Trial Sub-Area 3, the northern section of East Heathrow.

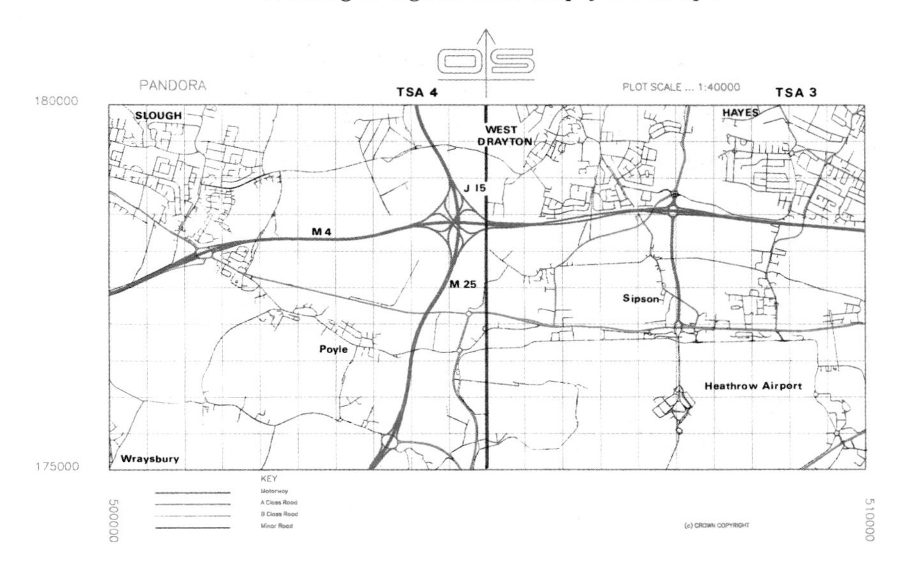

Figure 65.4. Plot showing the OS digital road network, centre-lines for Trial Sub-Areas 3 and 4: Heathrow Airport vicinity.

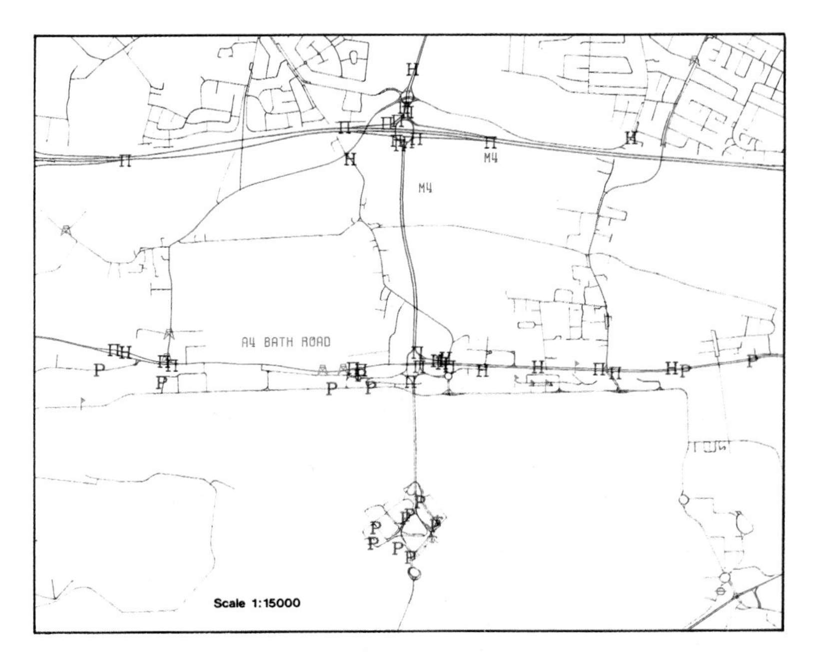

Key: H Hotel P Parking Π Direction sign ⊖ Station
Airport terminal Industrial site Car rental
—— Motorway —— 'A' Class road —— 'B' Class road
—— Unclassified road

Note: Other inventory items (e.g. tourist information, hospitals are not plotted as too numerous)

Figure 65.6. Plots of integrated data, pre-supply, for Trial Sub-Area 3, east Heathrow.

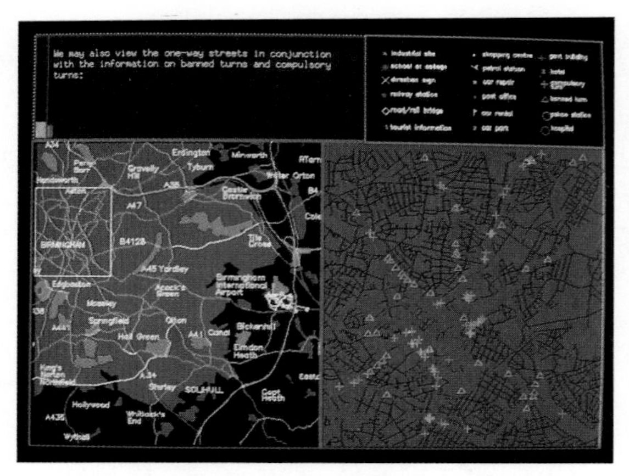

Figure 65.7: "Genamap" screen: a "window" placed on central Birmingham enables the location of one-way streets, banned turns and compulsory movements to be viewed at a larger scale.

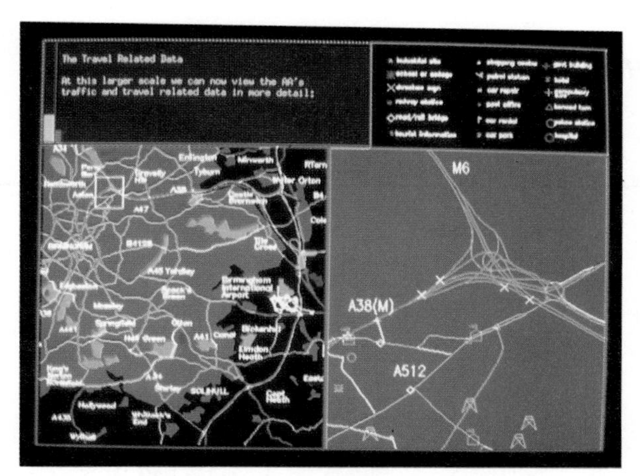

Figure 65.8: "Genamap" screen revealing detail in the vicinity of "Spaghetti Junction".

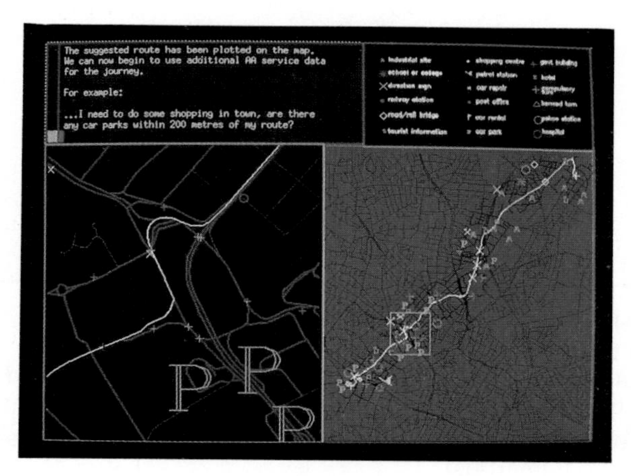

Figure 65.9: "Genamap" screen showing how AA data on car parks can be used in the vicinity of the route.

Field-trials: CARIN and TRAVELPILOT

The Automobile Association and Ordnance Survey collaborated in the selection of routes to be tested by the Philips and Bosch systems in field-trials. In addition to the three main inter-urban routes between London and Birmingham, 29 test routes were devised within the TSAs. CARIN field-trials were carried out in TSA2, central Birmingham during August 1990. Figure 11 shows the Philips test vehicle during the actual trials. Differences between the CARIN routes encountered in the field and the AA's recommended routes were generally due to data currency (i.e. road network change) and the omission of certain data types from the final CARIN Data File (CDF) data describing turns. A small number of deviations could be attributed to other data inconsistencies. Of the 16 test routes, 70 per cent of the routes selected by the CARIN Route planner were identical to the AA ones. In 12 per cent of the cases CARIN came up with an alternative, but good, route. In the remaining 18 per cent of cases the differences between the CARIN routes and the AA ones were attributable to the missing prohibited turn information, referred to above. Overall therefore, the field-trials were deemed to be successful. The TRAVELPILOT routeing simulations were undertaken in Germany using data for Central Birmingham (TSA2) and London-Heathrow (TSA 3, 4). The results were broadly similar to those found in the CARIN tests. Of the 20 test routes, the TRAVELPILOT Route-Planner gave the same (or

Figure 65.11. The Philips test vehicle during field-trials of CARIN in central Birmingham.

very nearly the same) route as the AA in 70 per cent of the cases; alternative, but good, routes were given in 20 per cent; and the remaining 10 per cent were quite different, for various reasons. In April 1991 TRAVELPILOT field-trials were undertaken in England to test the suitability of PANDORA data for map matching positioning purposes and the representation of the network on the TRAVELPILOT display. The results confirmed the high precision of data derived from large-scale mapping (1:10 000 up to 1:1 250) and highlighted problems associated with generalization in smaller-scale mapping (1:50 000). Bosch adjudged these with laboratory simulations and field-trials to be successful overall.

The GDF Draft Standard

Bosch and Philips are the joint authors of the draft GDF Standard that was issued as version 1.0. Prior to the PANDORA Project, only relatively small GDF test datasets had been created. The first edition of GDF represents a valuable statement of requirement, but its authors understand that it will need to be modified in the light of the experience gained in trying to meet it. It is a specific objective, both of the PANDORA and TFEDRM projects, that the existing draft GDF Standard is further developed and improved. The work undertaken in PANDORA in applying the standard has therefore been invaluable in this respect and recommendations for improvements to the GDF have been put forward. GDF Release 2.0 has recently been made available from Philips.

Legal protection of data providers

The AA and OS collaborated in the production of a deliverable (document reference D6273. 13) entitled ''Proposals for Legal Protection for Data Providers'' which was issued at the end of August 1990. The deliverable provides outline legal guidelines written in a style which is readily comprehensible to non-legally trained professionals working as data providers.

Supply of benchmark test data

A specific objective of the PANDORA project was ''to provide the UK element (Task 12) of the Benchmark Test put forward by CERCO Working Group VII''. This objective was achieved when a copy of the PANDORA data in GDF-EF format, together with supporting documentation, was supplied to the Task Force European Digital Road Map project in August 1990.

Project management and liaison activities

Management and QA

The management structure comprised a steering committee made up of representatives of each of the four partners, with a project manager and secretariat provided by MVA Systematica. Liaison and reporting to DRIVE Central Office (DRCO) was effected through the prime contractor via the project manager, who also orchestrated liaison and concertation with other DRIVE projects. Each of the 15 separate PANDORA Work Packages became the responsibility of a work package leader within a specific organization. Work package leaders were represented within a technical committee which reported to the steering committee via the project manager. Regular monthly and quarterly management reports were prepared and submitted to DRCO. As each work package was completed, the draft report was prepared and circulated to all participants for comment. Once comments had been incorporated into the text, the final report on the work package was sent to DRCO with copies to other DRIVE projects as appropriate. As an element of the overall project management and quality plan, each work package leader was required to circulate for comment a first draft of QA documents dealing with (1) the specification of the work package and (2) any controlled items, including deliverables. This process enabled any potential misunderstandings and technical problems to be resolved well in advance of the commencement of the major thrust of work on a work package or deliverable.

All DRIVE projects are subjected to an annual technical audit by an audit panel appointed by DRCO. The PANDORA project was audited in October 1989 and the result was to continue with the work as planned. A further and final audit recorded the success of the project in October 1990.

Liaison and concertation

PANDORA was only one of some 70 DRIVE projects running in parallel, and there was obviously a need for some form of inter-project liaison to be effected. In order for regular, periodic liaison between projects to take place, DRCO orchestrated large two-day gatherings, termed concertation meetings, comprising representatives from all DRIVE projects. PANDORA representatives attended every concertation meeting and actively participated in several workshop and group activities, as well as being rapporteurs. Specific additional liaison with the Task Force European Digital Road Map (TFEDRM-V1021) and the Road Information and Management Euro-System (RIMES-V1034) was effected through special meetings and by the exchange of Deliverables.

Conclusions

It has been shown in this chapter that the PANDORA Project was very successful. The following conclusions may be drawn:

(1) With one minor exception relating to AUTOGUIDE field-trials, all of the project's objectives, goals and deliverables, as originally defined in 1988, were successfully achieved.

(2) The five PANDORA participants worked well, both individually and collectively, thus engendering a spirit of genuine inter-organizational co-operation.

(3) Participation in the DRIVE programme is viewed positively by the participants.

(4) PANDORA was the first project to have produced a large (32 MBytes) dataset conforming to the draft GDF Standard 1.0 for digital road network data.

(5) Production of integrated large-scale digital road network data is a complex and time-consuming process and it is easy to underestimate the amount of work involved. Delays in production of data caused the main project timescale to slip by three months.

(6) The PANDORA data were intensively validated and shown to be of good quality and accuracy.

(7) Laboratory testing and field-trials using PANDORA data in the CARIN and TRAVELPILOT navigation systems demonstrated the successful use both of these navigation systems and of the data.

(8) Infrastructure-based systems which need transient and dynamic data as well as static data place additional, but quite separate, requirements for data updating

via mobile telecommunications. The provision of integrated road network data in GDF format will satisfy the main static data requirements of infrastructure-based systems. Further work is needed, in conjunction with the system providers, to determine how the requirements of infrastructure-based systems for mobile telecommunications data can be met.

(9) Issues associated with the long-term maintenance of digital road network data need to be the subject of further research and prototyping projects.

(10) The technical processes, framework and data model for the supply, integration and testing of digital road network data which have been developed and prototyped in the PANDORA project appear to form a sound basis for the creation of a much larger pan-European digital road network database.

(11) Work on the creation of a large-scale pan-European digital road network database needs to be progressed urgently as many existing and planned RTI functions and services (including vehicle navigation) need this database as a basis for locational referencing of data and for operation, monitoring and display purposes.

(12) However, creating and maintaining digital road network data is expensive and resource-consuming and suitable sources of funding for such a pan-European venture need to be established with the objective of having such a system, ideally by 1994.

(13) The project management and quality assurance procedures developed for the PANDORA project worked well and form a suitable template for adoption on all future DRIVE projects.

(14) Standards for digital road networks need to be extended to define how networks are numbered and referenced.

The current situation

As a direct result of the PANDORA and TFEDRM DRIVE projects, plus input from other European research projects (e.g. the EUREKA project CARMINAT and the European Transfer Format Working Group) and with assistance from MVA Systematica the draft GDF standard has been updated and issued as release 2.0, available from Philips. It is envisaged that the draft GDF standard (Wood, 1992) will ultimately be adopted by the CEN European standardization body via the activities of CEN technical committee 278.

Two rival commercial consortia for the creation of a digital road network for Europe have now emerged. These comprise (1) a Philips-led consortium called European Geographic Technologies and (2) an as yet unnamed consortium involving ETAK (US) and led by Bosch. Whether revenues from the several varied applications of digital road mapping (including vehicle navigation) are sufficient to support the costs of these two competing consortia remains an open question.

Acknowledgements

The author wishes to thank the other four organizations participating in this project and the DRIVE Central Office for their permission to publish this chapter. The participation of Messrs. A.B. Smith and S. Hartley (Ordnance Survey); S. Hoffman, R. Robbins and F.G. Lay (Automobile Association); W. Zechnau and H. Claussen (Robert Bosch GmbH); L. Heres (PhilipsBV); Gnd C. Quere'e and T. Wood (MVA Systematica) in the preparation of this paper and within the project is gratefully acknowledged.

Glossary

ALI-SCOUT/EURO-SCOUT: An infrastructure-based route guidance system using infrared for the vehicle/beacon communication link, being trialled in West Berlin by Siemens, Blaupunkt and the German Government.

AUTOGUIDE: The UK version of EURO-SCOUT (formerly ALI-SCOUT) to be piloted in London.

CEC: Commission of the European Communities.

CERCO: Comité Européen de ces la Responsables de la Cartographic Officielle. A board of all official cartographic organizations in Western Europe. A Council of Europe Study Group.

DEMETER: Digital Electronic Mapping of European TERritory. A EUREKA-supported project undertaken by Philips and Bosch, concerning the specification of data standards for digital road networks for navigation systems.

ECU: European Currency Unit. Used by European Community (EC) member states and roughly equivalent to one US dollar. EUREKA co-ordinated research and development programme within the EC to encourage collaboration between industries. One aspect is the solution of future transport problems using advanced control systems and strategies. Directed by European Technology Ministry.

EVA: An autonomous route guidance and information system utilizing CD-ROM developed by Robert Bosch GmbH, Mobile Communications Division, up to prototype stage.

PROMETHEUS: Programme for European Traffic with Highest Efficiency and Unprecedented Safety. A EUREKA project, running from 1986 to 1994, and being undertaken by a consortium of major European car manufacturers. Aims are to enhance future driving through information technology and micro-electronics, including vehicle navigation.

References

Belcher, P., and Catling, I., 1987, Electronic route guidance by AUTOGUIDE: The London demonstration, *Traffic Engineering and Control*, **28**(11), November.

Claussen, H., Heres, L., Lichtner, W. and Schlogl, D., 1988, Digital Electronic Mapping of European TERritory (DEMETER). International Symposium on Automotive Technology and Automation Conference, Florence.

Commission of the European Communities, 1988, Call for Tender for the DRIVE Programme, DRCALLI, Directorate General XIII, Brussels, Belgium, 20 May.

Divisional Standardization Department, 1988, Draft GDF Standard Release 1.0. Nederlandse Philips Bedrijven BV, Divisional Standardisation Department, Consumer Electronics, 5600 MD Eindhoven, The Netherlands.

Haywood, P. *et al.*, 1986, The National Transfer Format, Working Party to produce national standards for the transfer of digital map data, final draft (issue 1.3), Southampton: Ordnance Survey.

Hoffman, G., Sparmann, R., Von Tomkewitsch, R. and Zechnall, W., 1988, LISB research project: Guidance and information system Berlin. Paper presented at Street and Traffic 2000, Berlin.

Neffendorf, H., and McCallum, D., 1987, The geographic base for transport informatics. *Proceedings of the Information Technology in Transport and Tourism Seminar*, P302, pp 67–79, PTRC Transport and Planning 15th Summer Annual Meeting, University of Bath, September.

Supchowerskyj, W.E., Neukirchner, E. and Zechnall, W., 1988, Vehicle navigation systems status and developments in Europe: An overview, International Symposium on Automotive Technology and Automation, Florence.

Thoone, M.L.G., Driessen, L.M.H.E., Hermus, C.A.C.M. and Van Der Valk, K., 1987, The car information and navigation system CARIN and the use of compact disc interactive, SAE International Conress and Exposition, Detroit, Michigan, Technical Paper Series 870139.

Von Tomkewitsch, R., 1987, LISB Large-scale test Navigation and Information System Berlin, *Proceedings of the Information Technology in Transport and Tourism Seminar*, P302, pp. 145–155, PTRC Transport and Planning 15th Summer Annual Meeting, University of Bath, September.

Wood, T.F., 1990, The design and operation of a navigation database of road network attributes the PANDORA experience, PTRC 18th Summer Annual Meeting on European Transport and Planning, University of Sussex, September.

Wood, T.F., 1992, GDF: A Single European currency for Geographic Information, *GIS EUROPE Magazine*, forthcoming.

66

Developing spatial decision support systems for retail planning and analysis: Toyota GB—A case study

Dr Martin Clarke
GMAP Ltd

Abstract

This chapter describes how sophisticated analytical planning tools can be integrated with Geographical Information Systems technology to provide powerful predictive capabilities for organizations in retailing and related activities. Achieving this capability goes considerably beyond what is currently provided by most proprietary Geographical Information Systems and can lead to a more intelligent and structured approach to problem solving and strategic planning. Some of the prerequisites for developing a successful predictive modelling capability are outlined and, in particular, we emphasize the importance of understanding the real requirements of the end user. By establishing these needs in an iterative dialogue with the client, a set of priorities can be developed which drive the application, rather than being constrained by the technological constraints of a particular piece of packaged software. To illustrate the principal arguments of this chapter I describe the development of a decision support system for Toyota GB, describe the range of applications that it is put to and evaluate its usefulness.

The author is Managing Director of GMAP Ltd, a company owned by the University of Leeds, which specializes in the development of spatial decision support systems for a range of blue-chip clients in the UK, Europe and North America. He is also a senior lecturer at the University where he has undertaken innovative research on a range of spatial modelling methods that are at the heart of the systems that are developed for GMAP's clients.

Introduction

In the 1980s businesses of many different types began to recognize the importance of the geographical dimensions of their activities. Markets were not spatially homogeneous, retail centres contained different levels of competition in the relation to their catchment populations, the interaction patterns of customers were complicated and varied from region to region. At the same time significant changes were taking place within the business environment. Retail groups merged to form substantial organizations with ambitious growth plans, large edge-of-town and, latterly, out-of-town retail developments appeared on a regular basis, marketing and advertising grew, both in terms of expenditure and sophistication. To assist in the planning of this activity a number of tools were developed by market analysis organizations. The first wave of these tools consisted of geodemographic and related census data products and services (such as ACRON, PiN, MOSAIC and SuperProfiles). More recently Geographic Information Systems software packages (such as ARC/INFO and SPANS) have been developed to provide a more powerful approach to data analysis. In 1992, the Geo-Data and Information industry is a significant business sector in its own right. However, there are a number of general limitations to conventional approaches that fall under this heading, including:

(1) Most of what is offered are products that offer generic approaches that may or may not fit into the client's needs;

(2) Most GIS software has very restricted analytical capability (despite the claims made in some of the marketing material);

(3) Often the software is difficult to use and requires extensive training;

(4) The analysis that is possible is of limited interest and certainly rarely addresses more strategic/corporate issues;

(5) Software systems and data products are sold to anyone who cares to purchase them—inevitably this dilutes any competitive advantage that might be obtained.

To summarize, while there has been benefit gained from employing these products and systems, there is considerable scope for going much further beyond what is normally thought possible. This involves the development of methods of analysis and supporting information technology that can provide a truly predictive "what if?" capability. The rest of this chapter describes the principles behind achieving this capability and demonstrate an example of its application in the context of the decision support system that GMAP has developed for Toyota GB.

From descriptive to predictive analysis

Introduction

In most organizations planning involves a number of distinct stages that should take the form of a planning cycle. The first stage may involve a detailed descriptive analysis of the organization's products, its structure and competition, and its performance in the market at a number of scales *vis à vis* its competitors. Following the setting of a number of objectives (e.g. for growth in market share, profitability and so on) a number of alternative approaches will have to be identified for achieving them. The next step will be to evaluate the alternatives using a number of criteria (financial return, risk, etc.) and a set of recommendations made. Once the plan is implemented, a monitoring process should be established to ensure that the objectives are indeed being met and, if not, remedial action instigated to get the ship back on course. While recognizing that in practice the planning process can be more complex than this outline, the broad set of stages are probably common to most companies in the retail and related sectors.

The problems faced in undertaking this process are typically ones of information availability and access to tools that will identify and evaluate options. Information is needed for the descriptive analysis of existing circumstances, methods are needed for prescriptive and predictive analysis of the alternative courses of action. The argument presented in this paper is that most of what is available in the current market analysis industry is limited in use to descriptive analysis. However, tools developed within an academic environment are now being made available which allow organizations not only to obtain a better descriptive analysis of the existing situation but, crucially, also to be able to examine in detail the impacts of their plans, in terms of sales, market share, return on capital and other criteria. In other words, to both improve both business planning and performance.

Model-based methods for planning

Throughout the 1970s and 1980s significant developments took place in geographical research that produced new and improved methods for spatial analysis. For a variety of reasons there was very limited transfer of these methods into the commercial world. This limitation did not imply that there were no requirements for the method but, rather, ignorance on behalf of both academics responsible for the methodological developments and potential commercial users. Indeed, in many cases commercial organizations were using (and in some cases continue to use) very simple methods completely inappropriate for the complex systems they were trying to study (drawing circles around retail centres and counting the numbers of households in the circle is the favourite example!).

In recent years there has been more success in transferring modelling technology from the academic to the business environment, which in turn has helped focus the research agenda. Companies such as GMAP have been able to gain experience in technology transfer without having to compromise the academic credibility that underpins the methods they have developed. Alongside this increased confidence in model application we have also witnessed the wider availability of spatially referenced data (through computerized customer and transaction data; this is likely to increase with EPOS etc.) and the power of desktop computing and associated mapping and interface software.

Both of these developments make for better model performance and increased ease of use. Perhaps most importantly, however, there has been the pressure of a highly competitive retail environment which has generated a genuine requirement for competitive advantage. In the future, if not already, the development of a modelling capability will be a necessary, though not in itself a sufficient, criterion for competitive advantage. Readers disagreeing with this statement are probably unaware of the investments being made in geographical modelling by some of the UK's most successful retailers.

Given this background, what are the types of use to which modelling methods can be put? The following categories of use can be identified:

(1) *Providing a framework for data transformation*
Models provide a systematic accounting framework that establishes a consistent approach to the analysis of spatial data (see Figure 1). Consistency in this area

is not always achieved in data products—for example through the double counting of overlapping catchment area populations. More generally and importantly having a framework for data transformation can add value to data through the calculation of performance indicators, such as market share in a centre and market penetration in a residential district (e.g. postal sector). These data can then be used as a baseline for comparison with predictive model output to assess the impacts of proposed changes.

(2) *Synthesis and integration of data*
A problem with spatial data is that it usually comes at a wide variety of spatial scales and at different levels of aggregation. Models can be used to link and merge different data and to estimate the values of missing variables. This can both enhance the value of one-off market research data as well as give insights into markets and behaviour that are not possible from the basic data. One example that has been the subject of research at Leeds is the generation of individual and household income distributions at the enumeration district level. This data has obvious applications in target marketing and retail analysis where income may be a more powerful discriminator than conventional geodemographic clusters.

(3) *Updating information*
Census and other related data quickly ages from the time it was collected. In some cases, particularly where the focus is on spatial units smaller than local authorities, this "data ageing" will pose real problems to the planning analyst and there will be a need to employ demographic models to update the characteristics of the population.

(4) *Forecasting*
Updating takes us from the past to the present. There may also be a need to produce forecasts of how the characteristics of the population and other variables are likely to change at the small area level in the future. Forecasting will be of particular importance in planning services such as health care and education where demand for services is very much a function of the population in specific age cohorts. Again, models are needed for this purpose.

(5) *Impact analysis*
A common question facing many retailers is, what will happen if we take a particular course of action—such as open, close or refurbish an outlet? Simple but usually inappropriate methods for producing answers to this type of question have been used for some time. Only more recently have more powerful methods been applied that allow the full complexity of retail and related systems to be represented within a consistent accounting framework. The development of these spatial modelling tools also extends the range of questions that can be asked—such as what will be the impact of a competitor action or where should we target the marketing budget for a new store? Importantly, it is also possible to integrate financial modelling tools with spatial models to determine the return on the investment being considered and its associated risk.

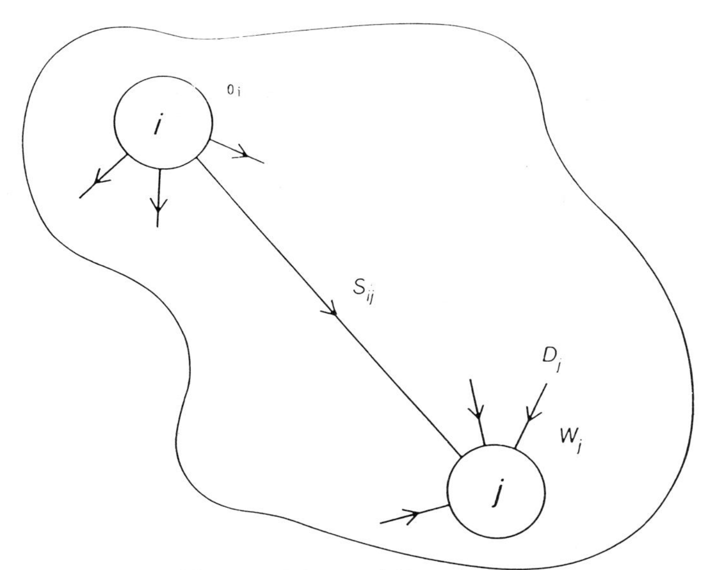

Figure 66.1. Modelling framework

Where

S_{ij} is the 'flow' of car purchases between origin zone i and dealer j

O_i is the demand in zone i

W_j is the attractiveness of dealer j

The basic accounting frameworks are:

$\underset{j}{S_{ij}} = O_i$

$\underset{i}{\epsilon S_{ij}} = D_i$ (total number of cars sold by dealer j)

$\underset{i}{\epsilon O_i} = \underset{j}{\epsilon D_i}$

(6) *Optimization*
Questions that are frequently asked include, "where is the best site for a new outlet?" or "what should the national network of outlets consist of?" To attempt to answer these types of questions optimisation methods can be employed. These methods typically involve identifying an objective function, such as maximize sales and an associated set of constraints (capital expenditure of x, no more than y outlines). Such methods can be computationally very demanding and often heuristic methods have to be employed.

The six uses of model-based methods identified above provide an illustrative but not exhaustive overview of what is possible. A recurring and justified question that potential users raise is, do they work? The answer is, provided that models are currently specified to reflect the complexity of the system being studied and take on board the substantial experience that now exists in model application, very impressive levels of performance can be achieved. The benchmark of model performance should not be 100 per cent perfection (nice this though this would be!) but relative to the current qualitative or quantitative methods that are used. Additionally, experience at GMAP suggested that mathematical models will almost always perform better than statistical models when the impacts of alternative decisions are being assessed.

Integrating models in decision support systems

Given the wide variation in needs and objectives that will exist between different organizations and the variety of different modelling tools that exist, it is unlikely that a packaging of modelling software in a proprietary system is a viable way forward. Our experience suggests that developing and customizing software to meet the requirements of clients both in terms of information systems and modelling is the preferred strategy. GMAP's approach is not to employ proprietary GIS software but to centre applications around a user interface from which information can be accessed, displayed and mapped and the modelling capability invoked. We have developed our own powerful graphics, mapping and relational database software which makes us independent of any software vendor or specific hardware platform. Each of the systems that have been produced for our clients has qualitatively different features and "look and feel".

A decision support system for Toyota GB

Introduction

Toyota GB is a very successful distributor of Toyota cars and light commercial vehicles in the UK. In the UK, as in most European markets, vehicles are sold to the public through a network of franchised dealers who are usually restricted to selling the products of a single manufacturer (though this is currently subject to some uncertainty following the publication of the Monopolies and Mergers Commission report in early 1992). There are over 8000 franchised dealers in the UK which sell in the order of two million vehicles annually. With a few exceptions, franchises are operated independently from manufacturers and distributors. In addition to new car sales dealers also generate revenue from parts and service business, financial service products and from second-hand car sales. Margins are typically very tight and large amounts of expenditure are directed at marketing and promotional activity both on a national and local canvas. Some £500 is spent under this banner for every new car sold in Britain.

Nationally, Toyota GB currently has 220 dealers selling over 40 000 cars which represents about 2 per cent of the market share. All of the Japanese importers are quota-restricted by agreement to a total of 11 per cent of the UK car market. As such there has been no great impetus to extend the dealer network, rather there has been a shift in model mix towards higher value vehicles, notably the Celica, MR2 and Supra. However, it has been clear for some time that Toyota Motor Corporation (TMC) has planned to develop a European production facility. In 1989 TMC announced that a factory would be constructed at Burniston, near Derby. European sourced vehicles are not part of the quota restriction in the UK or most other European countries. As such there now exists an opportunity to increase UK sales significantly beyond the current level. To achieve TGB and TMC objectives demands that detailed planning and analysis is undertaken. This was the main motivation for the development of the decision support system by GMAP. We now examine the set of objectives that it was designed to meet, its main characteristics and functionality and an evaluation of its use within TGB.

Objectives for TGB

GMAP commenced discussions with Toyota GB in 1987. As we have worked closely together since, different objectives have emerged in response to the developments in the car industry generally and TMC policy and product evolution specifically. However, the following objectives can be distilled from discussions over the last three years:

(1) The development of a system that would allow a detailed appraisal of Toyota and competitor performance in the UK at a variety of different geographical scales;

(2) To develop a method of systematically appraising the performance of individual Toyota dealers with respect to new car sales (by model segment) and parts and service business and that would allow the setting of realistic targets for future years' sales;

(3) To be able to identify opportunities for new dealer openings in terms of predicted business levels for new car sales and associated parts and service business. This would have to take account of the impacts on existing dealers, both Toyota and competitors and calculate the change in market penetration by postal district. In addition a method was needed for assigning the Areas of Responsibility (AoRs) for new dealers;

(4) To be able to assess the impact of dealer terminations in an analogous way to objective 3;

(5) To identify what the idealized dealer network would consist of under a number of different scenarios;

(6) To develop a strategy for extending the dealer network in response to the increased sales opportunity provided through European production;

(7) To develop a dealer strategy for assigning franchises for the luxury executive vehicle ''Lexus'' launched in 1990. This involved identifying dealers within the network who would achieve a minimum level of sales to warrant the award of the Lexus franchise of commercial criteria.

The majority of these objectives have or are currently being achieved. We now describe the main characteristics of the system developed for TGB.

Main characteristics of the TGB decision support system

The main characteristics of the decision support system can be summarized as follows:

(1) *Data sources*
The motor industry is data rich in that a large amount of data on new car registrations and vehicles on the road is captured by DVLC and made available to the Society of Motor Manufacturers and Traders who pass this information on to members on a syndicated basis. For TGB the following information is available:

- car registrations for each manufacturer by model segment for every postal district in the UK;
- for each Toyota dealership the source (postal district) of each registration by model segment;
- the total number of vehicles on the road by place of registration for each manufacturer by model segment (the car park).

In addition we assembled the following data from other sources:

- census by postal district;
- location of each of the 8000+ UK franchised dealers;
- drive times postal district dealer;
- sales data and targets for Toyota dealers.

This data has been assembled for 1987, 1988, 1989, 1990 and 1991 to provide not only a snapshot but a dynamic picture of performance within the car market.

(2) *System components*
The main decision support system developed has menu driven user-interface written in Clipper linked to the relational database. This system allows a wide variety of queries to be performed that produces output in tabular, graphical and mapped form. Time series data can also be displayed:

- an intelligent mapping system written in Clipper which has a high degree of functionality and also allows queries on the database to be performed through the map interface;
- a modelling system that allows a wide variety of scenarios to be run under different assumptions. The output from the model can be displayed in the same way as the baseline data and comparisons made. The model has been specifically designed for the application, and while the precise details are the subject of commercial confidence, it is based on substantial extensions of the family of spatial interaction models developed in Leeds (see Wilson and Bennett, 1985). The following factors have to be taken into account:

  — demand for new car sales by postal sector by model segment,
  — location of franchised dealers by manufacturer,
  — brand attractivity—the relative performance of the relationship between demand for new car sales and dealer location can be clearly seen in Figure 2; that is there is a strong distance-decay effect on sales as you

progress further from the dealer of different manufacturers in different model segments,

— dealer facilities by manufacturer,
— travel times between postal sectors and dealers.

We note that the use of standard statistical approaches, such as multiple regression, are singularly inappropriate for the type of problems addressed in this chapter. They fail to account properly for the complex pattern of consumer and competitor interaction in most markets and are not underpinned by what is known as double entry bookkeeping accounting principles. That is, if a new outlet is opened, its revenue must be generated by taking revenue from other outlets, assuming constant demand. To calculate this situation systematically, it is necessary to have an interaction approach that models every possible demand supply relationship. Regression models do not usually adopt this approach. Despite their dubious value they still seem to be adopted by many retailers and retail consultants to address location analysis problems.

In addition, other systems have been developed to meet specific requirements such as the Lexus study, where an optimization approach was needed.

An evaluation of the use of the decision support system

However good a decision support system that is developed to address a set of client objectives, it will only be as valuable as the use that is made of it. This requires that high levels of skill are demanded of the user—not in a computational or systems sense, but in an interpretative and planning sense. One of the problems of many proprietary GIS installations is that they need to be used by individuals who can program or

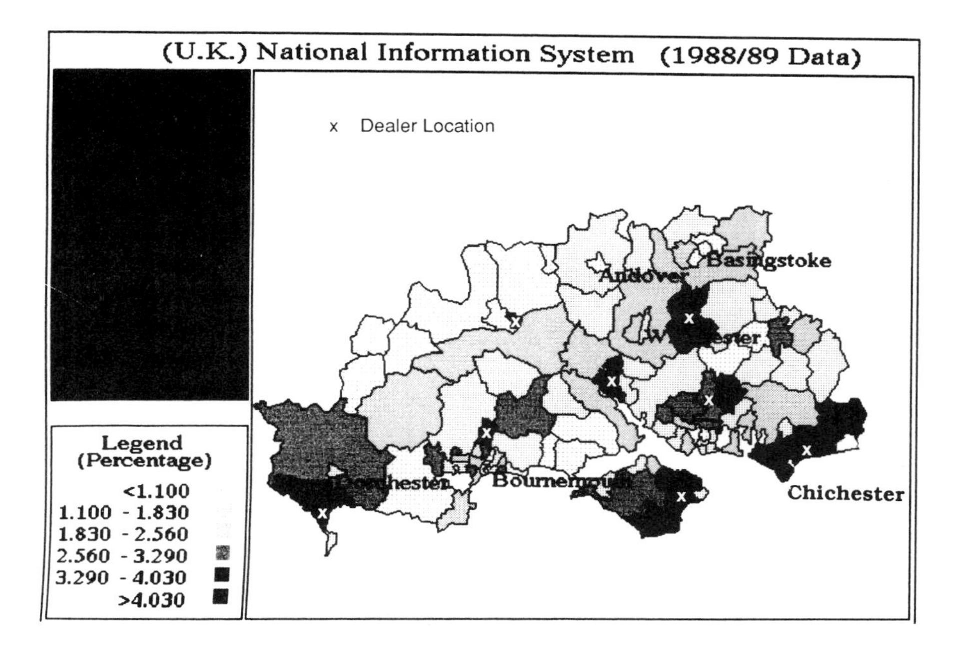

Figure 66.2. Market penetrations of a car manufacturer across Hants/Dorset postal districts.

understand structured relational database queries. What we have provided with the TGB system is a highly user friendly system that effectively provides "do it yourself" consultancy.

Since the installation of the first version of the decision support system, a number of different tasks have been undertaken by TGB. Some of them are the achievement of the business objectives described above. Furthermore, each individual dealer has had their performance assessed, their franchise territory appraised and their targets evaluated. Most importantly, a blueprint has now been developed for the future expansion of the franchised network. Any potential franchisee can be assessed in a short period of time and decisions made in a timely fashion. Of course, a decision support system does not make decisions; it merely helps support that process. The work with TGB has now been extended in a number of directions and we are working successfully with Toyota Belgium and discussing the wider potential of our approach with Toyota in the rest of Europe and North America.

Summary

The Information Age has brought vast amounts of data and technology into business. Most companies are not constrained by the amount of data they have access to or the computer power at hand. What is often missing is a framework for identifying business objectives and developing appropriate tools that will help in making more informed decisions that will help achieve these objectives. Many of the applications of GIS and related systems have focused on the technological aspects of the hardware or software. In our view information and technology utilization issues will be solved primarily with knowledge—which combines business expertise with technological expertise. We believe the success of the TGB/GMAP relationship has resulted from the combination.

Reference

Wilson, A.G. and Bennett, R.J., 1985, *Mathematics in Human Geography and Planning*, London: Riley.

67

A street-based Geographical Information System for local government revenue monitoring

Gary Higgs, David Martin and Paul Longley

Abstract

The UK has recently experienced significant changes to the basis of local government taxation. In 1989, the long-standing property-based rating system was replaced by the "Community Charge" which was designed to be levied upon adult individuals rather than property. In practice, however, this tax has been extensively hybridized in the second year of its operation, in a move originally intended to defuse the tension generated by losers in the switch from property to person-based local taxation. In its turn, this modified community charge is now to be superseded by a new hybrid tax, the "Council Tax". In view of local variability in household composition and property values, these changes may be expected to have had major redistributional effects between households at the local scale.

This chapter describes the development of a street-based Geographical Information System (GIS) for the city of Cardiff which has been devised in order to understand and monitor these effects. The construction of a street-level database has required management and extensive manipulation of the city's existing person and property-based records, neither of which are explicitly georeferenced. The research has addressed issues of confidentiality and spatial and temporal mismatches in the available data. This merged database has been created alongside a digital street network, providing a powerful analytical tool for the evaluation of recent change and for modelling the impact of future legislation. Street-based information reveals a high degree of variation in the changing tax burden below the level of "community" zones established for its administration.

Introduction

A number of studies have advocated the use of Geographical Information Systems (GIS) in urban policy-making ranging, for example, from the use of GIS as a decision support tool at the strategic level (e.g. Reeve and Wheeler, 1991; Worrall 1991a), through the use of GIS for targeting Urban Programme Resources (Hirschfield *et al.*, 1991) to the use of GIS in defining problems and preparing improvement plans for street networks (Yamanaka and Yoshikawa, 1991). Further examples of the use of GIS in urban policy-making are highlighted in texts by Huxhold (1990), Worrall (1990, 1991b) and Scholten and Stillwell (1990).

There is also a growing literature on the organizational and managerial implications of such studies for local government (e.g. Campbell and Masser, 1991; Coulson and Bromley, 1990) especially in relation to the shift away from the direct provision of services by local government in the 1990s. One of the major changes in the last ten years has been the change in the system of local government finance and taxation. This study reports the creation of a street-based GIS to begin to analyse some of these changes for the inner area of the city of Cardiff and to assess the redistributional effects on local government finance generation.

The study is a follow-up to a previous paper (Martin *et al.*, 1992) which highlighted variations in individual and household contributions under two regimes of local government finance between 1989-91; namely, the property-based rating system and the person-based community charge which replaced it in England and Wales on 1 April 1990. In common with a number of other cities in England and Wales (Burnett, 1989) this highlighted the importance of two main factors—the historical rateable value of the property and the number of adults in the household—in explaining the redistributive consequences of the community charge. The domestic rating system was based upon "rateable values" which were originally devised to represent the annual rent that a dwelling could command in a rent clearing market. By contrast, in its full-blooded form, the personal community charge is a flat-rate amount which is paid by everyone aged 18 and over. Payments under each form of taxation vary between local government jurisdictions as a result of different levels of government expenditure and variable levels of central government support. In practice, the underlying principles of the community charge (specifically the notion of a fixed flat rate charge which is paid, in whole or in part, by every eligible adult) have been tempered by political considerations, and it is appropriate to review these charges here using Welsh examples.

In the first year of operation of the Community Charge (1990/91), modest "transitional relief" arrangements were set in operation in order to cushion the heaviest losers from the initial shift from property towards person-based taxation. In England this relief was apportioned on a house by house basis, although in Wales a fixed level of relief was given to each chargepayer registered in a particular community. In 1990–91 320 out of 860 communities in Wales received some level of transitional relief, amounting to 35 per cent of the chargepayers (Welsh Office, 1990). In 1991–92 the financial commitment to transitional relief was stepped up, and there was some attendant increase in the number of Welsh communities that were to benefit, such that all residents in 613 of the 856 communities (some 67 per cent of all chargepayers) became recipients of transitional relief. The average reduction per receiving adult rose from £27 to £45 in the corresponding time period (Welsh Office, 1991).

A second, and possibly less considered, change to the Community Charge arose in the 1991 Budget, in which the Chancellor of the Exchequer announced a £140 blanket reduction in all charge bills in an attempt to defuse persistent protest against the charge regime (the loss of revenue was to be made good by a 2.5 percentage point increase in Value Added Tax). The compound effect of these changes was that established patterns of revenue raising became further obscured rather than clarified. In Cardiff, for example, the basic charge set for 1991–92 was £277.73. This was reduced by £140 in the March 1991 Budget. In addition, 17 out of 30 communities received reductions (61 per cent of charge payers) ranging from £5 (Butetown) to £86 (Adamsdown) and averaging £41 per receiving adult.

In our previous study, shifts in the spatial aspects of revenue generation within Cardiff's inner area (the aging Victorian core of the city) were examined at a variety of scales, from community level to 81 "House Condition Survey" (HCS) areas. This was achieved through the aggregation of street level data by interfacing a spreadsheet with a GIS (Higgs *et al.*, 1991). This chapter takes the analysis one stage further by examining the calculated revenue generated under the varying tax variations in household and individual revenue raising, highlighted in previous studies (e.g. Burnett, 1990 for Portsmouth; Paddison, 1989 for Glasgow), is analysed in further detail for the city of Cardiff in order to monitor the spatial effects of the particular community partitioning that has been used in Wales to administer community charge transitional relief.

Methodology

Two main data sets have been employed in this study—namely the publicly-available Electoral Register and the Rates Register for the city of Cardiff. The former has been used as surrogate data for the Community Charge Register which remains confidential below the community scale. A previous study found a high correspondence between the Community Charge and Electoral Registers at the community level for the 12 inner area communities of Cardiff (Higgs *et al.*, 1991). Street counts of electors from the 1991 Electoral Register were used in the study and data were input into a LOTUS 1-2-3 spreadsheet. Data were added on a street by street basis from the Rates Register (using a unique property reference number which incorporates its street identifier). For a number of street entries, data exist in the Rates Register and not the Electoral Register or vice versa. Possible reasons have been outlined in our previous paper (Higgs *et al.*, 1991) and include the renaming, misclassification, creation and demolition (in some cases) of streets. The Rates Register will also include entries for individual blocks of flats/maisonettes which cannot be distinguished from the Electoral Register.

For each street, the total revenue generated has been calculated for:

a) the *actual* rates payable in 1989-90;

b) the *notional* rates payable had this form of taxation existed in 1990-91 (calculated by estimating the rate poundage which corresponds to the community charge as levied for 1990-91);

c) the *notional* rates for 1991-92 (calculated as in (b), for 1991-92 community charges);

d) the *actual* community charge payable in 1990-91, allowing for minor transitional relief in two inner area communities (amounting to 7 per cent of the total number of people registered) and excluding any potential rebates/non -payments; and

e) the *community* charge payable in 1991-92, allowing for more widespread transitional relief and the blanket reductions announced in the 1991 budget.

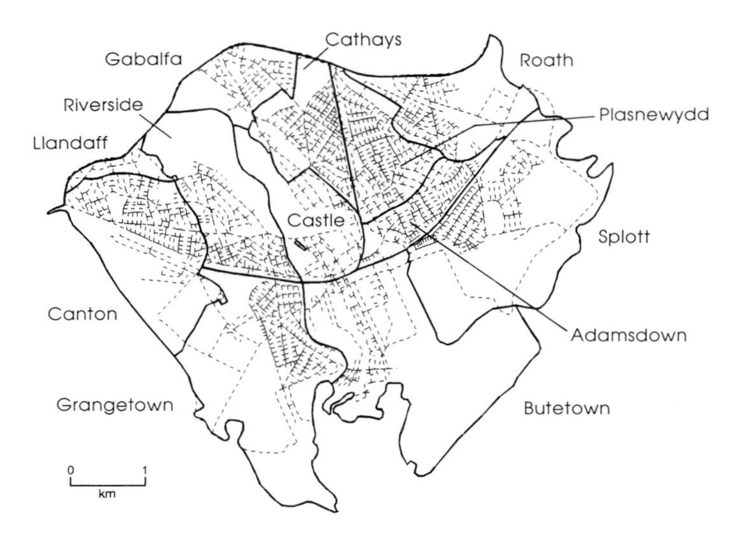

Figure 67.1. The digitized street network of Cardiff's Inner Area showing community areas

The aim of the present study has been to combine the database created in the spreadsheet with a computer (digitized) version of the street network using a GIS, in order to investigate in more detail shifts in the pattern of revenue raising in the city. We focus upon the inner area of Cardiff, which essentially comprises the late nineteenth and early twentieth-century urban core of the city and is an areal unit used by Cardiff City Council for some planning purposes. The street network of the inner area has been digitized in-house from Ordnance Survey (OS) 1:10 000 maps using pc ARC/INFO's ADS (Arc Digitizing Systems) module, since no other consistent and clean digitized data bases were available (Figure 1). Figure 2 illustrates the data linkage mechanism between the spreadsheet and the digitized street network. Each property in the Rates Register has an associated 13 digit code which is the unique property reference number (UPRN). This reference number includes codes which relate to the street and community in which the property fall, but no direct linkage exists to a grid reference or postal address. The first seven digits of the code represent the street index which has been used in the spreadsheet. Street based data from the Electoral Register has been input to the spreadsheet using the address link, as again this register contains no direct linkage to grid references or postal geography. During digitization, each street has been given a three figure code which has been entered into to act as the "link" between the digitized network and the database in the LOTUS 1-2-3 spreadsheet.

The relevant information from the LOTUS 1-2-3 spreadsheet is then exported using the Print–File facility within the package so as to produce an ASCII version of the database. This is imported into the INFO relational database by predefining field lengths/characteristics and by using the GET FROM command within INFO to access the data. The "relational join" between the database and the digitized street network is the three figure code common to both data sets. At the individual street level output from the analysis can be presented in pc ARC/INFO aggregated to the level of HCS area using the seven figure street-code which is present in the household survey data, or alternatively at the street level.

Results are presented below at the street-level for the differences between the notional rates to be paid in the financial year 1990–91 and the actual amount paid through the Community Charge (when only two communities had transitional relief arrangements in Cardiff). Our estimate of the distribution of the rates burden is obtained by assuming an identical contribution by the inner area to Cardiff's 1990–91 tax burden as was the case in the last year of operation of the rating system (1989–90). In the case of the Community Charge, we have assumed that every individual on the Electoral Register was liable for electoral *full* community charge, and that this was abated only by the small amount of transitional relief given to two communities. Our crude estimate of the total rates revenue generated by the inner area was 18.21 million pounds while our crude estimate of the inner area yield from the Community Charge was approximately 22.3 million pounds. This discrepancy arises for a number of reasons, including our failure to incorporate the reduced contributions of students, and may be the subject of refinement in future research.

Similar maps are also presented for the differences between the two regimes in the following financial year (1991–92) when more widespread relief arrangements existed. Differences in *total* amounts paid under the two regimes on a street basis are inevitably a function of the street length. These totals are therefore standardized by dividing these totals by the number of households and electors (as derived from the Electoral Registers) thus producing average and per-household estimates.

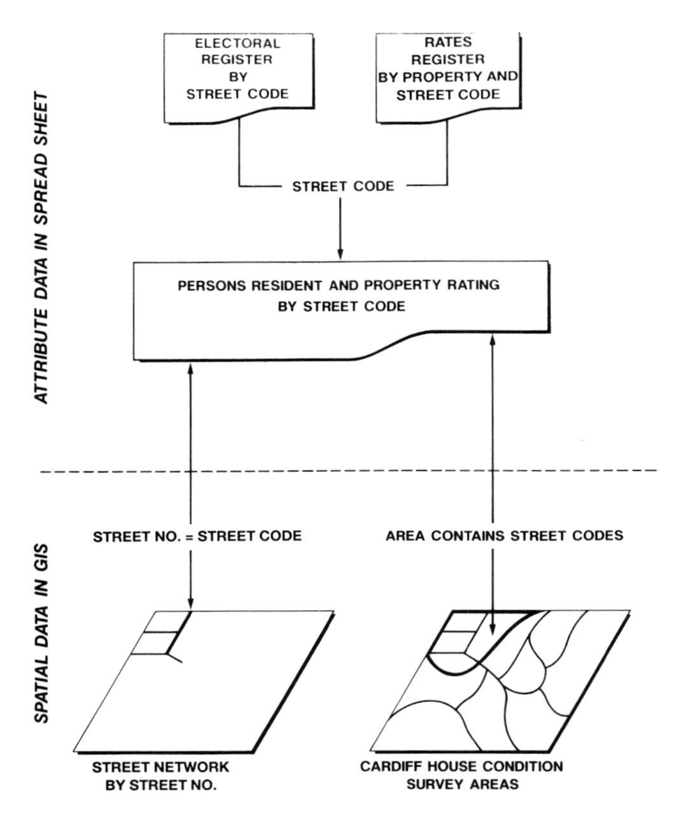

Figure 67.2. Data linkage between attributes in spreadsheets and street network in GIS.

Results

Differences in street revenues in 1990-91

Street revenue per household generated under the two regimes using the above techniques reveal a number of spatial trends. For example, an examination of so-called "losers" and "gainers" in the change between the two regimes in the first year of the community charge revealed that the majority the households in the inner area ended up paying more under the community charge than under the rates system. This general tendency does, however, conceal quite wide variations between the fortunes of individuals and households at the finer spatial scales. For example, on the one hand, residents of those properties (especially in the city centre) that had high ratable values gained from the change in the regime. Similarly those residents of properties in more suburban locations to the north of the city were proportionally better off under the new system. The heaviest losers, by contrast are concentrated among those residents of properties converted into flats and bedsits in parts of the city with high proportions of houses in multiple occupation (some of this disparity, however, may have been alleviated by the students' rebate system outlined above).

These patterns can be detected using our street level database. Figure 3a identifies those streets which lost out in the initial (1990–91) switch to the community charge, while Figure 3b identifies the minority of streets that gained from this charge. It is likely, in fact, that Figure 3b underestimates gainers since, for the reason outlined above, our Community Charge yield for the inner area is approximately 22 per cent higher than the yield from the rates. These maps nevertheless do serve to highlight those areas which benefited most from the taxation change, and as such represent a first step towards monitoring of changes in alternative taxation regimes.

Differences in street revenues in 1991–92

Figures 4a and 4b show the differences in revenue generated per household in the financial year 1991–92 after transitional relief and the post-1991 Budget arrangements. Again, apparent shifts have to be evaluated in view of data shortcomings and mismatches. For a variety of reasons outlined in our previous paper (Martin *et al.*, 1992) 26 streets appear in the Rates Register and not in Electoral Register, and similarly 49 streets are in the latter and not the former. Inevitably, therefore, the grand total revenues generated for the two regimes from the spreadsheet will be different; the total yield from the Rates System is estimated to be 8.85 million pounds and the total yield from the Community Charge is estimated to be 8.1 million pounds. Even allowing for these differences, a number of distinct trends can be noted. Figure 4b shows that there are more gainers than losers for 1991–92, largely because our analyses require the rates to raise a higher total revenue than the Community Charge. The extent of this gain, however, is also spatially variable across the city such that certain streets located in more densely populated areas such as Cathays, Canton and Grangetown are still relatively worse off under the Community Charge than under the rates, irrespective of the compensatory effect of the *areal* component of transitional relief. In contrast, properties in relatively affluent parts of the city (for example, Llandaff and North Roath) figure frequently amongst the largest gainers.

The overall impression from these maps, however, is the lack of consistent patterns under the post-transitional relief regime such that variations exist even between adjoining streets within communities. This would suggest the need for a more

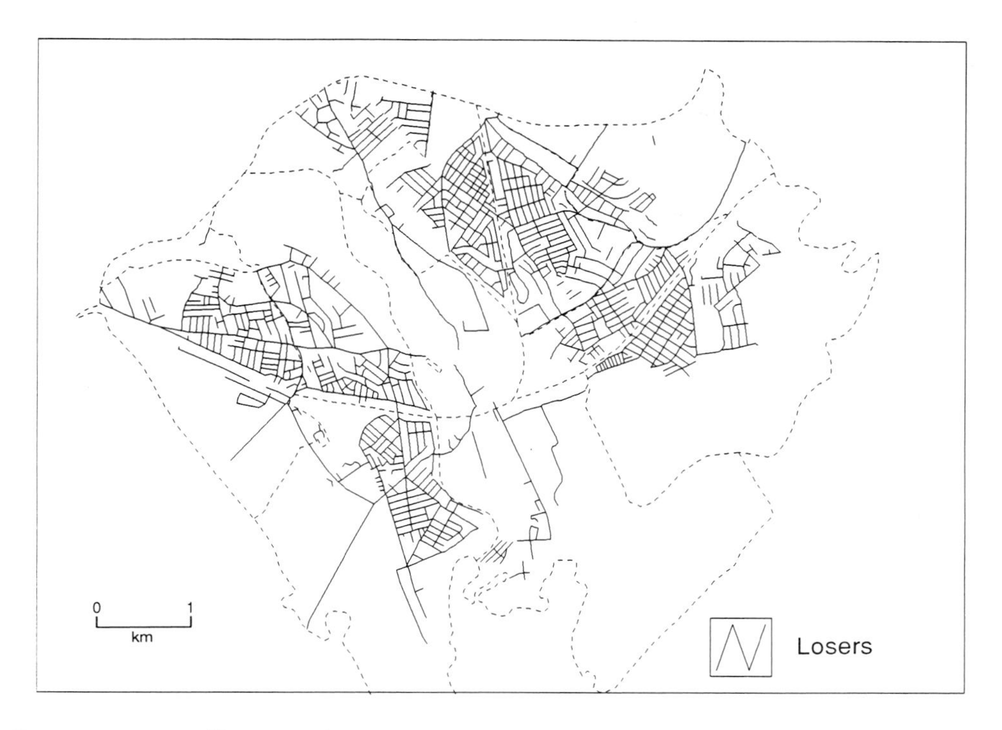

Figure 67.3a. Differences between Notional Rates and Community Charge for 1990-91 (per household): (a) losing streets.

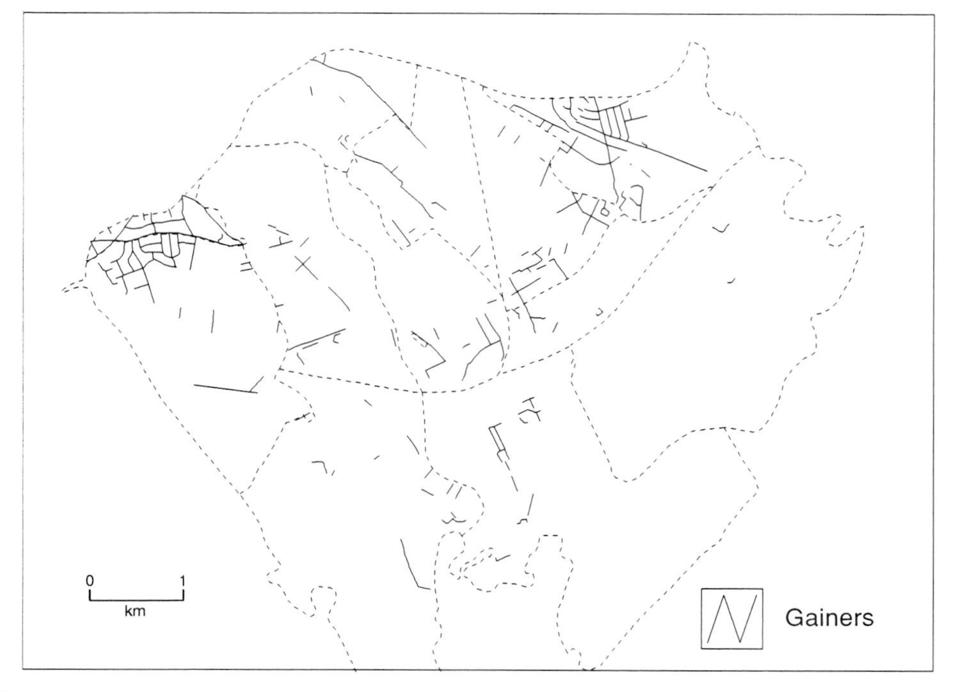

Figure 67.3b. Differences between Notional Rates and Community Charge for 1990-91 (per household): (b) gaining streets.

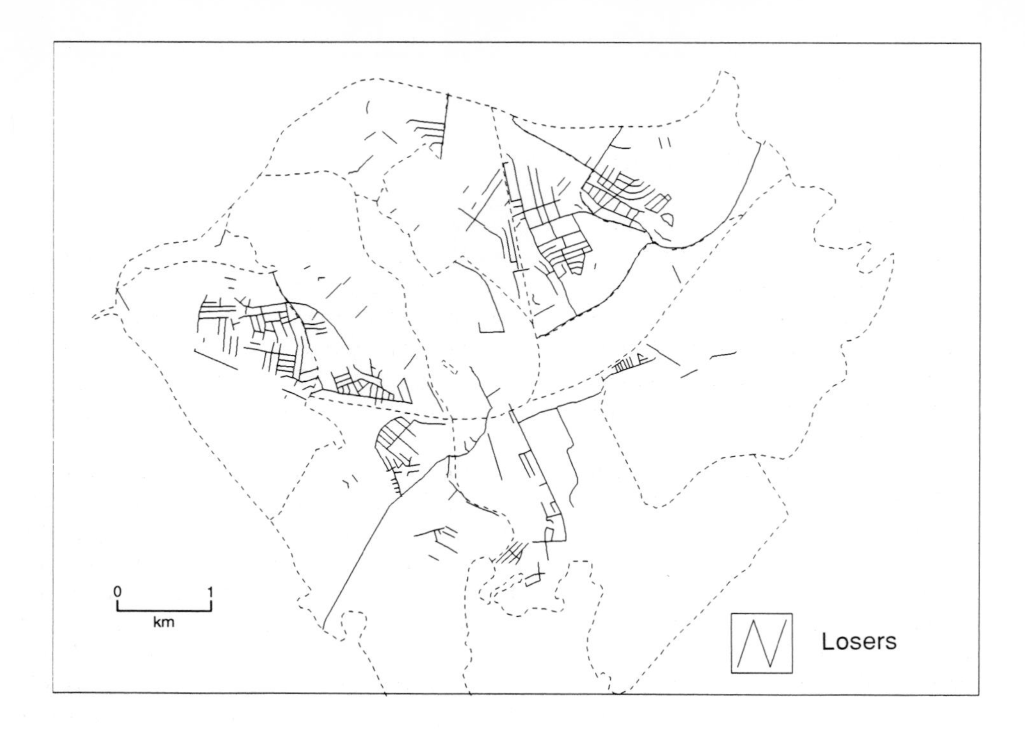

Figure 67.4a. Differences between Notional Rates and Community Charge for 1991-92 (per household)—post transitional relief, (a) losing streets.

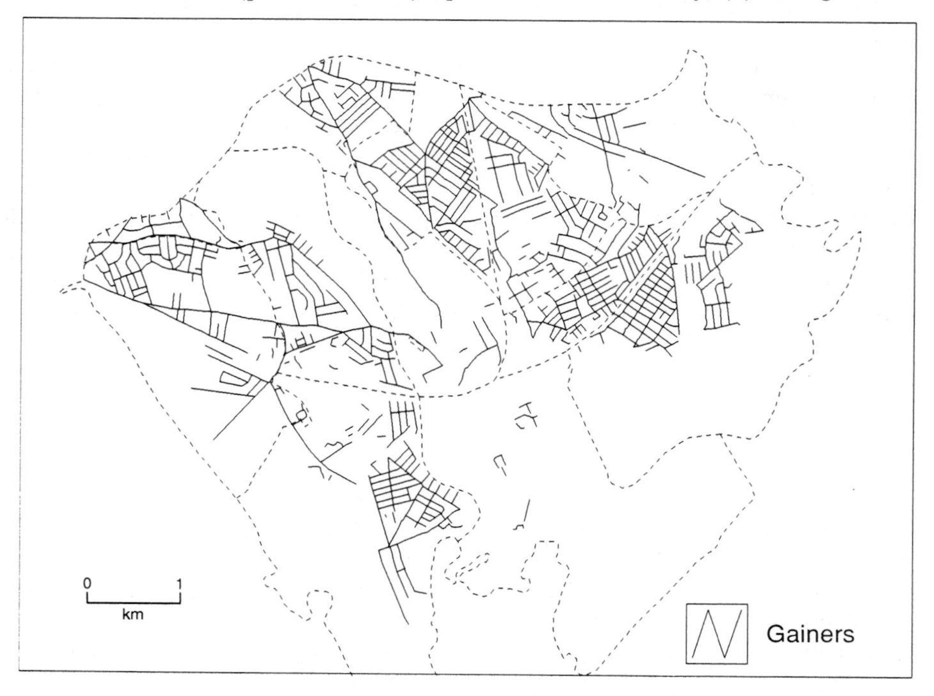

Figure 67.4b. Differences between Notional Rates and Community Charge for 1991-92 (per household)—post transitional relief, (b) gaining streets.

detailed case study approach at the individual property level as advocated by Paddison (1989) for example.

Conclusions and future work

This preliminary study has found a great deal of spatial variation in the impacts of the changing taxation regimes. The application of "blanket" coverage transitional relief arrangement within communities has not had a systematic effect upon the changed burden of taxation in the City. Residents in adjoining streets frequently fare significantly better or worse following the change from the rating system. These results mirror those emphasized in a previous paper (Higgs *et al.*, 1991) which utilized spatial aggregates of this data, namely communities and House Condition Survey (HCS) area (Keltecs, 1989).

More generally, the chapter has also demonstrated the capabilities of a combined GIS–spreadsheet approach to the analysis of policy issues in an urban environment, in particular in relation to the potential benefits to accrue from combining spatially referenced data (in this case the Electoral and Rates Registers). Furthermore, the graphics highlight the potential role of such systems in the monitoring of the spatial implications of policy, especially given the availability of data at suitable scales and in appropriate formats. The integration of previously disparate data sets using a combination of proprietary GIS software and customized GIS-type programs in this way can enhance the range and type of analysis undertaken in local government (Worrall and Rao, 1991). The Cardiff study with specific concern for local taxation services is a useful illustration of the issues which will face a local authority seeking to implement GIS solutions against a background of pre-existing poorly integrated administrative databases.

Once created, there exists the prospect of extending this database to assess a wide range of scenarios under planned or hypothetical local revenue raising regimes. A wide range of household characteristics might be included in order to model, for example, the geographical patterns of gainers and losers under a system of local income tax levied irrespective of residence attributes. Alternatively, the influence of different dwelling attributes upon gain or loss might be tackled. More generally, it is likely that future patterns of gain and loss, such as those anticipated under the Council Tax will be hybridized consequences of the interplay of household characteristics and dwelling attributes; we propose to investigate such issues in future research.

References

Burnett, A.D., 1989, The geography of the community charge/poll tax, Working Paper No. 6, Department of Geography, Portsmouth Polytechnic.

Burnett, A.D., 1990, The community charge (poll tax) in Portsmouth: The geography of winners and losers, Working Paper No. 10, Department of Geography, Portsmouth Polytechnic.

Campbell, H. and Masser, I., 1991, The impact of GIS on local government in Great Britain, in Rideout, T.W. (Ed.), *Geographical Information Systems in Urban and Rural Planning*, Institute of British Geographers, Planning and Environment Study Group, pp. 14–25.

Coulson, M. and Bromley, R., 1990, The assessment of user needs for corporate GIS: The example of Swansea Council, in Harts, J., Ottens, H.F.L. and Scholten, H.J. (Eds), *EGIS '90 Proceedings. First European Conference on Geographical Information Systems*, Amsterdam. Volume 1, Utrecht: EGIS Foundation, pp. 209–217.

Higgs, G., Longley, P. and Martin, D., 1991, An analysis of changing local taxation regimes using a street-level database, *Technical Reports in Geo-Information Systems, Computing and Cartography*, No. 34, University of Wales, Cardiff: Department of City and Regional Planning.

Hirschfield, A., Brown, J.B. and Marsden, J., 1991, Targeting Urban Programme Resources: A GIS-linked Decision Support System for St Helens, in: Rideout, T.W. (Ed), *Geographical Information Systems in Urban and Rural Planning*, Planning and Environment Study Group, Institute of British Geographers, pp. 41–57.

Huxhold, W.E., 1990, *An Introduction to Urban Geographical Information Systems*, Oxford University Press.

Keltecs, 1989, The Cardiff House Condition Survey, Phase 1: Inner Areas Final Report, Talbot Green: Keltecs Consulting Architects and Engineerings Ltd.

Martin, D., Longley, P. and Higgs, G., 1992, The geographical incidence of local government revenues: An intra-urban case study, Environment and Planning C, in press.

Paddison, R., 1989, Spatial Effects of the Poll Tax: A preliminary analysis, *Public Policy and Administration*, **4**(2), pp. 10–21.

Reeve, D. and Wheeler, R., 1991, Geographic Information Systems and local government policy making: The Kirklees Policy Mapping Project, *Local Government Policy Making*, **18** (3), pp. 41–49.

Scholten, H.J. and Stillwell, J.C.H., 1990, *Geographical Information Systems for Urban and Regional Planning*, Dordrecht, Netherlands: Kluwer Academic Publishing.

Welsh Office, 1990, Welsh local government Financial Statistics, No. 14, London: HMSO.

Welsh Office, 1991, Welsh local government Financial Statistics, No 15, London: HMSO.

Worrall, L. (Ed.), 1990, *Geographic Information Systems: Developments and Applications*, London: Belhaven Press.

Worrall, L., 1991a, GIS in decision making: A case study from the Telford Urban Policy Information Systems Project, in Klosterman, R.E. (Ed.) *Proceedings Second International Conference in Urban Planning Management*, Volume 2., pp. 107–23.

Worrall, L., 1991b, *Spatial Analysis and Spatial Policy using Geographic Information Systems*, London: Belhaven Press.

Worrall, L. and Rao, L., 1991, The Telford Urban Policy Information Systems Project, in Worrall, L. (Ed) *Spatial Analysis and Spatial Policy using Geographic Information Systems*, London: Belhaven Press, pp. 127–51.

Yamanaka, H. and Yoshikawa, K., 1991, Evaluation Systems of Local Streets Network based on Digital Town Map Database, paper presented at the Second International Conference on Computers in Urban Planning and Urban Management, Oxford, 6-8 July.

68

A comparative study of the performance of manual generalization and automated generalizations of line features

Zhilin Li
University of Southampton
and

Stan Openshaw
University of Newcastle

Abstract

This chapter presents the results of an empirical study of three different generalization procedures, namely, the traditional manual technique, Douglas-Peucker algorithm and an algorithm based on a natural principle. The performance of these three methods is compared for a small sample of linear features. The results suggest that the new method, derived from a natural principle (Li and Openshaw, 1990b), is surprisingly close to replicating the manual generalization process, at least in the example studies, and that it may well offer the basis for a new generation of automated line generalization procedures.

Introduction

Generalization is one of the fundamental processes in GIS (see Abler, 1987; Rhind, 1988). It occurs in many spatial data manipulations, especially those involving change in scale of map data and the overlay of map data sets acquired at different scales. A wide variety of algorithms have been developed (see McMaster, 1987, 1989; Nickerson, 1988). Unfortunately, most methods are based on algorithms which were developed for data reduction and then applied to line generalization. As a result, it has been concluded that the possibilities for an automated geometric solution to the line generalization problem is limited (Muller, 1990). However, it has been argued elsewhere by the authors (Li and Openshaw, 1990a) that there is a relatively straightforward natural principle that can be used to guide the generalization process and that algorithms based on this principle should perform better than data reduction algorithms in the generalization context. The purpose of this study is to measure the performance of a new automated line generalization procedure based on this principle (see Li and Openshaw, 1990b) and to compare it with both manual generalization and the widely used Douglas-Peucker algorithm (Douglas and Peucker, 1973).

The strategy involves digitizing some arbitrarily selected features from existing topographic maps at various scales and comparing the results of manual generalization with those produced by the two automated procedures. In the next section, a brief description of the test dataset is given. The third section outlines the measures of generalization effects that are used in this study, while the fourth compares the performance of the three different methods. The chapter finishes with some discussion and conclusions.

Description of test data set

A piece of river network with three line segments has been selected for this study as it provides a reasonable spread of complexity. There is nothing special about the choice. Figure 1 shows the three river segments digitized from four different map scales (1:25 000; 1:50 000; 1:250 000 and 1:625 000). The 1:10 000 scale is ignored because the 1:25 000 scale maps are basically the photographic reductions of the 1:10 000 scale maps and there is little extra to be gained from the digitizing effort. The selected features were digitized from OS topographic maps using ARC Digitizing System (ADS), which is part of ARC/INFO. Digitization was carefully performed to ensure that the line segments are recorded as accurately as possible in order to avoid too much information loss. It is not known what effects the data acquisition process has had on the subsequent analysis of generalization effects. It is probably small and in any case the effects are common to all methods.

Some measures of generalization effects

In seeking to understand the effects of generalization on the characteristics of linear features, it is useful to have some numerical measures which can be used to compare the three generalization techniques. McMaster (1986) describes 30 possible measures. However, Visvalingam and Whyatt (1990) pointed out that many of these measures are "inappropriate, misleading and questionable". In this chapter the following two measures are used to quantify the effects of generalization on the characteristics of linear features. They are the sinuosity ratio, and an error level for measuring displacement of line segments.

Sinuosity ratio and line length

Sinuosity ratio is a statistic which is designed to measure the wandering or meandering of a linear features. It is defined as follows (Unwin, 1981):

$$S = \frac{\text{Observed line length}}{\text{Straight-line distance from origin to end}} \quad (1)$$

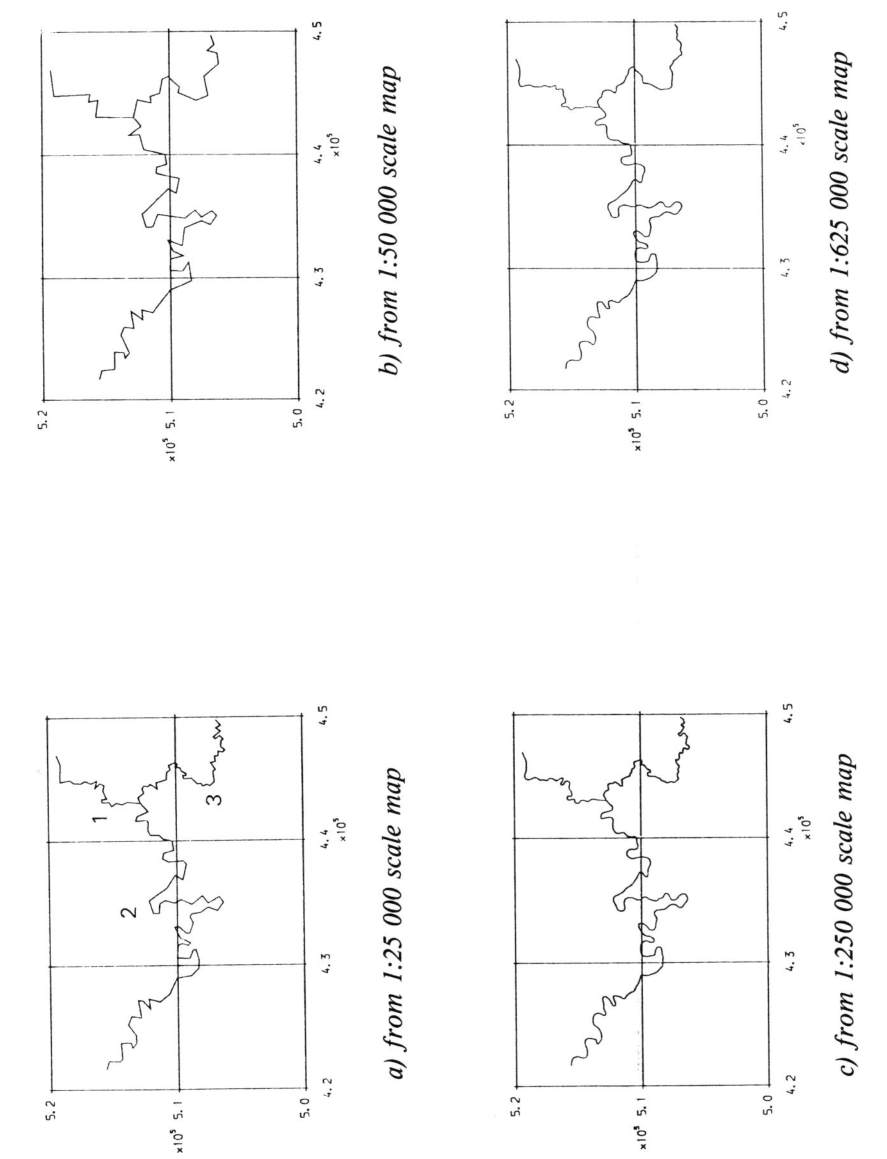

a) from 1:25 000 scale map

b) from 1:50 000 scale map

c) from 1:250 000 scale map

d) from 1:625 000 scale map

Figure 68.1. River segments digitized from various scale maps.

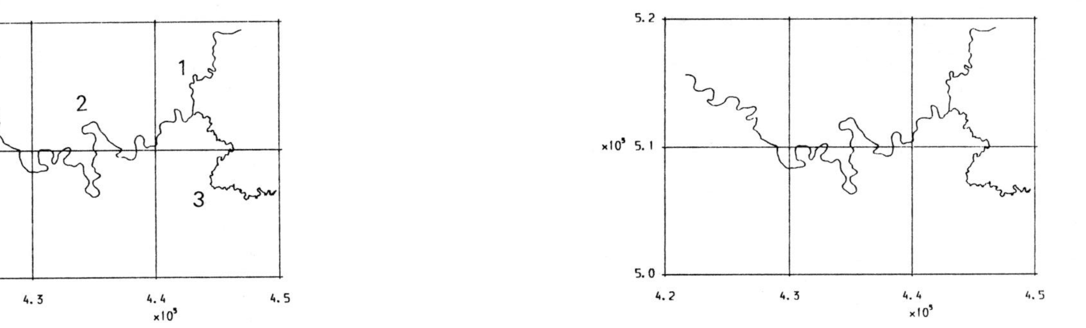

a) 1:250 000 scale by Douglas-Peucker algorithm

b) 1:625 000 scale by Douglas-Peucker algorithm

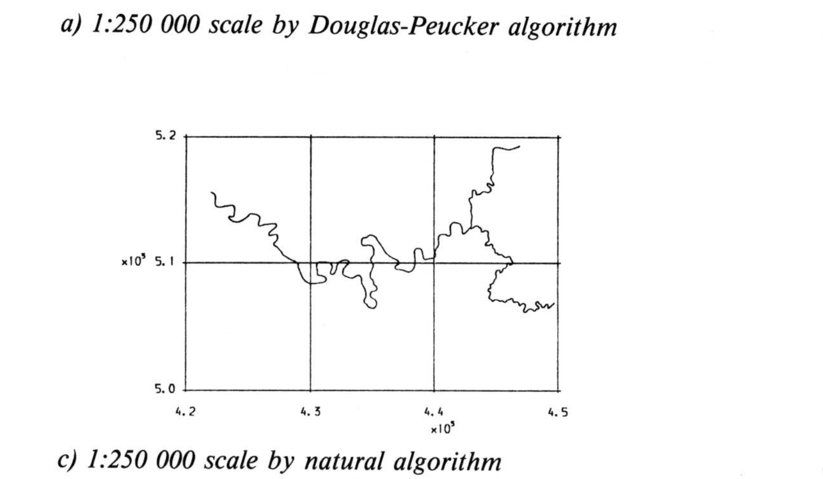

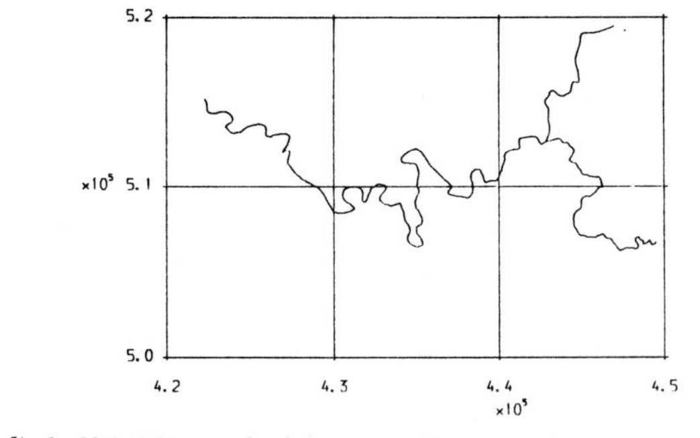

c) 1:250 000 scale by natural algorithm

d) 1:625 000 scale by natural algorithm

Figure 68.2. Examples of line generalization by automated techniques. (sequentially generalized from 1:25 000 to 1:50 000, then to the scale shown)

The larger the sinuosity ratio, the more complex or meandering the line. Sinuosity is not an absolute measure since the value of the ratio may vary with the positions of the two end points of a line segment. Nevertheless, it is a useful measure of the increase (or decrease) in line length due to generalization.

Error level for displacement of line segments

It is usually assumed that, after undergoing a generalization process, some parts of the features presented on smaller scale maps have been displaced from their original positions.

Locational errors created by the generalization process can be measured using the so-called "vector displacement". This is widely used to reassure the performance of data reduction algorithms. However, there are some problems in its applications. If the vector displacement is to be measured, then one must consider (a) how to find the corresponding points on two different lines to generate the vector and (b) if those points on spikes should be taken into account. In this study, a relative measure, an approximate value at map scale, e.g. 0.25mm, termed "error level" is used instead of an absolute value (e.g. 95m in terms of ground distance). For example, if the scale to which a map is to be generalized is 1:620 000, then 0.25mm at this map scale represents a value of about 156m in ground distance. Additionally, only those typical errors, which could be local maximas but occur most frequently, are measured.

An empirical comparison of automated and manual generalization

Two automated techniques are being tested here, the Douglas-Peucker algorithm and the new algorithm as recently developed by Li and Openshaw (1990b). The former is widely recognized as being the best among the existing algorithms for automated line generalization and it has been implemented in many GIS systems (e.g. ARC/INFO), while the latter is thought by the authors to be a very promising technique. The empirical results will test this conjecture.

It is noted that both algorithms have some critical parameters that need to be set. In the Douglas-Peucker algorithm, the tolerable distance is set at 0.5mm at map scale. This is the value of maximum permissible displacement error due to generalization by the algorithm. There are three variants of the new algorithm and here the raster-vector (without overlap) version (see Li and Openshaw, 1990b) is selected with fuzzy raster size set at 0.7mm at map scale.

An intuitive comparison

The features selected for this study have been described previously. Figure 68.1, in fact, shows the generalization of these river features by manual techniques. Figures 68.2a and b show the generalization of the same line features by Douglas-Peucker algorithm from initial scale of 1:25 000 to both 1:250 000 and 1:625 000 scales, While Figures 68.2c and d show the same features but this time generalized by the new algorithm.

In Figure 68.1, it is obvious that river features become increasingly less complex with decreasing map scale. A visual comparison with Figure 68.2c and d and Figure 68.1 suggests that there is a high degree of similarity between the features generalized

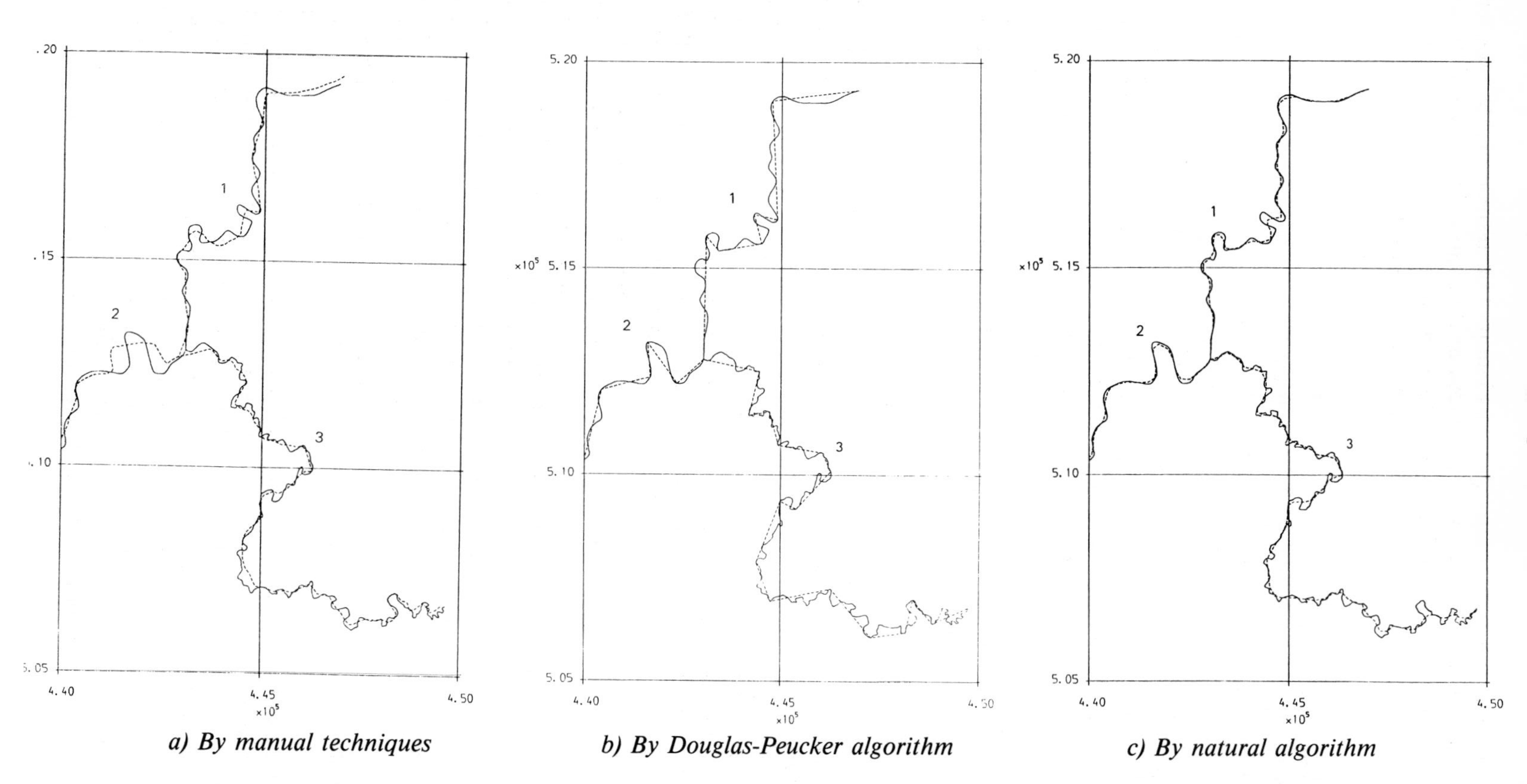

a) By manual techniques

b) By Douglas-Peucker algorithm

c) By natural algorithm

Figure 68.3. Example for the errors generated by generalization between 1:25 000 scale and 1:625 000 scale

by manual techniques and the new algorithm. By contrast, there is less apparent visual similarity between the features generalized by Douglas-Peucker algorithm and those by manual techniques. Indeed, the features in Figure 68.2a and b seem to represent a destruction of the nature of those river features shown in Figure 68.1. This is an intuitive impression and some quantitative results may be desirable.

Change in sinuosity ratio

The sinuosity ratio provides some general information about the relative complexity of those line segments that have the end points at the same positions (coordinates). However, it proved difficult to match the end points when the same feature was digitized from different scale maps. This introduces some uncertainty into the results. Table 68.1 gives the results for both manual and automated generalization. Surprisingly, there is not much difference for the sinuosity factors between manual generalization and the automated generalization carried out by the two selected algorithms. In some cases, the results by Douglas-Peucker algorithm are more close to those by manual generalization while in other cases the natural algorithm produces results closer to those by manual generalization. One concludes that sinuosity ratio is not very sensitive to the different characteristics of the generalized features which are visually apparent in Figure 68.2.

Error level for displacement by manual generalization

Figure 68.3 (a, b, c) show examples of displacement errors generated by the three generalization techniques. In extreme cases, small features may have been completely destroyed.

In map accuracy specification, there is a tolerable error for a given scale of map. The value varies with country. The US specification which is most widely quoted states: "95 per cent of all well-defined cultural and drainage features shall be plotted on the map in correct horizontal coordinate position within 1/50 inch (0.5mm), at publication scale". Of course, this value represents the accumulated effects of all the errors that may have been introduced at different stages of the entire mapping and generalization processes. The errors created by generalization alone should be smaller than this quantity. It is interesting, then, to compare features at two different scales to obtain the error level due to generalization and to then investigate how much this contributes in percentage terms to the tolerable mapping error as a whole.

Through the overlay of two features digitized from two maps with different scales, the discrepancies can be inspected and measured if the differences are large enough. Table 68.2 lists the error levels that have been measured by overlaying the three river segments. It can be seen that typical levels of error introduced by manual generalization of linear features is within a band of 0.35mm at map scale. This suggests that errors due to generalization account for up to 70 per cent of the total error budget.

Of course, these values also include some errors caused by map distortion and digitization. However, they do provide some information about the approximate total error levels. The largest error for the second river segment is created by cartographic exaggeration and/or displacement. If this part is excluded, then the error level is more or less the same as those of the other two. It may be interesting to note that the error level for the third river segment is no higher than the other two although this segment looks much more complex. That is, the error level due to generalization for more complex line segments is not necessarily higher than that in other segments. In fact, the error distribution for the complex lines is even more homogeneous.

Table 68.1. Sinuosity factors for features at various scales using different generalization methods

To Scale	From Scale	Segment	Manual	Douglas-Peucker	Natural
1: 50 000	1: 25 000	No. 1	1.668	1.650	1.667
		No. 2	2.620	2.606	2.627
		No. 3	2.511	2.544	2.604
1:250 000	1: 50 000	No. 1	1.535	1.593	1.592
		No. 2	2,534	2.513	2.561
		No. 3	2.232	2.163	2.079
	AG: 1:50K	No. 1	—	1.585	1.603
		No. 2	—	2,518	2.569
		No. 3	—	2.225	2.073
1:625 000	1:250 000	No. 1	1.330	1.372	1.354
		No. 2	2.316	2.342	2.356
		No. 3	1.800	1.826	1.715
	AG 1:250K	No. 1	—	1.442	1.427
		No. 2	—	2,429	2.219
		No. 3	—	1.9834	1.696

Note: In this table, 'AG' under the heading of 'from scale' means the automated generalized map.

Table 68.2. Error levels for displacement by manual generalization

From:	1:25 000						1:50 000				1:250 000	
To:	1:50 000		1:250 000		1:625 000		1:250 000		1:625 000		1:625 000	
error:	Max	Typi	Max	Typi	Max	Typi	Max	Typi	Max	Typi	Max	Typi
River No. 1	s	s	0.80	0.30	0.45	0.25	0.80	0.30	0.45	0.25	0.45	0.25
River No. 2	s	s	1.30	0.30	0.65	0.30	1.30	0.30	0.65	0.30	0.65	0.30
River No. 3	s	s	0.80	0.30	0.45	0.25	0.80	0.30	0.45	0.25	0.65	0.25

Note: 'S' denotes the errors are too small to be measured by inspection;
'Max' denotes the largest error (mm at map scale);
'Typi' means typical error level (mm at map scale)

For the Douglas-Peucker algorithm, the maximum error is controlled by the parameter given. It might be expected that the error level between successive scales (e.g. between 1:50 000 and 1:250 000) for Douglas-Peucker algorithm will be about 0.25mm at map scale, for a given tolerable distance of 0.5mm at map scale. It is also expected that the resulting error level by the natural algorithm will be smaller than 0.35mm at map scale for the fuzzy raster size of 0.7mm at map scale. It is also very important to investigate the error level between a given scale and other larger scale maps. For example, if the error level between 1:50 000 scale map and the generalized 1:250 000 scale map is 0.3mm at map scale, the error level between the 1:25 000 scale base map and the generalized 1:250 000 scale map could be quite different.

Table 68.3 shows the results for the error level generated by both the Douglas-Peucker algorithm and the natural algorithm. The symbol "D" in this table means "generalized by Douglas-Peucker algorithm from the immediate larger scale map which was digitized" and "DD" means the map was generalized from the immediate larger scale map which was already generalized by Douglas-Peucker algorithm. The "N" and "NN" have very similar meanings for the natural algorithm. From this table, it can be found that the actual errors for Douglas-Peucker algorithm are much larger than expected and much greater than the results for manual generalization.

Table 68.3 also shows the error levels generated by the natural algorithm. The errors introduced by this algorithm are very similar to those of the manual generalization procedure and smaller in many cases. In particular, the generalized 1:250 000 and 1:625 000 scale maps so accurately match the main shape of the original 1:25 000 scale map that the error levels are very small.

Change in fourier spectrum

The dataset used to generate these results has been very simplistic. However, the natural principle has also been applied to the somewhat more complex and perhaps more realistic geographical feature. If the coastline shown in Figure 68.4a is considered, then the Douglas-Peucker algorithm completely fails to produce a line which has any similarity to that produced manually (see Figure 68.4c) while the natural algorithm will still be capable of producing a very desirable result (see Figure 68.4d). If these findings are supported by further study, then it only remains to discover how best to implement this new method in a form suitable for application in both GIS software and, perhaps in graphics display hardware.

Concluding remarks

This chapter presents some results about the generalization for line features via three different methods. The results are important for three reasons. First, they show that the widely used Douglas-Peucker algorithm should be used for data reduction rather than line generalization; although both are clearly related and may not be separable activities. Indeed, if data reduction largely destroys the characteristics of a feature, then it can hardly be a useful cartographic function. Second, displacement of vectors due to manual generalization can itself account for around two-thirds of feature displacement permitted by some map accuracy standards. Tests showed that while the natural principle produced similar levels of displacement error to manual methods the Douglas-Peucker method typically produced higher levels of error than both.

Table 68.3. Error level generated by automated algorithms

Digitized Map	Generalized Map	River seg 1		River seg 2		River seg 3	
		Max	Typi	Max	Typi	Max	Typi
1: 25 000	D 1: 50 000	0.50	0.30	0.50	0.35	0.50	0.30
	D 1:250 000	0.50	0.30	0.62	0.37	0.50	0.37
	D 1:625 000	0.50	0.30	0.55	0.35	0.62	0.37
	DD 1:250 000	0.50	0.30	0.55	0.35	0.50	0.30
	DD 1:625 000	0.50	0.30	0.50	0.35	0.50	0.30
1: 50 000	D 1: 50 000	0.50	0.25	0.50	0.37	0.50	0.30
	D 1:250 000	0.50	0.25	0.50	0.37	0.50	0.35
	D 1:625 000	0.50	0.25	0.57	0.30	0.62	0.35
	DD 1:250 000	0.58	0.35	0.58	0.37	0.58	0.35
	DD 1:625 000	0.50	0.25	0.50	0.30	0.50	0.30
1:250 000	D 1:250 000	0.74	0.30	1.23	0.37	0.74	0.37
	D 1:625 000	0.50	0.25	0.50	0.30	0.50	0.35
	DD 1:250 000	0.70	0.25	1.23	0.40	0.68	0.30
	DD 1:625 000	0.50	0.25	0.57	0.25	0.74	0.25
1:625 000	D 1:625 000	0.50	0.25	0.57	0.25	0.52	0.35
	DD 1:625 000	0.50	0.25	0.62	0.25	0.42	0.25
1: 25 000	N 1: 50 000	s	s	s	s	s	s
	N 1:250 000	0.20	s	0.31	0.10	0.49	0.20
	N 1:625 000	0.74	0.25	0.62	0.25	0.49	0.27
	NN 1:250 000	0.18	s	0.25	s	0.50	0.25
	NN 1:625 000	0.42	0.15	0.49	0.15	0.62	0.25
1: 50 000	N 1: 50 000	0.30	0.15	0.50	0.15	0.30	0.15
	N 1:250 000	0.15	s	0.20	s	0.43	0.15
	N 1:625 000	0.74	0.25	0.65	0.25	0.50	0.35
	NN 1:250 000	0.20	s	0.20	s	0.60	0.30
	NN 1:625 000	0.42	0.15	0.49	0.15	0.62	0.25
1:250 000	N 1:250 000	0.76	0.10	1.00	0.25	0.69	0.20
	N 1:625 000	0.54	0.10	0.37	0.10	0.49	0.20
	NN 1:250 000	0.69	0.15	0.95	0.35	0.70	0.25
	NN 1:625 000	0.40	0.20	0.44	0.20	0.61	0.25
1:625 000	N 1:625 000	0.55	0.25	0.55	0.25	0.50	0.20
	NN 1:625 000	0.35	0.25	0.69	0.25	0.61	0.20

Note: 's' denotes the errors are too small to be measured by inspection;
'Max denotes the largest error (mm at map scale);
'Typi' means typical error level (mm at top scale).

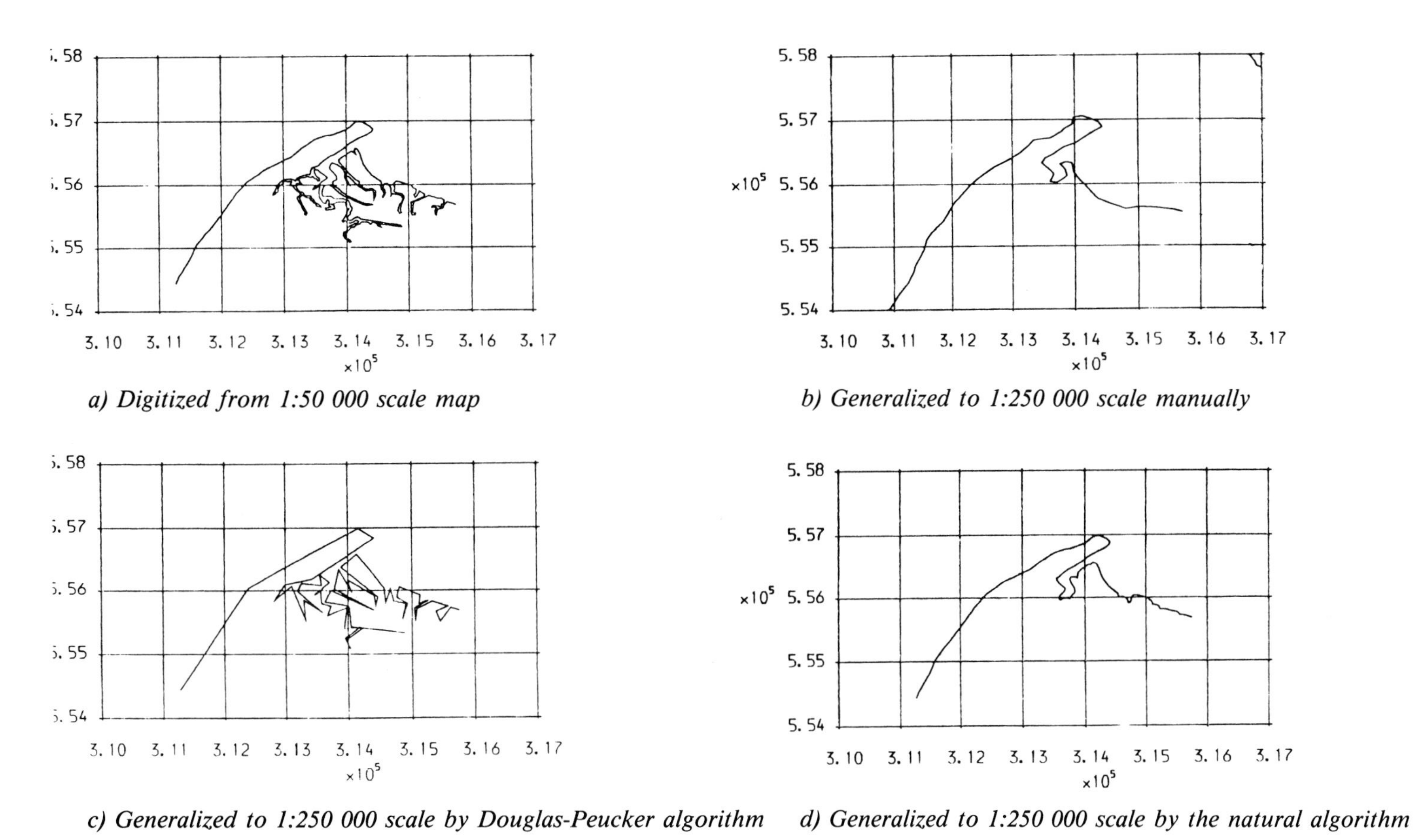

a) Digitized from 1:50 000 scale map

b) Generalized to 1:250 000 scale manually

c) Generalized to 1:250 000 scale by Douglas-Peucker algorithm

d) Generalized to 1:250 000 scale by the natural algorithm

Figure 68.4. Generalization of a piece of coastline by different methods

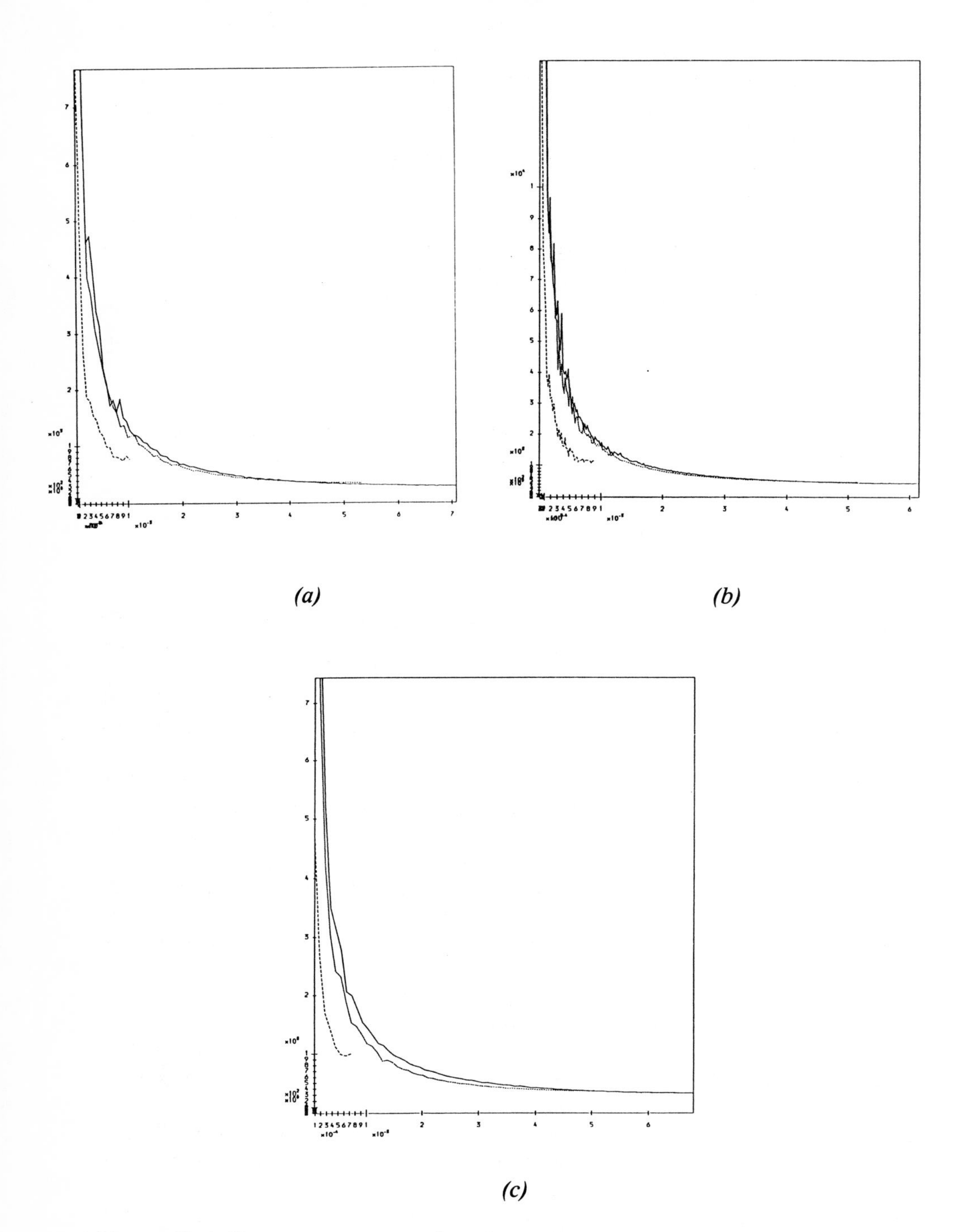

Figure 68.5: Frequency spectra for the river segments generalized by different techniques (for 1:625 000 scale). a) to c) for river segment 1 to 3

a: manual; b: natural algorithm; c: Douglas-Peucker algorithm

Third, it appears that the natural principle for automated generalization can produce results very similar to those produced by manual cartographic procedures and works effectively even on very sinuous lines which are typical of geographic features. These potentially important findings suggest that this method deserves closer scrutiny.

Of course, the magnitude of the errors introduced by these algorithms may be affected by the criteria being used. In order to provide a more comprehensive comparison, the frequency spectra are also examined. Figure 68.5 shows the examples of the spectra of the same features but generalized by three different methods. It can be seen that the spectra for both manual generalization and by the natural algorithm are very close together in all three cases, while Douglas-Peucker results are quite different. This again supports the view that the ones generalized by Douglas-Peucker algorithm have a very different nature and characteristics.

Acknowledgement

The authors would like to thank ESRC for funding the North East Regional Research Laboratory (NE.RRL). The assistance from colleagues in NE.RRL is highly appreciated.

References

Abler, R., 1987, The National Science Foundation National Center for Geographic Information and Analysis, *International Journal of Geographical Information Systems*, **1** (4), pp. 303–26.

Douglas, D. and Peucker, T., 1973, Algorithm for the reduction of the number of points required to represent a digitized line of its caricatore, *Canadian Cartographer*, **10** (2), pp. 112–22.

Li, Z. and Openshaw, S., 1990a, A natural principle for objective generalisation of digital map data and other spatial data, NE.RRL Report, University of Newcastle upon Tyne; Also accepted to be published in *Cartography and Geographic Information Systems* (formerly *The American Cartographer*).

Li, Z. and Openshaw, S., 1990b, Algorithms for automated like generalisation based on a natural principle, NE.RRL Report, University of Newcastle upon Tyne; Also accepted to be published in *International Journal of Geographical Information Systems*.

McMaster, R., 1986, A statistical analysis of mathematical measures for linear simplification, *The American Cartographer*, **13** (2), pp. 103–16.

McMaster, R., 1987, Automated line generalisation, *Cartographica*, **24**(2), pp. 74–111.

McMaster, R. (Ed.), 1989, Numerical Generalisation in Cartography, *Cartographica*, **40**(1), Monograph No. 40.

Muller, J., 1990, The removal of spatial conflicts in line generalisation, *Cartography and Geographic Information Systems* (formerly *The American Cartographer*), **17**(2), pp. 141–9.

Nickerson, B., 1988, Automated cartographic generalisation for linear features, *Cartographica*, **25**(3), pp. 15–66.

Rhind, D., 1988, A GIS research agenda. *International Journal of Geographical Information Systems*, **2**(1), pp. 23–28.

Unwin, D., 1981, *Introductory Spatial Analysis*. London and New York: Methuen.

Visvalingam, M. and Whyatt, J., 1990, The Douglas-Peucker algorithm for line simplification: Reevaluation through visualisation, *Computer Graphics Forum*, **9**(3), pp. 213–28.

White, E., 1985, Assessment of line-generalisation algorithms using characteristic points, *The American Cartographer*, **1**, pp. 17-27.

Part XII

Trade directories

Introduction

One of the primary aims of the *Yearbook* is to inform the user as fully as possible on many and varied aspects of GIS/LIS. With the help of Geobase Consultants a series of trade directories has been constructed broken down into the following twelve categories:

Within each category the entrants are listed alphabetically. In addition to brief tabular entries, certain entrants have elected to supply a more full description of their services and products. These have been written by the companies concerned; they do not constitute impartial reviews by the Editors or the AGI. The extended entries are indicated by page number in the final column of each table (headed *Ref.*), and follow the tabular entries in each category.

Companies or organizations wishing to be included in these directories in future editions of the *Yearbook* are welcome to contact.

Geobase Consultants Ltd.
28 Church Road,
Epsom,
Surrey KT17 4DX

GIS software & systems

Organisation, contact, telephone	Address	Products	Outstanding features	Ref	Surrogate Page
Action Information (Management) Limited John Page 0225 777288	Ashton Road, Hilperton, Wiltshire BA14 7SZ	MAPS IN ACTION. The map driven operational information system for decision makers. Easy manipulation and display of geo referenced data including AVLS reports in real time.	Stand alone on IBM PC compatible or as an intelligent terminal to a network or mainframe.	0	
Alper Systems Limited Clive Baker 0223 420464	Cambridge Science Park, Milton Road, Cambridge CB4 4WG	Alper GIS: state of the art full GIS: data conversion (automatic), user configurable Applications Builder with a wide range of standard applications for utilities and local government.	A widely used mixed raster/ vector, object based, seamless and distributed system with links to remote databases. Extremely easy to use and offers full PC access and networking.	0	
W.S. Atkins Management Consultants Angus Looney 0372 726140	Woodcote Grove, Ashley Road, Epsom, Surrey KT18 5BW	GEOSCOPE - a PC-based system running under Microsoft Windows which can store, merge, analyze and display geographic data in maps and reports.	An object-oriented GIS with intuitive help screens, Data can be dynamically exchanged between EXCEL and other Windows applications.	0	
Axis Software Systems Limited Neil Farmer 0604 720477	Park House, 88-102 Kingsley Park Terrace, Northampton, NN2 7HJ	Fully functional single / corporate GIS, raster and vector display concurrently, multiple overlay of data all accessed by Unix or PCDos processors on normal network.	Application modules customised for users, full polygon - polygon database searching for planning applications, land charges, contaminated land register. Also demographic analysis and database links.	0	
CACI Limited John Rae 071 602 6000	CACI House, Kensington Village, Avonmore Road, London W14 8TS	InSite, a PC based GIS designed to analyze people and markets for locational, marketing and 'resource allocation' applications.	PC based, stand alone or networked - accurate drive time polygons - easy to use report generator - powerful database manager - geographic and query based analysis.	0	

Organisation, contact, telephone	Address	Products	Outstanding features	Ref	Surrogate Page
CACI Limited Nigel Towell/ Kate Tylor 071 602 6000	CACI House, Kensington Village, Avonmore Road, London W14 8TS	ATLAS*GIS / ATLAS*PRO / ATLAS Mapmaker for Windows: enhanced mapping with some GIS functionality. High quality output chloropleth and dot mapping.	Assortment of packages to run under DOS / Windows / Macintosh. Supports wide variety of plotters and printers.	0	
CDR Mapping Limited Christopher Cooper 0433 621282	Hawthorn House, Edale Road, Hope, Sheffield S30 2RF	Supergazetteer is used for loading data into GIS. All data is stored in Oracle and is available for other packages. Full support services provided.	Map management of all NTF / OSTF mapbase using industry-standard SQL databases. Ideal for data loading and GIS use. Stand-alone PC to full DOS / UNIX networking.		
Claymore Services Limited Chris Whitt 0404 823097	Station House, Whimple, Exeter, Devon EX5 2QH	MAP91 is a windows based package designed specifically for mapping UK census data on a PC. Single user, network and site licences @ £750, £1500 and £5000 respectively.	MAP91 interfaces directly with digitized census area boundaries and census data files, as well as offering a dynamic link with windows spreadsheets and other packages.	0	
Conic Systems Limited Iain Cooke 031 667 2728	20a Mayfield Terrace, Edinburgh EH9 1SA	Locator - an object-oriented GIS for portable, pen computers. Integrates with office systems, so field personnel can use and gather spatial and other data.	Runs on all PenPoint hardware. Raster and vector display and capture. Full range of supporting software available - forms, images, diaries etc.	526	101
The Data Consultancy Bryan Wade 0734 588181	7 Southern Court, South Street, Reading, Berkshire RG1 4QS	MAPINFO the fully featured mapping and analysis system for PC's, Macintosh, SUN and HP systems. Supplied complete with The Data Consultancy GB Startup pack.	Map, browse, graph and query your data with ease. A cost effective solution with full support from our Reading office.	527	102
Digital Equipment Co. Limited Andrea White 0734 868711	Digital Park, Imperial Way, Reading, Berkshire RG2 0TE	Digital supports UNIX, VMS, OSF Motif, X-windows, POSIX, SQL, FDDI, Ethernet, X.25, TCP/IP, DECnet/OSI. Digital works closely with GIS vendors Smallworld, Laser-Scan, EDS, ESRI, SysScan, Pafec and APIC	Digital's track record in systems integration and networking provides a unique platform to address the major issue of GIS projects - making data/ information available to users and providing real application benefits.		0

Organisation, contact, telephone	Address	Products	Outstanding features	Ref	Surrogate Page
Earth Observation Sciences Mark Thomas 0252 811181	Branksome Chambers, Branksomewood Road, Fleet, Hampshire GU13 8JS	ILWIS: serious GIS and satellite image processing on a PC.	PC under DOS with maths co-processor. Super VGA up to 1024 x 768 x 256. HPGL plotters. Ink-jet printers. 4-button digitizer.	0	
ESRI (UK) Limited Carole Smith 0923 210450	23 Woodford Road, Watford, Hertfordshire WD1 1PB	ArcCAD is a new product which brings together in a unified environment, the market leading GIS software (ARC/INFO) to the world's number one CAD software (AutoCAD).	ArcCAD extends the AutoCAD data model to include spatial relationships, a topological database & selection, query & analysis tools, giving new power for the development of AutoCAD GIS applications.	530	105
ESRI (UK) Limited Carole Smith 0923 210450	23 Woodford Road, Watford, Hertfordshire WD1 1PB	ArcView is a powerful,new software tool that brings geographic information to the desktop. An easy-to-use & cost-effective query & display tool, it can be used stand-alone or with ARC/INFO GIS.	With ArcView's intuitive GUI, access to, visualization, printing & exporting of maps & tabular data are quickly learned. ArcView runs on PC's, Macintosh & UNIX workstations.	529	104
ESRI (UK) Limited Carole Smith 0923 210450	23 Woodford Road, Watford, Hertfordshire WD1 1PB	ESRI's ARC/INFO is recognized world-wide as the leading GIS and is in use in thousands of organizations including local & central government, oil companies, utilities, retailers, map makers, universities & many more.	ARC/INFO's extensive 'tool-box' of sophisticated geoprocessing tools, application language & open systems philosophy provides users with the flexibility to design & implement their own GIS solutions.	528	103
Eurecart Andy Wolfe 0621 851114	Ulting Lane, Hatfield Road, Langford, Essex CM9 6QA	APIC is an advanced object-oriented GIS, including development environment, customisation tools and a powerful query language.	APIC is an advanced object-oriented GIS proven in operational use around the world.	530	106
Genasys (UK) Limited Sara Thompson 061 232 9444	Enterprise House, Manchester Science Park, Lloyd Street North, Manchester M15 4EN	Genasys provides an integrated family of Open Systems products addressing GIS, Civil Engineering, Data Capture and Document Management.	Open Architecture; links to multiple external RDBMS; powerful applications development; OSF/ Motif graphical user interface; established worldwide customer base.	0	

Organisation, contact, telephone	Address	Products	Outstanding features	Ref	Surrogate Page
Geographic Management Systems Limited Bill Knowles 0454 618618	Woodlands Grange, Woodlands Lane, Almondsbury, Bristol BS12 4JY	PC-M.A.P (single screen) & GS-M.A.P (twin screen) are multi-application GIS developed specifically for the networked MSDos PC market running on IBM AT & PS/2 micro's.	Runs under Dos with bi-directional links to multiple external databases on remote hardware and operating systems (eg McDonnell Douglas, ICL, UNIX, Pick etc.)	531	107
GeoMEM Software Jim Tweedie / Marlon Binner 0250 872284	60 High Street, Blairgowrie, Perthshire PH10 6DF	MAPVIEWER and MAPVIEWER UK. Powerful thematic mapping package for displaying all kinds of business or scientific information geographically. Includes spreadsheet data entry system. Windows based.	Import/ export of map and data files. Displays many different map types. Creates multi-layers. Easy to use, with full tool-box facility. Exceptional value for money.	0	
GeoVision Systems Limited Walt Beisheim 0276 677707	80 Park Street, Camberley, Surrey GU15 3PG	GeoVision's all-relational GIS product line, VISION* supports all major open systems platforms and commercial relational database management systems (RDBMS).	Open systems / client server basis - all-relational GIS DBMS - comprehensive user tools - application development tools (4GL) - industry-specific applications - full integration of third party applications.	0	
GGP Systems Limited Tim Maxwell 081 656 8562	12 Vincent Road, Croydon, Surrey CR0 6ED	GGP is a PC Digital Mapping / GIS system. Purpose built in the UK for OS 1:1250 / 1:2500 mapping. Many Local Authority users.	Inexpensive, easy to install and use. Runs on networks, for example, Novell, PC-NFS. Includes links to databases and spreadsheets.	532	108
Gimms (GIS) Limited Marlene Ferenth 031 668 3046	30 Keir Street, Edinburgh EH3 9EU	GIMMS mapping and graphics software for thematic/ statistical mapping. DIGIT-II digitizing software for precision data capture.	GIMMS is available for most mainframe platforms, as well as workstations and PCs. GIMMS also supports a wide range of output devices. DIGIT-II is available for the PC platform and supports a wide range of digitizers.	0	
Glen Computing Limited Graham Bugler 0689 875577	309 High Street, Orpington, Kent BR6 0NN	GEO/SQL - powerful, C-based, object-oriented topologically structured GIS with interactive embedded SQL and Autocad user interface.	SQL interface, seamless topographical database, intelligent indexing, comprehensive management tools, menu driven.	532	109

Organisation, contact, telephone	Address	Products	Outstanding features	Ref	Surrogate Page
Graphical Data Capture Limited Peter C Klein 081 349 2151	262 Regents Park Road, London N3 3HN	MapInfo, the leading cost effective desktop GIS mapping system. Integrates MapData with Database. Has an Autocad link and import/ export of data. All fully supported.	IBM PCs and all compatibles; versions for Windows, Apple Macintosh, SUN and HP; networks; realtime electronic tracking. Additional map data available.	0	
IBM United Kingdom Limited Preet Dhillon 0926 332525	UK GIS Solution Centre, PO Box 31, Birmingham Road, Warwick CV34 5JL	IBM, along with Business Associates, offer an enterprise approach to GIS - addressing departmental needs (data capture / maintenance & asset management) & also integration of GIS with corporate information systems.	IBM provides integrated GIS solutions based on the RISC System/6000 range of technical workstation, covering data capture from existing records, data modelling, infrastucture management & corporate integration.	533	110
ICL (UK) Limited Tony Grimbley 0532 441111	ICL House, 5 South Parade, Leeds LS1 5QZ	PLANES centralized GIS providing data integration and application development. DINIS network analysis. PROSCAN and SYSTEM 9 GIS.	ICL Series 39 mainframe range from stand-alone departmental systems to large DP centre systems. ICL SUN workstation range. PC range.	534	111
Kingswood Limited CJ Greenwood 081 994 5404	19 Kingswood Road, London W4 5EU	MAPInfo and ATLAS PC products; routing software; TRUCKSTOPS 2 scheduling software; OPTISITE depot location software; QUICKADDRESS post code geocoding.	KINGSWOOD are specialists at integrating GIS applications with vehicle routing and scheduling packages to solve logistics and distribution problems.	0	
Laser-Scan Rob Walker 0223 420414	Cambridge Science Park, Milton Road, Cambridge CB4 4FY	HORIZON is a 3 dimensional GIS designed specifically to address issues of environmental monitoring, management and protection.	HORIZON combines graphic and alphanumeric data in a range of forms from a variety of sources. It also has a terrain modelling capability.	0	
Laser-Scan Rob Walker 0223 420414	Cambridge Science Park, Milton Road, Cambridge CB4 4FY	METROPOLIS is a GIS for urban applications including land and property management, networks and demographic analysis.	METROPOLIS can integrate digital map data in raster or vector form and has a relational database for holding attribute data.	0	

Organisation, contact, telephone	Address	Products	Outstanding features	Ref	Surrogate Page
Logica Bob Walters 071 637 9111	Medina House, Business Park 4, Randalls Way, Leatherhead, Surrey KT22 7TW	MapInfo is a low-cost desktop mapping system. Runs on IBM PC or compatible, Macintosh, SUN and HP. Reads / writes dBase, Foxbase or other DBF format files.	As strategic VAR, Logica offers map data, MapInfo training, technical support, applications development, customisation and systems integration.	579	701
Mapping Information Technology Unit Dr David Parker 091 230 4450	Nuventures Limited, 18 Windsor Terrace, Jesmond, Newcastle upon Tyne NE2 4HE	SLIMPAC: mapping information system for property terrier, control of building complexes (industrial, retail, residential, educational), archeological site, construction site.	AutoCAD with real time link to dBASE 4 with wide range of specialist mapping functions. Fully user customisable or supplied as turnkey system. PC 386 system required.	0	
Midlands Regional Research Laboratory Dr Alan Strachan 0533 523849	Bennett Building, University of Leicester, Leicester LE1 7RH	Distributors of GRASS GIS, Raster GIS, installation and support materials.	System runs on SUN workstations, ACORN R260.34, Silicon Graphics workstations.	0	
Pafec Limited - PAFEC GIS Amanda Ward 0602 357055	Strelley Hall, Nottingham NG8 6PE	PAFEC GIS is open, flexible and modular; object oriented; relational database links; UNIX workstations; raster and vector.	Easy to use and customise; fast database information retrieval, analyses and results displays; expert advice; quality and reliability.	0	
PAMAP Technologies Corporation Alison Malis / Ron King +1 604 381 3838	301-3440 Douglas Street, Victoria, British Columbia, Canada V8Z 3L5	PAMAP GIS Version 3, a fully integrated Natural Resource GIS delivering the benefits of vector and raster structures. Acknowledged by over 300 users in 20 countries as an outstanding natural resource product.	Latest release provides new features for external database links, enhanced 3D manipulation / analysis, user interface, combined raster / vector plotting. Integrates R-S data. Superior quality cartography.	0	
Remote Sensing Services Limited John Allan 0672 20226	Lychgate House, 24 High Street, Ramsbury, Wiltshire SN8 2PB	RSSL distribute ERDAS, a comprehensive raster GIS and remote sensing system. Supported on PCs and workstations, it provides full modelling and visualisation facilities.	Virtual disc roam; geographic linking of windows; virtual LUTs for windows; GIS modelling and image processing; extensive data format support.	0	

Organisation, contact, telephone	Address	Products	Outstanding features	Ref	Surrogate Page
SIAS Limited Sarah McRae 031 557 0200	4 Heriot Row, Edinburgh EH3 6HU	BROWSER (Crown copyright) provides a user-friendly graphical interface to spatial data which may be downloaded from other computers and databases. Map, graph and report formats.	Digital VAX and VAXstation (VMS) and DECstation (ULTRIX), IBM PC-AT and compatibles (DOS), IBM PS2 (DOS), Logicraft 386ware (DOS).	0	
Siemens-Nixdorf Information Systems Ltd. Denis Shaughnessy 0344 850576	Siemens-Nixdorf House, Oldbury, Bracknell, Berkshire RG12 4FZ	SICAD: a corporate GIS.	European market leader; production oriented; high level development environment; leading edge data structures and data management; integration with external systems.	0	
Smallworld John Banfield 0223 460199	Burleigh House, 13-15 Newmarket Road, Cambridge CB5 8EG	Smallworld GIS - integrated GIS technology providing distributed access to distributed data in a single environment	Hardware supporting UNIX/VMS and X__Windows in a networked workstation environment - SUN SPARC, DIGITAL DECSTATION, IBM RS6000.	535	112
Structural Technologies Limited David F Schindler 0527 854819	Woodside, The Slough, Studley, Warwickshire B80 7EN	STRUMAP: interactive PC based GIS system. Rule-based definition of topology for networks, areas and spaghetti models. Imports and exports OSTF, NTF1-4, DXF and GFIS.	Validates imported data against user criteria for both cartographic features and attributes. Automated and semi-automated capture features including network macros and line-following.	0	
SysScan UK Limited John Moss 0344 424321	SysScan House, Easthampstead Road, Bracknell, Berkshire RG12 1NS	SysScan mapping and GIS products include TELLUS for an open systems environment. PC-based TerraSoft and DNMS for utilities.	Openness, modularity and industry standards - easy to customise to your needs - easy to expand. Digital, SUN, HP and PC platforms.	536	113
Systems Distribution International Ltd. Dermot O'Beirne 081 877 3744	Unit 6 Garratt Court, Furmage Street, Wandsworth, London SW18 4DF	Geo/SQL is a complete GIS/FM/AM system. Its superior spatial database and SQL data capabilities make it an ideal solution for most applications.	Seamless spatial database. Standard SQL (ORACLE). ease of use and development. Excellent graphics and hardcopy. Flexible: PC, UNIX or both.	536	114

Organisation, contact, telephone	Address	Products	Outstanding features	Ref	Surrogate Page
Tenet Systems Limited Alan Taylor 0403 711555	Clothalls, West Grinstead, Horsham, West Sussex RH13 8NE	MapLink is an interactive, object-oriented graphical tool for creation and editing of map data for use in GIS applications.	MapLink imports, transforms and merges diverse map data formats with declutter and LoD control for GIS application development.	537	115
Trimble Navigation Limited Helen Knight 0256 760150	Meridian Office Park, Osborn Way, Hook, Hampshire RG27 9HX	Trimble Pathfinder GPS is the ultimate tool for simple and accurate field collection of position, time and attribute data for over 140 GIS systems.	6 channel GPS reveivers, accurate to 2 metres in differential operation. Up to 15000 position storage, optional hand held data collectors and bar code reader.	0	
Tydac Technologies Limited Andrew Day 0491 411366	Chiltern House, 45 Station Road, Henley-on-Thames, Oxfordshire RG9 1AT	SPANS - spatial analysis systems products: SPANS GIS and SPANS MAP used in: environmental management, resource development, business, local government.	Best user interface in the industry. Industry leader in data integration. Unparalleled analysis and modelling. Easily customised.	0	

CONIC SYSTEMS LIMITED (101)

Locator is designed for mobile users, and concentrates on their special needs. To synchronise with the office system, to annotate and correct information in the field, and to integrate maps into a total mobile information resource. Locator takes full advantage of the latest technology, being designed for the PenPoint operating system. Drag-and-Drop, hyperlinks, document embedding, an object-oriented programming interface, printing and faxing, and a highly gestural user interface give a more effective computing environment. Key features of Locator include:

- Quick and easy map manipulation using gestures.
- Full layer control.
- Handles vector and raster data, with simultaneous display.
- Effective access to related information from the map display.
- Pen Point documents and hyperlinks can be embedded anywhere on a map.
- Scalable markup and annotation.
- Accurate drawing facility for surveying and update of maps on site.
- Change management allowing modifications to be extracted and communicated between systems, even by modem.
- Support for visual comparison of different versions of a map.
- Facilities for communication with other systems, allowing upload and download of map and database information.
- Object-oriented programming interface for close integration with other applications.
- Use of data compression minimises hardware requirements.

THE DATA CONSULTANCY (102)

The Data Consultancy is an established provider of GIS related systems including drive time systems and geodemographic analysis systems and is an authorised value added reseller of MapInfo the leading desktop GIS and mapping system. The Data Consultancy also provide a range of other software products designed for market analysis and site assessment studies. Although available as separate systems each system interfaces to the others to provide a full range of GIS, market analysis and mapping capabilities.

- MAPINFO: the leading desktop GIS and mapping system which sets the standard for cost effectiveness. It is particularly suitable for marketing, site assessment, planning and business applications. Includes display, editing, overlay, analysis, query, searching, geocoding and digitising in one easy to use package. Fast display, analysis and querying of user data files in dBASE, Lotus and Excel format. Ideal for mapping and analysing Census data at all spatial levels, including enumeration districts.
- **DRIVETIME:** an extremely powerful and flexible interactive system to generate drive time isochrones around any location in Great Britain. Outputs maps and plots and generates boundary files suitable for input to mapping, GIS and other software.
- **ILLUMINE:** a full geodemographic analysis and reporting system giving direct access to the Census of Population small area statistics for all 130,000 Census enumeration districts and/or postcode sectors in Great Britain. Analyses can be undertaken for any area however defined. The modular design allows users to start with a simple entry level system with a few variables and add additional variables, population updates, neighbourhood profiling, expenditure estimates, retail business turnover potential, drive time isochrones etc. as and when required.
- **MARKETS:** our widely acclaimed and widely used store location model which simulates and predicts shopping patterns and expenditure flows. Model results include store or scheme turnover levels, turnover to floorspace ratios, trade draw rates and market penetration by drive time bands or zones.

The software can be purchased on its own or fully configured with U.K. boundaries and data. A wide range of databases, boundary files, maps and analyses can be supplied.

ESRI (UK) LIMITED (103)

The world's leading GIS, ARC/INFO has been proven as successful in providing GIS solutions to many thousands of users in a wide range of application areas. Its success is due to the quality and reliability of its technology, strong data model, openness and independence, easy-to-use graphical interface and extensive spatial analytical functionality.

The key features which go a long way to guarantee successful GIS solutions are:

- ARC/INFO with over 2,200 tried and tested commands offer users the widest range and most sophisticated collection of geoprocessing tools. Spatial analysis tools include topological map overlay, buffer generation and proximity analysis, spatial and logical query, surface analysis, network modelling, raster modelling, and tabular analysis.
- A comprehensive suite of tools for displaying maps and charts on screens and hard copy devices includes support for typeset quality typefaces, true colour and over 40 map projections. ARC/INFO has extensive capabilities for classifying and symbolising point, line, area and surface datasets.
- ARC/INFO'S ability to work in heterogenous hardware, networking and database environment (e.g. mainframe, mini, workstations and PC's running multiple databases). Organisations can utilise existing PC's and mainframes and add special purpose GIS workstations where and when high performance is really needed.
- Support for both centralised and distributed databases. ARC/INFO can provide access to either centralised maps and tabular databases, or alternatively data can be distributed at department level throughout the network. ARC/INFO's open architecture also means it is possible to link it to existing databases and software routines.
- Ability to create continuous map bases and provide tools for corporate access.
- ARC/INFO's sister products of ArcView and ArcCAD offer users low cost, easy access to their data.

  Arc View is an easy-to-use query and display tool. With the intuitive GUI users can quickly learn how to access and visualise both map and tabular data, compose their own maps and output to plotters or DTP systems. ArcView runs on PC's, Macintosh and UNIX workstations.

  ArcCAD links the market leading GIS software(ARC/INFO) to the world's number one CAD software (AutoCad). ArcCAD extends the AutoCAD data model to include spatial relationships, a topological database, and selection, query, and analysis tools, which give new power for the development of specialised GIS applications within AutoCAD.

ESRI(UK) has created a portfolio of turnkey applications covering Highway Information Systems, Land Charges, Land and Property Terrier, Census and Development Control.

ESRI (UK) LIMITED (104)

ArcView is a new and innovative addition to ESRI's ARC/INFO portfolio of GIS software It represents a new breakthrough in GIS software which will make GIS available to tens of thousands of users. In essence, ArcView is a simple to use interface to geographical databases. It shifts the emphasis away from the interface and onto the data themselves. As a result, new and casual users can examine their data quickly and easily without having to learn a complex access mechanism.

ArcView is available on workstations, PCs and compatibles, and Apple Macintoshes. It is inherently graphical in its operation. Users simply point at features using the mouse and click to select the objects. Thus to see some data, users click on the name of the theme in the table of contents and it will be drawn. To zoom in, select the 'zoom icon' from the palette of twelve icon tools, point at the area of interest and ArcView automatically zooms in.

Within ArcView there are extensive features for the display of all types of ARC/INFO vector, raster, grid and attribute data either on their own or overlaid. These data can be queried and selected sets can be generated based on both spatial and attribute logic. Maps and statistical data can be printed out or transferred to other desktop packages for further analysis.

ArcView has many possible uses in an organisation. New and casual users will find it a quick and enjoyable access mechanism to databases. Since ArcView is read-only there is no risk of corrupting the data. Widespread access to an organisation's valuable data can be easily achieved because ArcView is a networkable product offering true client-server computing. An organisation may choose to hold their data in a single centralised database, but provide access via a local area network and cheap PC or Macintosh terminals. ArcView is also an electronic publishing system. Imagine distributing your data to other people in your organisation, or selling it to members of the general public with a simple to use interface that is easy and fun to use.

ArcView is clearly a significant advance on current GIS access systems. It represents a major improvement in the way we interact with geography and at a price many can afford.

ESRI (UK) LIMITED (105)

ArcCad is a new product which brings together in a unified environment, the world's leading CAD system, AutoCad from Autodesk, and the world's leading GIS , ARC/INFO from ESRI. ArcCAD is an extension of the AutoCAD data model that enables users to perform topological and spatial analytical operations on the entities in an Auto CAD drawing database.

With ArcCAD, database entities can be created and edited in AutoCAD, using AutoCAD's powerful graphical tools. A topological coverage can then be created from the same entities using ARC/INFO's topological BUILD functions. Once topological coverages have been created, all the powerful ARC/INFO functions are available for data display, query and analysis. Objects can be for data display, query and analysis. Objects can be selected according to any combination of spatial, graphical and attribute criteria. Spatial analysis tools permit users to perform point-in-polygon, polygon overlay, buffering, dissolve and many more operations. ArcCAD also allows attributes to be added to objects and stored in dBASE format files. Relational database management system functionality is thus available to ArcCAD users.

ArcCAD customizations are undertaken using AutoCAD ADS and AutoLISP commands. Thus, experienced users and third party developers can create simple to use, yet sophisticated turn-key applications based on ArcCAD.

ArcCAD is fully integrated with the entire ESRI product range and can be networked to provide an integrated portfolio of products to meet user requirements. For example, databases can be created using ArcCAD, viewed using PC ArcView, and analyzed using PC or workstation ARC/INFO.

Initially available on PC's, ArcCAD provides all the menus and GIS functions necessary to create a comprehensive display, query and analysis environment within AutoCAD, ArcCAD requires AutoCAD 386 Release 11 and dBASE is optional.

EURECART (106)

APIC, supplied and supported in the UK by Eurecart, is an advanced object-oriented GIS proven in operational use in a range of organisations around the world.

APIC KEY BENEFITS:

- Single continuous database - the user is not limited to map sheets and can access the whole database area instantly.
- Object orientation - objects can have graphical appearance, text attributes, relationships with other objects and behaviour associated with them - the user has the ability to model the world as it really is.
- Powerful and flexible query language and customisation tools - enables the user to create and build customised applications to meet specific requirements.
- Fully distributed database capabilities - maximises efficient use of databases and communications.
- Full GIS functionality including - polygon processing, thematic mapping, comprehensive reporting, planning, facilities management and network analysis.

APIC is in operational use in over 60 sites in commercial companies, utilities and government organisations at locations ranging from Brazil and China to countries throughout Europe.

With APIC, Eurecart deliver complete GIS solutions - from consultation to defined user requirements through to customisation and integration with existing systems and procedures.

GEOGRAPHIC MANAGEMENT SYSTEMS LIMITED (107)

PC-M.A.P (single screen) and **GS-M.A.P** (twin screen) are a Multi Application Product family of GIS software developed specifically for the MS-DOS PC market. All **M.A.P** systems run on IBM AT and PS/2 micros with two way links to external databases on remote hardware and operating systems.

Graphical data can be used as a powerful enquiry medium linked to unlimited text for use in a wide variety of applications.

M.A.P Software Provides:

- 90% of large system functionality at a fraction of the cost.
- Menu driven system with minimal training requirements.
- Modular construction permitting system expansion when required.
- Single and twin screen options available plus networking options.
- Multiple simultaneous linkages to up to 15 external databases as well as its own in-built files.
- Photographs and scanned raster images (plans, drawings etc) - as backdrop maps or referenced to elements.
- Logic network for route and flow analysis using its own spatial link and node database.
- Output of thematic maps and attribute hardcopy to plotters, laser printers etc.
- Easy data entry via digitizers, scanners and mouse. Direct input from Ordnance Survey digital data, CAD and mapping systems plus satellite.
- Two-way search facility between graphical and textual information databases by creating search fences and search masks.
- Unrestricted zoom capability (9^9).
- Seamless map background.
- Polygon overlay and comparison from up to 255 layers with automatic text carry-over when elements move from layer to layer.
- Automated vehicle location and multi-tracking from satellite G.P.S.
- Full data handling including editing tools for comprehensive information management.

Applications Include:

Local Authority - Planning, Land Register, Terrier, Highways, Public Works and contracts.
Property and Assets - Retail, Commercial and Facilities Management.
Location Tracking - Vehicle, Aeroplane and Boats.
Emergency Planning - Police, Military and Emergency Planning.
Archiving - Storage, Retrieval and referencing of scanned plans, drawings and photographs.

GGP SYSTEMS LIMITED (108)

GGP is a PC Digital Mapping/GIS system, purpose built in the UK for OS 1:1250/2500 mapping. It has many Local Authority users and is also inexpensive, easy to install and use. It runs on networks eg. Novell, PC-NFS and includes links to databases and spreadsheets.

Additional Features:

Geocoding of locations of, for example, lamp columns. Automatic length & area calculations. Radial, Buffer and Point-in-polygon spatial analysis. Multiple data files and graphic overlays supported (over networks if required). Inbuilt Street Gazetteer using Post Office data (not provided). Wide range of printers, plotters and digitisers supported.

GGP single user price is £1500, annual customer support programme including GGP upgrades costs £400 per copy. Optional onsite training / consultancy @ £500 per day is available.

Please call for more details or demonstration disk. An evaluation copy of the complete system including user manual is available.

GLEN COMPUTING LIMITED (109)

Geo/SQL offers the complete Geographical Information System solution for the management and analysis of all types of spatially referenced data. The software combines a powerful method of indexing spatial objects using SQL, the structural Query language database management standard, thus providing a common SQL language for managing and querying both spatial and attributable data.

Geo/SQL has been designed to work on integrated partnerships with today's industry-standard software. As its graphics front-end, Geo/SQL uses the AutoCAD graphics environment. For its SQL back-end, Geo/SQL is designed to work with a variety of SQL database management systems.

Geo/SQL applications exist for Line-Plant Distribution and Management, Environmental Protection Analysis and Data Management, Emergency Planning and Impact Modelling Analysis, Oil and Gas Field Facility Management, Land Records Information, Highways Facility Management and Marketing. Worldwide Geo/SQL platforms already serve the following industries; Oil, Gas, Water, Highways, Local Government, Telecommunications, Cable T.V.

Geo/SQL operates directly from within either the Sun Unix or DOS environments. A full range of hardware platforms and peripherals can be supplied as part of a total package. This package backed up by support, training, installation and data capture services.

IBM UNITED KINGDOM LIMITED (110)

IBM's enterprise approach to GIS is designed to meet three key objectives:

- Provide the comprehensive functionality required for such diverse departmental applications as data capture, digital mapping, network/polygon modelling, spatial analysis and infrastructure management.
- Ensure that GIS is integrated within the overall IT architecture and no longer viewed as an isolated and specialist tool.
- Support established and emerging industry standards covering Open Systems, distributed relational database architecture, distributed computing environment and application development.

The integrated GIS architecture allows conventional maps and descriptive data for objects on the maps to be stored in a relational database and subsequently retrieved, viewed, updated, interrogated and analyzed by personnel throughout an organisation. A variety of terminals can be used - full function GIS workstations, general purpose IBM PS/2's and X-stations, or enquiry only, business terminals.

In order to provide a comprehensive GIS solution set, IBM has established alliances with a number of GIS Business Associates, who can provide complementary applications. These work with the IBM GIS data model, enabling IBM to provide turnkey solutions-covering consultancy, implementation services and tailored applications designed to streamline departmental operations.

IBM GIS is currently used throughout the world in industries as diverse as:

- Utilities
- Local and Central Government
- Communications
- Transportation
- Asset/Property Management
- Retail/Site location
- Health and Environmental monitoring
- Mining and Oil Exploration

There are well-established UK, European and Worldwide User Groups which provide a forum to exchange ideas and share knowledge.

IBM GIS - *PUTTING DATA ON THE MAP*

IBM GIS together with the RISC System/6000 workstation brings a new dimension to the world of GIS that can help enhance end-user satisfaction, enable growth and help ensure the success of your GIS project. To learn more about the IBM GIS solution, contact your local IBM representative or IBM's GIS solution Centre.

ICL (UK) Limited (111)

PLANES - The heart of ICL's strategic GIS development consists of three closely linked information systems. These cover property and land parcel related information, network related information together with the spatial representation of this and other data. This modular approach allows the processing of these different information types to be optimised, especially important when considering the very large geographically-based information sets that will be commonplace in the future, whilst allowing complete integration where necessary.

PLANES has been purchased by over 110 customers worldwide.

The ICL philosophy is one of integration and co-ordination and although PLANES provides full database facilities, great emphasis is placed on its ability to link to external databases and applications. This enables PLANES to act as the Geographic Information Manager, providing common access to all data in textual or graphical form and gaining the maximum benefit from the geographic content that is an integral part of up to 80% of an organisation's information.

PLANES **Property** is essentially concerned with address based information. It allows for a central directory of this information to be linked to the internal PLANES database, (Information Manager), external databases and external applications and to the Network or Spatial modules.

ICL also supplies **PAM**s (**P**LANES **A**pplication **M**odules) for the following Local Government business needs:-

Environmental Health; Estate Management; Housing Management; Planning; Land Terrier; Land Charges; Contaminated Land.

PLANES **Network** is concerned with network information such as highways or utility networks. It provides a central directory of all network elements and connectivity with flow and time stamping information and is linked to Information Manager, external databases, external applications and to the property and Spatial modules.

Both these modules allow the incremental development of functional applications which are automatically integrated with GIS. Applications may share common data and can be developed by users using 4GL facilities without the need for scarce computer staff expertise.

PLANES **SPATIAL** is concerned with user-specific geographic data and general topographic data. It provides a topologically-structured, object orientated database that allows the geographical objects to be linked to either Property, Network, Information Manager or external systems. Spatially-based enquiry to all PLANES modules and linked systems is available through PLANES Spatial Desktop Access an MS-DOS/Windows3 based application.

The ICL approach to GIS also includes the provision of related workstation based applications, working either standalone or linked to PLANES. These include data take-on and management systems for digital mapping via a range of third party solutions, network analysis through DINIS Electricity and Gas and structured data editing and spatial analysis systems.

ICL supplies System 9 GIS (from Computervision) in the Local Government market for GIS applications including Land Charges, Asset Management, and general map production, which can be provided in a UNIX / X-Terminal / PC environment.

In the Utilities market, ICL supplies the PRO SCAN II GIS system for Network management and mapping. This, integrated with DINIS provides a complete Network planning system.

ICL is committed to providing the hardware, software and services needed to make the best use of geographically related information and through this provide integrated, corporate-wide geographically-based, business information systems.

SMALLWORLD (112)

Smallworld Systems entered the GIS market in 1988; our aim was to create the most technically advanced GIS available and be a market leader within 5 years. Smallworld GIS is the result of a comprehensive worldwide market survey which ensured product development was firmly rooted in the market's requirements. Smallworld also started with a distinct advantage - they were unencumbered by an existing product.

Phenomenal growth has been assisted by a European network of partnerships and subsidiaries. Smallworld Systems GmbH was established in partnership with Gelsenwasserr, the largest German gas and water company; Smallworld Systems BV and Smallworld Svenska AB have opened in Holland and Sweden and the company has representation in France, Spain, Norway, Switzerland and Czechoslovakia.

Smallworld's biggest single asset has always been its people; the company's reputation for technological excellence and sales success ensures that, despite increasingly rapid growth, the calibre of staff remains unrivalled.

Consultancy, project management, systems integration, implementation, data conversion, training and after sales support allow our clients to derive the greatest benefit from their investment. A wide range of essential capabilities are already implemented in a Smallworld System; in addition specific applications have been developed with customers in gas, electricity, water supply, drainage, telecommunications, environmental, cadastral, local and regional government and facilities management.

Investment in software development tools and the use of standards, particularly Unix, X, TCP/IP and SQL, has resulted in the most easily customised system on the market. Currently running on high performance Unix workstations from Sun, Digital Equipment and IBM; the fundamental architecture frees users from dependence on any single supplier of hardware, or any one database. The system's unique features are based on its own powerful programming language - Smallworld Magik - an interactive object-oriented language for the implementation of large interactive systems. Smallworld Magik offers one environment and one language for system programming, applications development and customisation; applications are easily transferred between hardware platforms and programs are developed in a seamless environment. Developers using Smallworld Magik can prototype new applications very quickly and existing systems can be fully integrated appearing as part of one homogeneous system.

The Smallworld system is a continually evolving product with new applications regularly incorporated. Having developed a highly competitive, industry standard and hardware independent GIS platform, Smallworld is on target to become the market leader and support the future expansion of GIS worldwide.

SYSSCAN UK LIMITED (113)

Since the early eighties, SysScan's products have been recognised as pace setters in GIS. SysScan develops and markets Geographic Information Systems (GIS) and digital systems for spatial and map information. Our users range from small study groups to major mapping agencies, including well over 100 national mapping authorities and utilities across the world.

SysScan's digital mapping software combines maps of any scale with records and attribute data, and includes interactive editing and map management facilities.

TELLUS is a powerful and flexible GIS toolset. Its open, modular and object-oriented design makes Tellus easy to customise to your needs and easy to expand. Based on industry standard client/server technology, Tellus is designed for compatibilty and freedom of choice.

For utilities, SysScan's **DNMS** application software keeps all network data in Digital's widely used relational database, Rdb. **DNMS*CABLE** is SysScan's leading solution for the communications industry.

GEOREC automated data conversion software delivers maps or map related information, to SysScan and other vendors' GIS, dramatically faster than manual digitising methods. GEOREC takes scanned and vectorised input and delivers structured data via a process of pattern recognition algorithms.

For the PC user, **TERRASOFT** provides a low cost entry into all GIS applications.

SYSTEMS DISTRIBUTION INTERNATIONAL (114)

Geo/SQL has become the new standard for GIS and FM. The revolutionary design adapted by its developers, Generation 5 Technology Inc., allow for solutions to be achieved quickly and effectively.

Applications include Local and Central Government, Health, Industry, Utilities, Marketing, etc. As well as conventional GIS (eg. Contaminated Land, Farm Management, Water and Cable networks, Resource analysis, etc), it is particularly suited to Estates and Asset Management applications.

Geo/SQL provides full spatial referencing and analysis of data to meet the information systems needs of organisations. Through Geo/SQL, GIS technology can be introduced in a controlled manner. Users may develop their own applications, or use an SDI ready built solution as the basis for meeting urgent or long term operational needs. GIS operations can be expanded across the organisation as needs grow, and budgets and experience permit.

Geo/SQL's inherent flexibility is further enhanced by its open systems approach, permitting its use across PC and SUN-Unix based stand-alone and networked configurations. Its use of industry standard graphics and relational database technologies (such as AutoCAD and ORACLE) maximises use of existing skills and data sources and minimises training and implementation timescales. Full function GIS technology is therefore available at an acceptable cost.

TENET SYSTEMS LIMITED (115)

Tenet offers a family of open and extensible, object-oriented software tools used for development of real-time, interactive graphical applications for workstations. **GMS** is a tool for creation of the graphical user interface; **MapLink** is an interactive map editing tool, **GSQL** a graphical interface for relational databases and **Connex** a connectivity editing tool with **Envoy** addressing the management of interprocess communications. Together they form a comprehensive and adaptable environment for rapid development of interactive graphical applications. Tenet also provides software development and engineering services to help applications developers to apply and exploit these novel tools in their own systems.

GIS-linked application software & system

Organisation, contact, telephone	Address	Products	Outstanding features	Ref	Surrogate Page
W.S. Atkins Management Consultants Angus Looney 0372 726140	Woodcote Grove, Ashley Road, Epsom, Surrey KT18 5BW	GEOSCOPE - a PC-based system running under Microsoft Windows which can store, merge, analyze and display geographic data in maps and reports.	An object-oriented GIS with intuitive help screens, Data can be dynamically exchanged between EXCEL and other Windows applications.	0	
CDR Mapping Limited Christopher Cooper 0433 621282	Hawthorn House, Edale Road, Hope, Sheffield S30 2RF	Supergazetteer is used for loading data into GIS. All data is stored in Oracle and is available for other packages. Full support services provided.	Map management of all NTF / OSTF mapbase using industry-standard SQL databases. Ideal for data loading and GIS use. Stand-alone PC to full DOS / UNIX networking.	0	
Conic Systems Limited Iain Cooke 031 667 2728	20a Mayfield Terrace, Edinburgh EH9 1SA	Locator - an object-oriented GIS for portable, pen computers. Integrates with office systems, so field personnel can use and gather spatial and other data.	Runs on all PenPoint hardware. Raster and vector display and capture. Full range of supporting software available - forms, images, diaries etc.	526	101
The Data Consultancy Bryan Wade 0734 588181	7 Southern Court, South Street, Reading, Berkshire RG1 4QS	DRIVETIME the definitive PC system for generating accurate drivetime isochrones around any point in Great Britain. Supplied complete with road networks and a gazetteer of over 1000 places.	DRIVETIME gives complete control over the generation of drive time isochrones. Up to four hour and composite isochrones. Outputs in many formats.	527	102
Empress Software UK Dennis Flavell 0483 861990	Godalming Business Centre, Woolsack Way, Godalming, Surrey GU7 1XW	EMPRESS: an advanced high performance RDBMS and 4GL with query language, report writer, user defined operators, extended data types and programming interfaces.	Standard product supports special BULK data type suitable for 'object-orientated' co-ordinate data, SQL and custom query (eg spatial) extensions.	0	

Organisation, contact, telephone	Address	Products	Outstanding features	Ref	Surrogate Page
ESRI (UK) Limited Carole Smith 0923 210450	23 Woodford Road, Watford, Hertfordshire WD1 1PB	ESRI (UK) recognize the many requirements for turn-key solutions with the development of a portfolio of ARC/INFO based GIS application packages.	ESRI's Application Portfolio includes Highways Information System, Local Land Charges, Land & Property Terrier, Census, Development Control, Map Management System with others under development.	542	201
Eurecart Andy Wolfe 0621 851114	Ulting Lane, Hatfield Road, Langford, Essex CM9 6QA	APIC is an advanced object-oriented GIS, including development environment, customisation tools and a powerful query language.	APIC is an advanced GIS proven in operational use around the world, utilised for many applications improving business performance.	543	202
Fraser Williams (Commercial Systems) Ltd Sandra Bullock 071 240 8011	11 Maiden Lane, Covent Garden, London WC2E 7NA	SKYLINE is designed to manage all aspects of the property business for government agencies, local government, utilities, finance and insurance, leisure and retail, property investment companies.	Single comprehensive database using open systems allowing easy links to other software/ hardware and offers flexibility, portability and ease of use.	0	
Geographic Management Systems Limited Bill Knowles 0454 618618	Woodlands Grange, Woodlands Lane, Almondsbury, Bristol BS12 4JY	PC-M.A.P (single screen) & GS-M.A.P (twin screen) are multi-application GIS developed specifically for the networked MSDos PC market running on IBM AT & PS/2 micro's.	Runs under Dos with bi-directional links to multiple external databases on remote hardware and operating systems (eg McDonnell Douglas, ICL, UNIX, Pick etc.)	531	107
GeoVision Systems Limited Walt Beisheim 0276 677707	80 Park Street, Camberley, Surrey GU15 3PG	GeoVision's all-relational GIS product line, VISION* supports all major open systems platforms and commercial relational database management systems (RDBMS).	Open systems / client server basis - all-relational GIS DBMS - comprehensive user tools - application development tools (4GL) - industry-specific applications - full integration of third party applications.	0	
Glen Computing Limited Graham Bugler 0689 875577	309 High Street, Orpington, Kent BR6 0NN	GEO/SQL - powerful, C-based, object-oriented topologically structured GIS with interactive embedded SQL and Autocad user interface.	Applications in line plant management, facilities management, environmental management, retail and marketing and oil exploration.	532	109

Organisation, contact, telephone	Address	Products	Outstanding features	Ref	Surrogate Page
IBM United Kingdom Limited Preet Dhillon 0926 332525	UK GIS Solution Centre, PO Box 31, Birmingham Road, Warwick CV34 5JL	IBM, along with Business Associates, provide a comprehensive suite of GIS applications running on RISC System / 6000 workstations designed to cover most aspects of GIS usage - data capture - maintenance - analysis.	IBM's application development approach provides 'off-the-shelf' solutions covering basic GIS functionality and conforming to Open Systems standards along with a highly productive application development environment.	544	203
IMASS Limited Sue Gadd / Peter Taylor 091 213 5555	Information Management and Systems Solutions, Eldon House, Regent Centre, Gosforth, Newcastle upon Tyne NE3 3PW	iGIS product set running on Intergraph UNIX systems.	Includes facilities for water distribution and sewerage asset management, corporate gazetteer, customer services management.	0	
Laser-Scan Rob Walker 0223 420414	Cambridge Science Park, Milton Road, Cambridge CB4 4FY	VTRAK provides a semi-automatic means of digitising map based data by interactive line following from a raster image.	VTRAK can follow a variety of linestyles within binary, greyscale or colour raster data, and produces vector data in a choice of formats.	0	
Modulo 4 Limited Martin Horner 0532 771344	Suite 27A, Concourse House II, 432 Dewsbury Road, Leeds LS11 7DF	Land and property information systems for government, local government and commercial clients.	Land allocation, land administration, property management and asset registers running on ICL, Unisys and SUN systems.	572	605
MVA Systematica Keith Dugmore 0483 728051	MVA House, Victoria Way, Woking, Surrey GU21 1DD	SASPAC: software for the analysis of 1991 UK census Small Area Statistics and similar datasets. MATPAC: software for analysing 1991 census Workplace and Migration origin / destination statistics.	SASPAC has already been purchased by more than 250 organisations. Both SASPAC and MATPAC run on PC's and many mainframe machines.	0	
Norcom Technology Limited Stuart Duncan 0603 765252	7A Friars Quay, Norwich, Norfolk NR3 1ES	Real time data acquisition systems for hydrographic / environmental surveys, high accuracy positioning etc.	IBM PC compatible systems with comprehensive interfacing and display capabilities. Specialist post processing facilities with output to CAD / GIS system formats.	545	204

Organisation, contact, telephone	Address	Products	Outstanding features	Ref	Surrogate Page
Pinpoint Analysis Limited Martin Higgins 071 612 0568	Tower House, Southampton Street, London WC2E 7HN	GEOPIN, our PC ARC/Info based GIS, supports applications for marketing, retailing, network analysis and modelling.	Functions include interactive query, map and report display, with the ability to incorporate a wide range of datasets.	0	
Siemens-Nixdorf Information Systems Ltd. Denis Shaughnessy 0344 850576	Siemens-Nixdorf House, Oldbury, Bracknell, Berkshire RG12 4FZ	Applications for local government and utilities; includes planning, land charges, network management and analysis.	High degree of automation for productivity benefits; interworking of applications guaranteed through common data models; integration with external systems and applications.	0	
SysScan UK Limited John Moss 0344 424321	SysScan House, Easthampstead Road, Bracknell, Berkshire RG12 1NS	Applications include DNMS*CABLE for design, planning and management of cable TV / telephony networks and GEOREC for data conversion.	DNMS*CABLE uses Digital's leading edge relational database Rdb to allow for integration with existing databases.	536	113
Trimble Navigation Limited Helen Knight 0256 760150	Meridian Office Park, Osborn Way, Hook, Hampshire RG27 9HX	Trimble Pathfinder GPS is the ultimate tool for simple and accurate field collection of position, time and attribute data for over 140 GIS systems.	6 channel GPS reveivers, accurate to 2 metres in differential operation. Up to 15000 position storage, optional hand held data collectors and bar code reader.	0	
Tydac Technologies Limited Andrew Day 0491 411366	Chiltern House, 45 Station Road, Henley-on-Thames, Oxfordshire RG9 1AT	SPANS application development toolkit. Enables direct development of customised GIS applications to fulfil specific GIS needs.	Powerful GIS applications toolkit provides easy development environment. Enables users to develop third party applications.	0	

ESRI (UK) LIMITED (201)

A notable trend in GIS is the increase in inexperienced end users. For such uses, ESRI (UK) has invested many years of programmer and analyst effort to create a portfolio of turn-key applications. The main ARC/INFO-based applications available at present include:

HIGHWAYS - this allows county councils to map and analyse road inventory and condition survey data. It provides a dynamic link to Oracle's RMMS.
LAND CHARGES - provides a comprehensive answer to the needs of local government land charges departments. Users can select properties to search and can automate completion of CON29 and LLC1 forms.
LAND AND PROPERTY TERRIER - many organisations have a requirement to manage the property they own. This application allows users to look at the transactional history of ownership, to determine the amount of land currently owned and to manage letting and licensing.
CENSUS - the 1992 population census is one of the basic data sets used in many organisations. The census application provides quick and easy access to census data. Reports can be produced incorporating maps, tables and descriptive statistics.
MAP MANAGEMENT SYSTEM AND GAZETTEER - management of large vector and raster databases is a substantial task whether the data are in paper or digital format. The Map Management system controls updates and allows gazetteer and spatial selection of data.
In preparation - applications currently being developed include: Contaminated land register; Development and building control; and Planning.

EURECART (202)

APIC is an advanced, object-oriented GIS which is in operational use in over 60 sites in commercial companies, utilities and government organisations at locations ranging from Brazil and China to countries throughout Europe.

APIC is used to manage and coordinate activities in:

UTILITIES:

- Network management
- Scheme Management
- Customer services
- Emergency Planning
- Incident control
- Interruptions to supply

LOCAL AUTHORITIES:

- Traffic management
- Urban planning
- Urban service management
- Facility management
- Pollution control
- Demographics

TELECOMMUNICATIONS

PRIVATE SECTOR MARKETS

APIC is a continuous object-oriented GIS. This means APIC provides advanced application development facilities and easily integrates into existing systems and forms links with SQL databases and distributed databases.

A high-quality product such as APIC, is essential to achieve successful geographic business management. However, technology alone will not deliver the full benefits. That is why three of Europe's top business and technical suppliers: Eurecart, SEMA Group and CAPITA Group have come together in the APIC Partnership. We work with you to ensure that this technology delivers real benefit within your organisation, providing an environment where solutions are tailored to meet your specific requirements.

IBM UK LIMITED (203)

IBM's application software and systems are currently being used by industries as diverse as local and central government agencies, utility companies, transportation departments, petroleum companies, telecommunication firms, retail organisations and others seeking more efficient ways of managing their geographically related land and facilities data.

Applications are built using and object-based data model which takes graphic data (what it looks like - vector and/or raster), descriptive data (what it is), spatial data (where it is) and connectivity data (what it is linked to) and integrates it all within an industry standard relational data base model. The data is stored as a continuous map base with the ability for a user to specify the portion and type of data needed. Database retrievals can be area or network based. If required retrievals can be modified by user-definedqualification criteria - using layers, object types and attribute data fields.

This object-based data model offers four primary benefits to GIS users:

- It enables spatial information to be shared between different departments by eliminating the reliance on special purpose GIS structures.
- It permits integration of spatial information either other applications within an enterprise - such as work management systems, customer information databases, in-house publishing and corporate decision support applications.
- It delivers a true information system by enabling corporate users to access both spatial and non-spatial data, using general purpose terminals if required - thus reducing the cost of implementing a GIS.
- It uses industry standard data security, integrity, back-up and recovery features augmented by spatial indexing routines designed to enhance GIS performance.

TAILORED APPLICATIONS:

Also working with the IBM GIS data model are a number of function - specific departmental applications, developed by our GIS Business Associates worldwide. The application portfolio currently includes:

- Data Capture (covering both manual and assisted digitising).
- Map Management
- Network and Polygon modelling and analysis
- Transportation - infrastructure management, routing and analysis
- Site selection and catchment area analysis (using 1992 census data)
- Asset/property management
- Environmental monitoring and impact analysis

IBM GIS - *PUTTING DATA ON THE MAP*

This application portfolio is continually being extended and a current list can be obtained from your local IBM representative or from the UK GIS Solution centre.

NORCOM TECHNOLOGY LIMITED (204)

NORCOM TECHNOLOGY is an independent software and systems development company specialising in real time data acquisition, display and processing applications including navigation, hydrographic survey, and environmental survey.

The client base includes Ports and Harbours, contracting survey companies, Government institutions and Defence agencies

NORCOM's systems are designed for use on IBM PC compatible machines. Specialised interfacing cards allow data capture form a wide range of positioning systems(including GPS) and survey sensors. Data can be edited and selected before transfer to CAD or GIS systems.

DXF exchange format is commonly used, but translation routines can be developed to suit most formats.

Hardware platforms, peripherals & networks

Organisation, contact, telephone	Address	Products	Outstanding features	Ref	Surrogate Page
Camel Services Limited David Turnbull 0865 512678	66 Banbury Road, Oxford OX2 6PR	DITHER is a plotting accessory for AutoCAD and other vector-graphic users who need to add a colour wash to spot and full colour maps.	60 line / fill colours. Replot with new colours from the plot files. Fast, no bit-image conversion on the PC/WS. True / merged colour and selective erase.	0	
Gammadata Computer Limited Anne Chisholm 0256 817640	Worting House, Worting Road, Basingstoke, Hampshire RG23 8PY	GAMMACOLOR 300: modular, multi-functional colour printer; A3 and A4 300 dpi colour thermal transfer and A4 dye sublimation printers available.	Optional hardware and software modules allow the printing of screendumps, HPGL plots and Postscript files. Connection accessories and networking solutions also available.	0	
GeoVision Systems Limited Walt Beisheim 0276 677707	80 Park Street, Camberley, Surrey GU15 3PG	GeoVision's all-relational GIS product line, VISION* supports all major open systems platforms and commercial relational database management systems (RDBMS).	Open systems / client server basis - all-relational GIS DBMS - comprehensive user tools - application development tools (4GL) - industry-specific applications - full integration of third party applications.	0	
Glen Computing Limited Graham Bugler 0689 875577	309 High Street, Orpington, Kent BR6 0NN	Turnkey PC systems, peripheral equipment includes OCÉ, Calcomp, GTCO, Mutoh, HP, SPEA, TDS and Versatec.	Software products include GEO/SQL, AutoCAD, Oracle, CADOverlay, specialist data capture applications, links with industry standard peripherals.	532	109
IBM United Kingdom Limited Preet Dhillon 0926 332525	UK GIS Solution Centre, PO Box 31, Birmingham Road, Warwick CV34 5JL	IBM provides a broad range of workstations, PC's, peripheral devices & communications equipment - which support both IBM GIS and also a wide variety of third-party GIS software.	IBM offers a full & flexible range of workstations providing cost-effective solutions using IBM RISC System / 6000 technical workstations & the IBM PS/2 range using the OS/2 operating system.	549	301

Organisation, contact, telephone	Address	Products	Outstanding features	Ref	Surrogate Page
Kirstol Limited Peter Dickinson 061 338 7512	Cheethams Park Estate, Park Street, Stalybridge, Cheshire SK15 2BT	High resolution colour and monochrome large format scanners. Also raster output plotters and inkjet printers.	Real time colour separation and RGB 24 bit/pixel. Maximum true 1,016 dpi resolution, scan area 120cm x 120 cm.	100	302
Papergraphics Limited Ian W Baxter 0342 312488	Kings House, Cantelupe Road, East Grinstead, West Sussex RH19 3BE	A single source for all plotting supplies to support every plotter technology and application.	A range of application specific supplies pertinent to the GIS, seismic and mapping functions needs.	0	
Siemens-Nixdorf Information Systems Ltd. Denis Shaughnessy 0344 850576	Siemens-Nixdorf House, Oldbury, Bracknell, Berkshire RG12 4FZ	Full range of mainframe, server, workstation, PC and peripheral products.	Wide performance and capacity range; outstanding quality and reliability; full support for open systems standards.	0	
SUN Microsystems Limited Jonathan Cooper 0276 20444	Watchmoor Park, Riverside Way, Camberley, Surrey GU15 3YL	SUN SPARCstations and SPARCservers; extensive range of UNIX-based desktop workstations, network servers and connectivity solutions.	Market leading range of workstations and servers. Popular platform for GIS application developers and users, offering industry leading price / performance, application availability and flexibility.	0	
Taywood Data Graphics Pat Jarvis 081 575 9240	Greenford House, 309 Ruislip Road East, Greenford, Middlesex UB6 9BQ	ANA Tech 3640 high resolution scanner, driven by a PC, Apple Macintosh or UNIX workstation. Packaged solution covering scanned maps / drawings, optical disc and database software.	With a scanning solution provided in-house, and a full bureau service back-up, potential users can take full advantage of the capabilities of raster data.	594	906
Taywood Data Graphics Ralph Yardley / Kevin Ovington 0642 750515	Teesside Industrial Estate, Thornaby, Stockton-on-Tees, Cleveland TS17 9LT	ANA Tech 3640 high resolution scanner, driven by PC, Apple Macintosh or UNIX workstation. Packaged solution covering scanned maps / drawings, optical disc and database software.	With a scanning solution provided in-house, and a full bureau back-up, potential users can take full advantage of the capabilities of raster data.	594	906

Organisation, contact, telephone	Address	Products	Outstanding features	Ref	Surrogate Page
TDS Digitizers Limited Nick Walker 0254 676921	TDS House, Lower Philips Road, Blackburn, Lancashire BB1 5TH	TDS manufactures a range of high precision Mantissa digitizers - opaque and backlit - and supplies associated scanner and plotter products.	Mantissa - highest level of resolution, accuracy and performance available. Market leading A0 scanners. Revolutionary A0 inkjet raster plotter.	0	
Trimble Navigation Limited Helen Knight 0256 760150	Meridian Office Park, Osborn Way, Hook, Hampshire RG27 9HX	Trimble Pathfinder GPS is the ultimate tool for simple and accurate field collection of position, time and attribute data for over 140 GIS systems.	6 channel GPS reveivers, accurate to 2 metres in differential operation. Up to 15000 position storage, optional hand held data collectors and bar code reader.	0	
Zeh Graphic Systems (Europe) Limited Paul Popeck 0306 740105	Tillingbourne Court, Dorking Business Park, Station Road, Dorking, Surrey RH4 1HJ	Sophisticated colour raster plotting software available on UNIX and other systems. Designed for GIS and CAD users. Feature rich. Used internationally.	Supports all types of raster plotters. Montages raster with vector data. Hershey and Compugraphic font handling. Plot file management and scheduling.	0	
Carl Zeiss (Oberkochen) - Surveying Limited Ernie Wickens 0963 33425	The Stable House, Holbrook House, Holbrook, Wincanton, Somerset BA9 8BS	Comprehensive range of photogrammetric and surveying instruments with associated software.	Photogrammetric instruments may be used with Zeiss Phocus software or interfaced to a variety of GIS or CAD products from other manufacturers.	0	

IBM (UK) LIMITED (301)

RISC SYSTEM/6000 PRODUCTIVITY: The Performance Optimisation with enhanced RISC (POWER) architecture, gives the IBM RISC System/6000 outstanding power at an affordable price. It thus provides a means of distributing GIS data and processing power to a greater number of users at a lower cost per user - using the full power of distributed network computing.

OPEN SYSTEMS: The RISC System/6000 utilises AIX, IBM's enhanced version of the industry standard UNIX operating system, providing users with open architecture of the UNIX environment as well as interoperability with IBM systems Application Architecture (SAA) systems. AIX Version 3 conforms to all of today's important UNIX standards and as key industry standards evolve, AIX will continue to adhere to them - given IBM's commitment to Open Systems.

AN INVESTMENT FOR THE FUTURE: The RISC system/6000 family provides a powerful range of strategic workstation platforms, not only for GIS but also for other departmental and engineering applications. Thus users can run CAD, network analysis, text processing, publishing, spread sheets, project management, electronic mail and other applications alongside GIS - and also use the RISC System/6000 within a client/server environment, accessing host-based applications.

PERIPHERALS: IBM recognises that the peripherals are an important component of a GIS, as they provide the all important user interface. IBM therefore provides a range of input and output devices, together with support for other industry standard GIS peripherals, e.g. electrostatic plotters, high resolution digitisers and A0 scanners. Contact the IBM UK GIS Solution Centre for a current list of peripherals available for connection to IBM workstations.

NETWORKS: Maximum GIS benefit is associated with open access to information by users in various departments within an organisation - often involving the connection of terminals, workstations, PS/2's and host computers across distances that require local or wide area networks. These networks often need to work in a multi-vendor environment and IBM can provide a comprehensive set of network architectures and products to fulfil these requirements - supporting both Open System (OSI and OSF) and IBM SNA protocols. Practical demonstrations of this capability can be arranged at a number of "Connectivity Centres" based at IBM offices throughout the UK.

The use of networks can be further enhanced by employing a value-added-network of the type provided by IBM's Information Network Services. This can provide a valuable means of both developing network-based applications and of accessing other database information or integrating GIS with other business systems.

IBM GIS - *PUTTING DATA ON THE MAP*

To learn more about the power and flexibility of the IBM GIS solution, contact your local representative or IBM's UK GIS Solution Centre.

KIRSTOL LIMITED (302)

Kirstol Limited, distributor/agents for the Industrial Graphics Department of Dainippon Screen Manufacturing Co Ltd, Japan expand their established scanners and film making systems in self colour printing into the GIS fields.

Dainippon Screen's Large Format Scanner is complete with a unique Self Colour Discrimination Unit(SCD) and is particularly suited for the GIS, Self Colour Printing and related industries.

The system incorporating the ISC-1200PD Colour Scanner answers the need for higher quality input of colour images at high resolutions - up to 1000 lines/inch (40 lines/mm) and the facility to scan large format colour originals of up to 1200 mm x 1200mm. Large format colour scanning is available in full colour (24 bit RGB) or 8 bit colour coded data and 8 bit run length data, providing up to 255 user defined separated colours in a single scan, with high speed and quality data capture.

Remote sensed, mapping & GIS data products

Organisation, contact, telephone	Address	Products	Outstanding features	Ref	Surrogate Page
The Association for Geographic Information Maxine Allison 071 222 7000 /226	12 Great George Street, Parliament Square, London SW1P 3AD	Central government Datasets list.	A paper list of the titles of central government spatially referenced Datasets.	0	
The Association for Geographic Information Maxine Allison 071 222 7000 /226	12 Great George Street, Parliament Square, London SW1P 3AD	Central government Metadata directory.	A paper directory detailing central government spatially referenced datasets.	0	
The Association for Geographic Information Maxine Allison 071 222 7000 /226	12 Great George Street, Parliament Square, London SW1P 3AD	Central government Metadata disk.	Floppy disk documenting central government spatially referenced datasets.	0	
Automobile Association Ralph Robbins 0256 492906	Information Services, Fanum House, Basingstoke, Hampshire RG21 2EA	AA copyright data of GB, Europe & C. London. Structured place names gazetteers. Roadwatch dynamic traffic reports. Restricted road data. AA Milemaster calc. package. European postcodes.	GIS, vehicle scheduling & consumer location systems. Tourist & travel information. Route planning & scheduling systems. Databases levelled & feature coded, supplied as vectors or link/node.	0	
Bartholomew Kevin Dickson 031 667 9341	12 Duncan Street, Edinburgh EH9 1TA	A range of digital map databases including 1: 250,000 of Great Britain, 1: 14,000,000 of the World and 1: 5000 of London; postcode sector boundaries for Great Britain.	Bartholomew databases have application in the fields of marketing research, command and control, route planning, management support systems and many more.	0	

Organisation, contact, telephone	Address	Products	Outstanding features	Ref	Surrogate Page
CACI Limited John Rae 071 602 6000	CACI House, Kensington Village, Avonmore Road, London W14 8TS	Census data 1991, 1981 & 1971 - ACORN geodemographic classification - retail expenditure databases - indices of financial consumption & purchasing behaviour - current unemployment & population projections.	First census agency in the UK. First geodemographic classification in the UK.	0	
The Data Consultancy Bryan Wade 0734 588181	7 Southern Court, South Street, Reading, Berkshire RG1 4QS	Specialists in the provision of spatially referenced data sets including maps, boundaries, roads, demographics, income, expenditure, stores, centres and other data.	Data sets supplied in a wide range of formats for use with a wide variety of mapping, GIS and other software. A full data service.	555	401
Gazetteer Systems Robin A Hooker 0450 370959	1 Northcote Street, Hawick, Roxburghshire TD9 9QU	Index to the Ordnance Survey 1:25000 PATHFINDER™ second series map for the Great Britain. PATHFINDER™ is a registered trademark.	Four times more detailed than any currently available gazetteer for Great Britain. 100 metre accuracy. Versatile ASCII format.	0	
GeoMatrix Geoffrey Beacon 0742 724272	The Cooper Buildings, Sheffield Science Park, Arundel Street, Sheffield S1 2NS	Postline postal geography.	Postline: accurate and up to date; sector and district files; wide range of formats; highly competitive price.	0	
Geoplan (UK) Limited Andrew Kelly 0423 566755	14-15 Regent Parade, Harrogate, North Yorkshire HG1 5AW	Digitized postcode boundaries for the United Kingdom.	Royal Mail GEOPLAN is the definitive postcode boundary file. It is digitized from 1:50,000 scale maps and updated at six monthly intervals.	0	
Graphical Data Capture Limited Peter C Klein 081 349 2151	262 Regents Park Road, London N3 3HN	ED91, the 1991 census enumeration district boundary set for population mapping, also as ward, parish, health district, county boundaries plus Post Code/ ED tabulation.	Available at different filters (scales). Low cost and high accuracy. Use with and without census data. All OS fees paid. Available for most platforms.	0	

Organisation, contact, telephone	Address	Products	Outstanding features	Ref	Surrogate Page
Graphical Data Capture Limited Peter C Klein 081 349 2151	262 Regents Park Road, London N3 3HN	SlimNet simplified MapData for the UK based on 1:10000 scale and includes centreline of Motorways, A, B and C roads, roundabouts, bridges etc. located + gazetteer of road names.	SlimNet comes in three levels of density, in NTF, OSTF, ARC/Info and many other popular formats. Economical in price. Excellent thematic mapping and lots more.	0	
Kingswood Limited CJ Greenwood 081 994 5404	19 Kingswood Road, London W4 5EU	Distributors for AA data; GB - national coverage; central London to street level; European atlas plus postal areas; UK census and postal geography.	AA data can be provided in a format of the Client's own choice for use in industry standard packages or customised systems. It is accurate, current and affordable.	0	
MVA Systematica Keith Dugmore 0483 728051	MVA House, Victoria Way, Woking, Surrey GU21 1DD	ED-LINE: the authoritative national 1991 census digital enumeration district boundaries. Related products are WARD-LINE, and the simplified *MARKETEER boundaries for PC marketing systems.	ED-LINE is integrated with the Ordnance Survey's administrative boundaries, and is ideally suited for use with ADDRESS-POINT. ED-LINE has been purchased by more than 300 organisations.	0	
National Remote Sensing Centre Limited Dr Nick Veck 0252 541464	Delta House, Southwood Crescent, Southwood, Farnborough, Hampshire GU14 0NL	Suppliers of satellite and airborne imagery: GIS and image processing; consultancy and mapping services.	Priority is given to adopting a systematic approach to projects, ensuring high quality products and services which match customer requirements.	0	
Ordnance Survey Digital Sales 0703 792324	Romsey Road, Maybush, Southampton, Hampshire SO9 4DH	Digital data is available for most Ordnance Survey of Great Britain mapping products. A digital map data catalogue is available on request.		0	
Photogrammetric Data Services Limited Robert Finch 0273 464883	1 Ham Business Centre, Brighton Road, Shoreham-by-Sea, West Sussex BN43 6PA	Digital data capture using analytical stereo compilation devices, for 2D/3D production. Full range of aerial survey services provided with emphasis on photogrammetric applications. All markets supported.	In-house translation to a variety of output formats. International capability with projects in UK, Europe, Africa and the USA.	0	

Organisation, contact, telephone	Address	Products	Outstanding features	Ref	Surrogate Page
Ramtek (UK) Limited Jonathan Shears 0256 469541	Bushnell House, Manor Farm, Farleigh Road, Cliddesden, Basingstoke, Hampshire RG25 2JB	Unix and X-Windows based software for a broad range of satellite and aerial imagery integrated into a raster GIS environment for spatial analysis.	Tools for algorithm development, menu customisation; accelerated and high resolution versions; MOTIF / OPEN LOOK; any image size: 8 or 24 bit colour.	0	
Taywood Data Graphics Pat Jarvis 081 575 9240	Greenford House, 309 Ruislip Road East, Greenford, Middlesex UB6 9BQ	Sites of Special Scientific Interest (SSSI's) and Census enumeration districts (Ed-line) in vector form for a wide range of GIS systems. Ordnance Survey maps in full colour raster.	Data available now to support a wide range of GIS applications.	594	906
Trimble Navigation Limited Helen Knight 0256 760150	Meridian Office Park, Osborn Way, Hook, Hampshire RG27 9HX	Trimble Pathfinder GPS is the ultimate tool for simple and accurate field collection of position, time and attribute data for over 140 GIS systems.	6 channel GPS reveivers, accurate to 2 metres in differential operation. Up to 15000 position storage, optional hand held data collectors and bar code reader.	0	

THE DATA CONSULTANCY (401)

The Data Consultancy specialises in the provision of spatially referenced data sets for marketing and planning applications. Analysis and mapping of data is undertaken together with a full and comprehensive service of help and advice on the sources, collection and use of data. Survey research and cross analysis of data is also undertaken. Most data sets are spatially referenced to the national grid, administrative and postal geographies. Major data sets include:

- **MAP FILES:** roads, rivers, railways, town street maps, central London street map and european maps all in digital form.
- **BOUNDARY FILES:** U.K. administrative areas, counties, districts, local authority wards, health authorities enumeration districts; postcode areas, districts and sectors; drivetime isochrones and many others.
- **POINT FILES:** 1981 and 1991 enumeration district centroid, postcode sectors, local authority wards, shopping centres, retail stores, banks, building societies, garages and full gazetteers.
- **DEMOGRAPHIC DATA:** Census small area statistics at enumeration district, ward, postcode sector and other levels. Population updates, 1991 household counts at enumeration district, ward and postcode. Super Profile neighbourhood classification.
- **CONSUMER EXPENDITURE AND INCOME:** Consumer retail expenditure, retail business turnover potential and household income available at any spatial level down to enumeration district.
- **STORES:** U.K. hypermarkets and superstores, grocery stores, multiple retailers and service businesses.
- **SHOPPING CENTRES AND SCHEMES:** Details of all large managed shopping schemes together with proposed schemes and those with planning permission. Locations of all major shopping centres including town and city centres with cross references to Goads and Newmans data.

Data and analyses can be supplied with or without associated GIS software. See our full catalogue of Spatially Referenced Data sets for more details.

Project management/management contractor

Organisation, contact, telephone	Address	Markets supported	Outstanding features	Ref	Surrogate Page
W.S. Atkins Planning Consultants Keith Warren 0372 726140	Woodcote Grove, Ashley Road, Epsom, Surrey KT18 5BW	Systems design, software development for social, urban and transportation planning, health, water and waste management, mapping and survey.	Atkins GIS capability is supported by its 50 years of consultancy experience in technologies ranging from agriculture to urban and regional planning.	0	
CDR Mapping Limited Christopher Cooper 0433 621282	Hawthorn House, Edale Road, Hope, Sheffield S30 2RF	Supergazetteer is used for loading data into GIS. All data is stored in Oracle and is available for other packages. Full support services provided.	Map management of all NTF / OSTF mapbase using industry-standard SQL databases. Ideal for data loading and GIS use. Stand-alone PC to full DOS / UNIX networking.	0	
CSL Group Limited Steve Watson 071 982 0071	Consilium House, City Forum, City Road, London EC1V 2NY	All public sector organizations including central and local government, utilities, health authorities, emergency services and regulatory bodies.	Expert GIS implementation with experienced public sector consultants. Services: implementation plans, procurement support and project management.	0	
Data Base Builders Mr. Douglas Cross 0487 813745	1 Lawrence Road, Ramsey, Huntingdon, Cambridgeshire PE17 1UY	Management of GIS projects for utilities or local government is DBB's principal activity: we may act as manager, or team resource on large enterprises: and we may act for the vendor or the customer.	Specialist in vendor/ customer relations; contracts and specifications; data capture; staff selection; coordination of resources (including those of the contractor).	0	
ESRI (UK) Limited Carole Smith 0923 210450	23 Woodford Road, Watford, Hertfordshire WD1 1PB	ESRI (UK) provide a comprehensive range of GIS management services including turnkey applications, project management, systems design and implementation.	ESRI (UK)'s project management staff have extensive experience in implementing successful GIS solutions using PRINCE project management methodologies and SSADM system design.	0	

Organisation, contact, telephone	Address	Markets supported	Outstanding features	Ref	Surrogate Page
Fairbairn Services Limited Martin Dibnah 061 976 3536	Fairbairn House, Ashton Lane, Sale, Manchester M33 1WP	Utilities, central and local government, property owners, cable and telecommunications and general practice.	Control of full, pilot and feasibility projects from a background of engineering, computing, quality control and project management.	0	
Flynn & Rothwell Linda Elkins 0279 507346	Charrington House, The Causeway, Bishop's Stortford, Hertfordshire CM23 2ER	Consulting civil engineers offering GIS and mapping services to the water, waste-water, petroleum and solid-waste industries.	Environmental management, water cycle engineering, soil and ground water pollution assessment and hydrocarbon pollution remediation.	0	
Geobase Consultants Limited John Rowley 0372 745261	28 Church Road, Epsom, Surrey KT17 4DX	European, national and local government, environmental and other government agencies, utilities, GIS service and supply industry, transport infrastructure sector and private companies worldwide.	GCL can take full responsibility for projects from inception to completion or support a Client's project manager with 'second opinions' and expert advice.	570	603
GeoVision Systems Limited Walt Beisheim 0276 677707	80 Park Street, Camberley, Surrey GU15 3PG	Gas, electric and water utilities, telecommunications, transportation, governments, land-use and military.	Open systems / client server basis - all-relational GIS DBMS - comprehensive user tools - application development tools (4GL) - industry-specific applications - full integration of third party applications.	0	
Sir Alexander Gibb & Partners Limited Peter Beaumont 0734 261061	Earley House, 427 London Road, Reading RG6 1BL	International GIS and remote sensing. Project Management, database design, data acquisition, GIS analysis, image processing, cadastral mapping.		0	
Glen Computing Limited Graham Bugler 0689 875577	309 High Street, Orpington, Kent BR6 0NN	GEO/SQL application management and customisation in FM, line plant management, environment, oil, emergency planning and civil engineering.	Data capture, implementation strategy, project and quality management, customisation training, support and maintenance.	560	501

Organisation, contact, telephone	Address	Markets supported	Outstanding features	Ref	Surrogate Page
IBM United Kingdom Limited Preet Dhillon 0926 332525	UK GIS Solution Centre, PO Box 31, Birmingham Road, Warwick CV34 5JL	GIS design & implementation: covering pre-sales consultancy, data modelling, systems implementation, networking, corporate integration & overall management of projects using IBM RISC System / 6000 based GIS.	Local & central government, utilities, retail & distribution, transportation, cable & telecommunications, asset management.	560	502
IMASS Limited Sue Gadd / Peter Taylor 091 213 5555	Information Management and Systems Solutions, Eldon House, Regent Centre, Gosforth, Newcastle upon Tyne NE3 3PW	Utilities, local government and private sector.	Complete Project Management from selecting vendor through to implementation and data capture.	0	
Logica Bob Walters 071 637 9111	Medina House, Business Park 4, Randalls Way, Leatherhead, Surrey KT22 7TW	Utilities, oil, environment, transport, local / central government, defence, mapping agencies, space, remote sensing, retail and private sector in UK and overseas.	Logica has over 20 years' experience of project management. This includes managing subcontractors, data capture, training and software development.	579	701
Mason Land Surveys Bob Owen 0383 727261	Dickson Street, Dunfermline, Fife KY12 7SL	Experience with local authorities, district councils, development corporations and other organisations defining and implementing their GIS requirements.	Managing the customisation, initial data capture, systems and training for GIS. In-depth knowledge of systems: AutoCAD and Intergraph; experience in managing data capture contractors and suppliers of systems and data.	0	
Midsummer Computing Exchange Limited Steve Gill 0908 668866	Midsummer House, 429 Midsummer Boulevard, Central Milton Keynes, Buckinghamshire MK9 2HE	Central / local government and agencies, MoD, utilities, health and emergency services, transportation, environmental and geological.	Management of systems selection, procurement, installation, customisation, data conversion, training and staffing for GIS.	0	
Modulo 4 International Tony Hart 0532 771344	Suite 27A, Concourse House II, 432 Dewsbury Road, Leeds LS11 7DF	Public utilities, local government, national government and private sector in the UK and worldwide.	Procurement, implementation, training, and data acquisition for corporate or departmental systems, integration of disimilar systems.	572	605

Organisation, contact, telephone	Address	Markets supported	Outstanding features	Ref	Surrogate Page
MVA Systematica Doug McCallum 0483 728051	MVA House, Victoria Way, Woking, Surrey GU21 1DD	Central and local government, utilities, transport and distribution, automotive and informatics industry, emergency services / command and control, commercial and retail planning.	Feasibility studies, requirements / resources, systems design / selection, project / quality management, data take-on, property / network referencing, postcode / address matching, applications development.	0	
Posford Duvivier Tim Jeffries-Harris 0733 334455	Rightwell House, Bretton Centre, Peterborough, Cambridgeshire PE3 8DW	Services offered worldwide to public and private sectors, specialising in all aspects of environmental resource, and asset management.	Wide experience of IT consultancy and management from feasibility study, through functional specification, procurement, development and ongoing maintenance.	0	
QC Data David Roche 081 866 4400	Canada House Business Centre, 227 Field End Road, Eastcote, Ruislip, Middlesex HA4 9NA	Local and central government, telecommunications, gas, electrical and water utilities, mapping agencies and petroleum exploration in the UK, Europe and North America.	QC Data can resource 'turnkey' responsibility for entire projects or participate in any phase of GIS implementation. Project management staff are experienced in all aspects of GIS development.	592	904
Siemens-Nixdorf Information Systems Ltd. Denis Shaughnessy 0344 850576	Siemens-Nixdorf House, Oldbury, Bracknell, Berkshire RG12 4FZ	Utilities, central and local government.	59 strong project management group specializing in medium to large (£multi-million) projects using the 'PRINCE' (UK Government standard) methodology.	0	
Smith System Engineering Limited Dr Trevor Clifton 0483 505565	Surrey Research Park, Guildford, Surrey GU2 5YP	Requirements capture / specification, bid evaluation, implementation management and studies to HMG, emergency services (police) and standards bodies.	Data exchange standards (worked for AGI). Large IS where GIS is one of many technologies. 20 years of procurement experience.	0	
Terraquest Limited John Gannon 021 500 5818	Kenrick House, Union Street, West Bromwich, West Midlands B70 6DB	Services are provided primarily to public sector clients including central and local government and public sector industries.	TerraQuest have in-depth experience of all stages of GIS projects coupled with a complete range of appropriate technical skills.	0	

GLEN COMPUTING LIMITED (501)

In association with our sister company, Glen Surveys Ltd, we offer a unique range of hardware/software, data capture and computing services.

A high standard of services is assured through all stages of procurement, installation, training and maintenance, by our BS5750 Part I certified Quality Management System.

PROJECT MANAGEMENT AND GIS IMPLEMENTATION SERVICES

Glen Computing offers GIS solutions based on its core system Geo/SQL. We specialise in preparing a flexible configuration based on industry standards such as AutoCAD and SQL. Geo/SQL is available on PC and SUN UNIX platforms. The most important advantage of our approach is that Geo/SQL is designed to work in an integrated partnership with today's industry standard software, it does not force the user to rely on a set of proprietary graphics, database and management tools.

This flexibility also allows automatic input/integration of existing data into the new systems, minimising losses associated with data integrity.

DATA COLLECTION/CONVERSION

The Glen Group have a full range of data collection facilities including digitising, scanning, ground/aerial surveys, CAD coupled with extensive conversion / transfer facilities. We can supply structured or unstructured data in a range of industry standard CAD or GIS formats for direct input to end user systems.

IBM (UK) LIMITED (502)

IBM - *KNOW HOW*

IBM has provided solutions for the AM/FM/GIS market since the early 1970's. With this background, IBM can provide :

- The experience to assess your needs and recommend cost-effective solutions best suited to your business.
- The resources to either help you build the right application yourself or do it on your behalf.
- The expertise to work with you in order to ensure that GIS fits into your overall IT architecture, thus providing benefit to the whole organisation.

GIS projects often involve a number of sub-projects, each with a different supplier - with separate contracts and skill requirements. Under an IBM Systems Integration contract, there is no need to co-ordinate the services of a number of suppliers. IBM provides you with one source, one-stop shopping for expertise, hardware and software.

In IBM you have an experienced partner to help you with all aspects of GIS, large or small. Using established or innovative techniques, we can provide an effective coherent solution, combining management and technical aspects into successful GIS.

Independent GIS consultancy services

Organisation, contact, telephone	Address	Markets supported	Outstanding features	Ref	Surrogate Page
W.S. Atkins Planning Consultants Keith Warren 0372 726140	Woodcote Grove, Ashley Road, Epsom, Surrey KT18 5BW	Systems design, software development for social, urban and transportation planning, health, water and waste management, mapping and survey.	Atkins GIS capability is supported by its 50 years of consultancy experience in technologies ranging from agriculture to urban and regional planning.	0	
Chris Britton Consultancy Chris Britton 0483 68027	No 3 The Crossways, Onslow Village, Guildford, Surrey GU2 5QG	Highway management specialists serving local and central government in locational referencing, data capture, applications development and organisation / management studies.	High-calibre teams of Engineers, Managers and IT professionals assembled to review, problem-solve, project manage and assist implementation on any highways related system.	0	
Business Information Management Rob Mahoney 0273 515018	14 Kings Avenue, Denton, Newhaven, East Sussex BN9 0NA	Independent consultants to utilities, local government, retail and private sector. UK, Europe and beyond.	All GIS areas, awareness/ seminars, requirements study, strategy, market research, project management and implementation support.	569	601
Cambridge Computer Consultants (UK) Ltd. Colin Hookham 0480 69577	18 Oaklands, Fenstanton, Huntingdon, Cambridgeshire PE18 9LS	Various public and private sector markets.	Full range of GIS life-cycle services.	569	602
CDR Mapping Limited Christopher Cooper 0433 621282	Hawthorn House, Edale Road, Hope, Sheffield S30 2RF	Supergazetteer is used for loading data into GIS. All data is stored in Oracle and is available for other packages. Full support services provided.	Map management of all NTF / OSTF mapbase using industry-standard SQL databases. Ideal for data loading and GIS use. Stand-alone PC to full DOS / UNIX networking.	0	

Organisation, contact, telephone	Address	Markets supported	Outstanding features	Ref	Surrogate Page
Comsult DR Foster 0234 342401	67 Goldington Road, Bedford MK40 3NB	Utilities, local and central government.	Technical consultancy and bespoke software for a wide variety of applications. Emphasis on GIS and display systems.	0	
Concurrent Appointments Alan Carnell 0582 712976	27 Field Close, Harpenden, Hertfordshire AL5 1EP	Concurrent provides a permanent recruitment service to the GIS industry in the UK and Europe. The service covers technical, support, sales and marketing professionals.	Enquiries with CV welcomed from all levels of executive personnel. Enquiries with job specification are also welcomed from prospective employers.	0	
Coopers and Lybrand Nick Pearce 071 583 5000	Plumtree Court, London EC4A 4HT	Utilities, local and central government, private sector and vendors in UK, Europe and worldwide.	Totally independent consultancy on GIS, IT and other business areas. Executive awareness, strategy, system selection, implementation, project management and support.	0	
CSL Group Limited Steve Watson 071 982 0071	Consilium House, City Forum, City Road, London EC1V 2NY	All public sector organizations including central and local government, utilities, health authorities, emergency services and regulatory bodies.	Expert GIS consultancy with public sector focus. Services: product evaluation, GIS strategies, implementation reviews and system integration.	0	
Dan Rickman Associates Daniel Rickman 081 444 9116	74 Church Lane, East Finchley, London N2 0TE	Support from initial concept to final review: cost / benefit analysis, system selection, project management, business analysis: for utilities, government, retail, health and transport sectors.	Wide range of experience of GIS, including IT and Open Systems issues, data conversion and information management, links with image management and office systems.	0	
Data Base Builders Douglas Cross 0487 813745	1 Lawrence Road, Ramsey, Huntingdon, Cambridgeshire PE17 1UY	Tactical and project assistance offered to users and vendors from a decade of work on 'both sides of the fence'.	Support offered for the management of change: installations, staff issues, vendor/customer relations, purchase justifications, application specification and vendor knowledge.	0	

Organisation, contact, telephone	Address	Markets supported	Outstanding features	Ref	Surrogate Page
DWH Associates Limited Derek Hilder 0789 268102	20 Rother Street, Stratford upon Avon, Warwickshire CV37 6NE	Government and agencies, local government, utilities, health services, environmental bodies, police and emergency services.	Feasibility studies, requirements analysis, functional design, implementation support, post implementation reviews.	0	
Earth Observation Sciences Matthew Stuttard 0252 811181	Branksome Chambers, Branksomewood Road, Fleet, Hampshire GU13 8JS	Environmental - Remote sensing - Space sector (EC, ESA) - Central government - UK, Europe and International.	Consultancy in data management, GIS, remote sensing and applications. Software engineering to international standards.	0	
Fairbairn Services Limited Martin Dibnah 061 976 3536	Fairbairn House, Ashton Lane, Sale, Manchester M33 1WP	Utilities, central and local government, property owners, cable and telecommunications and general practice.	GIS, facilities and asset management systems. Independent hardware and software advice. Pilot and feasibility studies.	0	
Geobase Consultants Limited John Rowley 0372 745261	28 Church Road, Epsom, Surrey KT17 4DX	European, national and local government, environmental and other government agencies, utilities, GIS service and supply industry, transport infrastructure sector and private companies worldwide.	GCL provide independent services to identify, plan, specify, procure, implement, project manage and audit integrated GIS/IT solutions using recognized and GCL methodologies.	570	603
Gimms (GIS) Limited Ms Marlene Ferenth 031 668 3046	30 Keir Street, Edinburgh EH3 9EU	Customers have included local/ central government agencies, utilities and marketing agencies.	Our consultants offer unbiased professional advice on a variety of GIS applications including project management, software design/ implementation and benchmarking.	0	
Hoskyns GIS Bob Selley 0703 766777	Phi House, Enterprise Road, Chilworth Research Centre, Chilworth, Southampton, Hampshire SO1 7NS	All sectors.	Hoskyns offer a pragmatic approach to GIS consultancy at all stages based on a track record of successful implementations.	0	

Organisation, contact, telephone	Address	Markets supported	Outstanding features	Ref	Surrogate Page
IMASS Limited Peter Taylor / Sue Gadd 091 213 5555	Information Management and Systems Solutions, Eldon House, Regent Centre, Gosforth, Newcastle upon Tyne NE3 3PW	Utilities, local government and private sector.	GIS strategy and vendor selection, logical and physical systems design; asset survey; project management; tele-communications consultancy.	0	
Know Edge Limited Robin A. McLaren 031 443 1872	33 Lockharton Avenue, Edinburgh EH14 1AY	Worlwide markets: power and water utilities, local government, retail, property, environmental and land registration agencies.	The oldest, specialist GIS consulting company in the UK providing truly independent advice from strategy to implementation.	0	
KPMG Management Consulting Malcolm Leith 071 236 8000	8 Salisbury Square, London EC4Y 8BB	Central and local government, non-departmental government bodies, private sector including retail, property, distribution and financial services; in the UK, Europe and Worldwide.	Independent consultancy, business IT and GIS. Awareness seminars, strategy studies, requirement definition, systems evaluation and selection, customisation and systems development, project management.	571	604
John D. Leatherdale FRICS John Leatherdale 081 449 0123	10 Bartrams Lane, Hadley Wood, Barnet, Hertfordshire EN4 0EH	Independent consultant in GIS and digital mapping to users worldwide: government and local authorities, private sector, development agencies.	Feasibility studies, definition of user needs, implementation strategies, systems and data procurement, land information management.	0	
Lofthouse Associates Ian J Lofthouse 0695 576522	6 Peet Avenue, Ormskirk, Lancashire L39 4SH	All GIS market areas, local government in particular. Application integration.	Extensive IT experience, plus GIS development and implementation, plus business skills, provide unbiased analysis and guidance.	0	
Logica Bob Walters 071 637 9111	Medina House, Business Park 4, Randalls Way, Leatherhead, Surrey KT22 7TW	Utilities, oil, environment, transport, local / central government, defence, mapping agencies, space, remote sensing, retail and private sector in UK and overseas.	Logica provides independent consultancy support, enhanced by our practical experience of implementing GIS to provide successful business solutions.	579	701

Organisation, contact, telephone	Address	Markets supported	Outstanding features	Ref	Surrogate Page
Mason Land Surveys Bob Owen 0383 727261	Dickson Street, Dunfermline, Fife KY12 7SL	Consultancy services to local authorities, district councils and development corporations and especially the provision of IT strategies for smaller organisations starting in GIS.	Data capture and conversion advice, identification and installation of approriate software and hardware, staff training and database assembly.	0	
Midlands Regional Research Laboratory Dr Alan Strachan 0533 523849	Bennett Building, University of Leicester, Leicester LE1 7RH	Local and central government, demographics, training materials, health, marketing research.	Feasibility studies, spatial analysis, training, GIS applications.	0	
Midsummer Computing Exchange Limited Steve Gill 0908 668866	Midsummer House, 429 Midsummer Boulevard, Central Milton Keynes, Buckinghamshire MK9 2HE	Central / local government and agencies, MoD, utilities, health and emergency services, transportation, environmental and geological.	Feasibility studies, benchmarking, system selection, system analysis, implementation, pilot projects, customisation and system management.	0	
Modulo 4 International Tony Hart 0532 771344	Suite 27A, Concourse House II, 432 Dewsbury Road, Leeds LS11 7DF	Public utilities, local government, national government. Private sector in UK and worldwide.	Feasibility studies, project management, procurement assistance, systems and data analysis, GIS audit, specialist recruitment, market research.	572	605
MVA Systematica Doug McCallum 0483 728051	MVA House, Victoria Way, Woking, Surrey GU21 1DD	Central and local government, utilities, transport and districbution, automotive and informatics industry, emergency services / command & control, commercial and retail planning.	Feasibility studies, requirements / resources, systems design / selection, project / quality management, data take-on, property / network referencing, postcode / address matching, applications development.	0	
MVM Consultants Plc. David Rix 0272 252885	10-17 Park Place, Clifton, Bristol BS8 1JP	Local and central government; public utilities; transport and distribution; retail sector in the UK, Europe and Middle East, North America.	Project planning and analysis; feasibility studies; project implementation and management; systems installation and integration; systems design and development.	0	

Organisation, contact, telephone	Address	Markets supported	Outstanding features	Ref	Surrogate Page
P-E International Plc Guy Pullen 0784 434411	Park House, Wick Road, Egham, Surrey TW20 0HW	P-E support a diverse range of clients in central and local government, transport, retail and distribution, and the utilities - in the UK and overseas.	P-E provide GIS advice as part of developing business solutions, complemented by over 50 years' in helping clients to operate more successfully.	0	
PlanGraphics Inc. Dennis Kunkle +1 502 223 1501	202 West Main Street, Suite 200, Frankfort, Kentucky 40601, United States of America	PlanGraphics designs and implements GIS for public and private clients, including local and regional governments and all types of utility agencies.	Plan Graphics' clients receive the benefit of more than 13 years' GIS consulting experience in all phases of system design and implementation.	0	
Posford Duvivier Tim Jeffries-Harris 0733 334455	Rightwell House, Bretton Centre, Peterborough, Cambridgeshire PE3 8DW	Services offered worldwide to public and private sectors, specialising in all aspects of environmental resource, and asset management.	Indpendent services available for projects of all scales covering general IT implementations as well as specialist GIS applications.	0	
Scott Wilson Kirkpatrick Stephen Vincent / T Beaumont 0256 461161	Scott House, Basing View, Basingstoke, Hampshire RG21 2JG	Worldwide systems implementation and GIS applications in earth resources, water resources and transportation.	Practical applications experience combined with objective systems analysis. Regional offices throughout the UK.	573	606
SIAS Limited Sarah McRae 031 557 0200	4 Heriot Row, Edinburgh EH3 6HU	Central and local government organisations and private sector. Concentrating on highways, transportation and utilities.	Application design, implementation and consultancy. INGRES, X and MS Windows specialists, SSADM experience. Custom software in C and Fortran. Graphical interfaces to spatial data.	0	
Simon Petroleum Technology Cathy Woodhead, Garry Green 0492 581811	Llandudno, Gwynedd LL30 1SA	GIS in petroleum, environment and natural resources worldwide. Remote sensing, image processing, interpretations and database construction.	ARC/Info, ERDAS, ORACLE, Apple Macintosh, Intergraph and a suite of seismic interpretation and mapping software. Resources survey, evaluation and exploration.	573	607

Organisation, contact, telephone	Address	Markets supported	Outstanding features	Ref	Surrogate Page
Smith System Engineering Limited Dr Trevor Clifton 0483 505565	Surrey Research Park, Guildford, Surrey GU2 5YP	Requirements capture / specification, bid evaluation, implementation management and studies to HMG, emergency services (police) and standards bodies.	Data exchange standards (worked for AGI). Large IS where GIS is one of the technologies. 20 years of procurement experience.	0	
Ian Stratford and Associates Ian Stratford 0737 244522	Woodcroft, 2 Alders Road, Reigate, Surrey RH2 0ED	Central and local government, utilities, transport and highways, environmental agencies, private sector and system vendors in UK, Europe and Overseas.	GIS and CAD consultancy. Feasibility / requirements analysis / functional design / implementation / project management and support / QA / market research.	0	
Summerside Services Limited Martin C Rickman 0480 497160	Atlas House, 3-4 Free Church Passage, St. Ives, Cambridgeshire PE17 4AY	Local and central government, environmental agencies, national mapping agencies, utilities and private sector in the UK, Europe and worldwide.	Mapping / GIS awareness training, implementation strategy planning, data capture strategy, project and quality management for all LIS / GIS and mapping related applications.	573	608
Taywood Data Graphics Pat Jarvis 081 575 9240	Greenford House, 309 Ruislip Road East, Greenford, Middlesex UB6 9BQ	Utilities, local, county and central government, defence, architectural, engineering, environmental bodies in UK and Europe.	Extensive hands-on experience in a wide range of data capture and GIS systems provide the base for pilot projects and consultancy in systems selection.	594	906
Terraquest Limited John Gannon 021 500 5818	Kenrick House, Union Street, West Bromwich, West Midlands B70 6DB	Services are provided primarily to public sector clients including central and local government and public sector industries.	Terraquest have exceptional skills knowledge and experience supported by close involvement with eight major implementation projects.	0	
Trimble Navigation Limited Helen Knight 0256 760150	Meridian Office Park, Osborn Way, Hook, Hampshire RG27 9HX	Trimble Pathfinder GPS is the ultimate tool for simple and accurate field collection of position, time and attribute data for over 140 GIS systems.	6 channel GPS reveivers, accurate to 2 metres in differential operation. Up to 15000 position storage, optional hand held data collectors and bar code reader.	0	

Organisation, contact, telephone	Address	Markets supported	Outstanding features	Ref	Surrogate Page
WRc Stuart Goodwin 0793 511711	Frankland Road, Blagrove, Swindon SN5 8YF	Water industry, National Rivers Authority, government, private sector, environmental and overseas.	Business / environmental services; consultancy for all aspects of GIS; bureau spatial analysis using in-house GIS.	0	
Rob Walker Associates Rob Walker 0954 51003	64 Histon Road, Cottenham, Cambridgeshire CB4 4UD	GIS users, particularly in the utilities and public sector, systems suppliers and data providers in the UK and overseas.	Specification of user requirements and system functionality, management of data conversion projects, consultancy on systems and standards.	0	

BUSINESS INFORMATION MANAGEMENT (601)

Business Information Management provides a wide range of services to meet the ever growing market demand for independent GIS consultancy.

The company has provided services to a variety of organisations in both the public and private sectors and has a proven track record in delivering genuinely impartial advice to clients.

The company's philosophy is always to work closely with its clients ensuring that they receive the best advice and assistance in defining their business needs and the most appropriate solution within an agreed time frame.

A modular approach has been developed to all GIS services that allows the client to select any combination of our services appropriate to their needs. Joint project modules are an important aspect of our services that allow our clients staff to work alongside Business Information Management to develop their own in house skills.

A selection of our services are as follows:

- Executive overviews
- Awareness seminars
- Feasibility studies
- Requirement studies
- Market research
- Project support
- Digital map use and management

Business Information Management takes a pride in its independence and the services it provides.

CAMBRIDGE COMPUTER CONSULTANTS (U.K.) LIMITED (602)

Cambridge Computer Consultants provides consultancy services for all stages of GIS procurement and implementation, including:

- Feasibility
- Cost / benefit / risk
- Business analysis
- Requirements specification
- System selection + benchmarking
- Data conversion planning
- Implementation management
- Training

The company delivers technical and management expertise acquired from eight years of GIS project experience in the following market sectors:

- Local Government
- Electricity
- Telecommunications
- Water
- Central Government (PRINCE / SSADM)
- Environment
- Transport

GEOBASE CONSULTANTS LIMITED (603)

Geobase Consultants is a **leading** independent consultancy firm specialising in Geographic Information and Systems. We justify ourselves as **leading** because we have taken a high profile role for providing **Standards** to the geographic information market on a National and a European basis. We do this because the surest way of protecting our client's investment is to ensure that standards are specified wherever possible.

Geobase Consultants provides a broad range of services to users and intending users of geographic information. These services cover the entire life-cycle of a system procurement from first thoughts through planning and implementation to a post-implementation review. Our services our tailored to individual client needs but we do have a preference for working closely with the Client's personnel to ensure an element of 'technology transfer'.

We also service the needs of the institutions which support the users as well as the industry's system suppliers and data providers. Our services are broadly as follows:

- Education: awareness seminars at business or technical levels
- Education: briefings / lectures on the role of Standards
- Preliminary and feasibility studies
- Statements of requirement
- Advice on technical issues such as mapping technology
- Analysis of suppliers ability to meet requirements
- Procurement management, documentation and services
- Implementation and data take-on management
- Project management and audits
- Market profiling including review of inhibiting factors
- Product (application or 'toolbox') planning and design

During the previous year we have participated in major consultancy assignments for local authorities providing them with:

- A statement of requirements for a corporate Land and Property GIS with applications for Terrier, Legal and Property Services, corporate property gazetteer
- A feasibility report examining the benefits of integrating Highways Information Management with GIS
- A feasibility report indicating how to expand departmental GIS initiatives into a Council-wide strategy whilst avoiding unnecessary centralisation.

In much the same time-scale we have prepared a major report for the Association for Geographic Information concerning a proposed strategy for standards, prepared two British Standards (for the National Transfer Format and for the Local Government Management Board's Street Gazetteer). We have also been involved in specifying data conversion programmes for a Regional Water Authority and advising a data provider organisation.

No job is too small - or too large for us! No job is too 'high level' or too 'technical' for us to be involved: we have appropriate skills available to suit all requirements!

KPMG MANAGEMENT CONSULTING (604)

KPMG Management Consulting is one of the UK's largest providers of professional services to both the public and private sectors. KPMG's GIS group offers a range of services designed to assist clients with development and implementation of GIS.

Our approach to GIS provides a systematic evaluation of the opportunities presented by GIS within the framework of an organisations corporate objectives and overall business development plan. We can assist with the examination of the implications of GIS on established techniques to resolve key issues including:

- Cost justification of GIS
- Phased implementation management
- Merging of GIS and global IT strategies
- Data ownership and data management
- Human resources and management of change

We have a strong, multidisciplinary team of consultants who are able to assist at every stage of evaluation, planning and implementation of GIS and our range of services include:

Awareness	• executive briefings	
	• management seminars	
Feasibility /strategy	• business strategy	• IT/GIS strategy
	• data audit	• Cost benefit analysis
	• EEE reviews	• organisation review
	• IT existing/required	• systems integration
Implementation	• specifications	• ITT
	• business process design	• selection
	• benchmarks	• project management
	• customisation	• change management
	• training	
Continuing Strategic Advice	• monitoring	
	• horizon technology	
	• review	

MODULO 4 GROUP (605)

The Modulo 4 Group is committed to the wider application of GIS technology in the mainstream IT environment. It specialises in Land, Property, Plant and Asset information needing spatial referencing using addresses, parcels and networks - often with an underlying base-map. Modulo 4 consultants have worked on a wide range of applications with most of the leading GIS software and on every continent.

Modulo 4 International (M4I) is an independent consultancy offering a range of services to clients in the Public Utilities, National and Local Governments, Commerce and Industry

- feasibility studies / pilot projects
- systems and data analysis
- procurement, data acquisition and project management
- GIS Audit and value added development
- market research and specialist recruitment
- education and training

M4I consultants have managed large and complex 'working' GIS and have been actively involved in the development and marketing of some of the world's leading products. They have also been involved, since the earliest days of digital mapping, with the specification, acquisition, quality control and restructuring of spatial data and its attributes.

M4I consultants play leading roles in AGI, RICS, BCompS, BCartS, BSi and other industry groupings, professional bodies and learned societies. M4I is proud to be a member of the British Consultants Bureau and is very experienced in the management of GIS projects including the personnel, equipment, institutional and integration problems which inevitably arise. M4I has a wealth of general IT and business experience to set GIS in its proper context. M4I carry out Cost Benefit Analysis, Functional Design, System Design and Detailed specification.

MODULO 4 LIMITED (M4L) develop specialised software for GIS and GIS related projects. M4L staff have worked with ICL, UNISYS, IBM, DEC, ESRI, Laser-Scan, Genasys, Alper, Axis and other systems. Applications have included Land Allocation, Land Charges, Utility Plant Networks, Emergency Services, Military Simulation, Radio Planning and Environmental Impact. M4L have recently worked in UK, Australia, Europe and Africa. M4L are organised into product Business Units for different proprietary software and are equipped with a networked workstation and PC environment with on-line access to mainframe systems.

SCOTT WILSON KIRKPATRICK (606)

Scott Wilson Kirkpatrick are leading international Consulting Engineers, Transportation and Environmental Planners, with 18 offices in the UK and offices in over 30 countries worldwide. The Information Systems Division concentrates on the practical application and integration of information systems, both as part of engineering projects and as independent GIS consultants. Specialist professional staff provide a comprehensive service covering all aspects of GIS consultancy, from system evaluation, design, and implementation to the interpretation of remotely sensed data.

During the past five years, work has focused on earth resources applications in rural development, water resources applications in hydrology and hydrogeology, and transportation applications in road maintenance. In many cases this has involved the integration of remote sensing with GIS.

The experience gained designing and implementing solutions to practical problems has also been combined with objective systems analysis to develop an increasing workload in system evaluation and implementation for local and regional government departments in the UK.

A strong involvement with development and research has also been maintained, resulting in the development of Scott Wilson Kirkpatrick's own Geographic Information System Toolkit **(GIST)**, which supplements the in-house facilities which include **SPANS** and **ILWIS** GIS software and a **GEMS35** image processing system.

SIMON PETROLEUM COMPANY (607)

The acquisition of the Robertson Group plc by Simon Engineering plc has created a new and important name in the international petroleum industry: Simon Petroleum Technology

Already one of the world's largest, most experienced and progressive petroleum data acquisition, processing, services and consulting groups, Simon Petroleum Technology has an unrivalled range of technical capability encompassing virtually all facets of the petroleum exploration and production process. The company is dedicated to the development of innovative technology and its practical application in order to generate the cost-effective solutions that our clients need in today's increasingly complex search for, and exploitation of, hydrocarbons.

SUMMERSIDE SERVICES LIMITED (608)

Contract Cartographic Staff : We provide fully trained, experienced computer mapping technicians to work within the client's own premises and in-house digital facilities on part or full project basis. This service ensures an efficient and cost-effective mapping project solution.

Project Management and Consultancy : We offer project and facilities management services and consultancy for all mapping related projects.

Training : Specialised staff training is provided in digital mapping and GIS applications for technicians, supervisors and managers.

GIS implementation services

Organisation, contact, telephone	Address	Markets supported	Outstanding features	Ref	Surrogate Page
W.S. Atkins Planning Consultants Keith Warren 0372 726140	Woodcote Grove, Ashley Road, Epsom, Surrey KT18 5BW	Systems design, software development for social, urban and transportation planning, health, water and waste management, mapping and survey.	Atkins GIS capability is supported by its 50 years of consultancy experience in technologies ranging from agriculture to urban and regional planning.	0	
CDR Mapping Limited Christopher Cooper 0433 621282	Hawthorn House, Edale Road, Hope, Sheffield S30 2RF	Supergazetteer is used for loading data into GIS. All data is stored in Oracle and is available for other packages. Full support services provided.	Map management of all NTF / OSTF mapbase using industry-standard SQL databases. Ideal for data loading and GIS use. Stand-alone PC to full DOS / UNIX networking.	0	
CSL Group Limited Steve Watson 071 982 0071	Consilium House, City Forum, City Road, London EC1V 2NY	All public sector organizations including central and local government, utilities, health authorities, emergency services and regulatory bodies.	Expert GIS implementation with a public sector focus. Services: implementation plans, professional and technical support, systems integration.	0	
ESRI (UK) Limited Carole Smith 0923 210450	23 Woodford Road, Watford, Hertfordshire WD1 1PB	ESRI (UK) offer a comprehensive range of GIS implementation services to support our customers covering the management, development, installation and support of ARC/INFO based GIS solutions.	ESRI (UK) staff have extensive experience in GIS and specialise in project management, system design, application consultancy, software development, database design, user support & training.	0	
Fairbairn Services Limited Martin Dibnah 061 976 3536	Fairbairn House, Ashton Lane, Sale, Manchester M33 1WP	Utilities, central and local government, property owners, cable and telecommunications and general practice.	Pilot and feasibility studies undertaken. From format and model specification, through implementation and quality procedures to data conversion.	0	

Organisation, contact, telephone	Address	Markets supported	Outstanding features	Ref	Surrogate Page
Geobase Consultants Limited John Rowley 0372 745261	28 Church Road, Epsom, Surrey KT17 4DX	European, national and local government, environmental and other government agencies, utilities, GIS service and supply industry, transport infrastructure sector and private companies worldwide.	GCL project manage and organize services to implement an integrated GIS/IT installation including designing, planning and contracting data take-on, customization, training and hand-over.	0	603
GeoSearch Inc. Richard Serby +1 719 260 7087	P.O. Box 62178, Colorado Springs, CO 80920, United States of America	Worldwide personnel recruitment/ placement. GIS, GPS, photogrammetry and related sciences.	Database of 10,000+ candidates specialized in GIS, GPS, photogrammetry and related sciences. Candidates submitted according to SKILL, EXPERIENCE, WAGE/ SALARY and LOCATION.	0	
GeoVision Systems Limited Walt Beisheim 0276 677707	80 Park Street, Camberley, Surrey GU15 3PG	Gas, electric and water utilities, telecommunications, transportation, governments, land-use and military.	Open systems / client server basis - all-relational GIS DBMS - comprehensive user tools - application development tools (4GL) - industry-specific applications - full integration of third party applications.	0	
Sir Alexander Gibb & Partners Limited Peter Beaumont 0734 261061	Earley House, 427 London Road, Reading RG6 1BL	International GIS and remote sensing. Project Management, database design, data acquisition, GIS analysis, image processing, cadastral mapping.		0	
Glen Computing Limited Graham Bugler 0689 875577	309 High Street, Orpington, Kent BR6 0NN	GEO/SQL application management and customisation, services include support, training, installation, development, maintenance and data capture.	Hardware/ software system supply, integration and analysis of peripherals and spatial/ relational data requirments.	560	501
Haiste Limited James Gossage 0532 620000	Newton House, Newton Road, Leeds LS7 4DN	Utilities, environmental agencies, water industry in UK, Europe and overseas.	Facilities include scanning and colour electrostatic plotting service. Custom GIS software written. Training and support.	0	

Organisation, contact, telephone	Address	Markets supported	Outstanding features	Ref	Surrogate Page
IMASS Limited Sue Gadd / Peter Taylor 091 213 5555	Information Management and Systems Solutions, Eldon House, Regent Centre, Gosforth, Newcastle upon Tyne NE3 3PW	Utilities, local government and private sector.	Develop GIS applications, customize existing iGIS products (see GIS-linked application software and systems entry); data survey and capture.	0	
Kember Smith Limited Malcolm Smith 071 721 7222	Blackfriars Foundry, 156 Blackfriars Road, London SE1 8EN	Specialist on site support service for imaging and GIS systems. Completely independent services include selection advice, installation and support.	Wide range of scanners, optical systems, jukeboxes and software supported or tailored to needs. Independent of hardware and software sales. Consultancy and bureau services also provided.	0	
Logica Bob Walters 071 637 9111	Medina House, Business Park 4, Randalls Way, Leatherhead, Surrey KT22 7TW	Utilities, oil, environment, transport, local / central government, defence, mapping agencies, space, remote sensing, retail and private sector in UK and overseas.	Logica designs and builds customised information systems, integrates GIS with other corporate systems and provides training and support.	579	701
Mason Land Surveys Bob Owen 0383 727261	Dickson Street, Dunfermline, Fife KY12 7SL	Experience with local authorities, district councils and development corporations defining objectives and specifications in relation to asset and facilities management.	Full implementation services including data capture, training and long or short term support including staff secondment, feasibility and pilot studies. Ordnance Survey Superplan Agent.	0	
Midsummer Computing Exchange Limited Steve Gill 0908 668866	Midsummer House, 429 Midsummer Boulevard, Central Milton Keynes, Buckinghamshire MK9 2HE	Central / local government and agencies, MoD, utilities, health and emergency services, transportation, environmental and geological.	Acceptance testing, application, graphics and database design, customisation, client training, technical staffing and data conversion.	0	
Modulo 4 International Tony Hart 0532 771344	Suite 27A, Concourse House II, 432 Dewsbury Road, Leeds LS11 7DF	Public utilities, local government, national government and private sector in the UK and worldwide.	Implementation of digital mapping and GIS products with corporate IT systems. ICL PLANES and Unisys ARGIS implementations with SUN or PC workstations.	572	605

Organisation, contact, telephone	Address	Markets supported	Outstanding features	Ref	Surrogate Page
MVA Systematica Doug McCallum 0483 728051	MVA House, Victoria Way, Woking, Surrey GU21 1DD	Central and local government, utilities, transport and distribution, automotive and informatics industry, emergency services / command & control, commercial and retail planning.	Feasibility studies, requirments / resources, systems design / selection, project / quality management, data take-on, property / network referencing, postcode / address matching, applications development.	0	
QC Data David Roche 081 866 4400	Canada House Business Centre, 227 Field End Road, Eastcote, Ruislip, Middlesex HA4 9NA	Local and central government, telecommunication, gas, electrical and water utilities, mapping agencies and petroleum exploration in the UK, Europe and North America.	QC Data provide cost effective implementation services using a full range of technical staff and working closely with clients during all phases of project implementation.	592	904
Science Systems Limited David Wintle 0272 717251	23 Clothier Road, Brislington, Bristol BS4 5PS	Software development and consultancy services to cover the complete project lifecycle for the utilities, aerospace, communications, environment, energy and defence markets.	Applications on whole range of mid-range computers, workstations and PCs. Specialists in TT&C ground control systems for remote sensing satellites.	0	
Siemens-Nixdorf Information Systems Ltd. Denis Shaughnessy 0344 850576	Siemens-Nixdorf House, Oldbury, Bracknell, Berkshire RG12 4FZ	Utilities, local government.	Siemens Nixdorf has an exceptional record of achievement in large GIS projects world-wide. Proven capability in all services, including training, consultancy, application development and systems integration.	0	
Terrahunt Geoscience Limited Derek Morris 0727 822287	Herts Business Centre, Alexander Road, London Colney, Hertfordshire AL2 1JG	Electrostatic colour plotting bureau service (large format A0 plus).	Choice of any 15 colours from 256; fast turnaround; fixed price; Xerox 8944 - 4XRE electrostatic plotter.	0	
Terraquest Limited John Gannon 021 500 5818	Kenrick House, Union Street, West Bromwich, West Midlands B70 6DB	Services are provided primarily to public sector clients including central and local government and public sector industries.	TerraQuest have evolved a complete range of implementation skills and facilities developed on eight major GIS projects.	0	

Organisation, contact, telephone	Address	Markets supported	Outstanding features	Ref	Surrogate Page
Trimble Navigation Limited Helen Knight 0256 760150	Meridian Office Park, Osborn Way, Hook, Hampshire RG27 9HX	Trimble Pathfinder GPS is the ultiate tool for simple and accurate field collection of position, time and attribute data for over 140 GIS systems.	6 channel GPS reveivers, accurate to 2 metres in differential operation. Up to 15000 position storage, optional hand held data collectors and bar code reader.	0	
Tydac Technologies Limited Andrew Day 0491 411366	Chiltern House, 45 Station Road, Henley-on-Thames, Oxfordshire RG9 1AT	Tydac Technologies provide implementation, system design and bespoke training for the SPANS GIS system.		0	
WRc Stuart Goodwin 0793 511711	Frankland Road, Blagrove, Swindon SN5 8YF	Water industry, National Rivers Authority, government, private sector, environmental and overseas.	Business/environmental services; consultancy for all aspects of GIS; bureau spatial analysis using in-house GIS.	0	

LOGICA PLC (701)

Logica is a leading independent computer software, systems and consultancy company with an international capability and reputation. The Company's 3500 staff operate from offices in twelve countries. Logica has a global client base covering a range of market sectors and we are dedicated to the provision of high quality, state of the art information technology solutions. Our work revolves around the skill of our own technical specialists in harnessing the power of new digital technologies to meet client needs, and to enable systems to work more effectively and efficiently.

Logica's reputation is based largely on its success as a systems integrator, that is, in taking a system approach, looking at a business requirement and incorporating whatever elements (hardware, software, communications, etc.) are required into a unified whole.

As well as systems integration, Logica staff are also available to provide strategic and technical consultancy, and application specific technical support.

Logica's involvement in a project which includes GIS would, then, take the form of:

- the development of a strategy for geographic information
- analysis of requirements
- procurement consultancy
- assistance in data capture
- system design and build
- post-design services

The nature of the GIS work which Logica has carried out includes:

- development of a water and sewerage asset maintenance system
- comparison of GIS products on both commercial and technical grounds.
- development of a map management system to manage the procurement of maps along with the implementation of operational procedures
- definition of a data architecture which provides the basis of a system to control the collection of spatial data, analysis and dissemination of spatial data.

GIS can provide significant benefit when integrated with other corporate computer systems by providing additional decision support and operational capabilities. Thus the future of GIS depends on systems integration companies such as Logica, who have the technical skills allied to specialist understanding of the business sector and the capability to integrate them into a successful solution.

Applied GIS services

Organisation, contact, telephone	Address	Markets supported	Outstanding features	Ref	Surrogate Page
Aegis Survey Consultants Limited Nicholas Taylor 0992 814717	Waltham End, London Road, Abridge, Essex RM4 1UX	Location and mapping of utilities in 2D and 3D. AEGISMAP/ LANDMARK - buried services interrogation software. Land, building and engineering surveys.	Buried/ visible services referenced spatially in 3D. Multiple X-sections on mainframes and portables. Data transferred into major CAD and GIS systems.	0	
W.S. Atkins Planning Consultants Keith Warren 0372 726140	Woodcote Grove, Ashley Road, Epsom, Surrey KT18 5BW	Systems design, software development for social, urban and transportation planning, health, water and waste management, mapping and survey.	Atkins GIS capability is supported by its 50 years of consultancy experience in technologies ranging from agriculture to urban and regional planning.	0	
Binnie & Partners Ian Bush 0737 774155	Grosvenor House, 69 London Road, Redhill, Surrey RH1 1LQ	GIS & 3D surface modelling and analysis services applied to environmental and civil engineering projects. Extensive experience in the field of water engineering.	Hydrodynamic modelling for river basin and coastal zone management studies, contingency planning for hydraulic hazards, environmental impact assessment studies and visualisation.	584	801
CACI Limited John Rae 071 602 6000	CACI House, Kensington Village, Avonmore Road, London W14 8TS	Location analysis, impact studies, customer profiling and market targetting for retail, finance, health, government and FMCG:	ACORN geodemographic classification. Analysis of 1991 census. Demographic and market forecasts.	0	
CDR Mapping Limited Christopher Cooper 0433 621282	Hawthorn House, Edale Road, Hope, Sheffield S30 2RF	Supergazetteer is used for loading data into GIS. All data is stored in Oracle and is available for other packages. Full support services provided.	Map management of all NTF / OSTF mapbase using industry-standard SQL databases. Ideal for data loading and GIS use. Stand-alone PC to full DOS / UNIX networking.	0	

Organisation, contact, telephone	Address	Markets supported	Outstanding features	Ref	Surrogate Page
GeoSearch Inc. Richard Serby +1 719 260 7087	P.O. Box 62178, Colorado Springs, CO 80920, United States of America	Worldwide personnel recruitment/ placement. GIS, GPS, photogrammetry and related sciences.	Database of 10,000+ candidates specialized in GIS, GPS, photogrammetry and related sciences. Candidates submitted according to SKILL, EXPERIENCE, WAGE/ SALARY and LOCATION.	0	
GeoVision Systems Limited Walt Beisheim 0276 677707	80 Park Street, Camberley, Surrey GU15 3PG	Gas, electric and water utilities, telecommunications, transportation, governments, land-use and military.	Open systems / client server basis - all-relational GIS DBMS - comprehensive user tools - application development tools (4GL) - industry-specific applications - full integration of third party applications.	0	
Sir Alexander Gibb & Partners Limited Peter Beaumont 0734 261061	Earley House, 427 London Road, Reading RG6 1BL	International GIS and remote sensing. Project Management, database design, data acquisition, GIS analysis, image processing, cadastral mapping.		0	
Hunting Land and Environment Limited Graham Deane or M. Whitelegge 0442 231800	Thamesfield House, Boundary Way, Hemel Hempstead, Hertfordshire HP2 7SR	UK and overseas. Remote sensing and GIS consultancy for all resource survey applications. Digital mapping for all sectors.	Integration and analysis of spatial data for environmental monitoring / land use planning. Remote sensing / GIS training including 3 month resource surveys course.	0	
MVA Systematica Doug McCallum 0483 728051	MVA House, Victoria Way, Woking, Surrey GU21 1DD	Central and local government, utilities, transport and distribution, automotive and informatics industry, emergency services / command & control, commercial and retail planning.	Requirements analysis, systems design, data capture / integration, property / network referencing, postcode / address matching, applications development, new product R&D, census / geo-demographic services.	0	
Pinpoint Analysis Limited Martin Higgins 071 612 0568	Tower House, Southampton Street, London WC2E 7HN	Suppliers of data services ranging from geodemographics to digitised grid references. Sectors supported include local government, utilities, retail and financial.	Datasets are software independent with most systems and character formats supported.	0	

Organisation, contact, telephone	Address	Markets supported	Outstanding features	Ref	Surrogate Page
Posford Duvivier Tim Jeffries-Harris 0733 334455	Rightwell House, Bretton Centre, Peterborough, Cambridgeshire PE3 8DW	Services offered worldwide to public and private sectors, specialising in all aspects of environmental resource, and asset management.	Applications of GIS are focused on the coastal and offshore environment, including resource management, civil engineering works and environmental management tools.	0	
Remote Sensing Services Limited John Allan 0672 20226	Lychgate House, 24 High Street, Ramsbury, Wiltshire SN8 2PB	Data supply: satellite and image photography; consultancy: remote sensing and GIS projects; GIS support services: digitising and data conversion.	RSSL have access to both raster and vector GIS facilities, ensuring that projects of any type and size can be undertaken.	0	
Silsoe College Dr Chris Bird 0525 60428	Cranfield Institute of Technology, Silsoe, Bedfordshire MK45 4DT	A consultancy and academic research institution that serves the rural sector with regard to land resource planning and environmental monitoring.	Extensive experience in remote sensing and GIS applications both in Europe and developing countries.	0	
Terraquest Limited John Gannon 021 500 5818	Kenrick House, Union Street, West Bromwich, West Midlands B70 6DB	Services are provided primarily to public sector clients including central and local government and public sector industries.	TerraQuest provide data processing, bureau services and remote facilities to support land and property requirements.	0	
Trimble Navigation Limited Helen Knight 0256 760150	Meridian Office Park, Osborn Way, Hook, Hampshire RG27 9HX	Trimble Pathfinder GPS is the ultimate tool for simple and accurate field collection of position, time and attribute data for over 140 GIS systems.	6 channel GPS reveivers, accurate to 2 metres in differential operation. Up to 15000 position storage, optional hand held data collectors and bar code reader.	0	
Anthony Walker and Partners Edward J Sharkey 0564 795005	47a High Street, Henley in Arden, Solihull, West Midlands B95 5AA	AWP carry out environmental assessments for motorway / trunk road schemes, power line studies, mineral sites and development site location studies.	AWP have expertise in using DTMs for visual assessment, in generating photomontages and in using sieve analysis techniques in digital mapping.	0	

Organisation, contact, telephone	Address	Markets supported	Outstanding features	Ref	Surrogate Page
WRc Stuart Goodwin 0793 511711	Frankland Road, Blagrove, Swindon SN5 8YF	Water industry, National Rivers Authority, government, private sector, environmental and overseas.	Business / environmental services; consultancy for all aspects of GIS; bureau spatial analysis using in-house GIS.	0	
Posford Duvivier Tim Jeffries-Harris 0733 334455	Rightwell House, Bretton Centre, Peterborough, Cambridgeshire PE3 8DW	Services offered worldwide to public and private sectors, specialising in all aspects of environmental resource, and asset management.	Applications of GIS are focused on the coastal and offshore environment, including resource management, civil engineering works and environmental management tools.	0	
Remote Sensing Services Limited John Allan 0672 20226	Lychgate House, 24 High Street, Ramsbury, Wiltshire SN8 2PB	Data supply: satellite and image photography; consultancy: remote sensing and GIS projects; GIS support services: digitising and data conversion.	RSSL have access to both raster and vector GIS facilities, ensuring that projects of any type and size can be undertaken.	0	
Silsoe College Dr Chris Bird 0525 60428	Cranfield Institute of Technology, Silsoe, Bedfordshire MK45 4DT	A consultancy and academic research institution that serves the rural sector with regard to land resource planning and environmental monitoring.	Extensive experience in remote sensing and GIS applications both in Europe and developing countries.	0	
Terraquest Limited John Gannon 021 500 5818	Kenrick House, Union Street, West Bromwich, West Midlands B70 6DB	Services are provided primarily to public sector clients including central and local government and public sector industries.	TerraQuest provide data processing, bureau services and remote facilities to support land and property requirements.	0	
Trimble Navigation Limited Helen Knight 0256 760150	Meridian Office Park, Osborn Way, Hook, Hampshire RG27 9HX	Trimble Pathfinder GPS is the ultimate tool for simple and accurate field collection of position, time and attribute data for over 140 GIS systems.	6 channel GPS reveivers, accurate to 2 metres in differential operation. Up to 15000 position storage, optional hand held data collectors and bar code reader.	0	0

BINNIE AND PARTNERS. (801)

Binnie and Partners are consulting engineers specialising in the water sector. We offer GIS and 3D surface modelling services applied to a wide range of environmental and civil engineering projects.

We carry out hydrodynamic, water quality, sediment and wave modelling for river, lake, esturial and coastal regions. Recent work includes river basin management plans, coastal stability and pollutant dispersion studies and the effects of esturial barrages.

River modelling is carried out to evaluate flood alleviation schemes, the effects of developments, mining settlement, and contingency planning for major hazards such as dam failure and uncontrolled floods.

GIS data collection/conversion services

Organisation, contact, telephone	Address	Markets supported	Outstanding features	Ref	Surrogate Page
Atkins AMC WM McKay 0672 63922	Fordbrook Business Centre, Marlborough Road, Pewsey, Wiltshire SN9 5NU	Data capture of fore and background mapping for GIS by digitizing, land survey, photogrammetry and CAD. Also industrial, architectural and archeological surveying, data conversion and terrain modelling.	AutoCAD, Intergraph, GDS and MOSS formats supported including DXF, IGDS, SIF, MOSS, GDS, OSTF together with most raster formats.	0	
W.S. Atkins Planning Consultants Keith Warren 0372 726140	Woodcote Grove, Ashley Road, Epsom, Surrey KT18 5BW	Systems design, software development for social, urban and transportation planning, health, water and waste management, mapping and survey.	Atkins GIS capability is supported by its 50 years of consultancy experience in technologies ranging from agriculture to urban and regional planning.	0	
B.K.S. Surveys Limited Anne Deehan 0265 52311	47 Ballycairn Road, Coleraine, County Londonderry BT51 3HZ	BKS offer a worldwide spatial data service using photogrammetric and interactive graphics workstations operating under a BS 5750 Quality Assurance scheme.	An experienced team offer a cost-effective and powerful database creation facility using Microstation (Intergraph), Kork KDMS, Moss, AutoCAD and Synercom Informap.	591	901
Cad-Capture Limited Simon Watts 0254 583534	Whitebirk Estate, Blackburn, Lancashire BB1 5UD	Scanning services for raster & vector GIS or CAD systems. Specialising in small scale & contour mapping. Clients include Ordnance Survey, County Councils & Public Utilities.	Any GIS or CAD system: OSTF, NTF, TIFF, DXF, IGES, SIF, IRF, GENIO, via ½″ magnetic tape, floppy or optical disk.	0	
Data Base Builders Douglas Cross 0487 813745	1 Lawrence Road, Ramsey, Huntingdon, Cambridgeshire PE17 1UY	GIS data capture and conversion advice offered and contracts arranged, for utilities and local government.	Specialist data capture advice regarding text and graphics, raster and vector; also in primary data capture (GPS - global positioning systems).	0	

Organisation, contact, telephone	Address	Markets supported	Outstanding features	Ref	Surrogate Page
Engineering Surveys Group Bill Stedman 0932 348981	Incorporating Clyde Surveys, Rosemount House, Rosemount Avenue, West Byfleet, Surrey KT14 6NP	Digital aerial and ground surveys. Field data capture service, digital mapping, data conversion, scanning and format translation for utilities, environment and communication.	Services are quality assured to BS 5750 and undertaken by professionally qualified staff using leading edge technology and equipment.	0	
Fairbairn Services Limited Martin Dibnah 061 976 3536	Fairbairn House, Ashton Lane, Sale, Manchester M33 1WP	Utilities, central and local government, property owners, cable and telecommunications and general practice.	Data conversion for most CAD and GIS. Service includes engineering support, quality assurance and project management.	0	
Gardline David Pettit 0493 850723	Admiralty Road, Great Yarmouth, Norfolk NR30 3NG	Services to utilities, government and other GIS users including map/ records conversion; data structuring and translation; consultancy; project management, quality control and assurance.	High quality and cost effective data capture solutions provided by highly skilled staff and the latest technology. Compatibility with most GIS formats.	591	902
Sir Alexander Gibb & Partners Limited Peter Beaumont 0734 261061	Earley House, 427 London Road, Reading RG6 1BL	International GIS and remote sensing. Project Management, database design, data acquisition, GIS analysis, image processing, cadastral mapping.		0	
Glen Computing Limited Graham Bugler 0689 875577	309 High Street, Orpington, Kent BR6 0NN	Digitizing, scanning, ground survey, photogrammetric mapping, modelling, CAD, data conversion, and data transfer services.	Output formats supported: GEO/SQL, AutoCAD, GDS, Intergraph, GENIO, OSTF, Eclipse, Panterra, DXF, IGES, GXF, ASCII, RLC, TIFF.	560	501
Graphical Data Capture Limited Peter C Klein 081 349 2151	262 Regents Park Road, London N3 3HN	Contract digitizing specialists for the OS, ARC/Info, utilities, terriers, property registers, networks and zones of all kinds. High Quality Assurance is an integral production feature.	Data supplied OSTF, NTF, 3D models, MapInfo, MOSS and others. NJUG13 supported. 20 years experience. Maps into Magtape is our business!	0	

Organisation, contact, telephone	Address	Markets supported	Outstanding features	Ref	Surrogate Page
Grove Consultants Limited A.H. Palmer 081 846 2468	28 Hammersmith Grove, London W6 7EN	Digital data capture for GIS and record purposes from existing maps and plans or direct from land and building surveys. A0 colour plotting. BS 5750 approved company.	Data supplied in OSTF, AUTOCAD DXF, MOSS, GDS. NJUG 13 testing.	0	
IMASS Limited Peter Taylor / Sue Gadd 091 213 5555	Information Management and Systems Solutions, Eldon House, Regent Centre, Gosforth, Newcastle upon Tyne NE3 3PW	Asset survey; data quality control; digitizing; scanning; project design and management.	End-user systems include Intergraph Microstation, UNIX / FRAMME, IRAS environments.	0	
Laser-Scan Rob Walker 0223 420414	Cambridge Science Park, Milton Road, Cambridge CB4 4FY	A variety of map projects are carried out for governmental and commercial organisations, both in the UK and overseas.	A range of services from digitising to complete map production, and from ad hoc consultancy to total project management are offered.	0	
Loy Surveys Limited Mr Jim Loy 0800 833312	UK Support Office, 1 Paisley Road, Renfrew PA4 8JH	GIS 'Get it Surveyed'. Approved supplier to government bodies. Scottish Agent: Ground Modelling. Software products ECLIPSE / PanTerra. Main digital formats supported.	Chartered Land Surveyors. BS 5750 Quality Assured Firm. Member of Survey Association. Nationwide Service. Fully automated field to finish.	0	
Mason Land Surveys Bob Owen 0383 727261	Dickson Street, Dunfermline, Fife KY12 7SL	Digitising and raster scanning, database assembly, linking to graphics and validation, land and aerial surveys, extensive reprographics including A0 colour plotting.	MLS use Intergraph, AutoCAD, MOSS and GRADE and can output to other major systems: GDS, OSTF, NTF and Mapdata. Ordnance Survey Superplan Agent.	0	
McLintock Limited Stephen Crowder 0865 749957	Gresham Telecomputing House, 244 Barns Road, Oxford OX4 3RW	Specialists in the public sector with Prime, Digital, ROCC, ICL, SUN, Tektronix hardware and GDS, ARC/Info software.	Our range of services includes: graphics capture, text capture, data conversion, source data integration, data collection consultancy.	0	

Organisation, contact, telephone	Address	Markets supported	Outstanding features	Ref	Surrogate Page
Midsummer Computing Exchange Limited Maralyn Gill / Matt Tuohy 0908 668866	Midsummer House, 429 Midsummer Boulevard, Central Milton Keynes, Buckinghamshire MK9 2HE	Data conversion with / without database records, vector processing for GIS, raster scanning, 3D modelling, client training and support staff supplied.	Data transfer in industry standard formats including DXF, OSTF, SIF, NTF, MOSS, ASCII, Intergraph, TIFF, VBIT, LRD, CALS and IGDS.	0	
National Remote Sensing Centre Limited Dr Nick Veck 0252 541464	Delta House, Southwood Crescent, Southwood, Farnborough, Hampshire GU14 0NL	Suppliers of satellite and airborne imagery; GIS and image processing; consultancy and mapping services.	Priority is given to adopting a systematic approach to projects, ensuring high quality products and services which match customer requirements.	0	
Pafec Limited - RAVEN Amanda Ward 0602 357055	Strelley Hall, Nottingham NG8 6PE	Image scanning and data capture. Vectorisation, including text and symbol recognition. Also, bureau and consultancy service.	Raster redlining, editing / cut and paste; fast vectorising; hardware independent; output in a variety of formats.	0	
Plowman, Craven & Associates Mark Phillips 0582 765566	141 Lower Luton Road, Batford, Harpenden, Hertfordshire AL5 5EQ	Digital mapping; terrain modelling; survey data processing; data format translation; attribute data capture.	Full input and output from and to: McDonnell Douglas GDS; Intergraph IGDS/SIF; DXF; IGES and others.	592	903
QC Data David Roche 081 866 4400	Canada House Business Centre, 227 Field End Road, Eastcote, Ruislip, Middlesex HA4 9NA	Local and central government, telecommunication, gas, electrical and water utilities, mapping agencies and petroleum exploration in the UK, Europe and North America.	QC Data provides a high quality, cost-effective data conversion service using several conversion technologies and supporting Intergraph, ARC/Info, Unisys ARGIS, AutoCAD, Oracle, Informix and Ingres.	592	904
Reprocart BV Cartography Service Centre E. Korver +31 10 461 88 83	Ceintuurbaan 213 a, 3051 KC Rotterdam, The Netherlands	High resolution scanning and plotting. Data capture by automatic vectorisation and manual digitisation. Inkjet proofing up to A0.	Inhouse programming DIGISYS mapping software. Services at IOTEC Brussels and Tennyson Graphics Melbourne.	0	

Organisation, contact, telephone	Address	Markets supported	Outstanding features	Ref	Surrogate Page
Sterling Surveys Ltd Peter Wightman 0428 604911	Grove House, Headley Road, Grayshott, Hindhead, Surrey GU26 6LE	Topographic surveys; measured building surveys; inshore hydrographic surveys; data capture and output; digitising existing plans.	Data output to any commonly used format.	0	
Structural Technologies Limited David F Schindler 0527 854819	Woodside, The Slough, Studley, Warwickshire B80 7EN	Conversion in any direction between OSTF, NTF, DXF, GFIS and maintaining topological and relational information where possible.	All OS datasets supported including OSBASE, administrative areas, 250k and OSCAR. Also ED-Line. We can carry out conversions in-house or appropriate software.	0	
Summerside Associates Limited David Ellis 0480 497160	Atlas House, 3-4 Free Church Passage, St. Ives, Cambridgeshire PE17 4AY	Bureau data capture service: large or small scale background or foreground mapping. Provision of structured data with attributes utilising Laser-Scan LITES tablet based system or on screen VTRAK raster to vector.	Data provided to OS88 NJUG or equivalent standard. Data conversion to / from most standards including NTF Level 3, SIF, OSTF, DXF, Infomap, MOSS, GFIS, ARC/Info, GGP API or plain ASCII.	593	905
Summerside Services Limited Martin C Rickman 0480 497160	Atlas House, 3-4 Free Church Passage, St. Ives, Cambridgeshire PE17 4AY	For data capture and system operations: provision of contract staff and services including mapping technicians and project management staff.	Comprehensively trained computer mapping technicians and professional staff experienced in Laser-Scan LITES and VTRAK, GFIS, Infomap, Intergraph, ARC/Info, Scitex, MapData and Viewmap.	573	608
Survey & Development Services Ltd John McCreadie 0506 825121	3 Hope Street, Bo'Ness, West Lothian EH51 0AA	Vector digitising or raster scanning of existing maps and record drawings.	Output to all major data formats such as OSTF, NTF Level 3, MOSS, DXF and others.	0	
Taywood Data Graphics Pat Jarvis 081 575 9240	Greenford House, 309 Ruislip Road East, Greenford, Middlesex UB6 9BQ	Comprehensive scanning and digitising services. Scanning available in colour, greyscale and binary. Foreground data conversion of graphics and attributes. Provision of digital SSSI data and ED-line.	Raster, vector and attribute data to any system: eg ARC/Info, AutoCAD, GDS, GeoVision, GFIS, Intergraph, MapData, Scitex, Smallworld, Synercom, DXF, NTF level 3, OSTF, SIA and SIF.	594	906

Organisation, contact, telephone	Address	Markets supported	Outstanding features	Ref	Surrogate Page
Taywood Data Graphics Ralph Yardley / Kevin Ovington 0642 750515	Teesside Industrial Estate, Thornaby, Stockton-on-Tees, Cleveland TS17 9LT	Volume production capabaility operating 24 hour per day. Prompt turn around and competitive pricing. Full range of scanning, digitising, auto-vectorizing and line following services.	Raster, vector and attribute data to any system: ARC/Info, AutoCAD, GDS, GeoVision, GFIS, Intergraph, MapData, Scitex, Smallworld, Synercom, DXF, NTF level 3, OSTF, SIA and SIF.	594	906
Terraquest Limited John Gannon 021 500 5818	Kenrick House, Union Street, West Bromwich, West Midlands B70 6DB	Services are provided primarily to public sector clients including central and local government and public sector industries.	TerraQuest provide comprehensive field and documentary research, data preparation and loading services. Over 100 projects have been completed successfully.	0	
Trimble Navigation Limited Helen Knight 0256 760150	Meridian Office Park, Osborn Way, Hook, Hampshire RG27 9HX	Trimble Pathfinder GPS is the ultimate tool for simple and accurate field collection of position, time and attribute data for over 140 GIS systems.	6 channel GPS reveivers, accurate to 2 metres in differential operation. Up to 15000 position storage, optional hand held data collectors and bar code reader.	0	

B.K.S. SURVEYS LIMITED (901)

We are one of Europe's leading spatial data suppliers with an established reputation for quality in over 40 markets. This quality has been achieved by building our skills in the production of digital spatial information onto our cartographic foundations. Technical direction is supplied by our established team of specialists, who have on average served our customers for over 18 years.

Several propriety systems are used in the development of our spatial information e.g. Microstation (Intergraph), Kork KDMS, Moss, AutoCAD and Synercom Informap.

Our resources include 9 Photogrammetric workstations and 11 Interactive Graphics workstations which are networked so as to produce a cost-effective and powerful database creation facility. Three large, high accuracy plotters, supported by pen and electrostatic capability are used for hard copy representation.

We are accredited under the BS5750 quality assurance scheme and are continuing to seek to improve our operations.

To our customers we offer a quality spatial data service based upon our experience, skills, resources and representation.

GARDLINE DIGITAL MAPPING SERVICES (902)

Gardline provides a range of digital mapping/record conversion services. We offer considerable expertise and advanced technical solutions developed from performing both large conversion contracts for the Utilities and small specialised projects for other geographic information users. Over the last 12 months we have conducted numerous contracts varying considerably in their requirements, including:

- Map scales 1:500 - 1:250,000
- Cable mains conversion
- Structured data
- Topological networks
- Polygon formation
- Multiple attribute coding
- 3D elevation data for digital terrain models
- Output to Intergraph / Laser-Scan / AutoCAD / OSTF / NTF

Whatever the size or complexity of your data capture requirement, Gardline provides the quality and cost effective answer.

PLOWMAN CRAVEN AND ASSOCIATES (903)

Plowman, /Craven and Associates have been providing surveying and mapping services worldwide for over 27 years. Digital products have been produced for the last 14 years and our dedicated development team is always up-to-date with the latest techniques and equipment.

Of specific interest to GIS users we run 8 interactive graphic workstations for the preparation of digital maps and plans, with associated attribute data. Data input can be direct from field survey instruments, analytical photogrammetric workstations and digitising workstations.

Data translations are achieved by our own software and we have extensive experience with many formats reflecting the variety of clients we serve.

Current projects for GIS include Property Information for Developers, Statutory Authorities, Government departments and the creation of database for space audits of large buildings.

For property departments we present digital floor plans and geographical location along with a database of maintenance and condition survey information.

Space audit information is collected with our own data-logging software and transferred directly to a relational database. These databases are used by Architects and Building Surveyors in their analysis of existing building space.

QC DATA (904)

QC Data is an international company with eight offices in North America and Europe and production centres in Cork Ireland, Calgary, Denver and Houston. With a corporate commitment to become an industry leader, **QC Data** provides a full spectrum of GIS solutions specialising in conversions, database management and related support services. A dedicated research and development effort has been established to facilitate **QC Data**'s goal of continually meeting industry requirements.

QC Data has established a successful track record in providing GIS solutions to **Local and Central Government, Utilities and the Petroleum Industry**. By recruiting highly trained technicians and industry specialists, combined with a continual investment in state-of-the-art conversion technology, **QC Data** has earned a reputation for providing cost-effective and high quality solutions to complex GIS projects.

QC Data has developed considerable expertise in data conversion providing both scanning and digitising input systems. Utilising internally developed proprietary systems, **QC Data** consistently maintains its position in a highly competitive market. The firm's background in the conversion industry has provided hands-on experience with most of the leading hardware platforms and software environments. In addition, **QC Data**, has developed a number of proprietary data translation programmes that facilitate the transfer of data from one format to another.

In summary, **QC Data** offers solution to address all aspects of GISdevelopment including:

- **CONVERSION SERVICES**
- **DATABASE DEVELOPMENT AND MANAGEMENT**
- **DOCUMENT IMAGE MANAGEMENT**
- **PROJECT IMPLEMENTATION**

SUMMERSIDE ASSOCIATES LIMITED (905)

Summerside Associates currently provide quality GIS data capture and support services to an impressive list of clients including Government agencies, major utility and private sector companies both in the UK and overseas. Particular emphasis is placed on combining the best people, quality training, exceptional technical resources and management expertise into a full service package that puts the interests of the customers first.

COMPUTER MAPPING BUREAU : Captures background topographic mapping to meet the standards of the Ordnance Survey of Great Britain, together with the quality assurance testing to NJUG13 requirements. The bureau particularly specialises in the capture and conversion of "foreground" mapping into structured digital records. All types of digitising and attribute capture are undertaken using modern Digital workstations, ALTEK digitising tablets and Laser-scan software products including the semi-automatic VTRAK raster to vector conversion package. Output can be simple map feature data or topologically structured into network, polygons or surfaces. The data may be transferred to the customer using National Transfer Format Level 3 (and now BS 7567 : 1992) or most other third party transfer formats.

TAYWOOD DATA GRAPHICS (906)

Data in whatever form is a key asset of any business. Converting this data currently on paper or film to a digital format is the speciality of Taywood Data Graphics. Maps, engineering drawings, site plans etc, can now be readily transferred by a variety of different techniques.

Maps at 1:2500 scale are digitized both for Ordnance Survey and a wide range of utility companies. These are all issued with an NJUG 13 quality assurance test certificate. Maps at other scales can be readily undertaken, and we have recently completed a 1:1 000 000 scale map for part of Russia.

Prior to volume data conversion of your records, it is crucial that a pilot project is undertaken. Pilots have been completed for British Telecom, EME, Norweb, and Yorkshire Water.

There are a number of ways of producing vector data, ie.

- manual digitising
- line-following
- screen tracing of raster
- automatic vectorization

We have the equipment and expertise to undertake any of these. Typically we use manual digitising where a symbol library and extensive levelling is used, raster scanning and line-following for contour and utility companiesmaps, screen tracing for drawing restoration, and auto vectorization for certain maps, engineering drawings and building floor plans.

An alternative to vector digitising is raster scanning which provides an excellent reference or archive base. Maps can be scanned at up to 1,000 dpi in colour, grey-scale or binary and output to all computer systems that handle raster data.

Colour scanning of large format maps is one of the latest services, and the colour raster data can be output in a very wide range of formats and media. The colour quality on screen that can be achieved especially starting with Ordnance Survey maps is very high.

Aerial photographs in colour and black and white, and also colour photographs can also be scanned and included in your GIS for reference.

Data supply is now an increasing part of our business. We distribute the vector Sites of Spatial Scientific Interest data set for the UK., and are the digitising partner in the successful ED-line consortium comprising Ordnance Survey, MVA Systematica, L.R.C and ourselves. To date over 300 sales of the 1991 census enumeration district boundaries have been sold, and the data set for England and Wales is now complete and validated. This data is available in a number of formats for both PC and workstation GIS.

With 40 large digitising tables, 7 scanning systems, 8 line-following work-stations we can readily complete projects of one drawing to many thousands of maps.

Value added/other network services

Organisation, contact, telephone	Address	Markets supported	Outstanding features	Ref	Surrogate Page
Comsult DR Foster 0234 342401	67 Goldington Road, Bedford MK40 3NB	Specialists in digitizing large scale Ordnance Survey maps. Fifteen years experience of mapping work.	OSTF and NTF formats. BS 5750 certification.	0	
Concurrent Appointments Alan Carnell 0582 712976	27 Field Close, Harpenden, Hertfordshire AL5 1EP	Concurrent provides a permanent recruitment service to the GIS industry in the UK and Europe. The service covers technical, support, sales and marketing professionals.	Enquiries with CV welcomed from all levels of executive personnel. Enquiries with job specification are also welcomed from prospective employers.	0	
GeoVision Systems Limited Walt Beisheim 0276 677707	80 Park Street, Camberley, Surrey GU15 3PG	Gas, electric and water utilities, telecommunications, transportation, governments, land-use and military.	Open systems / client server basis - all-relational GIS DBMS - comprehensive user tools - application development tools (4GL) - industry-specific applications - full integration of third party applications.	0	
IMASS Limited Sue Gadd / Peter Taylor 091 213 5555	Information Management and Systems Solutions, Eldon House, Regent Centre, Gosforth, Newcastle upon Tyne NE3 3PW	IMASS operates extensive communications facilities in the North East of England (telephone, radio, WANS and LANS) and has carried out consultancy and implementation in the UK and abroad.	Integration of Intergraph UNIX and Novell networks.	0	

Education/training/information services

Organisation, contact, telephone	Address	Markets supported	Outstanding features	Ref	Surrogate Page
The Association for Geographic Information Maxine Allison 071 222 7000 /226	12 Great George Street, Parliament Square, London SW1P 3AD	AGI's UK central government Metadata service.	Telephone enquiries about central government data availability.	0	
University of Edinburgh Bruce M Gittings 031 650 2523	Department of Geography, Drummond Street, Edinburgh EH8 9XP	Diploma and MSc. in GIS. First in UK with large dedicated laboratory. M.Phil. and Ph.D research in GIS/ remote sensing and other areas at this major research centre.	9/12 month modular courses to suit all backgrounds. Wide range of software in use. Priority rated for ESRC and NERC support. Mature applicants wishing to consolidate skills are welcomed.	0	
Geobase Consultants Limited John Rowley 0372 745261	28 Church Road, Epsom, Surrey KT17 4DX	Geographic information and systems supply and support community in the UK, Europe and Worldwide.	Based on Market Research experience, analysis of common user needs and a leading role in Standards for the geographic information market, GCL offers marketing and information support services.	570	603
Sir Alexander Gibb & Partners Limited Peter Beaumont 0734 261061	Earley House, 427 London Road, Reading RG6 1BL	International GIS and remote sensing. Project Management, database design, data acquisition, GIS analysis, image processing, cadastral mapping.		0	
GIS Europe Stan Abbott 0969 667566	Leading Edge Press and Publication Limited, The Old Chapel, Burtersett, Hawes, North Yorkshire DL8 3PB	GIS EUROPE is the magazine read by Europeans in all sectors of the geographical information systems (GIS) community.		0	

Organisation, contact, telephone	Address	Markets supported	Outstanding features	Ref	Surrogate Page
Kingston University Seppe Cassettari 081 547 2000 /2508	School of Geography, Penrhyn Road, Kingston-upon-Thames, Surrey KT1 2EE	BSc GIS (3 years full time or part-time mode). HND GIS (2 years full-time). GIS professional training programme (short course or distance learing modes).	A comprehensive range of education and training courses in GIS technology, its applications and the broader management and business issues.	0	
Richard Langrish Associates Richard Langrish 081 394 0092	33 Dorling Drive, Epsom, Surrey KT17 3BH	Richard Langrish Associates serves product and service companies operating in the geographic information and other technical markets in the UK and Europe.	Richard Langrish Associates offers a full contract publishing facility for house magazines and newsletters along with a stage-management service for User group and promotional seminars.	600	1101
University of Leicester Dr Peter Fisher 0533 523839	Department of Geography, University of Leicester, Bennett Building, University Road, Leicester LE1 7RH	Diploma and MSc in GIS.	Diploma 9 months, MSc 12 months. In collaboration with Department of Computer Studies. ESRC, NERC and related. Studentships available. ARC/Info, IDRISI, ERDAS and SPANS.	0	
The Local Government Management Board Tony Black 0582 451166	Arndale House, The Arndale Centre, Luton, Bedfordshire LU1 2TS	Local Government in Great Britain.	A range of publications relating to the implementation of GIS in Local Authorities.	0	
Longman Group UK Limited Kim McDonald 0279 623290	Longman House, Burnt Mill, Harlow, Essex CM20 2JE	Publishers of a comprehensive range of texts and reference books of value to all GIS professionals and academics and a must for libraries.	The range is lead by 'Geographical Information Systems: Principles and Applications' edited by DJ Maguire, MF Goodchild and DW Rhind.	0	
Luton College of Higher Education Ron Beard 0582 34111	Park Square, Luton, Bedfordshire LU1 3JU	BSc Mapping Science: either full time or part time. BTEC HND Land Administration (Geographical techniques): sandwich course.	Well-established with major digital mapping and GIS elements. HND involves one year work placements supported by a wide range of organisations.	0	

Organisation, contact, telephone	Address	Markets supported	Outstanding features	Ref	Surrogate Page
Modulo 4 International Tony Hart 0532 771344	Suite 27A, Concourse House II, 432 Dewsbury Road, Leeds LS11 7DF	Public utilities, local government, national government and private sector in the UK and worldwide.	General GIS education and training. Bespoke training for existing or newly implemented systems. ICL PLANES and Unisys ARGIS courses for users and developers.	572	605
University of Newcastle-upon-Tyne Dr David Parker 091 222 6445	Department of Surveying, Newcastle-upon-Tyne NE1 7RU	BSc in Mapping Information Science. GIS masters degrees by research. Customised short courses for industry.	Specialism in property and utility applications based on large scale Ordnance Survey mapping. Facilities include Smallworld GIS, SLIMPAC, ARC/Info, GRASS, AutoCAD, ERDAS.	0	
Polytechnic Huddersfield, Manchester Dr James Petch 061 745 5670	Polytechnic, Salford University, Diploma Office, Peel Building, University of Salford, Salford M5 4WT	Postgraduate distance learning diploma in GIS. 18 month course for professional development, 10 full teaching packs. W/E and two weeks residential.		0	
GIS Unit University of Salford James Petch 061 745 5261	Department of Geography, University of Salford, Salford M5 4WT	Distance learning Diploma in GIS; approved training centre for System 9, SPANS GIS, EASI/PACE image processing. Bespoke courses 1 day - 1 week.		0	
Siemens-Nixdorf Information Systems Ltd. Denis Shaughnessy 0344 850576	Siemens-Nixdorf House, Oldbury, Bracknell, Berkshire RG12 4FZ	Utilities, central and local government.	Siemens Nixdorf provides a comprehensive training service, and, as a result of over 15 years experience in GIS implementation, is able to offer high level advice and guidance.	0	
Silsoe College Dr Chris Bird 0525 60428	Cranfield Institute of Technology, Silsoe, Bedfordshire MK45 4DT	A university institution providing both long and short course training for recent graduates or experienced planners in the rural sector.	MSc courses in Applied Remote Sensing, Information Technology and Land Resource Management. Short courses in remote sensing, GIS and land information databases.	0	

Organisation, contact, telephone	Address	Markets supported	Outstanding features	Ref	Surrogate Page
Summerside Services Limited Martin C Rickman 0480 497160	Atlas House, 3-4 Free Church Passage, St. Ives, Cambridgeshire PE17 4AY	Training services: comprehensive training in digital mapping and GIS applications for technicians, supervisors and managers.	Hands-on training on modern workstations using our in-house bespoke training programmes. (Supported by part-time academic units for suitable candidates.)	573	608
University College of Swansea Martin G Coulson 0792 295546	Department of Geography, Singleton Park, Swansea SA2 8PP	Swansea Geography Department specialises in geographic data handling with BSc, MPhil and PhD programmes. Mature students are welcome on all courses.	Courses include GIS, remote sensing and population data handling. LEA's may give discretionary awards for the 9 month Diploma in Topographic Science.	0	
Tydac Technologies Limited Andrew Day 0491 411366	Chiltern House, 45 Station Road, Henley-on-Thames, Oxfordshire RG9 1AT	Training on SPANS GIS: introductory courses; advanced SPANS courses; customised and in-house training.	Training given by external fully qualified lecturers and tutors.	0	

RICHARD LANGRISH ASSOCIATES (1101)

Richard Langrish Associates provide a complete publishing and consultancy service for organizations wishing to produce their own magazines and newsletters for employee or customer distribution.

We undertake the production of special one-off publishing projects, as well as providing a full contract publishing facility. Our services include all aspects of design and production, editorial and advertisement sales.

Part XIII

AGI membership directory

Council and committee members

Council

Gurmukh Singh **(Chairman)**, CAM Ltd, 126 Cornwall Road, London SE1 8TQ
Tel: 071-928-2433. Fax: 071-928-2366

Michael Blakemore **(Vice Chairman)**, NOMIS Unit 3P, Mountjoy Research Centre, University of Durham, Durham DM1 3SW
Tel: 091-374-2468/2490. Fax: 091-384-4971

Nick Pearce **(Vice Chairman)**, Coopers & Lybrand Deloitte, SA22 Plumtree Court, London EC4 4HT Tel: 071-212-4808/2959. Fax: 071-212-2850

Thomas Waugh **(Honorary Treasurer)**, GIMMS, University of Edinburgh, Drummond Street, Edinburgh EH3 9EU
Tel: 031-650-1000 x 4276. Fax: 031-668-2104

Dorothy Salathiel (Co-opted Member), Room P1/168B, Department of the Environment, 2 Marsham Street, London SW1P 3EB Tel: 071-276-4166

Frank Bennion, British Gas, West Midlands (MC76), 5 Wharf Lane, West Midlands B91 2JP Tel: 021-705-6888 x 23309. Fax: 021-704-3770

Michael Brand, Ordnance Survey, Northern Ireland, Colby House, Stranmillis Court, Belfast BT9 5BJ Tel: 0232-661-244 x 202. Fax: 0232-683-211

Guiseppe Cassettari, Kingston Polytechnic, Geography Department, Penrhyn Road, Kingston upon Thames, Surrey KT1 2EE
Tel: 081-549-1366 x 2510. Fax: 081-547-7419

Andy Coote, Ordnance Survey, Information & Computer Services, Romsey Road, Maybush, Southampton SO9 4DH Tel: 0703-792-382

Martin Coulson, University College Swansea, Natural Science Building, Singleton Park, Swansea SA2 8PP Tel: 0792-295-546. Fax: 0792-205-556

Kevin Crawley, SEEBOARD plc, Electric House, Wellesly Road, Croydon, Surrey CR9 2AR Tel: 081-760-0299 x 5200. Fax: 081-681-2862

Per Dirdal, Alper Systems Ltd, Cambridge Science Park, Milton Road, Cambridge CB4 4FQ Tel: 0223-420-464. Fax: 0223-420-324

Sara Finch, Logica Industry Ltd, Randalls Research Park, Randalls Way, Leatherhead, Surrey KT22 7TW Tel: 071-637-9111

Ian Gilfoyle, Cheshire County Council, Commerce House, Hunter Street, Chester CH1 2QP Tel: 0244-603-101. Fax: 0244-603-802

Bruce Gittings, University of Edinburgh, Geography Department, Drummond Street, Edinburgh EH8 9XP Tel: 031-650-1000 x 2258. Fax: 031-556-0544

Roy Haines-Young, University of Nottingham, Geography Department, University Park, Nottingham NG7 2RD Tel: 0602-484-848 x 3381. Fax: 0602-421-676

Phil Jeanes, PA Consulting Group, Cambridge Laboratory, Melbourn, Royston, Hertfordshire SG8 6DP Tel: 0763-261-222. Fax: 0763-260-023

Alistair Keddie, IMASS, Eldon House, Regent Centre, Gosforth, Newcastle upon Tyne NE3 3PW Tel: 091-213-5555, Fax: 091-213-0526

John Leonard, Ordnance Survey, Marketing Department, Romsey Road, Maybush, Southampton SO9 4DH Tel: 0703-792-558, Fax: 0703-792-660

Alastair Macdonald, Ordnance Survey, Romsey Road, Maybush, Southampton SO9 4DH Tel: 0703-792-550, Fax: 0703-792-888

Ian Masser, University of Sheffield, Town & Regional Planning, Western Bank, Sheffield S10 2TN Tel: 0742-768-555 x 6179

Michael Nicholson, Property Intelligence Ltd, Ingram House, 13-15 John Adam Street, London WC2N 67D Tel: 071-839-7684, Fax: 071-839-1060

John Rowley (Co-opted Member), Geobase Consultants, 28 Church Road, Epsom, Surrey KT17 4DX Tel: 0372-745-261, Fax: 0372-726-617

Nigel Sheath, ICL (UK) Ltd, 127 Hagley Road, Birmingham B16 8LD
Tel: 021-456-1111 x 2090, Fax: 021-455-0358

Peter Smith, Controller of Management Services, HM Land Registry, Lincolns Inn Fields, London WC2A 3PH Tel: 071-405-3488, Fax: 071-242-0825

Alan Summerside, Summerside Associates, 3 & 4 Free Church Passage, St. Ives, Cambs PE17 4AY Tel: 0480-497-160, Fax: 0480-497-148

David Unwin, University of Leicester, Department of Geography, University Road, Leicester LE1 7RH Tel: 0533-523-824, Fax: 0533-522-200

Less Worrall, Wreckin Borough Council, Principal Policy Planner, PO Box 213, Mainslee House, Telford TF3 4LD Tel: 0952-202-431, Fax: 0952-291-060

Peter Woodsford (ex-officio), Laser Scan Ltd, Cambridge Science Park, Milton Road, Cambridge CB4 4FY Tel: 0223-420-414, Fax: 0223-420-044

Management

Gurmukh Singh (**Chairman**), CAM Ltd, 126 Cornwall Road, London SE1 8TQ
Tel: 071-836-1511, Fax: 071-497-8610

Mike Blakemore (**Senior Vice-Chairman**), NOMIS, Unit 3P, Mountjoy Research Centre, University of Durham, Durham DM1 3SW
Tel: 091-3774-2468/2490, Fax: 091-384-4971

Nick Pearce (**Junior Vice-Chairman**), Coopers & Lybrand Deloitte, SA22 Plumtree Court, London EC4 4HT Tel: 071-212-4808, Fax: 071-212-2850

Dorothy Salathiel, Room P1/168B, Department of the Environment, 2 Marsham Street, London SW1P 3EB Tel: 071-276-4166

Thomas Waugh (**Honorary Treasurer**), GIMMS, University of Edinburgh, Drummond Street, Edinburgh EH3 9EU
Tel: 031-650-1000 x 4276, Fax: 031-668-2104

Kevin Crawley, SEEBOARD Plc, Electric House, Wellesly Road, Croydon, Surrey CR9 2AR Tel: 081-760-0299 x 5200, Fax: 081-760-0290 & 0339

Alastair Macdonald, Ordnance Survey, Romsey Road, Maybush, Southampton SO9 4DH Tel: 0703-792-550, Fax: 0703-792-889

Conference

Alistair Keddie (**Chairman**), GIS Mapping Services Manager, Northumbrian Water Group Plc, IMASS PO Box 4, Regent Centre, Gosforth, Newcastle upon Tyne NE3 3PX Tel: 091-213-555, Fax: 091-213-526

Nick Pearce, Principal Associate, Coopers & Lybrand Deloitte, SA22 Plumtree Court, London EC4A 4HT Tel: 071-212-4808/2559, Fax: 071-822-2850

Peter Clegg, Information Technology, Sheffield City Council, Town Hall, Surrey Street, Sheffield S1 2HH Tel: 0742-735-372, Fax: 0742-735-007

Gurmukh Singh, CAM Ltd, 126 Cornwall Road, London SE1 8TQ
Tel: 071-928-2433, Fax: 071-928-2366

Caroline Prescot, Westrade Fairs, Rickmansworth, Hertfordshire WD3 1DD
Tel: 0923-778-311, Fax: 0923-776-820

David Green, University of Aberdeen, Department of Geography, Elphistone Road, Aberdeen AB9 2UF Tel: 0224-272-324, Fax: 0224-487-048

Kevin Crawley, General Manager, SEEBOARD Plc, Electric House, Wellesley Road, Croydon, Surrey CR9 2AR Tel: 081-760-0299, Fax: 081-760-0290

Richard Sterling, BKS Surveys Ltd, Ballycairn Road, Coleraine BT51 3HZ
Tel: 0265-52311/7, Fax: 0265-57637

John Leonard, Ordnance Survey, Romsey Road, Maybush, Southampton SO9 4DH
Tel: 0703-792-558, Fax: 0703-792-888

Per Dirdal, Alper Systems Ltd, Cambridge Science Park, Milton Road, Cambridge CB4 4FQ Tel: 0223-420-464, Fax: 0223-420-324

Ian Homer, Northern Electric, Carliol House, Newcastle upon Tyne, NE99 1SE
Tel: 091-401-7451, Fax: 091-385-7255

Chris Corbin, IT Southern Ltd, Southern House, Lewes Road, Brighton, East Sussex BN1 9PY Tel: 0273-600-444, Fax: 0273-675-299

Ursula Cooke, IMASS Ltd, Eldon House, Regent Centre, Gosforth, Newcastle upon Tyne NE3 3PW Tel: 091-213-555, Fax: 091-213-526

Andrew Larner, AGI, 12 Great George Street, London, SW1P 3AD
Tel: 071-222-7000, Fax: 071-222-9430

Membership

Phil Jeanes, PA Consulting Group, 123 Buckingham Palace Road, London W1
Tel: 071-730-9000

Peter Woodsford, Laser-Scan Ltd, Cambridge Science Park, Milton Road, Cambridge CB4 4FY Tel: 0223-420-414, Fax: 0223-420-044

Warwick Brown, Regional Civil Engineering Department, British Rail (SR), Room 13/9, Southern House, Wellesley Grove, Croydon BR9 1DY
Tel: 081-666-6558, Fax: 081-666-6843

David Hendley, DH Associates Ltd, 10 Challoners Way, East Molesey, Surrey KT8 0DW Tel: 081-979-2745

Dorothy Pugh, Department of Property, Cambridgeshire County Council, Shire Hall, Castle Hill, Cambridge CB3 0DP Tel: 0223-317-454, Fax: 0223-317-341

Jonathan Budd, English Nature, Northminster House, Northminster, Peterborough P1 1VA Tel: 0733-340-345, Fax: 0733-898-341

Seppe Cassettari, Kingston Polytechnic, Geography Department, Penrhyn Road, Kingston upon Thames, Surrey KT1 2EE
Tel: 081-549-1366 x 2510, Fax: 081-549-7796

Promotions & Publicity

Michael Nicholson (**Chairman**), Property Intelligence Ltd, Ingram House, 13-15 John Adam Street, London WC2N 6LD Tel: 071-839-7684, Fax: 071-839-1060

Roland Cunningham, ICL Ltd, Observatory House, Windsor Road, Slough, Berkshire SL1 2EY Tel: 0753-516-000, Fax: 0753-516-778

Per Dirdal, Alper Systems Ltd, Unit 302 Cambridge Science Park, Milton Road, Cambridge CB4 4FQ Tel: 0223-420-464, Fax: 0223-420-324

Gurmukh Singh, CAM Ltd, 126 Cornwall Road, London SE1 8TQ
Tel: 071-836-1511, Fax: 071-497-8610

Sara Finch, Logica Industry Ltd, Randalls Research Park, Randalls Way, Leatherhead, Surrey KT22 7TW Tel: 071-637-9111, Fax: 071-383-0530

Terry Robinson, Glenmoor, Cheddar Road, Axbridge, Somerset BS26 2DL
Tel: 0934-732-421

Alastair Macdonald, Ordnance Survey, Romsey Road, Maybush, Southampton SO9 4DH Tel: 0703-792-558, Fax: 0703-792-889

Standards Steering/Executive Committee

John Rowley (**Chairman**), Geobase Consultants, 28 Church Road, Epsom, Surrey KT17 4DX Tel: 0372-745-261/0372-723-417 (mobile—0860-0865-880-236),
Fax: 0372-726-617

Kate De Groot (Group "C"), Logica Communications Ltd, 68 Newman Street, London W1A 4SE Tel: 071-637-9111 (home: 081-440-5925), Fax: 071-383-0530

Chris Gower (**Chairman Group "A"**), Berkshire County Council, Computer Manager, Department of Highways & Planning, Shire Hall, Shinfield Park, Reading RG2 9XG
Tel: 0734-875-444/234-780, Fax: 0734-310-268

Nick Green, Digital Equipment Company Ltd, The Crescent, Jays Close, Basingstoke, Hampshire RG22 4BS Tel:, Fax:

Mandy Lane (Group "B"), Institute of Terrestrial Ecology, Merlewood Research Station, Grange over Sands, Cumbria LA 6JU Tel: 0539-532-264, Fax: 0539-534705

Tom Fulton (Group "B"), British Telecom, NNE46/6th Floor, Commercial Union House, 124 Saint Vincent Street, Glasgow G2 5ER
Tel: 041-248-3480, Fax: 041-221-5406

John Turner (Group "B"), Technical Systems Consultancy, Ordnance Survey, Romsey Road, Maybush, Southampton SO9 4DH
Tel: 0703-792-514, Fax: 0703-792-404

Rob Walker (**Chairman Group "C"**), Laser Scan Ltd, Cambridge Science Park, Milton Road, Cambridge CB4 4FY Tel: 0223-420-414, Fax: 0223-420-044

Richard Ley, Systems & Techniques Unit, Royal Engineers Survey, Elmwood Avenue, Feltham, Middlesex TW13 7AH Tel: 081-890-3622 x 4377, Fax: 081-890-3622 x 4177

Thomas Waugh, University of Edinburgh, Drummond Street, Edinburgh EH3 EU
Tel: 031-667-1011-x 4276, Fax: 031-668-2104

Bryan Pyne, Advanced Technology Marine Systems, 5 May Cottages, Monkswell Lane, Chipstead, Coulsdon, Surrey CR5 3SX
Tel: 0737-833-499 (home and work, answering machine)

Mike Isherwood, London Fire & Civil Defence Authority, 8 Albert Embankment, London SE1 7SD Tel: 071-587-4023, Fax: 071-587-4042

Mike Stanbridge, Manager, Geographic Systems Support, BAeSEMA, 20/26 Lambs Conduit Street, London WC1N 3LF Tel: 071-404 0911, Fax: 071 405 9469

Dan Rickman, 74 Church Lane, East Finchley, London N2 0TE Tel: 081-444-9116

Peter Woodsford, Laser-Scan Ltd, Cambridge Science Park, Milton Road, Cambridge CB4 4FY Tel: 0223-420-414, Fax: 0223 420 044

Fred Mitchell, BSI, 2 Park Street, London W1A 2BS
Tel: 071-629-9000, Fax: 071-603-2084

Tony Black, LGMB c/o Arndale House, Arndale Centre, Luton, Beds LU1 2TU

Dorothy Pugh, Cambridgeshire County Council, Department of Property, Shire Hall, Castle Hill, Cambridge BG3 0DP Tel: 0223 317 454, Fax: 0223 317 341

P. Clark, Royal Geographical Society, 1 Kensington Gore, London SW7 2AR

John Moore, CAM Ltd, 126 Cornwall Road, London SE1 8TQ

Chris Drinkwater, Hydrographic Office, Ministry of Defence, Taunton TA1 2SN

M.S. Sowton, Bourne Cottage, Twyford, Winchester SO21 1NX

Robin Waters, Modulo 4, Suite 27a Concourse House, 2 432 Dewsbury Road, Leeds LS11 7DF Tel: 0532 771 344, Fax: 0532 700 090 x 124

Amanda Ellis, Land Surveyors Division, RICS

Sponsor members

Mr P Dirdal, Alper Systems Ltd, Milton Road, Cambridge, CB4 4FQ

Mr J N Murray, Blenheim Online, Blenheim House, Pinner, Middlesex, HA5 2AE

Mr P Todd, CMG Information Services Public Sector Ltd, Tothill Street, London, SW1H 9NB

Mr N Pearce, Coopers & Lybrand Deloitte, Plumtree Court, London, EC4 4HT

Mr J M Custance, Department of the Environment, LGS Division/Environment Protection Statistics, Room B/245, Romney House, Marsham Street SW1

Ms A White, Digital Equipment Company UK, Marketing Manager for GIS, Hampshire House, Wade Road, Basingstoke, RG24 0PL

Mr P Paisley, Doric Computer Systems, GIS Division, 23 Woodford Road, Watford, WD1 1PB

Mr S East, EDS United Kingdom, Wells Court, Albert Drive, Sheerwater, Woking, Surrey, GU21 5RN

Mr C Thompson, Geonex (UK) Ltd, Unit 4 + & Barwell Business Centre, Arthur Road, Barwell, Hinckley, Leicestershire

Mr P Dhillon, IBM (UK) Ltd, PO Box 31, Birmingham Road, Warwick, CV34 5JL

Mr N Sheath, ICL (UK) Ltd, 127 Hagley Road, Birmingham, B16 8LD

Mr C Branch, Intergraph (UK) Ltd, Government & Utilities Group, Great Western Way, Swindon, Wiltshire, SN5 7XP

Mr C E H Corbin, IT Southern Ltd, Marketing & Sales, Southern House, Lewes Road, Brighton, BN1 9PY

Mr M Jackson, Laser—Scan Ltd, Milton Road, Cambridge, CB4 4FY

Dr J Penreath, National Rivers Authority, Information Systems, 30-34 Albert Embankment, London, SE1 7TL

Mr J P Leonard, Ordnance Survey, Maybush, Southampton, SO9 4DH

Mr M Cleghorn, Prime Computer (UK) Ltd, 373/399 London Road, Camberley, Surrey, GU13 3HR

Mr P D B Gilbert, RICS Land Surveyors Division, 12 Great George Street, Parliament Square, London, SW1P 3AD

Mr T A Boley, SEEBOARD plc, Corporate Strategy Director, Grand Avenue, Hove, East Sussex, BN3 2LS

Mr T Venediger, Siemens plc, 2 Hanworth Road, Feltham, Middlesex, TW13 5BA

Mr J Moss, SysScan (UK) Ltd, SysScan House, Easthampstead Road, Bracknell, RG12 1NS

Mr R Mason, Taywood Data Graphics, 309 Ruislip Road East, Greenford, Middlesex, UB6 9BQ

Mr A Black, The Local Government Management Board Arndale House, The Arndale Centre, Luton, Beds, LU1 2TS

Mr M Kowalski, Unisys Ltd, Government & Public Services, Central Park, New Lane, Leeds, LS11 5EB

Corporate members

Mr F Gonzalez, Torralba Agencia de Medio Ambiente Sevicion de Evaluacion de Recursos Naturales, Eritana 2, 41013 SPAIN

Mr R A Sage, Anglian Water Regional Systems, Chivers Way, Histon, Cambridge

Dr J H Ashford, Ashford Associates Ltd, 72 Harrow Lane, Maidenhead, Berkshire, SL6 7PA

Mr J C Gyngell, Avon County Council Planning Department, PO Box 46, Middlegate (Floor 11), Whitefriars, Lewins Mead, Bristol BS99 7EV

Mr N A C Farmer, Axis Software Systems Ltd, 88-102 Kingsley Park Terrace, Northampton, NN2 7HJ

Mr R A Blunt, Baymont Engineering Company, President 14100 – 58th Street, N Rubin ICOT Center, Clearwater 34620, Florida, USA

Mr D K Lukes, Bedfordshire County Council, Planning Department, County Hall, Cauldwell Street, Bedford MK42 9AP

Mr C Gower, Berkshire County Council, Shire Hall, Shinfield Park, Reading RG2 9XD

Dr W A Swann, Beverley Group Limited, Woodlands Grange, Woodlands Lane, Almondsbury, Bristol BS12 4JY

Mr I G Bush, Binnie & Partners, MOSS Department, Grosvenor House, 69 London Road, Redhill, Surrey RH1 1LR

Dr J Raper, Birkbeck College, Geography Dept, 7-15 Gresse Street, London W1P 1PA

Mr T Wales, Birmingham City Council, Management & Information Services, 9th Floor 1 Victoria Square, Birmingham B1 1BD

Mr R H Blackwood, BKS Surveys Ltd, Computing Department, 47 Ballycairn Road, Coleraine, Northern Ireland, BT51 3HZ

Mr R B Brender, Brender Management Services Ltd, 1 Coldbath Square, Farringdon, London EC1R 5HL

Mr F C Baldwin, Brentwood District Council, Engineering Services, Council Offices, Ingrave Road, Brentwood, Essex CM15 8AY

Mr H D N Ditoos-Nahapatian, Bristol Water Company, Operations Technical Services, PO Box 218, Bridgwater Road, Bristol BS99 7AU

Mr P Finney, British Aerospace SEMA Ltd, Systems Division, Warton Aerodrome, Preston, Lancs PR4 1AX

Mr C Beattie, British Cartographic Society, c/o 13 Sheldrake Gardens, Hordle, Lymington, Hampshire SO41 1FJ

Mr V Lavan, British Coal, Information Technology Dept, Coal House, Cannock, Staffordshire WS11 3HZ

Mr M Ives, British Gas PLC, National Digital Records Project, 5th Floor Staines House, 158 High Street, Staines, Middlesex TW18 4AZ

Mr K Becken, British Geological Survey, Information and Central Services Directorate, Keyworth, Nottinghamshire NG12 5GG

Mr E Green, Broadlands District Council, Information Services, Thorpe Lodge, Yarmouth Road, Thorpe St Andrew, Norwich, Norfolk NR7 0DU

Mr B Dodds, Buckinghamshire County Council, County Engineer's Department Room 508, County Hall, Aylesbury, Buckinghamshire HP20 1UY

Mr K Dugmore, CACI Ltd, Public Services, Regent House, 89 Kingsway, London WC2B 6RM

Mr G Singh, CAM Ltd, 126 Cornwall Road, London, SE1 8TQ

Mr C Hookham, Cambridge Computer Consultants UK Ltd, 18 Oaklands Fenstanton, Huntingdon Cambs, PE18 9LS

Ms D P Pugh, Cambridgeshire County Council, Property Department, Shire Hall, Castle Hill, Cambridge CB3 0AP

Ms C Klavinskis, CES Ltd, 5 Tavistock Place, London, WC1H 9SS

Mrs J Dickerson, Cheltenham Borough Council, Architecture & Planning, PO Box 12, Municipal Offices Parade, Cheltenham, Gloucestershire GL50 1PP

Mr I Gilfoyle, Cheshire County Council, County Planning Officer, Commerce House, Hunter Street, Chester CH1 2QP

Mr M Whalley, City of Stoke on Trent, IT Services Department, Swift House, Glebe Street, Stoke on Trent ST4 1HP

Mr R H Hollingsworth, Colne Valley Water Company, GIS Manager Operations, Blackwell House, Aldenham Road, Watford, Hertfordshire WD2 2EY

Mr G B Paterson, Computer Corp of America (Int), 36-38 Market Street, Maidenhead, SL6 8A9

Mr A Ellis, Cornwall County Council, I T Officer, County Hall, Truro, Cornwall TR1 3AY

Mr Wier-Wierzbowski, Cotswold District Council, Housing & Environmental Services, Querns Business Centre, Whitworth Road, Cirencester, Glos GL7 1RT

Mr J G Warde, Crown Estate Office, Information Systems Branch, 16 Carlton House Terrace, London SW1Y 5AH

Mr E L Cripps, Cumbria County Council, Corporate Information Unit, Corporate Services Dep, The Courts, Carlisle CA3 8NA

Mr N Sandford, Cyngor Sir Powys C C, Planning Department, County Hall, Llandrindod Wells, Powys LD1 5LG

Mr D Rickman, Dan Rickman Associates, 74 Church Lane, London N2 0TE

Mr C G Oliver, Dartford Borough Council, Development Services Officer, Civic Centre, Dartford, Kent DA1 1DR

Mr P Ayscough, Department of Trade & Industry, Information Technology Division, Kingsgate House, Room 836, 66-74 Victoria Street, London SW1E 6SW

Mr G K Sutton, Dept of the Environment (NI), Water Service, Northland House, 3 Frederick Street, Belfast BT1 2NS

Ms L Drysdale, Design Computer Aids Ltd (DeCAL), Business Development Manager, 16/2 Timberbush Leeth, Edinburgh EH6 6QH

Mrs A M Bosworth, Desktop Engineering Systems, Pembroke House, 5/9 Pembroke Road, Ruislip, Middlesex HA4 8NQ

Mr M D MacDonald, Devon County Council, Chief Executive's (R&I), County Hall, Topsham Road, Exeter EX2 4QD

Gen R Wood, Directorate of Military Survey, Elmwood Avenue, Feltham, Middlesex TW13 7AH

Mr M S Parkin, Dorset County Council, Transportation & Engineering, County Hall, Dorchester, Dorset DT1 1XT

Mr H B O'Neill, Dublin Corporation C A M P, 16/19 Wellington Quay, Dublin 2, Ireland

Mr G O'Brien, Dublin Institute of Techology, Computing and Information Technology, Bolton Street, Dublin 1

Mr M Bosworth, Dudley Metropolitan District Council, Public Works, Council House, Mary Steven's Park, Stourbridge, Dudley, West Midlands DY8 2AA

Mr W S Dalrymple, Dumbarton District Council, Planning & Development Dept, 69 Glasgow Road, Dumbarton G82 1RE

Mr G L Mann, Dumfries & Galloway Regional Council, Physical Planning Council Offices, English Street, Dumfries DG1 2DD

Mr C J Collins, Durham County Council, Treasury Department, County Hall, Durham DH1 5UE

Mr D W Hilder, DWH Associates Ltd, 20 Rother Street, Stratford-upon-Avon, Warwickshire CV37 6NE

Ms R K Frean, Earth Observation Sciences, Branksome Chambers, Branksomewood Road, Fleet, Hampshire GU13 8JS

Mr R Fisher, East London Polytechnic, Land Surveying Dept, Longbridge Road, Dagenham, Essex RM8 2AS

Mr J M T Tebbs, East Midlands Electricity, Operations & Telecoms Dept, PO Box 4, North P D O, 398 Coppice Road, Arnold, Nottingham NG5 7HX

Mr K J Thompson, East Sussex County Council, County Management Services, PO Box 5, County Hall, Lewes, East Sussex BN7 1SF

Mr I Sayers, Elstree Computing Ltd, Elsmere House, 12 Elstree Way, Borehamwood, Herts WD6 1NF

Ms B Knox, English Heritage, Fortress House, 23 Savile Row, London W1X 2HE

Ms S McLean, ESRC S East Regional Research Lab, Department of Geography, London School of Economics, Houghton Street, London WC2A 2AE

Mr D Glading, Essex County Council, Planning Department, County Hall, Chelmsford, Essex CM1 1LF

Mr A W F Wolfe, Eurecart, Ulting Lane, Hatfield Road, Langford, Essex CM9 6QA

Mr M D Dibnah, Fairbairn Services Ltd, Fairbairn House, Ashton Lane, Sale, Cheshire M33 1WP

Mr D J R Stewart, Forestry Commission, Business Systems Division, 231 Corstorphine Road, Edinburgh EH12 7AT

Mr J P Doig, Friends of the Earth, GIS Section, 26-28 Underwood Street, London N1 7JQ

Mr R J Smith, Gedling Borough Council, Deputy Planning & Estate Officer, Civic Centre, Arnold Hill, Arnold, Nottingham NG5 6LU

Mr D A Orr, General Register Office for Scotland, Census Department, Ladywell House, Ladywell Road, Edinburgh EH12 7TF

Mr J R Rowley, Geobase Consultants Ltd, 28 Church Road, Epsom, Surrey KT17 4DX

Mr R Talbot, Geobuild Projects Ltd, 4 Grange Road, Burley in Wharfedale, Ilkley, West Yorkshire LS29

Mr J R M Turner, Geografix Ltd, Hurricane Way, Norwich NR6 6EW

Mr W J Knowles, Geographic Management Systems, Clyde House, Reform Road, Maidenhead, Berks SL6 8BU

Mr A Thomas, Geographical Association, 343 Fullwood Road, Sheffield S10 3BP

Mr E A Guiton, Geomatrix Ltd, Cooper Buildings, Sheffield Science Park, Arundel Street, Sheffield S1 2NS

Ms M A Ferenth, Gimms Ltd, 30 Keir Street, Edinburgh, EH3 9EW

Dr H D Parker, GIS World Inc, 2629 Redwing Road, Suite 280, Ft Collins, CO 80526, USA

Ms C S McAllister, Glasgow District Council, Planning Department, 231 George Street, Glasgow, G1 1RX

Mr G S Bugler, Glen Computing Services, 309 High Street, Orpington, Kent, BR6 0NN

Miss J C Battersby, Glenigan Ltd, Marketing Department, 41-47 Seabourne Road, Bournemouth, Dorset, BH5 2HU

Mrs I B Page, Gordon District Council, Planning Dept, Gordon House, Blackhall Road, Inverurie, Aberdeenshire AB51 9WA

Mr P M Klein, Graphical Data Capture Ltd, 262 Regents Park Road, London, N3 3HN

Mr A Giddins, Graphics Management Ltd, Director, 9 Braefoot Court Law, Strathclyde, ML8 5HY

Mr P J Smith, H M Land Registry, Plans Practice HQ Division, Lincoln's Inn Fields, London, WC2A 3PH

Mr R W Martin, H M Land Registry, Computer Services Division, Drake's Hill Court, Burrington Way, Plymouth PL5 3LP

Mr J Gossage, Haiste Ltd, Computer Applications Dept, Newton House, Newton Road, Leeds LS7 4DN

Mr J K Conway, Halcrow Fox & Associates, 44 Brook Green, Vineyard House, Hammersmith, London W6 7BY

Mr R Emmens, Hampshire County Council, County Planning Officer, The Castle, Winchester, Hampshire, SO23 8UD

Ms Y E Griffiths, Harrogate Borough Council, IT Division, Dept of Corporate Services, Council Offices, Crescent Gardens, Harrogate HG1 2SG

Mr D Goddard, Hart District Council, Civic Offices, Harlington Way, Fleet, Hampshire GU13 8AE

Mr G C Steeley, Hertfordshire County Council, Planning & Estates Department, County Hall, Hertford, SG13 8DN

Mr P Winter, Hertsmere District Council, Chief Executives Department, Civic Offices, Elstree Way, Borehamwood, Hertfordshire WD6 1WA

Mr Alexander Highland, Regional Council IT Department ☿, Regional Buildings, Glenurquhart Road, Inverness, IV3 5NX

Mr G J Tucker, Horsham District Council, Directorate of Planning & Law, New Park House, North Street, Horsham, West Sussex RH12 1RL

Mr M Porter, Hoskyns Group Plc, GIS Division, Technology House, Victoria Road, Winchester SO23 7DU

Mr I H S Stratford, Howard Humphreys IT Applications, Thorncroft Manor, Dorking Road, Leatherhead, Surrey KT22 8JB

Miss J C White, Hunting Technical Services Ltd, Remote Sensing & Info Sciences, Thamesfield House, Boundary Way, Hemel Hempstead HP2 7SR

Mr M J Sebbage, Hydrographic Society U K Branch, c/o Southampton Inst of H E, Newtown Road, Southampton SO3 9ZL

Mr D Huckle, Inland Revenue Valuation Office, New Court, Carey Street, London WC2A 2JE

Mr R V Moore, Institute of Hydrology, Catchment Characteristics, Maclean Building, Crowmarsh Gifford, Wallingford, Oxfordshire OX10 8BB

Mr C H Jackman, Intera Information Technologies, Highland Farm, Greys Road, Henley on Thames, Oxon RG9 4PS

Mr J M Watson, Isle of Man Government, Local Government & The Environment, Government Offices, Bucks Road, Douglas, Isle of Man

Mr P W Warne, ITNet, IS/GIS, 21 William Street, Birmingham B15 1LH

Mr R Orr, John Bartholomew & Son, Cartographic Director, 12 Duncan Street, Edinburgh EH9 1TA

Mr M R Ward, JPB Surveys Ltd, 53 Moor Street, Brierley Hill, West Midlands DY5 3SP

Mr R Laming, Kent County Council, Information Systems Department, Springfield, Maidstone, Kent ME14 2LX

Mr R K Moore, Kerrier District Council, Treasurer's Department, Council Offices, Dolcoath Avenue, Camborne, Cornwall TR14 8RY

Mrs Belshaw, Kilmarnock & Loudoun District Council, Ref F256/B, PO Box 13, Civic Centre, Kilmarnock KA1 1BY

Mrs S P Lloyd, Knowsley Metropolitan Borough Council, Planning Department, Municipal Buildings, PO Box 26 Archway Road, Huyton, Merseyside L36 9FB

Mr M Leith, KPMG Peat Marwick Development & Infrastructure Practice, 8 Salisbury Square, London EC4Y 8BB

Mr J Whitaker, Lancashire County Council, County Planning Department, East Cliff, Preston, Lancashire PR1 3EX

Ms S Finch, Logica Industry Limited, Randalls Research Park, Randalls Way, Leatherhead, Surrey KT22 7TW

Mr B Capper, London Borough of Barking & Dagenham, Borough Treasurer's Dept, Civic Centre, Dagenham, Essex RM10 7BH

Mr P Cridland, London Borough of Barnet, Technical Services Directorate, Engineering Services Division, Barnet House 1255 High Road, Whetstone, London N20 0EJ

Mr P W Wakeham, London Borough of Croydon, IT Dept Room 11 13, Taberner House, Park Lane, Croydon, Surrey CR9 3JS

Mr P J Morley, London Borough of Hounslow, Director of Engineering, Civic Centre, Lampton Road, Hounslow, Middlesex TH3 4DN

Mr P Bone, London Borough of Newham, Environment & Planning Dept, Town Hall Annexe, Barking Road, London E6 2RP

Mr R A Jones, London Borough of Southwark, Engineering & Public Works, Municipal Offices, Larcom Street, London SE17 1RY

Ms J Heynat, London Docklands Development Corporation, Executive Office, Thames Quay, 191 Marsh Wall, London E14 9TJ

Dr M C Isherwood, London Fire & Civil Defence Authority, 8 Albert Embankment, London SE1 7SD

Mr J C Hollis, London Research Centre, Deputy Director, Parliament House, 81 Black Prince Road, London SE1 7SZ

Mr D T Edridge, London Underground Ltd, Information Technology, 1st Floor East 55, Broadway, London SW1H 0BD

Miss V Lawrence, Longman Group UK Ltd, Longman Scientific & Technical, Longman House, Burnt Mill, Harlow, Essex CM20 2JE

Mr A Johnston, Lothian Regional Council, Systems Planning, Computer Services, Computer Centre, Warriston's Close (323 High St) Edinburgh EH1 1PG

Mr M Russell, Luton Borough Council, Town Hall, George Street, Luton LU1 2BQ

Mr R Beard, Luton College of Higher Education, Luton College, Applied Sciences, Park Square, Luton Beds LU1 3JU

Dr C H Osman, Macaulay Land Use Research Institute, Land Use Division, Craigiebuckler, Aberdeen AB9 2QJ

Mr M G Lafferty, Malvern Hills, District Council, Council House, Avenue Road, Malvern, Worcestershire WR14 4AX

Miss A Williams, Manpower (UK) Ltd, Business Development Department, International House, 66 Chiltern Street, London WC

Mr K J A Mason, Mason Land Surveys Ltd, Dickson Street, Elgin Street, Industrial Street, Dunfermline, Fife KY12 7SL

Mr T B Cater, Mercury Communications Ltd, Central Services, Mercury House, Waterside Park, Longshot Lane, Bracknell Berks

Mr J Willis, Merseyside Information Service, Suite 301, Royal Liver Building, Pierhead, Liverpool L3 1JH

Mr M F Biscoe-Taylor, Meta Generics, The Jeffreys Building, St John's Innovation Park, Cowley Road, Cambridge CB4 4WS

Mr C T Thomas, Mid Glamorgan County Council, Highways & Transportation, County Offices, Greyfriars Road, Cardiff CF1 3LJ

Mr Scoggins, Mid-Suffolk District Council, Chief Planning Officer, Council Office, High Street, Needham, Market, Ipswich PI6 8DL

Dr A J Strachan, Midland Regional Research Lab, Bennett Building, University of Leicester, Leicester LE1 7RH

Mrs K E Andres, Milton Keynes Borough Council, Forward Planning Department, Civic Offices, 1 Saxon Gate East, Central Milton Keynes, Bucks MK9 3HQ

Mrs B A Bond, Ministry of Defence, Hydrographic Department, Creechbarrow Road, Taunton, Somerset TA1 2DN

Mr G S Craine, Moss Systems Ltd, Engineering Department, Moss House, North Heath Lane, Horsham, West Sussex RH12 4QE

Mr A McNeill, Motherwell District Council, Planning Department, Civic Centre, Motherwell ML1 1TW

Mr M Q S Green, Museum of London, Dept of Urban Archaeology, 150 London Wall, London EC2Y 5HN

Mr D M W Logie, MVA Systematica, MVA House, Victoria Way, Woking, Surrey GU21 1DD

Mr R J Markham, MVM Consultants plc, 13 Great George Street, Bristol, Avon BS1 5RR

Mr A W Sheppard, N Health & Social Services, BD Planning Department, County Hall, 182 Galgorm Road, Ballymena, N Ireland BT42 1QB

Mr M B Riley, N Hertfordshire District Council, IT Manager Finance Department, Town Lodge, Gernon Road, Letchworth, Hertfordshire SG6 3HN

Prof P Stringer, N Ireland Reg Research Lab, 105 Botanic Avenue, Belfast BT7 1NN

Mrs K Walton, Nat Council for Ed Technology Unit, 6 Sir William Lyons Road, Science Park, University of Warwick, Coventry CV4 7EZ

Mr S Merrett, National Grid Company, I T Department, 185 Park Street, London SE1 9DY

Ms J Blackmore, National Joint Utilities Group, 30 Millbank, London SW1P 4RD

Ms D Marriott, Nature Conservancy for Scotland, Computing Dept, 12 Hope Terrace, Edinburgh EH9 2AS

Dr J Hellawell, Nature Conservancy Council, Monitoring & Carto Services, Northminster House, Peterborough, Cambridgeshire PE1 1UA

Mr D Margetts, NERC Scientific Services, Polaris House, North Star Avenue, Swindon, Wiltshire SN2 1EU

Mr I C Mercer, Nextbase Ltd, Headline House, Chaucer Road, Ashford, Middlesex TW15 2QT

Mr J Wilson, NI Housing Executive, Housing Centre, 2 Adelaide Street, Belfast BT2 8PB

Mr M Nasir, Nixdorf Computer (M) Sdn Bhd No, 36-40 Medan Setia, 2 Plaza Damansara, 50490 Kuala Lumpur, Malaysia

Mr M J Blakemore, NOMIS, Unit 3P Mountjoy Research Centre University of Durham, Durham DM1 3SW

Mr B Peacock, North East Water, GIS Department, PO Box 10, Allendale Road, Newcastle upon Tyne NE6 2SW

Mr AS Yardy, Northamptonshire County Council, Planning & Transportation Dept, Northampton House, Northampton NN1 2HZ

Mr I R Homer, Northern Electic plc, Northern Electric House, Station Road, New Penshaw, Tyne & Wear DH4 7LA

Mr J McCleary, Northern Ireland Electricity, Engineering Services, P O Box 2, Danesfort, 120 Malone Road, Belfast BT9 5HT

Mr A B Robertson, Northumbria Police, Computer Services, Police Headquarters, North Road, Pontleand, Newcastle upon Tyne N620 0BL

Mr A R Keddie, Northumbrian Water Group/IMASS Ltd, Information Technology Consultancy, Northumbria House, Regent Centre, Gosforth, Newcastle-upon-Tyne NE3 3PX

Mr R McMilland, NORWEB Plc, Construction Dept, HQ Talbot Road, Manchester M16 0HQ

Mr B J Bull, Nottinghamshire County Council, Planning & Transportation, Trent Bridge House, Fox Road, West Bridgford NG2 6BJ

Mr J M Dixie, O P C S, Census Division, Segensworth Road, Titchfield, Fareham, Hampshire PO15 5RR

Mr S Turner, Oracle Corporation UK Ltd, Broad Quay House, Prince Street, Bristol BS1 4DJ

Mr R A Kirwan, Ordnance Survey, Phoenix Park, Dublin, Ireland

Mr M J D Brand, Ordnance Survey, N I Department of Environment, Colby House, Stranmillis Court, Belfast BT9 5BJ

Orkney Islands, Council Planning, Department County Offices, Kirkwall, Orkney KW15 1NY

Mr A J Walker, Oxford City Council Planning, Clarendon House, 52 Cornmarket Street, Oxford OX1 3HD

Mrs A M Morris, Oxford Institute of Retail Management, Templeton College, Kennington, Oxford OX1 5NY

Mr P J Jeanes, P A Consulting Group, Computer Aided Engineering, 123 Buckingham Palace Road, London SW1W 9SR

Mr G N Pullen, P E International Plc, I T Division, Park House, Wick Road, Egham, Surrey TW20 0HW

Mr R Finch, Photogrammetric Data Services, 1 Ham Business Centre, Brighton Road, Shoreham By Sea, Sussex BN43 6PA

Mr J Farrow, Photogrammetric Society, Dept of Photogrammetry & Surveying, University College London, Gower Street, London WC1E 6BT

Mr J C Antenucci, Plangraphics Inc, 202 West Main Street, Suite 200, Frankfort, Kentucky, USA 40601

Mr M Phillips, Plowman Craven & Associates, R & D Department, 141 Lower Luton Road, Batford, Harpenden, Hertfordshire AL5 5EQ

Mr H Beeching, Poole Borough Council, Borough Engineer, Municipal Buildings, Civic Centre, Poole BH15 2RY

Mr M J L Nicholson, Property Intelligence Ltd, Ingram House, 13-15 John Adam Street, London WC2N 67D

Mrs D K Grieco, QC Data Limited, Sales/Marketing, Second Floor, 100 Park Lane, London W1Y 4AR

Ms F A Holdsworth, RAC Motoring Services, Information Services Dept, RAC House, PO Box 100, South Croydon CR2 6XW

Mr J Cogle, Registers of Scotland, Meadowbank House, 153 London Road, Edinburgh, Scotland EH8 7AU

Prof P M Mather, Remote Sensing Society, University of Nottingham, Department of Geography, University Park, Nottingham NG7 2RD

The Manager, Robertson McCarta, Cartography Department, Tyn-y-Coed, Llandudno LL30 1SA

Mr D Lowe, Royal Borough of Kensington, D B S I T, The Town Hall, Hornton Street, London W8 7NX

Mr R Davey, Royal Borough of Kingston upon Thames, Chief Executives Office, Office Automation Bureau, Guildhall, Kingston upon Thames, Surrey KT1 1EU

Ms C Shardin, Royal Mail Postcodes, 3-4 St Georges Business Centre, St Georges Square, Portsmouth PO1 3AX

Mr I Gilfoyle, Royal Town Planning Institute, Member Services, 26 Portland Place, London W1N 4BE

Ms Y Rendall, Sandwell Metropolitan Borough Council, IT Department, Sandwell Council House, Oldbury, Warley, West Midlands B69 3DF

Dr C L Garner, Scottish Homes, Research Manager, Thistle House, 91 Haymarket Terrace, Edinburgh EH12 5HE

Mr G Yarnell, Scottish Hydro Electric Plc, Commercial Division, Peasiehill Road, Arbroath, Angus DD11 2NJ

Mr T M Cox, Sefton Metropolitan Borough Council, Planning Department, Vermont House, 375 Stanley Road, Bootle L49 1RG

Mr T Barnett, Sevenoaks D C, Central Services Dept, Argyle Road, Sevenoaks, Kent TN13 1HG

Mr R Wantling, Sheffield City Council, Design and Building Services, Town Hall, Sheffield S1 1HH

Mr J Cornick, Shetland Islands Council, Finance Dept, Breiwick House, 15 South Road, Lerwick ZE1 ORB

Mr E Jones, Shropshire County Council, County Property & Planning Services Dept, The Shirehall Abbey, Foregate, Shrewsbury SY2 6ND

Dr T R Clifton, Smith Associates Ltd, 40 Priestley Road, Chancellor Court, Surrey Research Park, Guildford, Surrey GU2 5YP

Mr A S Miller, Soc of Surveying Technicians, Drayton House, 30 Gordon Street, London WC1H 0BH

Dr R J Jones, Soil Survey & Land Research Centre, Computing & Information Systems, Silsoe Campus, Silsoe, Bedford MK45 4DT

Mr J H Mclintock, South Oxfordshire District Council, Info Technology Services, PO Box 17, Council Offices, Crowmarsh, Wallingford, Oxford OX10 8NP

Mr C P Thomas, South Wales Electricity Engineering, Director, St Mellons, Cardiff CF3 9XW,

Mr B Hooper, South West Water Services Ltd, Peninsular House, Rydon Lane, Exeter DEVON

Mr R Freeston, Southern Electric Information Systems, Bartons Road, Havant, Hampshire PO9 5JB

Mr R A Webb, Southern Water, Guildbourne House, Worthing, Sussex BN11 1LD

Mr S N M Mackie, Spa Marketing Systems, 1 Warwick Street, Leamington Spa, Warwickshire CV32 5LW

Mr S R Moore, Sterling Surveys Ltd, Grove House, Headley Road, Grayshott, Hindhead, Surrey GU26 6LE

Mr G Thorley, Strathclyde Regional Council, Chief Executive's Department, 20 India Street, Glasgow G2 4PF

Mr D F Schindler, Structural Technologies Ltd, Head Office, Woodside, The Slough, Studley, Warwickshire B80 7EN

Mr P G Wiseman, Suffolk County Council, Highways Department, St Edmund House, County Hall, Ipswich IP4 1LZ

Mr R A Summerside, Summerside Associates, 3 & 4 Free Church Passage, St Ives PE17 4AY

Mr J M Whatmore, Sunderland Polytechnic, Environmental Technology, Benedict Building, St Georges Way, Sunderland, Tyne & Wear SR2 7BW

Mr J Callaghan, Survey International Manager, Dickens House, Enterprise Way, Off Maulden Road, Flitwick, Beds MK45 5BY

Mr M C May, Swale Borough Council, Information Technology, Swale House, East Street, Sittingbourne, Kent ME10 3HT

Mr E A Fearnley, Swansea City Council, Information Technology, The Guild Hall, Swansea, West Glamorgan SA1 4PN

Mr D McGrane, Systems Distribution International Ltd, 36 Lad Lane, Dublin 2, Ireland

Mr R G Anderson, Tameside Metropolitan Borough Council, Policy Services Department, Council Offices, Room 6 14 Wellington Road, Ashton under Lyme OL6 6DL

Mr D F Pigram, Tandridge District Council, Directorate of Planning Council Offices, Station Road, East Oxted, Surrey RH8 0BT

Miss M T Sharp, Tayside Regional Council, Planning Department, Tayside House, Crichton Street, Dundee DD1 3RB

Ms D M Freeman, The Advisory Unit, Computers in Education, 126 Great North Road, Hatfield, Herts AL9 5JZ

Mr S Liddiard, Thurrock Borough Council, Management Services, Civic Offices, New Road, Grays, Essex RM17 6SL

Dr D P Best, Touche Ross Management Consultants, Peterborough Court, 133 Fleet Street, London EC4A 2TR

Dr I Heywood, Tydac Technologies Ltd, Managing Director, Chiltern House, 45 Station Road, Henley on Thames, Oxon RG9 1AT

Mr S Kitching, Tyne & Wear, Research & Intelligence Unit, Room 394 Civic Centre, Newcastle upon Tyne NE1 8QN

Mr D P Chapman, University College London, Dept of Photogrammetry & Surveying, Gower Street, London WC1E 6BT

Mrs B A Morris, University of Edinburgh, GISA Department of Geography, Drummond Street, Edinburgh EH8 9XP

Mr B M Gittings, University of Edinburgh, Department of Geography, Drummond Street, Edinburgh EH8 9XP

Dr D Parker, University of Newcastle, Department of Surveying, The Old Forge Building, Newcastle upon Tyne NE1 7RU

Dr R H Haines-Young, University of Nottingham, Inst of Eng & Space Geodesy, University Park, Nottingham NG7 2RD

Mr R Parry, University of Reading, Geography Dept, Whiteknights, Reading RG6 2AB

Prof F I Masser, University of Sheffield, Town & Regional Planning, Western Bank, Sheffield S10 2TN

Prof A M Hay, University of Sheffield, Geography Department, Sheffield S10 2TN

Dr M J Clark, University of Southampton, Geodata Institute, Southampton SO9 5NH

Mr W I Wallace, Vale Royal Borough Council, Estates Section, Wyvern House, The Drumber, Winsford, Cheshire CW7 1AH

Mr R D Peel, ViewMAP Spatial Systems Ltd, Clifton Heights, Triangle, West Bristol BS8 1EJ

Dr Cromar, Wakefield Metropolitan District Council, Chief Planning Officer, Town Hall, Wood Street, Wakefield WF1 2HQ

Mr D Addyman, Warwickshire County Council, Planning & Transportation, PO Box 43, Shire Hall, Warwick CV34 4SX

Mr P J Turton, Water Research Centre, Asset Management Department, PO Box 85, Frankland Road, Blagrove, Swindon SN5 8YR

Mr C V Sale, Waveney District Council, Computer Centre, Town Hall, High Street, Lowestoft NR32 1HS

Mr I Jones, Welsh Water, SW Division, Hawthorne Rise, Haverford, West Dyfed SA51 2BH

Ms E Simmons, Welwyn & Hatfield Borough Council, Planning & Development Dept, Council Offices, The Campus, Welwyn Garden City, Hertfordshire AL8 6AE

Dr A S Walker, Wild Leitz UK Ltd, Photogrammetry & Systems, Davy Avenue, Knowlhill, Milton Keynes MK5 8LB

Mr A Trigg, Wiltshire County Council, Planning & Highways Department, County Hall, Trowbridge, Wiltshire BA14 8JJ

Mr S Lacey, Winchester City Council, IT Section City Offices, Colebrook Street, Winchester, Hants SO23 9LJ

Ms G Kenyon, Windsor & Maidenhead, Royal Borough Council, Finance Directorate/ Audit and Review, Town Hall, St Ives Road, Maidenhead, Berkshire SL6 1RF

Mr R Lewis, Wolverhampton Metropolitan Borough Council, Technical Services, Civic Centre, St Peters Square, Wolverhampton WV1 1RP

Mr P Curran, Xerox Engineering Systems, 5 Oxford Road, Newbury, Berks RG13 1QD

Mr A F Scholfield, Yorkshire Electricity Energy, Supply Division, Scarcroft, Leeds LS14 3HS

Mr C A Tunley, Yorkshire Water Services Ltd, Technology & Services Dept, Broadacre House, Vicar Lane, Bradford BD1 5PZ

Mr D Peterson, ZS Associates, Apex Plaza, Forbury Road, Reading, Berks RG1 1ZS

Individual members

Mr D Marsden, Heathwall Pumping Station, 60 Nine Elms Lane, London SW8 5DA

J Aldridge, c/o American Embassy – Box 1224, Grosvenor Square, London W1A 1AE

Mr D J Unwin, University Road, Leicester LE1 7RH

Mr J Curzon, 26 Axminster Walk, Bramhall, Stockport, Cheshire SK7 2QJ

Mr M G Coulson, Natural Science Building, Singleton Park, Swansea SA2 8PP

Dr B Kelk, Keyworth, Notts NG12 5GG

Mr E J Stephens, Antrobus House, 18 College Street, Petersfield, Hants GU31 4AD

Mr A J Roberts, PO Box 3764, Doha Qatar

Mr I M Ashworth, 14 Crutchfield Lane, Walton-on-Thames, Surrey KT12 2QZ

Mr P E Clegg, Cliffe House, Cottage High, Bradfield, Sheffield S6 6LJ

Mr K W Robinson, 65 School Drive, Newton, Longville, Milton Keynes MK17 0DD

Dr G A Cassettari, Penrhyn Road, Kingston upon Thames, Surrey KT1 2EE

Mr J D Leatherdale, 10 Bartrams Lane, Hadley Wood, Barnet, Hertfordshire EN4 0EH

Mr J R Nelson, Taberner House, Park Lane, Croydon CR9 1JT

Ms S Courtney, Middle House, The Green, Offham, West Malling, Kent ME19 5NN

Mr P G Coles, Council House, Room 75, Earl Street, Coventry CV1 5RQ

Mr A M Coote, Romsey Road, Maybush, Southampton SO9 4DH

Mr R P Mahoney, 14 Kings Avenue, Denton, Newhaven, East Sussex BN9 0NA

Mr B F Wade, 7 Southern Court, South Street, Reading RG1 4QS

Mr R Buxton, Plumtree Court, London EC4A 4HT

Ms L Clifford, Tangier Lane, Eton, Windsor, Berkshire SL4 6BB

Mr J G Loy, 1 Paisley Road, Renfrew PA4 8JH

Dr T C Bailey, Laver Building, Exeter University, Exeter, Devon EX4 4QE

Prof D Rhind, Winchester Mead, 1 Cold Harbour Close, Wickham, Hants PO17 5PT

Prof P F Dale, Longbridge Road, Dagenham, Essex RM8 2AS

Mr D R J Vaughan, 106 Gloucester Place, London W1H 3DB

Mr R W Laing, Sapphire Court, 274-276 High Street, Slough SL1 1NB

Mr K O'Hara, The Old Police Station, Newnham, Gloucester GL14 1AA

Mr R G Robbins, Fanum House, Basing View, Basingstoke, Hampshire RG21 2EA

Mr R F Oxley, Hornsey Town Hall, Crouch End, Broadway, London N8 9JJ

Ms J C Danczak, 104 St John Street, London E1M 4EH

Mr D Forrest, Glasgow G12 8QQ

Mr D A Wilkinson, James Weir Building, Montrose Street, Glasgow G1 1XJ

Dr J C Budd, Northminster House, Northminster, Peterborough PE1 1VA

Mr D H McPherson, Taunton TA1 2DN

Mr D W Hutchinson, Room A4/8 Government Buildings, Leatherhead Road, Chessington KT9 2LU

Mr R A McLaren, 33 Lockharton Avenue, Edinburgh EH14 1AY

Mr D C Hughes, 82 Melbourne St, North Adelaide, SA 5006, Australia

Mr J W Woodhams, Heathwall Pumping Station, 60 Nine Elms Lane, London SW8 5DA

Mr N E Pallister, Manley House, Kestrel Way, Sowton, Exeter, Devon EX2 7LQ

Mr N Holmes, County Hall, Atlantic Wharf, Cardiff CF1 5UW

Mr J A Polwin, The Civic Centre, Lampton Road, Hounslow, Middlesex TW3 4DN

Mr P A E Eames, Cory House, The Ring, Bracknell, Berkshire RG12 1AX

Mr I T Logan, 10 Horsebridge Way, Rownhams, Southampton, Hampshire SO1 8AZ

Dr J Cordingley, District Headquarters, Highfield, Cliftonville Road, Northampton NN1 5DN

Mr K J Maddock, 89 Lansdowne Place, Hove, East Sussex BN3 1FN

Mr C A W J Nicklin, 8 Ugg Mere Court Road, Ramsey-St-Marys, Huntingdon, Cambs PE17 1RQ

Mr M Cory, Romsey Road, Maybush, Southampton SO9 4DH

Mrs J W Thomson, High Cross, Madingley Road, Cambridge CB3 0ET

Mr P Mingins, 83 Pheasant Way, Beeches Park, Cirencester, Glos GL7 1BJ

Mr C S Bradford, 28 Crichton Street, Dundee DD1 3RQ

Mr R W Groom, 5 Rose Avenue, Hazelmere, Buckinghamshire HP15 7PR

Mrs B Ashton, Civic Offices, Guildhall Square, Portsmouth PO1 2BQ

Mr M N Forbes, Computer Team, 9-15 New Park Road, London, SW2 4DU

Mrs K M De Groot, 23 Evelyn Close, Egley Road, Woking, Surrey

Mr B P F Heran, Park House, 116 Park Street, London W1Y 4NN

Mr J R Tyler, 49 Stonegate, York, YO1 2AW

Mr J Matthews, Templar House, 81-87 High Holborn, London WC1V 6NU

Ms C M Owen, Stokes House, 17-25 College Square , East Belfast BT9 6AT

Mr R P E Cunningham, Observatory House, Windsor Road, Slough, Berkshire SL1 2EY

Dr A C H Bird, Silsoe, Bedfordshire MK45 4DT

Mr J N Rae, 112 Ingram Street, Glasgow G1 1ET

Col J P Elder, Elmwood Avenue, Feltham, Middlesex TW13 7AE

Mr A G B Ayres, Masefield House, Gatwick Airport, West Sussex RH6 0HZ

Mr I W Evans, Derlwyn, Lon Y Wern, Tregarth, Bangor, Gwynedd

Mrs J M McMorrow, St Helens Road, Ormskirk, Lancashire L39 4QP

Mr A S MacDonald, Romsey Road, Maybush, Southampton SO9 4DH

Mr R A Longhorn, 1 Potters Cross, Wootton, Bedfordshire MK43 9JG

Mr D W Moat, Council Offices, Wellington Road, Ashton under Lyne, Tameside OL6 6DL

Mr J W Ramster, Pakefield Road, Lowestoft, Suffolk NR33 0HT

Major A J M Vickers, 62 Craven Road, Newbury, Berkshire RG14 5NJ

Mr E B Howard, Herikebrink 177544ER, Enschede, The Netherlands

Mr P J Dowie, Drummond Street, Edinburgh, Scotland EH8 9XD

Mr A K Cooper, PO Box 3950001, Pretoria, South Africa

Mr J J Murray, 24 Cambridge Place, Hills Road, Cambridge

Dr C D Wellings, 1 Station Road, Pangbourne, Berkshire RG8 7AY

Mr F Bennion, 5 Wharf Lane, Solihull, West Midlands B91 2JP

Mr I D B Ballantyne, Butlers Leap, Rugby, Warwickshire CV21 3RQ

Mr W B Christian, 4 Boulton Avenue, New Ferry, Wirral L62 1DT

Mr A R Bromley, Council Offices, Nottingham Road, Melton Mowbray, Leicestershire LE13 0UL

Mr K M Atterbury, 4 Marine Drive, Edinburgh, Scotland EH5 1YB

Miss S E Kerr, 12-22 Linenhall Street, Belfast, Northern Ireland BT28 BS

Mr I Archibald, 10 Charlotte Square, Edinburgh EH2 4DR

Mr P Entwistle, PO Box 13, County Hall, North Humberside HU17 9AB

Mr A A Britten, Kingfisher House, Goldhay Way, Orton Goldhay, Peterborough PE2 0ZR

Mr K A Jones, Mucklow Hill, Halesowen B62 8BP

Mr S Hawkins, 46 Fortescue Road, Colliers Wood, London SW19 2EB

Mr G J Bardill, 200 Lichfield Lane, Mansfield, Nottinghamshire NG18 4RG

Mr D Y K Ng, Conway Mansion, 29 Conduit Road, Block F 16th Floor, Hong Kong

Mr M D Procter, Alston Farm, Alston Lane, Churston, Devon TQ5 0HT

Ms G P McIntyre, Lovat Bank Silver Street, Newport, Pagnell, Buckinghamshire MK16 OEQ

Mr H A C Thomson, Tayside House, 28 Crichton Street, Dundee DD1 3RD

Mr P Southorn, 21 Mowbray Road, Northallerton, N Yorkshire DL6 1QT

Mr S W Trow, 29 John Street, Sunderland SR1 1JT

Mr T J Perkins, 42 South Oswald Road, Edinburgh EH9 2HH

Mr D J Hendley, 10 Challoners Way, East Molesey, Surrey KT8 0DW

Mr D P Hughes, 21 Baden Powell Drive, Colchester, Essex CO3 4SL

Prof P W J Batey, PO Box 147, Liverpool L69 3BX

Mr R S Morgan, 20 India Street, Glasgow G2 4PF

Ms A W Carruthers, 30 Keir Street, Edinburgh EH3 9EU

Mr G W Carver, Telford House, Tothill Street, London SW1H 9NB

Mr J Turner, Romsey Road, Maybush, Southampton SO9 4DH

Mr E Hakki, 85 Queen Victoria Street, London EC4V 4AB

Mr S A Matthews, UCLA, Los Angeles, California 90024, USA

Mr A N Sinclair, Plumtree Court, London EC4A 4HT

Dr P D Marsh, Walford Manor, Baschurch, Shrewsbury SY4 2HH

Mr G W Webb, Telford House, Tothill Street, London SW1H 9NB

Dr P C Walker, St Andrews Hospital, North Side, Yarmouth Road, Norwich NR7 055

Mr H M H Lee, Silwood Park, Buckhurst Road, Ascot, Berkshire SL5 7QW

Dr N J Eales, Summer Fields, Sketty Road, Swansea SA2 0LH

Mr T C Waugh, Drummond Street, Edinburgh EH3 9EU

Mr G Levett, Consilium House, City Forum, City Road, London EC1V 2NY

Mr P Shand, High Winds, Cassington, Oxford OX8 1DL

Prof J Beaumont, Claverton Down, Bath BA2 7AY

Dr T J Nash, Hawthorn House, Edale Road, Hope, Sheffield S30 2RF

Mr D Keith, Pegasus House, 375 West George Street, Glasgow G2 4NN

Mr J S Voller, 33 The Paddock, Busby, Glasgow

Mr T S Fulton, Commercial Union House, 124 St Vincent Street, Glasgow G2 5ER

Mr M Bell, W5B Warton, Preston, Lancashire PR4 1AX

Mr N J Pearce, 17 Fox Brook, Wootton Bassett, Swindon, Wiltshire SN4 8QD

Mr G T Robinson, 15 Belsize Park, London NW3 4ES

Mr P J Churchward, Suite 27A/34, Concourse House, II432 Dewsbury Road, Leeds LS11 7DF

Mr P Turner, Suite 27A/34, Concourse House, II432 Dewsbury Road, Leeds LS11 7DF

Mr H McCracken, 20 India Street, Glasgow G2 4PF

Mr N D J Edmead, Silver Birches, Gentles Lane, Upper Passfield, Nr Liphook, Hampshire GU30 7RY

Mr J D Steel, 70 Smallfield Road, Horley, Surrey RH6 9AT

Dr J R Cuthbert, Room 5155, New St Andrews House, Edinburgh EH1 3SX

Mr H M Fingland, 13-15 New Market Road, Cambridge CB2 1OA

Mr M Bedford, 24 Repton Court, Repton Close, Basildon, Essex SS13 1LN

Mr P W Madeley, 24 Repton Court, Repton Close, Basildon, Essex SS13 1LN

Mr C J Parker, Pwll Monyn, Betws Road, Llanrwst, Gwynedd LL26 0PT

Mr N P Mackay , 8 Blyth Close, Manchester Road, London E14 9DU

Dr R Barr, Manchester M13 9PL

Miss S C Cox, 4 Greenway, Crosby, Liverpool L23 9PX

Mr M P Adam, 49 Histon Road, Cambridge CB4 3JD

Mr P G Davies, The Kilnhurst Business Centre, Victoria Street, Kilnhurst, Rotherham S62 5SQ

Miss E Simmons, Council Offices, Welwyn Garden City, Hertfordshire AL8 6AE

Mr M Fisher, PO Box 1732, Riyadh 11441, Saudi Arabia

Mr J Davie, Scott House, Basing View, Basingstoke RG21 2JG

Col P R Wildman, Dawson Building, Elmwood Avenue, Feltham, Middlesex TW13 7AF

Mr R M Hirst, Woodcote House, Ashley Road, Epsom, Surrey KT18 5BW

Mr C N Thompson, Burgh House, Burgh by Sands, Carlisle CA5 6AN

Mr P R Jones, Electricity House, PO Box4, Vale, Guernsey, Channel Islands

Mr C P Richardson, 51 York Road, Headington, Oxford OX3 8NR

Mr F G Cole, 197 East Clyde Street, Helensburgh, Dumbartonshire, Scotland GA4 7AJ

Mr I H Townend, Swindon, Wiltshire SN4 OQD

Mr A J Lodwick, 17 Clifton Park Road, Caversham, Reading, Berkshire RG4 7PD

Mr A Metcalfe, Council Offices, Farnborough Road, Farnborough, Hampshire GU14 6AH

Mr P H Neal, 87 Melville Road, Maidstone, Kent ME15 7UT

Mr G Hamilton, Steers House, Canning Place, Liverpool L1 8JA

Ms A Tait, Flat 12, Kingsleigh Walk, Westmoreland Road, Bromley, Kent BR2 0YE

Mr R Beard, Park Square, Luton, Beds LU1 3JU

Mr J Buchanan, Rodborough Manor, Bear Hill, Rodborough, Nr Stroud, Gloucestershire GL5 5DH

Mr T J Jeffries , Harris Rightwell House, Bretton Centre, Peterborough, Cambridgeshire PE3 8DW

Mr J B Noon, 3/4 West Powburn, West Savile Terrace, Edinburgh EH9 3EW

Mr D Noble, St James's Building, Oxford Street, Manchester M1 6FL

Mr F M A Hewlett, Miller's Way, 1A Shepherd's Bush Road, London W6 7NA

Mr I Storey, Trawsgoed Mansion, Aberystwyth, Dyfed SY23 4HT

Mr G L Ashton, 200 Lichfield Lane, Berry Hill, Mansfield NG18 4RG

Mr A F Wright, 2 Ridgacre Lane, Quinton, Birmingham B32 1ES

Mr D S Stewart, 23 London Road, Biggleswade, Bedfordshire SG18 8ER

Mr R B Coote, Vine House, 41 Portsmouth Road, Cobham, Surrey KT11 1JQ

Ms K Howard, PO Box 12, Richard Fairclough House, Knutsford Road, Warrington WA4 1HG

Mr R G W Griffiths, Hafod Elfyn, Penrhos Road, Bangor, Gwynedd LL57 2LQ

Dr P G Gretton-Watson, 49 Berners Street, London W1P 4AQ

Mr F Stuart-Brown, 41 Walton Park, Pannal, Harrogate HG3 1EJ

Ms J M Plant, 8 St Bride Street, London EC4A 4DA

Mr C Craig, 650 Aztec West, Bristol BS12 4SD

Mr N C R Taylor, Waltham End, London Road, Abridge, Essex RM4 1UX

Mr A T Mitchell, 231 Corstorphine Road, Edinburgh EH12 7AT

Mr J Falconer, Phoenix House, 202 Elder Gate, Milton Keynes MK9 1BE

Mr D Popovich, 95 St Bedes Crescent, Cambridge CB1 3TZ

Mr R N Vango, 92-94 Church Road, Mitcham, Surrey CR4 3TD

Dr P A Waters, 2 Prospect House, Prospect Avenue, Kingsdown, Bristol BS2 8EA

Mr T W Robinson, Glenmoor, Cheddar Road, Axbridge, Somerset BS26 2DL

Mr C M McLundie, Granton House, 4 Marine Drive, Edinburgh EH5 1YB

Mr D M Pettit, Admiralty Road, Great Yarmouth, Norfolk NR30 3NG

Mr J W Cadoux-Hudson, Mayfield, Buckland Road, Childswickham, Broadway WR12 7HH

Dr C N Edmonds, Cranford, Kenilworth Road, Blackdown, Royal Leamington Spa, Warwickshire CV32 6RG

Mr A Carnell, 27 Field Close, Harpenden, Hertfordshire AL5 1EP

Mr M P Wilkins, Borough House, Newark Road, Peterborough PE1 5YJ

Mr J Day, 81 Addison Road, Caterham, Surrey CR3 5LU

Ms C Kenyon, 4 Strode Road, Tottenham N17 6TZ

Mr V J Abbott, Drake Circus, Plymouth PL4 8AA

Miss J E Rigby, 6 The Hawthorns, East Boldon, Tyne and Wear NE36 0DP

Dr A A Lovett, University Plain, Earlham, Norwich, Norfolk NR4 7TJ

Mr J H Williams, Bangor, Gwynedd LL57 2UW

Mr M Calvert, RM511, Hydebank, 4 Hospital Road, Belfast BT8 8JP

Mr J Hutchinson, London Square, Cross Lanes, Guildford, Surrey GU1 1UJ

Mr D P Giles, Burnaby Building, Burnaby Road, Portsmouth PO1 3ZL

Mr P F Raven, Capenhurst, Chester CH1 6ES

Mr C W Decker, Terriers House, 201 Amersham Road, High Wycombe, Bucks HP13 5AJ

Mr J Scanlan, PO Box HM 1384, Hamilton, HMFX, Bermuda

Dr A M Murphy, St Andrews House, (SA42) St Andrews Street, London EC4A 4HT

Mr D I Wilson, Clumber Park, Worksop, Notts S80 3BE

Mr M P Butterworth, Elettra Avenue, Waterlooville, Hampshire PO7 7XS

Mr M J Stanbridge, "Rotherfield", Grange Road , St Cross, Winchester, Hants SO23 9RT

Mr R Stephenson, 22 Sunderland Road, Ealing, London W5 4JY

Prof R Dunn, Whiteknights, Reading RG6 2AF

Dr A C Kemp, Thorncroft Manor, Dorking Road, Leatherhead, Surrey KT22 8JB

Mr A Dickinson, Crowmarsh Gifford, Wallingford, Oxfordshire OX10 8BA

Mr H Orenstein, PO Box 1807, London NW4 2PF

Brig M P B G Wilson, Elmwood Avenue, Feltham, Middlesex TW13 7AE

Mr P M Gibbs, PO Box 5575, Ruwi, Sultanate of Oman

Mr P Dickinson, Cheethams Park Estate, Park Street, Stalybridge, Cheshire SK15 2BT

Mr E Parsons, Penrhyn Road, Kingston upon Thames, Surrey KT1 2EE

Mr R A Newell, 21 Hartington Villas, Hove, East Sussex BN3 6HF

Maj Gen R Wood, Elmwood Avenue, Feltham, Middx TW13 7AH

Mr J Vivian, 54 Church Street, Kidlington, Oxford OX5 2BB

Mr G R W Bown, 7 Commercial Road, Gloucester GL1 1NW

Dr J R Petch, Salford, M5 4WT

Mr G R Young, Rightwell House, Bretton Centre, Peterborough PE3 8DW

Mr J C Rowland, Woodhill House, Westburn Road, Aberdeen AB9 2LU

Mr P J Dowie, Kingfisher House, Goldhay Way, Orton Goldhay, Peterborough PE2 0ZR

Mr R A Hogg, Consort House, 12 West George Street, Glasgow G2 1HN

Ms S Stone, 9 St Leonards Place, York YO1 2ET

Mr M G Probert, Room C 341, Romsey Road, Maybush, Southampton SO9 4DH

Mr B Ahmad, No 12, Loron 51A/227C46100, Petaling Jaya, Selangor, W Malaysia

Ms F K McCready, 5-U Verner Street, Belfast BT7 2AA, Northern Ireland

Mr J Humphrey, Kingston Bridge House, Church Grove, Kingston upon Thames, Surrey KT1 4AG

Mr K Roy , Killingworth, Newcastle upon Tyne NE99 1LH

Mr I Clark, Milton Gate, 1 Moor Lane, London EC2Y 9PB

Mr J R Irwin, Civic Offices, Guildhall Square, Portsmouth PO1 2AU

Ms M Chung, MSIRI Reduit, Mauritius, Indian Ocean

Mr K F Clarke, Fallowfield, Glatton, Cambridgeshire PE17 5RU

Mr M Warboys, Keele, Staffs ST5 5BG

Mr S Asbury, 6285 Barfield Road, Atlanta, Georgia 30328, USA

M G J Eadie, Wallingford, Oxfordshire OX10 8BA

Mr J R ChantadaFrei, Rosendo Salvado, 2 4 F15701, Santiago de Compostela, SPAIN

Ms D Salathiel, Room P1/168B2, Marsham Street, London SW1

Mr R Green, The Tannery, Westgate, Chichester, West Sussex PO19 3RJ

Mr C J Cooper, Hawthorn House, Edale Road, Hope, Sheffield S30 2RF

Mr R Yunos, No 12 Lorong, SIA/227C, Petaling Jaya, 46100 Selangor, Malaysia

Mr Z Alias, 12 Lorong SIA/227C46100, Petaling Jaya, Selangor, Malaysia

Mr S Ward, St James Court, Great Park Road, Almondsbury, Bristol BS12 4QJ

Mrs L Longman, St James Court, Great Park Road, Almondsbury, Bristol BS12 4QJ

Mr C R Britton, 3 The Crossways, Onslow Village, Guildford, Surrey GU2 5QG

Mr A S Shanks, PO Box 141, Dhahran Airport, 31932 Saudi Arabia

Dr L Worrall, PO Box 213, Mainslee House, Telford TF3 4LD

Mr M Mitchell, PO Box 612, London SW11 3JX

Mr T W Ng, Room 5165/F, Murray Building, Garden Road Central, Hong Kong

Mr J R Lillywhite, 1 Northern Road, Cosham, Portsmouth PO6 3AA

Mr B D Pyne, 5 May Cottages, Monkswell Lane, Chipstead, Coulsdon, Surrey

Mr D Gliddon, Northgate Road, Farnborough, Hants GU14 6TW

Mr I H Kelly, 14 Broom Field, Lightwater, Surrey GU18 5QN

Mr R Deakin, 262 Ermin Street, Stratton, St Margaret, Swindon, Wilts SN3 4LW

Mr S Emm, 1 Fox Spring Crescent, Edinburgh EH10 6NB

Mr C Sabel, 19 Silver Street, Cambridge CB3 9EP

Mr G R Mears-Young, University of Leeds, 3F Springfield House, Hyde Terrace, Leeds LS2 9LN

Mr D Wilbie-Chalk, The Seminarium, Well Close, Rothbury, Morpeth, Northumberland NE69 7NZ

Mr W J Chaloner, "Faygate", 72 Liverpool Road, Walmer, Deal, Kent CT14 7LR

Mr E Lewis, Crud-Yr-Awel, Ffos-Y-Ffin, Aberaeron, Dyfed SA46 0HD

Mr P J Wootton, Southampton Depot, Western Esplanade, Southampton, Hants SO1 0EU

Mr T P Dawson, 30 New Bridge Street, Exeter, Devon EX4 3AH

Dr M L C Ferreira, Rua Alfredo Lopes, 1717Vila Elizabeth, CEP 13560, Sao Carlos, SPBrazil

Ms S Benson, PO BOX 231028, Omuramba Road, Windhoek, Namibia

Mr N Rowan, Room 451 Dundonald House, Stormont Estate, Belfast BT4 3SB

Mr P J Halls, Heslington, York YO1 5DD

Mr G K Hayhurst, 27 Willow Court, Abbey Road, Macclesfield SK10 3PD

Dr R W Alexander, Cheyney Road, Chester CH1 4BJ

Miss L Hampshire, Block B Elmbridge Court, Cheltenham Road, Gloucester GL3 1AG

Mr S J H Dawson, PO Box 41, Northern Road, Portsmouth, Hampshire PO6 3AU

Mr R S Greenly, 39 Thanes Street, Windsor, Berkshire SL4 1PR

Ms C R Rees, Park Square, Luton LU2 8LE

Mr C Wai Pun, Flat C 22 Floor, Kenyon Court, 50 Bonham Road, Hong Kong

Mr S B Thompson, 34/41 Station Road, Ashford, Kent TN23 1PP

Mr D A Cross, 1 Lawrence Road, Ramsey, Huntingdon, Cambs PE17 1UY

Mr I Cooke, 20a Mayfield Terrace, Edinburgh EH9 1SA

Mr J H Irving, 8 Tenby Avenue, Harrow, Middlesex HA3 8RX

Mr G L Radford, 24 Brynteg, Llandegfan Menai Bridge, Gwynedd LL59 5TY

Mr R S Douglass, Grosvenor HouseFore Street, St Stephen in Brannel, St Austell PL26 7NN

Mr R J Stapleton, The Tannery, Westgate, Chichester, W Sussex PO19 3RJ

Dr D B Kidner, Llantwit Road, Treforest, Mid Glamorgan CF37 1DL

Mr R M Crooks, Room A908 Government Buildings, 98-121 Epsom Road, Guildford GU1 2LD

Student members

Mr T C Payne, 126 Pembroke Road, Seven Kings Ilford, Essex
Mr F McLeod, Longbridge Road, Dagenham, Essex RM8 2AS
Mr L J Finniear, 67 Forest Road, Loughborough LE11 3NW
Mr J H Davies, Horn Hill Road, W Addersbury, Nr Banbury, Oxfordshire OX17 BHN
Miss C J Sayer, 91 Hazel Bank Road, Chertsey, Surrey KT16 8PB
Mr D B Cross, 33 Sykes Street, New Hey Rochdale, Lancashire OL16 4JP
Miss A K Bartlett, 16 Campion Road, Abingdon, Oxon OX14 3TQ
Mr G J Morris, 10 Church Way, Yealmpton, Devon PL8 2LA
Mr M F Fishwick, 8 Swallowfield Gardens, Theale, Berkshire RG7 5AD
Ms J L Wright, 1 Charleville Circus, Sydenham, SE26 6NR
Ms L Daldini, Guildbourne House, Chatsworth Road, Worthing, BN11 7LD
Mr J D S Stocks, Erw Wen Fron Park Avenue, Llanfairfechan, Gwynedd LL33 0AS
Mr G D Mowl, Lipman Building, Sandyford Road, Newcastle upon Tyne NE1 8ST
Mr M H W Hobbs, The University Kent, CT2 7NF
Mr A J Moore, 3 Crossgreen Place, Uphall West Lothian EH52 6TD
Mr F J Piccinini, 6 Endsleigh Gardens, London WC1H 0ED
Mr S C Kane, 6E Hawthorn Place Uphall, Broxburn West Lothian, EH52 5BX
Mr APC Tear, 4/1 Romero Place, Edinburgh EH16 5BJ
Mrs J McLaren, 33 Lockharton Avenue, Edinburgh EH14 1AY
Mr RB Williams, 18 Chester Court, Albany Street, London NW1 4BU
Dr C Mathers, Emmanuel College, Cambridge CB2 1JE
Mr G F Wade, 4 Sefton Terrace, Deganwy, Gwynedd
Miss E Baker, 39 Fir Street, Gospel End, West Midlands DY3 4AD
Mr N Mark, 26 Strode Road, London E7 0DU
Mr X Lopez, Birkbeck College, 7-15 Gresse Street, London W1P 1PA
Mr B E Baker, 43 Sherrards Park Road, Welwyn Garden City, Herts AL8 7LD
Mr R J Abrahart, 41 Grangeway Road, Wigston Fields, Leicester LE8 1JF

Index to advertisers